ELEMENTARY ALGEBRA

p. 92   Integers

p. 222   Binomials - Special Products

p. 279   Perfect Square Trinomials

Formulas from Geometry

Formulas for Area (A), Perimeter (P), Circumference (C), and Volume (V):

Square

$A = s^2$

$P = 4s$

Rectangle

$A = lw$

$P = 2l + 2w$

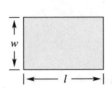

Circle

$A = \pi r^2$

$C = 2\pi r$

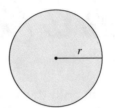

Triangle

$A = \dfrac{1}{2}bh$

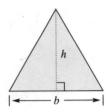

Trapezoid

$A = \dfrac{1}{2}h(b_1 + b_2)$

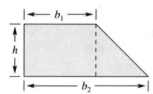

Parallelogram

$A = bh$

$P = 2a + 2b$

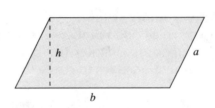

Pythagorean Theorem

$a^2 + b^2 = c^2$

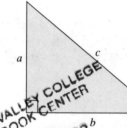

Cube

$V = s^3$

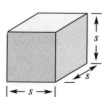

Rectangular Solid

$V = lwh$

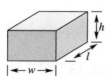

Circular Cylinder

$V = \pi r^2 h$

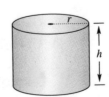

Sphere

$V = \dfrac{4}{3}\pi r^3$

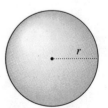

Could You Use Some Help In Your Algebra Course?

You can find it in both the **Study Guide** and the **Student Solutions Guide** accompanying this text.

The **Study Guide** (18765-8) includes:
- ▲ a review of key concepts for each section
- ▲ examples for review, with detailed solutions
- ▲ additional exercises to reinforce each topic
- ▲ cumulative chapter tests, with answers

In the **Student Solutions Guide** (1866-6), you'll find:
- ▲ detailed answers to all odd-numbered text exercises
- ▲ step-by-step solutions to Chapter Test questions from the text
- ▲ computer-generated graphics to clarify the presentation

You can get both the **Study Guide** and the **Student Solutions Guide** at your bookstore. It's an effective way to give yourself the advantage in algebra.

Many other learning aids are available with this text; you'll find them listed in the preface. If you have any questions regarding these extraordinary supplements, check with your textbook department.

D. C. Heath and Company
125 Spring Street
Lexington, MA 02173

Elementary Algebra

ROLAND E. LARSON   **ROBERT P. HOSTETLER**
The Pennsylvania State University
The Behrend College

with the assistance of DAVID E. HEYD
The Pennsylvania State University
The Behrend College

D. C. Heath and Company
Lexington, Massachusetts   Toronto

Address editorial correspondence to:

D. C. Heath
125 Spring Street
Lexington, MA 02173

Cover and title page: Paul Klee (1879–1940). *Polyphon gefasstes Weiss, 1930.140*. Paul Klee Foundation, Museum of Fine Arts Berne. © 1991, Copyright by COSMOPRESS, Geneva, Switzerland.

Paul Klee created *Polyphonically Enclosed White* by layering different-sized blocks of color on top of one another.

Technical art: Folium, Inc.; Illustrious, Inc.

Photo credits: p. 1 Marsha Goldberg; p. 38 Gallery of Prehistoric Art; p. 58 Charles Ficke; p. 100 Tedd Benson/Benson Woodworking Co. Inc., Alstead, NH; p. 144 Charles Ficke; p. 181 NASA; p. 208, p. 248 Tedd Benson/Benson Woodworking Co. Inc., Alstead, NH; p. 298 H. Armstrong Roberts/H. Abernathy; p. 341, p. 418 Tedd Benson/Benson Woodworking Co. Inc., Alstead, NH; p. 441 Foto Marburg/Art Resource, New York; p. 462 Tedd Benson/Benson Woodworking Co. Inc., Alstead, NH; p. 485 The Bettmann Archive; p. 505 Residence at Marblehead, MA, 1984–1987, designed by Robert A. M. Stern Architects, NY; p. 515 The Bettmann Archive.

Published simultaneously in Canada.

Printed in the United States of America.

International Standard Book Number: 0-669-18763-1

Library of Congress Catalog Card Number: 90-86147

10 9 8 7 6 5 4 3 2

PREFACE

Elementary Algebra provides comprehensive coverage of the topics required in a beginning course in algebra. The text has two basic goals: First, to help students develop proficiency in algebra; and second, to show students how algebra can be used as a modeling language for real-life problems. To support these two functions, the text has several key pedagogical features.

- **Problem-Solving Process** One general approach to problem-solving is stressed throughout. Students are taught to form a verbal model as the first step in solving any application.

- **Applications** Numerous applications are integrated throughout every section of the text both as solved examples and as exercises. As a result, students are constantly using and reviewing their problem-solving skills. The text applications cover a wide range of relevant topics, and many use real data.

- **Discussion Problems** Appearing at the end of each section, the discussion problems offer students the opportunity to think, talk, and write about mathematics in a different way. Students are encouraged to draw new conclusions about the concepts presented and to develop a sense of how each topic studied fits into the whole concept of algebra.

- **Exercise Sets** The exercise sets contain numerous computational and applied problems dealing with a wide range of topics. These problems are carefully graded to increase in difficulty as the student's problem-solving skills develop. Each pair of consecutive problems is similar, with the answer to the odd-numbered one given in the back of the text. Exercise sets appear at the end of each text section; an additional set of review exercises is given at the end of each chapter. Other chapter review features are summaries, chapter tests, and cumulative tests. The opportunity to use calculators—to show pattern, to experiment, to calculate, and to create graphic models—is available with selected topics.

These and other features of the text are described in greater detail on the following pages.

Features of the Text

The features of this text are designed to help students improve their skills and acquire an understanding of mathematical concepts. The functional use of four colors strengthens the text as a pedagogical tool.

Chapter Opener

A list of the topics to be covered and a brief chapter overview provide a survey of the contents of each chapter, showing students how the topics fit into the overall development of algebra.

Section Topics

Each section begins with a list of important topics that are covered in the section. These topics are also the subsection titles and can be used for easy reference and review by students.

Definitions and Rules

All of the important rules, formulas, and definitions are boxed for emphasis. Each is also titled for easy reference.

Notes

Notes appear after definitions and examples. They are designed to give additional insight, to point out common errors, and to describe generalizations.

Math Matters

Each chapter contains a Math Matters box that discusses an interesting historical note or mathematical problem. Some of the Math Matters features pose questions—in such cases, the answers are given in the back of the text.

CHAPTER FIVE

Polynomials

5.1 Adding and Subtracting Polynomials
5.2 Multiplying Polynomials
5.3 Dividing Polynomials
5.4 Negative Exponents and Scientific Notation

SECTION 5.1 Adding and Subtracting Polynomials   **209**

SECTION 5.1

Adding and Subtracting Polynomials
Basic Definitions • Adding Polynomials • Subtracting Polynomials

Basic Definitions

Remember that the *terms* of an algebraic expression are those parts separated by addition. An algebraic expression whose terms are all of the form ax^k, where a is any real number and k is a nonnegative integer, is called a **polynomial in one variable**, or simply a **polynomial**. Here are some examples of polynomials in one variable.

$$2x + 5, \quad x^2 - 3x + 7, \quad \text{and} \quad x^3 + 8$$

In the term ax^k, a is called the **coefficient** of the term and k is called the **degree** of the term. Since a polynomial is an algebraic sum, the coefficients take on the signs between the terms. For instance,

$$x^4 + 2x^3 - 5x^2 + 7 = (1)x^4 + 2x^3 + (-5)x^2 + (0)x + 7$$

has coefficients 1, 2, −5, 0, and 7. For this polynomial, the last term, 7, is called the **constant term**. We usually write polynomials in the order of descending powers of the variable. This is called **standard form**. For instance, the standard form of $3x^2 - 5 - x^3 + 2x$ is

$$-x^3 + 3x^2 + 2x - 5. \qquad \text{Standard form}$$

The **degree of a polynomial** is the degree of the term with the highest power, and the coefficient of this term is called the **leading coefficient** of the polynomial. For instance, the polynomial

$$-3x^4 + 4x^2 + x + 7$$

is of fourth degree, and its leading coefficient is −3.

Definition of a Polynomial in x	Let $a_n, a_{n-1} \ldots, a_2, a_1, a_0$ be real numbers and let n be a *nonnegative integer*. A **polynomial in x** is an expression of the form $$a_n x^n + a_{n-1}x^{n-1} + \cdots + a_2 x^2 + a_1 x + a_0$$ where $a_n \neq 0$. The polynomial is of **degree n**, and the number a_n is called the **leading coefficient**. The number a_0 is called the **constant term**.

NOTE: The following are *not* polynomials for the reasons stated.

$2x^{-1} + 5$ 　　　　　　Exponent in $2x^{-1}$ is *not* nonnegative

$x^3 + 3x^{1/2}$ 　　　　　Exponent in $3x^{1/2}$ is *not* an integer

---

SECTION 5.2 Multiplying Polynomials   **221**

...nd $(x - 3)^3$ as follows.

MATH MATTERS

The Spectrum

The spectrum is the visible result produced when light is resolved into its various wavelengths or frequencies. For instance, when light is passed through a prism, as shown in the accompanying figure, the light is separated into a rainbow of colors ranging from red to violet.

The frequencies and wavelengths of the colors of the rainbow are as follows.

Color of Light	Frequency (10^{14} hertz)	Wavelength (10^{-7} meters)
Red	4.284–4.634	6.470–7.000
Orange	4.634–5.125	5.850–6.470
Yellow	5.125–5.215	5.750–5.850
Green	5.215–6.104	4.912–5.750
Blue	6.104–7.115	4.240–4.912
Violet	7.115–7.495	4.000–4.240

Problem-Solving Process

Students are taught the following strategies—in keeping with the spirit of NCTM standards—for solving applied problems. (1) Construct a verbal model; (2) Label variable and constant terms; (3) Construct an algebraic model; (4) Using the model, solve the problem; and (5) Check the answer in the original statement of the problem.

Examples

Each of the more than 760 text examples was carefully chosen to illustrate a particular concept or problem-solving technique. Examples are titled for easy reference.

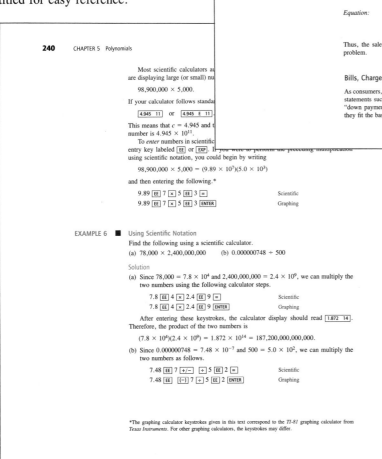

170 CHAPTER 4 Applications of Linear Equations

Keep these models in mind as you study the following examples.

EXAMPLE 3 ■ Finding the Discount Rate

During a midsummer sale, a lawn mower listed at $199.95 is on sale for $139.95. What is the discount rate?

Solution

Verbal model: Discount = Discount rate · List price

Labels: Discount = 199.95 − 139.95 = $60
List price = $199.95
Discount rate = p (percent in decimal form)

Equation: $60 = p(199.95)$

$$\frac{60}{199.95} = p$$

$$0.30 \approx p$$

Thus, the discount rate is 30%. Check this solution in the original statement of the problem.   ■

EXAMPLE 4 ■ Finding the Sale Price

A drugstore advertises 40% off on all summer tanning products. A bottle of suntan oil lists for $3.49. What is the sale price?

Solution

Verbal model: Sale price = List price − Discount

Labels: List price = $3.49
Discount rate = 0.4 (percent in decimal form)
Discount = 0.4(3.49) (dollars)
Sale price = x (dollars)

Equation: $x = 3.49 − (0.4)(3.49)$
$\approx 3.49 − 1.40$
$= \$2.09$

Thus, the sale price is $2.09. Check this solution in the original statement of the problem.   ■

Bills, Charges, and Payments

As consumers, we pay a wide variety of fees, bills, and charges that include descriptive statements such as "parts plus labor," "daily rate plus mileage," "food plus tips," or "down payment plus so much a month." Study the following examples to see how they fit the basic verbal model introduced in this section.

240 CHAPTER 5 Polynomials

Most scientific calculators au[...] are displaying large (or small) nu[...]

$98,900,000 \times 5,000.$

If your calculator follows standa[...]

4.945 11   or   4.945 E 11

This means that $c = 4.945$ and t[...] number is 4.945×10^{11}.

To *enter* numbers in scientific[...] entry key labeled EE or EXP. I[...] using the preceding multiplication[...] using scientific notation, you could begin by writing

$98,900,000 \times 5,000 = (9.89 \times 10^7)(5.0 \times 10^3)$

and then entering the following.*

9.89 EE 7 × 5 EE 3 =   Scientific
9.89 EE 7 × 5 EE 3 ENTER   Graphing

EXAMPLE 6 ■ Using Scientific Notation

Find the following using a scientific calculator.
(a) $78,000 \times 2,400,000,000$ (b) $0.000000748 \div 500$

Solution

(a) Since $78,000 = 7.8 \times 10^4$ and $2,400,000,000 = 2.4 \times 10^9$, we can multiply the two numbers using the following calculator steps.

7.8 EE 4 × 2.4 EE 9 =   Scientific
7.8 EE 4 × 2.4 EE 9 ENTER   Graphing

After entering these keystrokes, the calculator display should read 1.872 14. Therefore, the product of the two numbers is

$(7.8 \times 10^4)(2.4 \times 10^9) = 1.872 \times 10^{14} = 187,200,000,000,000.$

(b) Since $0.000000748 = 7.48 \times 10^{-7}$ and $500 = 5.0 \times 10^2$, we can multiply the two numbers as follows.

7.48 EE 7 +/− ÷ 5 EE 2 =   Scientific
7.48 EE (−) 7 ÷ 5 EE 2 ENTER   Graphing

*The graphing calculator keystrokes given in this text correspond to the *TI-81* graphing calculator from *Texas Instruments*. For other graphing calculators, the keystrokes may differ.

Calculators

Two types of calculators are discussed: scientific and graphing calculators. Sample keystrokes are given for both types. The graphing capability of graphing calculators and computer graphing software is discussed. However, coverage of this material is optional.

Discussion Problems

Discussion problems appear at the end of each section. They encourage students to think, talk, and write about mathematics, individually or in groups.

DISCUSSION PROBLEM ■ Factoring by Grouping

Some third-degree polynomials can be factored by grouping and some cannot. For instance, the polynomial

$$x^3 - 3x^2 - 2x + 6$$

can be factored by grouping, whereas the polynomial

$$x^3 - 3x^2 - 2x - 6$$

cannot be factored by grouping. Find several other third-degree polynomials, some of which can be factored by grouping and some of which cannot. ■

Warm-Up | *The following warm-up exercises involve skills that were covered in earlier sections. You will use these skills in the exercise set for this section.*

In Exercises 1–10, find the product.

1. $2(5 - 15)$ 2. $-3(8 + 6)$

3. $12(2x - 3)$ 4. $7(4 - 3x)$

5. $-6(10 - 7x)$ 6. $-2y(y + 1)$

7. $-3t(t + 2)$ 8. $8xy(xy - 3)$

9. $(2 - x)(2 + x)$ 10. $(x + 4)^2$

In Exercises 17–60, factor the expression by removing any common f
of the expressions have no common factor.)

17. $3x + 3$ 18. $5y + 5$ 19.
21. $8t - 16$ 22. $3u + 12$ 23.
25. $24y^2 - 18$ 26. $7z^3 + 21$ 27.
29. $25u^2 - 14u$ 30. $36t^4 + 24t^2$ 31.
33. $7s^2 + 9t^2$ 34. $12x^2 - 5y^3$ 35.
37. $-10r^3t^2 - 7rs^2$ 38. $-9a^2b^4 + 12a^2b$ 39.
41. $10abc + 10a^2bc$ 42. $21x^2y^2z - 35xz$ 43.
45. $100 + 75z - 50z^2$ 46. $42t^3 - 21t^2 + 7$ 47.
49. $x(x - 3) + 5(x - 3)$ 50.
51. $t(s + 10) - 8(s + 10)$ 52. $y(q - 5) - 10(q - 5)$
53. $a^2(b + 2) - a(b + 2)$ 54. $x^3(x + 4) + x^2(x + 4)$
55. $z^3(z + 5)^2 + z^2(z + 5)$ 56. $(a + b)(c + 7) - (a + b)(c + 7)^2$
57. $y^3(y - 8)^2 + y^2(y - 8)^3$ 58. $(x + y)(x - y) - x(x - y)$
59. $5u^2v^4 + 5u^2v^2 + 5uv^2$ 60. $11x^5y^3 - 22xy^2 + 11y^2$

In Exercises 61–70, factor the given expression by grouping.

61. $x^2 + 10x + x + 10$ 62. $x^2 - 5x + x - 5$ 63. $y^2 - 4y + 2y - 8$ 64. $y^2 + 3y + 3y + 9$
65. $x^3 + 2x^2 + x + 2$ 66. $x^3 - 5x^2 + x - 5$ 67. $t^3 - 3t^2 + 2t - 6$ 68. $3s^3 + 6s^2 + 2s + 4$
69. $z^3 + 3z^2 - 2z - 6$ 70. $4u^3 - 2u^2 - 6u + 3$

In Exercises 71–74, factor a negative real number from the given polynomial and then use the Commutative Property to write the polynomial factor with a positive leading coefficient.

71. $5 - 10x$ 72. $3 - x$ 73. $4 + 2x - x^2$ 74. $18 - 12x - 6x^2$

In Exercises 75–80, rewrite the given polynomial by factoring out the indicated fraction.

75. $\frac{1}{2}x + \frac{3}{4} = \frac{1}{4}(\underline{\hspace{1cm}})$ 76. $\frac{2}{3}x - \frac{1}{6} = \frac{1}{6}(\underline{\hspace{1cm}})$

77. $\frac{7}{8}x + \frac{5}{16} = \frac{1}{16}(\underline{\hspace{1cm}})$ 78. $\frac{5}{12}u - \frac{5}{8} = \frac{1}{24}(\underline{\hspace{1cm}})$

79. $2y - \frac{1}{5} = \frac{1}{5}(\underline{\hspace{1cm}})$ 80. $3z + \frac{3}{4} = \frac{1}{4}(\underline{\hspace{1cm}})$

6.1 EXERCISES

In Exercises 1–16, find the greatest common factor of the given expressions.

1. 24, 90   2. 20, 45   3. 18, 150, 100
4. 60, 80, 90   5. $z^2, -z^6$ 6. t^4, t^7
7. $2x^2, 12x$ 8. $36x^4, 18x^3$ 9. u^2v, u^3v^2
10. $r^6s^4, -rs$ 11. $9yz^2, -12y^2z^3$ 12. $-15x^6y^3, 45xy^3$
13. $28a^4b^2, 14a^3b^3, 42a^2b^5$ 14. $16x^2y, 12xy^2, 36x^2y^2$
15. $14x^2, 1, 7x^4$ 16. $5y^4, 10x^2y^2, 15xy$

Warm-Up

Each section (other than Section 1.1) contains a set of 10 warm-up exercises that allows students to practice the previously learned skills that are necessary to master the "new skills" presented in the section. All warm-up exercises are answered in the back of the text.

Exercises

The nearly 6,000 exercises include both computational and applied problems covering a wide range of topics. These are designed to build competence, skill, and understanding. Each exercise set is graded in difficulty to allow students to gain confidence as they progress. Answers to odd-numbered exercises are in the back of the text.

---

366 CHAPTER 8 Graphs and Linear Equations

In Exercises 37–42, use the vertical line test to determine whether y is a function of x.

37. $y = x - 2$

38. $x + y = 4$

39. $y = (x - 2)^2$

40. $x - y^2 = 0$

41. $x^2 + y^2 = 9$

42. $y = x^3 - x$

In Exercises 43–48, sketch the graph of the equation. Use the vertical line test to determine whether y is a function of x.

45. $y = x^2$ **46.** $y = x$

50. *Total Cost* The inventor of a new game estimates that the variable cost for producing the game is \$0.95 per unit and the fixed costs are \$6,000. Let C represent the total cost of producing x games. Write an equation that gives the total cost C in terms of the number of games produced x and sketch the graph of the equation.

SECTION 3.4 Inequalities **139**

45. $-4 < \dfrac{2x - 3}{3} < 4$ **46.** $0 \le \dfrac{x + 3}{2} < 5$ **47.** $6 > \dfrac{x - 2}{-3} > -2$ **48.** $-2 < \dfrac{x - 4}{-2} \le 3$

49. $\dfrac{3}{4} > x + 1 > \dfrac{1}{4}$ **50.** $-1 < -\dfrac{x}{3} < 1$

In Exercises 51–56, use inequality notation to denote the given statement.

51. x is nonnegative. **52.** P is no more than 2. **53.** y is more than -6. **54.** t is less than 8.

55. z is at least 3. **56.** x is a real number greater than or equal to -2 and less than 5.

In Exercises 57–60, write a verbal statement describing the set of real numbers satisfying the given inequality.

57. $x \le 10$ **58.** $z > 8$ **59.** $-\dfrac{3}{2} < y \le 5$ **60.** $-3 \le t < 3.8$

61. *Planet Distances* Mars is further from the Sun than Venus, and Venus is further from the Sun than Mercury (see figure). What can be said about the relationship between Mars's and Mercury's distances from the Sun?

Sun   Mercury   Venus   Earth   Mars

62. *Budget for Trip* Suppose you have \$2,500 budgeted for a trip. The transportation for the trip will cost \$900. To stay within your budget, all other costs must be no more than what amount?

63. *Annual Operating Cost* A utility company has a fleet of vans. The annual operating cost C (in dollars) per van is

$$C = 0.32m + 2,300$$

where m is the number of miles traveled by a van in a year. What number of miles will yield an annual operating cost that is less than \$10,000?

64. *Profit* The revenue for selling x units of a product is

$$R = 115.95x.$$

The cost of producing x units is

$$C = 95x + 750.$$

In order to obtain a profit, the revenue must be greater than the cost. For what values of x will this product produce a profit?

65. *Telephone Cost* The cost of a long-distance telephone call is \$0.46 for the first minute and \$0.31 for each additional minute. The total cost of the call cannot exceed \$4. Find the interval of time that is available for the call.

66. *Sides of a Triangle* The lengths of the sides of the triangle in the accompanying figure are a, b, and c. Find an inequality that relates $a + b$ and c.

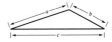

67. *Comparing Distances* Suppose you live three miles from college and your parents live two miles from you (see figure). Let d represent the distance between your parents' house and the college. Write an inequality involving d.

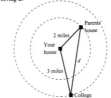

Graphics

The ability to visualize problems is a critical skill that students need in order to solve them. To encourage the development of this skill, the text has an abundance of figures, many computer-generated for accuracy.

Applications

Real-world applications are integrated throughout the text in both examples and exercises. This offers students a constant review of problem-solving skills and emphasizes the relevance of the mathematics. Many of the applications use real data, and all are titled for reference.

64. *Scientists* The accompanying table shows the number of women scientists in the United States in 1978 and 1988. (*Source:* U.S. Bureau of Labor and Statistics.)
(a) Use the table to find the total number of computer scientists (men and women) in 1988.
(b) Use the table to find the total number of physical scientists in 1978.

Women Scientists in the United States

Field	1978 Number	1978 %	1988 Number	1988 %
Computer Science	40,200	23%	218,700	31%
Biology	30,000	18%	89,200	30%
Mathematics	13,100	24%	44,900	27%
Physical Science	18,500	9%	46,500	15%
Environmental Science	7,200	10%	12,300	11%

Photo Enlarged In Exercises 65 and 66, Figure (a) was put into a photocopier and reduced or enlarged to produce Figure (b). Estimate the percentage of reduction or enlargement.

65. (a)   (b)

66. (a)   (b)

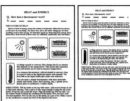

SECTION **Business and Consumer Problems**
Introduction • Markup and Discounts • Bills, Charges, and Payments •

our applications of linear equations have been limited to either s or problems that fit the percent equation model. In this section ications of linear equations to include common consumer problems repair bills, discounts, markup, salaries, admission fees, car rent- goal is for you to gain confidence in setting up and solving such ering how much alike these problems are. Many of the applications the algebraic structure given in the following verbal model.

Base amount + Percent · Base amount

In Exercises 51–54, solve the given equation and round your answer to two decimal places. (A calculator may be helpful.)

51. $0.234x + 1 = 2.805$ **52.** $275x - 3130 = 512$ **53.** $\frac{x}{3.155} = 2.850$ **54.** $\frac{2x}{3.7} = \frac{3}{4}$

In Exercises 55–66, solve the given word problem.

55. Find a number such that the sum of that number and 45 is 75.

56. The sum of four times a number and 16 is 100. Find the number.

57. The sum of two consecutive odd integers is 72. Find the two integers.

58. The sum of three consecutive even integers is 192. Find the three integers.

59. *Dimensions of a Tennis Court* The length of a rectangular tennis court is six feet more than twice the width (see figure). Find the width of the court if the length is 78 feet.

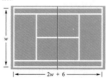

60. *Dimensions of a Sign* A sign has the shape of an equilateral triangle (see figure). The perimeter of the sign is 225 centimeters. Find the length of the sides of the sign. (An equilateral triangle is one whose sides have the same length.)

YIELD

61. *Cost of Housing* Suppose you budget 30% of your annual after-tax income for housing. If your after-tax income is $32,500, what amount can you spend on housing?

62. *Retirement Plan* Suppose you budget $7\frac{1}{2}$% of your gross income for an individual retirement plan. If your annual gross income is $29,800, how much would you put in your retirement plan each year?

63. *Original Price* A coat sells for $250 during a 20% storewide clearance sale. What was the original price of the coat?

64. *Buy Now or Wait?* A sales representative indicates that if a customer waits another month for a new car that currently costs $17,800, the price will increase by 6%. However, the customer will pay an interest penalty of $450 for the early withdrawal of a certificate of deposit if the car is purchased now. Determine whether the customer should buy now or wait another month.

65. *Height of a Fountain* Water is forced out of a fountain with an initial velocity of 28 feet per second (see figure). The velocity of the water stream is then given by $v = -32t + 28$, where t is the time in seconds. The height of the water is given by $h = -16t^2 + 28t$. What is the maximum height of the fountain? (*Hint:* Find the time when the velocity is zero, and then find the height for that time.)

Maximum height

66. *Hours of Labor* The bill (including parts and labor) for the repair of an automobile was $357. The cost for parts was $285. How many hours were spent repairing the car if the cost of labor was $32 per hour?

Geometry

Geometric formulas and concepts are reviewed throughout the text. For reference, common formulas are given inside the front and back covers.

End-of-Chapter Study Aids

A Chapter Summary outlines all of the skills that are presented in the chapter. For review, section and exercise references are given for the major topics.

A set of Review Exercises at the end of each chapter gives students an opportunity for additional practice. Each set of review exercises includes both computational and applied problems covering a wide range of topics.

A Chapter Test allows students to assess their own level of success.

Cumulative Tests appear after Chapters 4, 7, and 11. These tests help students to judge their mastery of previously covered concepts.

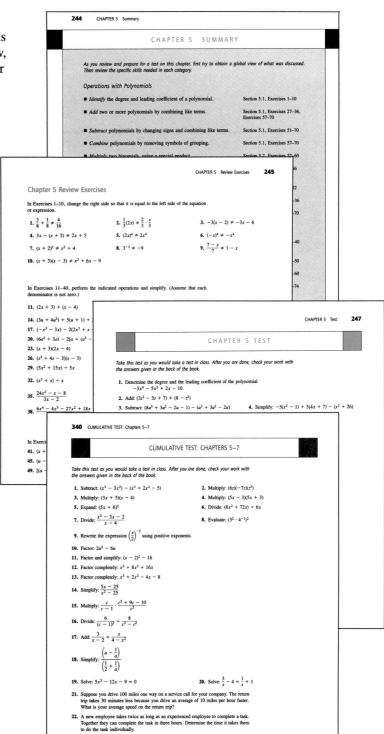

244 CHAPTER 5 Summary

CHAPTER 5 SUMMARY

As you review and prepare for a test on this chapter, first try to obtain a global view of what was discussed. Then review the specific skills needed in each category.

Operations with Polynomials

- *Identify* the degree and leading coefficient of a polynomial.   Section 5.1, Exercises 1–10
- *Add* two or more polynomials by combining like terms.   Section 5.1, Exercises 27–36, Exercises 57–70
- *Subtract* polynomials by changing signs and combining like terms.   Section 5.1, Exercises 51–70
- *Combine* polynomials by removing symbols of grouping.   Section 5.1, Exercises 57–70
- *Multiply* two binomials, using a special product   Section 5.2, Exercises 52–60

CHAPTER 5 Review Exercises   **245**

Chapter 5 Review Exercises

In Exercises 1–10, change the right side so that it is equal to the left side of the equation or expression.

1. $\frac{3}{8} + \frac{1}{8} \neq \frac{4}{16}$ 2. $\frac{1}{3}(2x) \neq \frac{2}{3} \cdot \frac{x}{3}$ 3. $-3(x - 2) \neq -3x - 6$

4. $3x - (x + 5) \neq 2x + 5$ 5. $(2x)^4 \neq 2x^4$ 6. $(-x)^4 \neq -x^4$

7. $(x + 2)^2 \neq x^2 + 4$ 8. $3^{-2} \neq -9$ 9. $\frac{7 - x}{7} \neq 1 - x$

10. $(x + 3)(x - 3) \neq x^2 + 6x - 9$

In Exercises 11–40, perform the indicated operations and simplify. (Assume that each denominator is not zero.)

11. $(2x + 3) + (x - 4)$

14. $(3u + 4u^2) + 5(u + 1) +$

17. $(-x^3 - 3x) - 2(2x^3 + x$

20. $(6a^3 + 3a) - 2[a + (a^3 -$

23. $(x + 3)(2x - 4)$

26. $(s^3 + 4s - 3)(s - 3)$

29. $(5x^2 + 15x) \div 5x$

32. $(x^2 + x) \div x$

35. $\frac{24x^2 - x - 8}{3x - 2}$

38. $\frac{6x^4 - 4x^3 - 27x^2 + 18x}{}$

CHAPTER 5 Test   **247**

CHAPTER 5 TEST

Take this test as you would take a test in class. After you are done, check your work with the answers given in the back of the book.

1. Determine the degree and the leading coefficient of the polynomial $-3x^4 - 5x^2 + 2x - 10$.

2. Add: $(3z^2 - 3z + 7) + (8 - z^2)$

3. Subtract: $(8u^3 + 3u^2 - 2u - 1) - (u^3 + 3u^2 - 2u)$ 4. Simplify: $-5(x^2 - 1) + 3(4x + 7) - (x^2 + 26)$

In Exerci

41. $(x +$

45. $(u -$

49. $[(a -$

340 CUMULATIVE TEST: Chapters 5–7

CUMULATIVE TEST: CHAPTERS 5–7

Take this test as you would take a test in class. After you are done, check your work with the answers given in the back of the book.

1. Subtract: $(x^3 - 3x^2) - (x^3 + 2x^2 - 5)$ 2. Multiply: $(6z)(-7z)(z^2)$

3. Multiply: $(3x + 5)(x - 4)$ 4. Multiply: $(5x - 3)(5x + 3)$

5. Expand: $(5x + 6)^2$ 6. Divide: $(6x^2 + 72x) \div 6x$

7. Divide: $\frac{x^2 - 3x - 2}{x - 4}$ 8. Evaluate: $(3^2 \cdot 4^{-1})^2$

9. Rewrite the expression $\left(\frac{x}{2}\right)^{-2}$ using positive exponents.

10. Factor: $2u^2 - 6u$

11. Factor and simplify: $(x - 2)^2 - 16$

12. Factor completely: $x^3 + 8x^2 + 16x$

13. Factor completely: $x^3 + 2x^2 - 4x - 8$

14. Simplify: $\frac{5x - 25}{x^2 - 25}$

15. Multiply: $\frac{c}{c - 1} \cdot \frac{c^2 + 9c - 10}{c^3}$

16. Divide: $\frac{6}{(c - 1)^2} \div \frac{8}{c^3 - c^2}$

17. Add: $\frac{3}{x - 2} + \frac{x}{4 - x^2}$

18. Simplify: $\dfrac{\left(a - \frac{1}{a}\right)}{\left(\frac{1}{2} + \frac{1}{a}\right)}$

19. Solve: $5x^2 - 12x - 9 = 0$ 20. Solve: $\frac{3}{x} - 4 = \frac{1}{x} + 1$

21. Suppose you drive 100 miles one way on a service call for your company. The return trip takes 30 minutes less because you drive an average of 10 miles per hour faster. What is your average speed on the return trip?

22. A new employee takes twice as long as an experienced employee to complete a task. Together they can complete the task in three hours. Determine the time it takes them to do the task individually.

SUPPLEMENTS

Elementary Algebra by Larson and Hostetler is accompanied by a comprehensive supplements package for maximum teaching effectiveness and efficiency.

- *Student Solutions Guide* by Carolyn F. Neptune, Johnson County Community College

- *Study Guide* by Jay Wiestling, Palomar College

- *Complete Solutions Guide* by Carolyn F. Neptune, Johnson County Community College

- *Instructor's Guide* by Robert P. Hostetler and Ann R. Kraus, The Pennsylvania State University

- *Test Item File* by Norman B. Patterson

- *Elementary Algebra* Videotapes by Dana Mosely, Valencia Community College

- *Elementary Algebra* TUTOR by Timothy R. Larson, Paula M. Sibeto, Kristin Winnen-Smith, and John R. Musser.

- Test-Generating Software

- Classroom Demonstration Software by Timothy R. Larson, Paula M. Sibeto, Kristin Winnen-Smith, and John R. Musser.

- Gradebook Software

This complete supplements package offers ancillary materials for students and instructors and for classroom resources. Each item is keyed directly to the textbook for ease of use. For the convenience of software users, a technical support telephone number is available with all D. C. Heath software products: (617)860-1218. The components of this comprehensive teaching and learning package are outlined in the diagram on the following page.

ELEMENTARY ALGEBRA

FOR INSTRUCTORS

Printed Ancillaries

Complete Solutions Guide
- Solutions to all text exercises

Test Item File
- Printed test bank
- Over 3,500 test items
- Open-ended and multiple-choice test items
- Available as a computerized test bank

Instructor's Guide
- Chapter commentary with general teaching strategies
- Section topics
- Sample tests
- Transparency masters

Software

Computerized Testing Software
- Test-Generating software
- Over 3,500 test items
- Also available as a printed test bank

Classroom Demonstration Software
- Additional examples with step-by-step solutions
- Interactive, menu-driven resource bank covering major topics
- Animated color graphics

FOR STUDENTS

Printed Ancillaries

Student Solutions Guide
- Solutions to all odd-numbered text exercises

Study Guide
- Section summaries
- Additional examples with solutions
- Starter exercises with answers

Software

TUTOR
- Interactive tutorial software follows section by section with text
- Diagnostic feedback
- Additional examples
- Chapter self-tests

Videotapes

Videotapes
- Comprehensive coverage
- For media/resource centers
- Additional explanation of important concepts, sample problems, and applications

CLASSROOM RESOURCES

Printed Ancillaries

Instructor's Guide
- Chapter commentary with general teaching strategies
- Section topics
- Sample tests
- Transparency masters

Software

Classroom Demonstration Software
- Additional examples with step-by-step solutions
- Interactive, menu-driven resource bank covering major topics
- Animated color graphics

Gradebook software
- Grade management system

Videotapes

Videotapes
- Comprehensive coverage
- For media/resource centers
- Additional explanation of important concepts, sample problems, and applications

ACKNOWLEDGMENTS

We would like to thank the many people who have helped us prepare the text and supplements package. Their encouragement, criticisms, and suggestions have been invaluable to us.

Reviewers: Linda Crabtree, Longview Community College; John Decoursey, Vincennes University; Kenneth Grace, Anoka-Ramsey Community College; Steve E. Green, Tyler Junior College; Robert J. Kosanovich, Ferris State University; Carolyn F. Neptune, Johnson County Community College; Richard Semmler, Northern Virginia Community College; Melissa J. Simpkins, Columbus State Community College; Dave Van Langeveld, Clearfield High School; Gerry Vidrine, Louisiana State University; Beverly Weatherwax, Southwest Missouri State University; and Martin Wells, East High School.

Focus Group Participants: Ruth A. Koelle, Roger Williams College; Jeri V. Love, Florida Community College at Jacksonville; Maureen A. McCarthy, Santa Rosa Junior College; Hank Martel, Broward Community College; Ellen Milosheff, Triton College; Charlotte Newsom, Tidewater Community College; and Keith Wilson, Oklahoma City Community College.

A special thanks to all of the people at D. C. Heath and Company who worked with us in the development and production of the text, especially Ann Marie Jones, Mathematics Acquisitions Editor; Cathy Cantin, Developmental Editor; Kathleen A. Savage, Production Editor; Cornelia Boynton, Designer; Carolyn Johnson, Editorial Associate; Lisa Merrill, Production Coordinator; Gary Crespo, Art Editor; and Billie L. P. Ingram, Photo Researcher.

Several other people worked on this project. David E. Heyd assisted us in writing the text and solving the exercises; Carolyn F. Neptune wrote the *Student Solutions Guide* and the *Complete Solutions Guide*; Jay Wiestling wrote the *Study Guide*; Norman B. Patterson wrote the *Test Item File* and proofed the manuscript; and Helen Medley proofed the manuscript. The following people also worked on the project: Linda L. Kifer, Linda M. Bollinger, Timothy R. Larson, Louis R. Rieger, Paula M. Sibeto, Kristin Winnen-Smith, Laurie A. Brooks, Richard J. Bambauer, Patricia S. Larson, Nancy K. Stout, and John R. Musser.

On a personal level, we are grateful to our wives, Deanna Gilbert Larson and Eloise Hostetler, for their love, patience, and support. Also, a special thanks goes to R. Scott O'Neil.

If you have suggestions for improving the text, please feel free to write to us. Over the past two decades we have received many useful comments from both instructors and students, and we value these very much.

Roland E. Larson
Robert P. Hostetler

CONTENTS

HOW TO STUDY ALGEBRA

After years of teaching and guiding students through algebra courses, we have compiled the following list of suggestions for studying algebra. These study tips may take some time and effort—but they work!

Making a Plan

Make your own course plan right now! Determine the number of hours you need to spend on algebra each week. Write your plans on your calendar or some other schedule planner, and then *stick to your plan*.

Preparing for Class

Before attending class, read the portion of the text that is to be covered. This takes a lot of self-discipline, but it pays off. By going to class prepared, you will be able to benefit much more from your instructor's presentation. Algebra, like most other technical subjects, is easier to understand the second or third time you hear it.

Attending Class

Attend every class. Arrive on time with your text, a pen or pencil, paper for notes, and your calculator.

Participating in Class

As you are reading the text before class, write down any questions that you have about the material. Then, ask your instructor during class.

Taking Notes

Take notes in class, especially on definitions, examples, concepts, and rules. Then, as soon after class as possible, read through your notes, adding any explanations that are necessary to make your notes understandable *to you*.

Doing the Homework

Learning algebra is like learning to play the piano or learning to play basketball. You cannot become skilled by just watching someone else do it. You must also do it yourself. A general guideline is to spend two to four hours of study outside of class for each hour in class. When working exercises, your ultimate goal is to be able to solve the problems accurately and quickly. When you start a new exercise set, however, understanding is much more important than speed.

Finding a Study Partner

When you get stuck on a problem, it may help to try to work with someone else. Even if you feel you are giving more help than you are getting, you will find that an excellent way to learn is by teaching others.

Building a Math Library

Start building a library of books that can help you with this and future math courses. You might consider using the *Student Solutions Guide* or the *Study Guide* that accompanies the text. Also, since you will probably be taking other math courses after you finish this course, we suggest that you keep the text. It will be a valuable reference book.

Keeping Up with the Work

Don't let yourself fall behind in the course. If you think that you are having trouble, seek help immediately. Ask your instructor, attend your school's tutoring services, talk with your study partner, use additional study aids such as videos or software tutorials—but do something. If you are having trouble with the material in one chapter of your algebra text, there is a good chance that you will also have trouble in later chapters.

Getting Stuck

Everyone who has ever taken a math course has had this experience: You are working on a problem and cannot see how to solve it, or you have solved it but your answer does not agree with the answer given in the back of the book. People have different approaches to this sort of problem. You might ask for help, take a break to clear your thoughts, sleep on it, rework the problem, or reread the section in the text. The point is, try not to get frustrated or spend too much time on a single problem.

Keeping Your Skills Sharp

Before each exercise set in the text we have included a short set of *Warm-Up Exercises*. These exercises will help you review skills that you learned in previous exercises. These sets are designed to take only a few minutes to solve. We suggest working the entire set before you start each new exercise set. (All of the Warm-Up Exercises are answered in the back of the text.)

Checking Your Work

One of the nice things about algebra is that you don't have to wonder whether your solution is correct. You can tell whether it is correct by checking it in the original statement of the problem. If, in addition to your "solving skills," you work on your "checking skills," you should find your test scores improving.

Preparing for Exams

Cramming for algebra exams seldom works. If you have kept up with the work and followed the suggestions given here, you should be almost ready for the exam. At the end of each chapter, we have included three features that should help as a final preparation. Read the *Chapter Summary*, work the *Review Exercises*, and set aside an hour to take the sample *Chapter Test*.

Taking Exams

Most instructors suggest that you do *not* study right up to the minute you are taking a test. This tends to make people anxious. The best cure for anxiousness during tests is to prepare well before taking the test. Once the test has begun, read the directions carefully, and try to work at a reasonable pace. (You might want to read the entire test first, then work the problems in the order with which you feel most comfortable.) Hurrying tends to cause people to make careless errors. If you finish early, take a few moments to clear your thoughts and then take time to go over your work.

Learning from Mistakes

When you get an exam back, be sure to go over any errors that you might have made. Don't be too quick to pass off an error as just a "dumb mistake." Take advantage of any mistakes by hunting for ways to continually improve your test-taking abilities.

WHAT IS ALGEBRA?

To some, algebra is manipulating symbols or performing mathematical operations with letters instead of numbers. To others, it is factoring, solving equations, or solving word problems. And to still others, algebra is a mathematical language that can be used to model real-world problems. In fact, algebra is all of these!

As you study this text, it is helpful to view algebra from the "big picture"—to see how the various rules, operations, and strategies fit together.

The rules of arithmetic form the foundation of algebra. These rules are generalized through the use of symbols and letters to form the basic rules of algebra, which are used to *rewrite* algebraic expressions and equations in new, more useful forms. The ability to rewrite algebraic expressions and equations is the common skill involved in the three major components of algebra—*simplifying* algebraic expressions, *solving* algebraic equations, and *graphing* algebraic functions. The following chart shows how this elementary algebra text fits into the "big picture" of algebra.

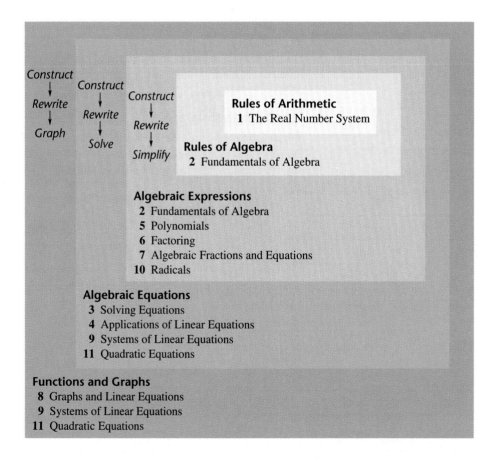

Construct
↓
Rewrite
↓
Graph

Construct
↓
Rewrite
↓
Solve

Construct
↓
Rewrite
↓
Simplify

Rules of Arithmetic
1 The Real Number System

Rules of Algebra
2 Fundamentals of Algebra

Algebraic Expressions
2 Fundamentals of Algebra
5 Polynomials
6 Factoring
7 Algebraic Fractions and Equations
10 Radicals

Algebraic Equations
3 Solving Equations
4 Applications of Linear Equations
9 Systems of Linear Equations
11 Quadratic Equations

Functions and Graphs
8 Graphs and Linear Equations
9 Systems of Linear Equations
11 Quadratic Equations

ELEMENTARY ALGEBRA

CHAPTER ONE

The Real Number System

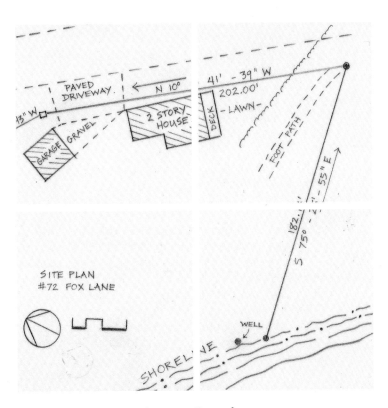

Chapter Overview

*This chapter sets the stage for our study of algebra by reviewing the sets, operations, properties, and algorithms of arithmetic. These are the basic ingredients of what is called a **mathematical system**. A clear understanding of arithmetic is the first step in learning the concepts and applications of algebra.*

Real Numbers: Order and Absolute Value

Sets and Real Numbers   •   The Real Number Line   •   Ordering Real Numbers   •   Absolute Value

Sets and Real Numbers

The ability to communicate precisely is an essential part of a modern society, and it is the primary goal of this text. Specifically, this text concerns the language we use to communicate ideas involving numbers.

The formal term that we use in mathematics to talk about a collection of objects is the word **set**.* For instance, the set

$$\{1, 2, 3\}$$

contains the three numbers 1, 2, and 3. Note that in this text we always use a pair of braces { } to list or describe the members of a set. Parentheses ( ) and brackets [ ] are used to represent other ideas.

The set of numbers that we use in arithmetic is called the set of **real numbers**. The term *real* distinguishes real numbers from *imaginary* numbers—a type of number that is used in some mathematics courses. We will not study imaginary numbers in this course.

The set of real numbers has many important **subsets**, each with a special name. For instance, the set

$$\{1, 2, 3, 4, \ldots\}$$

is called the set of **natural numbers** or **positive integers**. Note that the three dots indicate that the pattern continues. For instance, the set also contains the numbers 5, 6, 7, and so on. Every positive integer is a real number, but there are many real numbers that are not positive integers. For example, the numbers -2, 0, and $\frac{1}{2}$ are real numbers, but they are not positive integers.

Positive integers can be used to describe many things that we encounter in everyday life. For instance, you might be taking four classes this term, or you might be paying $180 a month for rent. But even in everyday life, positive integers cannot describe some concepts accurately. For instance, you could have a zero balance in your checking account, or the temperature could be $-10°$ (ten degrees below zero). To describe such quantities we need to expand the set of positive integers to include **zero** and the

---

*Whenever we formally introduce a mathematical term in this text, the word will occur in boldface type. Be sure you understand the meaning of each new word; it is important that each word becomes part of your mathematical vocabulary.

negative integers. The expanded set is called the set of **integers**, which we write as follows.

$$\underbrace{\{\ldots,\, -3,\, -2,\, -1,}_{\substack{\text{Negative} \\ \text{integers}}}\,\, \overbrace{0,}^{\text{Zero}}\,\, \underbrace{1,\, 2,\, 3,\, \ldots\}}_{\substack{\text{Positive} \\ \text{integers}}}$$

The set of integers is a subset of the set of real numbers. In other words, every integer is a real number.

Even with the set of integers, there are still many quantities in everyday life that we cannot describe accurately. The cost of many items is not in whole-dollar amounts, but in parts of dollars, such as \$1.19 or \$39.98. You might work $8\frac{1}{2}$ hours, or you might miss the first half of a movie. To describe such quantities, we expand the set of integers to include **fractions**. The expanded set is called the set of **rational numbers**. In the formal language of mathematics, we say that a real number is rational if it can be written as a ratio of two integers. Thus, $\frac{3}{4}$ is a rational number; so is 0.5 (it can be written as $\frac{1}{2}$); and so is every integer.

Each of the sets of numbers mentioned—the natural numbers, integers, and rational numbers—is a subset of the set of real numbers, as shown in Figure 1.1. As implied by this figure, some real numbers are not rational. A real number that is not rational is called **irrational** and it cannot be written as the ratio of two integers. One example of an irrational number is $\sqrt{2}$ (read as the positive square root of 2). We will discuss this type of real number in Chapter 10.

FIGURE 1.1   Subsets of Real Numbers

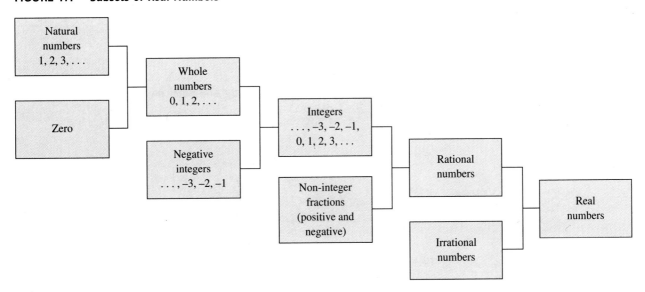

EXAMPLE 1   ■   Classifying Real Numbers

Determine which numbers in the following set are (a) natural numbers, (b) integers, and (c) rational numbers.

$$\left\{\tfrac{1}{2}, -1, 0, 4, -\tfrac{5}{8}, \tfrac{4}{2}, -\tfrac{3}{1}\right\}$$

Solution

(a) Natural numbers: $\left\{4, \tfrac{4}{2} = 2\right\}$

(b) Integers: $\left\{-1, 0, 4, \tfrac{4}{2} = 2, -\tfrac{3}{1} = -3\right\}$

(c) Rational numbers: $\left\{\tfrac{1}{2}, -1, 0, 4, -\tfrac{5}{8}, \tfrac{4}{2}, -\tfrac{3}{1}\right\}$ ■

The Real Number Line

The picture we use to represent the real numbers is called the **real number line**. It consists of a horizontal line with a point (the **origin**) labeled as 0. Numbers to the left of 0 are **negative** and numbers to the right of 0 are **positive**, as shown in Figure 1.2. Zero is neither positive nor negative. Thus, when we want to describe numbers that may be positive *or* zero, we use the term **nonnegative**.

FIGURE 1.2   The Real Number Line

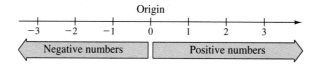

NOTE: Each point on the real number line corresponds to exactly one real number and each real number corresponds to exactly one point on the real number line, as shown in Figure 1.3.

FIGURE 1.3

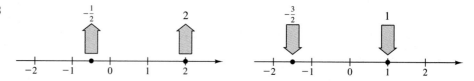

Each point on the real number line corresponds to a real number.

Each real number corresponds to a point on the real number line.

Ordering Real Numbers

The real number line provides us with a way of comparing any two real numbers. For instance, if we choose any two (different) numbers on the real number line, then one of the numbers must be to the left of the other number. We say that the number to the left is **less than** the number to the right, or that the number to the right is **greater than** the number to the left. For example, from Figure 1.4 we can see that -3 is less than 2 because -3 lies to the left of 2 on the number line. We denote a "less than" comparison by the **inequality symbol** $<$. That is, "-3 is less than 2" is denoted by

$$-3 < 2.$$

Similarly, we use the inequality symbol $>$ to denote a "greater than" comparison. For instance, "2 is greater than -3" is denoted by

$$2 > -3.$$

The inequality symbol $\leq$ means **less than or equal to**, and the inequality symbol $\geq$ means **greater than or equal to**.

FIGURE 1.4

-3 lies to the left of 2.

When you are asked to **order** two numbers, you are simply being asked to say which of the two numbers is greater.

EXAMPLE 2 ■ Ordering Integers

Place the correct inequality symbol ($<$ or $>$) between the two numbers.

(a) 3 ___ 5   (b) -3 ___ -5

(c) 4 ___ 0   (d) -2 ___ 2

Solution

See Figure 1.5.

(a) $3 < 5$, because 3 lies to the *left* of 5.

(b) $-3 > -5$, because -3 lies to the *right* of -5.

(c) $4 > 0$, because 4 lies to the *right* of 0.

(d) $-2 < 2$, because -2 lies to the *left* of 2.

FIGURE 1.5

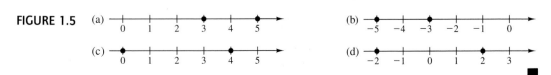

EXAMPLE 3   ■   **Ordering Fractions and Decimals**

Place the correct inequality symbol ($<$ or $>$) between the two numbers.

(a) $\dfrac{1}{3}$  $\dfrac{1}{5}$ (b) $-\dfrac{3}{2}$  $\dfrac{1}{2}$

(c) -3.1 ▨ 2.8 (d) -1.09 ▨ -1.90

Solution

See Figure 1.6.

(a) $\dfrac{1}{3} > \dfrac{1}{5}$, because $\dfrac{1}{3} = \dfrac{5}{15}$ lies to the *right* of $\dfrac{1}{5} = \dfrac{3}{15}$.

(b) $-\dfrac{3}{2} < \dfrac{1}{2}$, because $-\dfrac{3}{2}$ lies to the *left* of $\dfrac{1}{2}$.

(c) $-3.1 < 2.8$, because -3.1 lies to the *left* of 2.8.

(d) $-1.09 > -1.90$, because -1.09 lies to the *right* of -1.90.

FIGURE 1.6

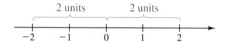

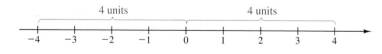

■

Absolute Value

Two real numbers are said to be **opposites** of each other if they lie the same distance from, but on opposite sides of, zero. For example, -2 is the opposite of 2, and 4 is the opposite of -4, as shown in Figure 1.7.

FIGURE 1.7

2 units	2 units

-2 — -1 — 0 — 1 — 2

-2 is the opposite of 2.

4 units	4 units

-4 — -3 — -2 — -1 — 0 — 1 — 2 — 3 — 4

4 is the opposite of -4.

Parentheses are useful for denoting the opposite of a negative number. For example, we interpret $-(-3)$ to mean the opposite of -3, which we know to be 3. That is,

$$-(-3) = 3. \qquad \text{The opposite of } -3 \text{ is } 3$$

For any real number, its distance from zero (on the real number line) is called its **absolute value**. We use a pair of vertical bars, $|\ \ |$, to denote absolute value.

$$|5| = \text{distance between 5 and 0} = 5$$
$$|-8| = \text{distance between } -8 \text{ and } 0 = 8$$

Since opposite numbers lie the same distance from 0 on the real number line, they have the same absolute value. Thus,

$$|5| = 5 \quad \text{and} \quad |-5| = 5$$

or we can write this more simply as $|5| = |-5| = 5$.

NOTE: The absolute value of a real number is either positive or zero. Moreover, zero is the only real number whose absolute value is 0. That is, $|0| = 0$.

In mathematics, we use the word **expression** to mean a collection of numbers and symbols such as $3 + 5$ or $|-4|$. When you are asked to **evaluate** an expression, you are being asked to find the *number* that is equal to the expression.

EXAMPLE 4 ■ Evaluating Absolute Value

Evaluate the following expressions.

(a) $|-10|$ (b) $\left|\dfrac{3}{4}\right|$ (c) $|-3.2|$ (d) $-|-6|$

Solution

(a) $|-10| = 10$, because the distance between -10 and 0 is 10.

(b) $\left|\frac{3}{4}\right| = \frac{3}{4}$, because the distance between $\frac{3}{4}$ and 0 is $\frac{3}{4}$.

(c) $|-3.2| = 3.2$, because the distance between -3.2 and 0 is 3.2.

(d) $-|-6| = -(6) = -6$.

Note in part (d) that $-|-6| = -6$ does not contradict the fact that the absolute value of a real number cannot be negative. ■

EXAMPLE 5 ■ Comparing Absolute Values

Place the correct symbol ($<$, $>$, or $=$) between the two numbers.

(a) $|-9|$ ▨ $|9|$ (b) 0 ▨ $|-5|$

(c) -4 ▨ $-|-4|$ (d) $|12|$ ▨ $|-15|$

Solution

(a) $|-9| = |9|$, because both are equal to 9.

(b) $0 < |-5|$, because $|-5| = 5$ and 0 is less than 5.

(c) $-4 = -|-4|$, because both numbers are equal to -4.

(d) $|12| < |-15|$, because $|12| = 12$ and $|-15| = 15$ and 12 is less than 15. ■

DISCUSSION PROBLEM ■ Comparing Opposites and Absolute Value

Write a short paragraph describing the difference between the *opposite* of a real number and the *absolute value* of a real number. ■

1.1   EXERCISES

In Exercises 1–8, write the real numbers shown by the points on the real number line and place the correct inequality symbol ($<$ or $>$) between the two numbers.

1.

2.
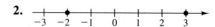

3.
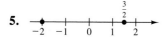

4.

5.

6.

7.

8.

In Exercises 9–20, show each real number as a point on the real number line and place the correct inequality symbol between the two numbers.

9. $\dfrac{1}{3}$ ▨ 4

10. 6 ▨ -2

11. 4 ▨ $-\dfrac{7}{2}$

12. 2 ▨ $\dfrac{3}{2}$

13. -8 ▨ -10

14. $-\dfrac{7}{2}$ ▨ $-\dfrac{7}{3}$

15. -4.6 ▨ 1.5

16. 28.60 ▨ -3.75

17. $\dfrac{7}{16}$ ▨ $\dfrac{5}{8}$

18. $-\dfrac{3}{8}$ ▨ $-\dfrac{5}{8}$

19. 0 ▨ $-\dfrac{7}{16}$

20. 0 ▨ $\dfrac{7}{4}$

In Exercises 21–24, determine the distance between the indicated point and zero on the real number line and write an appropriate absolute value expression for the distance.

21.
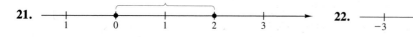

22.
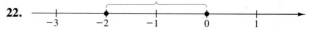

23.
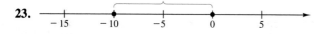

24.

In Exercises 25–28, find the opposite of the given number.

25. 5   **26.** 2   **27.** -3 **28.** -7

In Exercises 29–38, evaluate the expression.

29. $|7|$ **30.** $|76.3|$ **31.** $|-3.4|$ **32.** $|-16.2|$

33. $\left|-\dfrac{7}{2}\right|$ **34.** $\left|-\dfrac{9}{16}\right|$ **35.** $-|4.09|$ **36.** $-|-43.8|$

37. $-|-23.6|$ **38.** $-|91|$

In Exercises 39–50, place the correct symbol ($<$, $>$, or $=$) between the two real numbers.

39. $|-15|$ ▨ $|15|$ **40.** $|525|$ ▨ $|-525|$

41. $|-4|$ ▨ $|3|$ **42.** $|16|$ ▨ $|-25|$

43. $|32|$ ▨ $|-50|$ **44.** $|1026|$ ▨ $|800|$

45. $\left|\dfrac{3}{16}\right|$ ▨ $\left|\dfrac{3}{2}\right|$ **46.** $\left|-\dfrac{7}{8}\right|$ ▨ $\left|\dfrac{4}{3}\right|$

47. $-|-48.5|$ ▨ $|-48.5|$ **48.** $-|-64|$ ▨ $|-50|$

49. $|-3.1|$ ▨ $-|-3.1|$ **50.** $|-4.9|$ ▨ $|-10.2|$

In Exercises 51–54, show the given numbers on the real number line. (*Hint:* Start by writing each number in decimal form. A calculator might help with the fractions.)

51. $\dfrac{5}{2}, \dfrac{3}{2}, -2, -|-3|$ **52.** $3.7, \dfrac{16}{3}, |-1.9|, -\dfrac{1}{2}$

53. $-4, \dfrac{7}{3}, |-3|, 0$ **54.** $|-2.3|, 3.2, -2.3, -|3.2|$

In Exercises 55–60, determine whether the statement is true or false. If false, give an example that shows it is false.

55. The absolute value of any real number is positive.

56. The absolute value of a number is equal to the absolute value of its opposite.

57. The absolute value of a rational number is a rational number.

58. A given real number corresponds to exactly one point on the real number line.

59. The opposite of a positive number is a negative number.

60. Every rational number is an integer.

Introduction

In this section we discuss the four operations of arithmetic on the set of integers: addition, subtraction, multiplication, and division. As we proceed through this chapter, try to keep in mind the overall picture of a *mathematical system* (refer to the illustration in "What Is Algebra?" at the beginning of the text), and note the particular parts discussed in each section.

Addition of Integers

We summarize the operation of **addition** of integers as follows.

Addition of Integers	1. To add two integers *with like signs*, add their absolute values and attach the common sign to the result.
	2. To add two integers *with unlike signs*, subtract the smaller absolute value from the greater absolute value and attach the sign of the integer with the greater absolute value.

The result of adding two or more numbers is called the **sum** of the numbers, and each individual number is called a **term.** For instance, the sum of 3 and 5 is 8, because $3 + 5 = 8$.

EXAMPLE 1 ■ Adding Integers

(a) $-84 + 14 = -70$ (b) $-15 + 72 = 57$

(c) $-138 + (-62) = -200$ (d) $27 + (-52) + 13 = -12$ ■

To add two or more positive integers, it is often convenient to use the **carrying algorithm** (or procedure) with a vertical arrangement of the numbers. For instance, to evaluate the sum $148 + 62 + 536$, the carrying algorithm has the following form.

$$
\begin{array}{r}
^{1\,1} \\
148 \\
62 \\
+536 \\
\hline
746
\end{array}
$$

At the beginning of the chapter, we discussed the concept of a mathematical system. You now have an example of such a system. It consists of a set of numbers (the integers), an operation (addition), and algorithms for addition. The fourth part of this system (properties of addition) will be discussed in Section 1.5.

Subtraction of Integers

To **subtract** one number from another, add the opposite of the number being subtracted to the other number. Here are some examples.

Subtraction	*Related Addition*
$3 - 8$	$3 + (-8) = -5$
$10 - (-13)$	$10 + (13) = 23$
$-5 - 12$	$-5 + (-12) = -17$
$7 - 5$	$7 + (-5) = 2$

The expression "$7 - 5$" can be read as "7 minus 5" or "5 subtracted from 7."

Subtraction of Integers	To subtract one integer from another, change the sign of the integer being subtracted, and add it to the other integer. Another way of saying this is: to subtract an integer, add its opposite.

The result of subtracting one number from another number is called the **difference** of the two numbers. For instance, the difference of 12 and 9 is 3, because $12 - 9 = 3$.

EXAMPLE 2 ■ Subtracting Integers

Evaluate the following.

(a) $-14 - 6$ (b) $-25 - (-37)$ (c) $-4 - (-17) - 23$

Solution

(a) $-14 - 6 = -14 + (-6) = -20$
(b) $-25 - (-37) = -25 + (37) = 12$
(c) $-4 - (-17) - 23 = -4 + (17) + (-23) = -10$ ■

EXAMPLE 3 ■ Comparing Verbal Forms with Symbolic Forms

Verbal Form	*Symbolic Form*
(a) Negative five	-5
(b) Seven subtracted from twelve	$12 - 7$

	Verbal Form	*Symbolic Form*
(c)	Eleven minus six	$11 - 6$
(d)	The opposite of negative eight	$-(-8) = 8$
(e)	Minus four	-4

Note in parts (c) and (e) that the word *minus* is used in two ways in mathematics. It can denote subtraction, as in "eleven minus six," or it can denote a negative number, as in "minus four." ■

For subtraction problems involving only two integers, we can use the following **borrowing algorithm**.

$$
\begin{array}{r}
3\ 10\ 15 \\
\cancel{4}\ \cancel{1}\ \cancel{5} \\
-2\ 7\ 6 \\
\hline
1\ 3\ 9
\end{array}
$$

Multiplication of Integers

Multiplication of two integers can be described as repeated addition (or subtraction). Here are some examples.

Multiplication	*Repeated Addition*
$3 \times 5 = 15$	$\underbrace{5 + 5 + 5}_{\text{Add 5 three times}} = 15$
$4 \times (-2) = -8$	$\underbrace{(-2) + (-2) + (-2) + (-2)}_{\text{Add } -2 \text{ four times}} = -8$

Multiplying two integers is referred to as finding the **product** of two integers, and the integers are called **factors** of the product. We denote multiplication in a variety of ways. For instance,

$$7 \times 3, \quad 7 \cdot 3, \quad 7(3), \quad (7)3, \quad \text{and} \quad (7)(3)$$

all denote the product of "7 times 3," which we know is 21.

Multiplication of Integers

1. To multiply two integers *with like signs*, find the product of their absolute values and attach a *positive* sign. (In practice, when *no* sign is attached, the result is assumed to be positive.)

2. To multiply two integers *with unlike signs*, find the product of their absolute values and attach a *negative* sign.

The product of zero and any other number is zero. For instance, if we multiply 0 and 4, we obtain $(0)(4) = 0$.

EXAMPLE 4 ■ **Multiplying Integers**

Find the following products.

(a) $-6 \cdot 9$ (b) $(-5)(-7)$ (c) $3(-12)$

(d) $-12 \cdot 0$ (e) $5(-3)(-4)(7)$

Solution

(a) $-6 \cdot 9 = -54$ (b) $(-5)(-7) = 35$

(c) $3(-12) = -36$ (d) $-12 \cdot 0 = 0$

(e) To find the product of more than two numbers, we find the product of their absolute values. If there is an even number of negative factors, then the product is positive. If there is an odd number of negative factors, then the product is negative. In this case there are two negative factors, so the product must be positive, and we write

$$5(-3)(-4)(7) = 420.$$ ■

Be careful to properly distinguish between expressions like $3(-5)$ and $3 - 5$, or $-3(-5)$ and $-3 - 5$. The first of each pair is a multiplication problem while the second is a subtraction problem.

Multiplication	*Subtraction*
$3(-5) = -15$	$3 - 5 = -2$
$-3(-5) = 15$	$-3 - 5 = -8$

To multiply two integers having two or more digits, we suggest the **vertical multiplication algorithm** demonstrated in the next example. The sign of the product is determined by the usual multiplication rule.

EXAMPLE 5 ■ **Using the Vertical Multiplication Algorithm**

Multiply -34 and 78 using the vertical multiplication algorithm.

Solution

Using the vertical algorithm with the absolute values of the factors, we obtain the following.

```
      78
  ×   34
  ------
     312   ⟸  Multiply 4 times 78
     234   ⟸  Multiply 3 times 78
  ------
    2652   ⟸  Add columns
```

Now, since the factors have unlike signs, it follows that

$$-34 \times 78 = -2,652.$$ ■

Division of Integers

Just as we used addition to describe subtraction of two integers, we can use multiplication to describe the **division** of two integers. Consider the following illustrations.

Multiplication	*Division*
$12 = 3 \cdot 4$	$\dfrac{12}{4} = 3$
$15 = 5 \cdot 3$	$\dfrac{15}{3} = 5$
$15 = (-5)(-3)$	$\dfrac{15}{-3} = -5$
$-15 = 5(-3)$	$\dfrac{-15}{-3} = 5$

Division is denoted by the symbols $\div$ or $/$, or by a horizontal line. For example,

$$30 \div 6 = 30/6 = \frac{30}{6} = 5.$$

The result of dividing one number by another is called the **quotient** of the two numbers. Using the symbol $30 \div 6$, we call 30 the **dividend** and 6 the **divisor**. Using the symbols $30/6$ or $\frac{30}{6}$, we call 30 the **numerator** and 6 the **denominator**.

Division of Integers

1. To divide two integers *with like signs*, find the quotient of their absolute values and attach a *positive* sign (or no sign).

2. To divide two integers *with unlike signs*, find the quotient of their absolute values and attach a *negative* sign.

NOTE: Because the division of two integers can be rewritten as a multiplication problem, the rule of signs for division is the same as it is for multiplication.

EXAMPLE 6 ■ Dividing Integers

Perform the following divisions.

(a) $\dfrac{-42}{-6}$ (b) $36 \div (-9)$ (c) $\dfrac{0}{-13}$ (d) $-105 \div 7$

Solution

(a) $\dfrac{-42}{-6} = 7$ because $-42 = 7(-6)$.

(b) $36 \div (-9) = -4$ because $(-4)(-9) = 36$.

(c) $\dfrac{0}{-13} = 0$ because $(0)(-13) = 0$.

(d) $-105 \div 7 = -15$ because $(-15)(7) = -105$.   ■

In Example 6(c), note that there is nothing wrong with dividing 0 by a number that is not zero. The result of such a division is always zero. For instance,

$$\frac{0}{3} = 0 \quad \text{and} \quad \frac{0}{-4} = 0.$$

Division by zero, on the other hand, is not defined. For instance, the following expressions have no meaning.

$$\frac{3}{0} \quad \text{and} \quad \frac{-4}{0} \qquad \text{Division by zero is undefined}$$

Thus, whenever you are working with a division problem, you should be sure to check that the denominator (the divisor) is not zero.

When doing division problems involving large integers, the **long division algorithm** described in the following example can be used.

EXAMPLE 7   ■   Long Division Algorithm

Use the long division algorithm to find the following quotient.

$$\frac{351}{13}$$

Solution

$$
\begin{array}{r}
27 \\
13\overline{)351} \\
26 \\
\hline
91 \\
91 \\
\hline
\end{array}
$$

Hence, the solution is

$$\frac{351}{13} = 27.$$ ■

Applications

To complete this section we look at an application that illustrates the use of all four arithmetic operations: addition, subtraction, multiplication, and division.

EXAMPLE 8   ■   An Application: Stock Purchase

Suppose on Monday you bought $500 worth of stock in a company. During the rest of that week, you recorded the gains and losses, as shown in Table 1.1.

TABLE 1.1

Tuesday	Wednesday	Thursday	Friday
Gained $15	Lost $18	Lost $23	Gained $10

(a) What was the value of the stock at the close of Tuesday?
(b) What was the value of the stock at the close of Wednesday?
(c) What was the value of the stock at the end of the week?
(d) What would the total loss have been if Thursday's loss had occurred on each of the four days?
(e) What was the average daily gain (or loss) for the four days recorded?

Solution

(a) Since the original value of the stock was $500, and the stock gained $15 by the close of Tuesday, its value at the close of Tuesday was

$$500 + 15 = \$515.$$

(b) Using the result of part (a), the value at the close of Wednesday was

$$515 - 18 = \$497.$$

(c) The value of the stock at the end of the week was

$$500 + 15 - 18 - 23 + 10 = \$484.$$

(d) The loss on Thursday was $23. If this loss had occurred each day, the total loss would have been

$$4(23) = \$92.$$

(e) To find the average of the four gains and losses, we add and divide by 4. Thus, the average is

$$\text{Average} = \frac{15 + (-18) + (-23) + 10}{4} = \frac{-16}{4} = -4.$$

This means that during the four days, the stock had an average loss of $4 per day.   ■

DISCUSSION PROBLEM   ■   Using a Calculator

Use a calculator to perform each of the operations in the examples in this section.*
Write a short paragraph describing the difference between the calculator key marked

---

*The graphing calculator keystrokes given in this text correspond to the *TI-81* graphing calculator from *Texas Instruments*. For other graphing calculators, the keystrokes may differ.

$\boxed{+/-}$ and the calculator key marked $\boxed{-}$. (The first of these keys is called the **change sign key** and the second is called the **subtraction key**.) Illustrate this difference by showing which keys you would use to perform the following operation.

$$-3 - 4$$

Some calculators (including most graphing calculators) have a key labeled $\boxed{(-)}$ rather than a change sign key. If you have access to both a scientific calculator and a graphing calculator, explain how the key $\boxed{(-)}$ is different from a change sign key.   ■

Warm-Up

The following warm-up exercises involve skills that were covered in earlier sections. You will use these skills in the exercise set for this section.

In Exercises 1–4, show each real number as a point on the real line and place the correct inequality symbol ($<$ or $>$) between the real numbers.

1. -2.5 ▨ -4 **2.** $\dfrac{3}{16}$ ▨ $\dfrac{3}{8}$

3. -3.1 ▨ 2.7 **4.** 4.3 ▨ -1

In Exercises 5 and 6, evaluate the given expression.

5. $-|-0.75|$ **6.** $|25.2|$

In Exercises 7–10, place the correct symbol ($<$, $>$, or $=$) between the real numbers.

7. $\left|\dfrac{7}{2}\right|$ ▨ $|-3.5|$ **8.** $\left|\dfrac{3}{4}\right|$ ▨ $-|0.75|$

9. $|-25|$ ▨ $-|20|$ **10.** 0 ▨ $|-3|$

1.2   EXERCISES

In Exercises 1–10, find the sum.

1. $2 + 7$ **2.** $10 + (-3)$ **3.** $-6 + 4$ **4.** $(-8) + (-3)$

5. $-1 + 0$ **6.** $-3 + 0$ **7.** $14 + (-14)$ **8.** $(-8) + 8$

9. $(-14) + 13$ **10.** $(-20) + 19$

In Exercises 11–20, find the sum.

11. $-10 + 6 + 34$ **12.** $-15 + (-3) + 8$ **13.** $0 + 2 + 4$ **14.** $6 + 0 + 2$

15. $32 + (-32) + 16$ **16.** $-312 + (-564) + (-100)$ **17.** $49 + (-|-17|)$

18. $|-10| + |35|$ **19.** $550 + (-1,625) + (-4,060) + 7,132$ **20.** $-730 + 1,820 + 3,150 + (-10,000)$

In Exercises 21–26, find the difference by first writing the given subtraction problem in its related addition form.

21. $12 - 9$ **22.** $8 - 2$ **23.** $-10 - (-13)$ **24.** $-4 - (-1)$

25. $4 - (-4)$ **26.** $9 - (-6)$

In Exercises 27–40, evaluate the given expression.

27. $55 - 20$ **28.** $39 - 13$ **29.** $72 - 85$ **30.** $400 - 525$

31. $1,000 - (-500)$ **32.** $2,500 - (-600)$ **33.** $-210 - 400$ **34.** $-110 - (-30)$

35. $|15| - |-7|$ **36.** $|-100| - |25|$ **37.** $-2 - 4 - 6$ **38.** $-3 - 5 - 1$

39. $-2 - (-3) - (-4)$ **40.** $10 - (-3) - (-8)$

In Exercises 41–44, write each multiplication as a repeated addition problem, and find the product.

41. $3 \cdot 2$ **42.** $4 \cdot 4$ **43.** $4 \times (-3)$ **44.** $5 \times (-2)$

In Exercises 45–56, find the product.

45. 7×30 **46.** $0 \cdot 20$ **47.** $4(-8)$ **48.** $10(-25)$

49. $(-6)(-9)$ **50.** $(-20)(-8)$ **51.** $(310)(-32)$ **52.** $(-500)(-6)$

53. $5(-3)(6)$ **54.** $-7(3)(-1)$ **55.** $(142)(217)$ **56.** $(156)(393)$

In Exercises 57–66, perform the indicated division, if possible. If not possible, state the reason.

57. $27 \div 9$ **58.** $35 \div 5$ **59.** $72 \div (-12)$ **60.** $(-28) \div 4$

61. $\dfrac{-81}{-3}$ **62.** $\dfrac{0}{8}$ **63.** $\dfrac{8}{0}$ **64.** $\dfrac{-125}{-25}$

65. $\dfrac{-58}{2}$ **66.** $\dfrac{48}{4}$

In Exercises 67–76, perform the indicated division using the long division algorithm.

67. $1,440 \div 45$ **68.** $1,312 \div 16$ **69.** $2,750 \div 25$ **70.** $22,010 \div 71$

71. $\dfrac{2,209}{47}$ **72.** $\dfrac{8,700}{116}$ **73.** $\dfrac{44,290}{515}$ **74.** $\dfrac{33,511}{47}$

75. $\dfrac{169,290}{162}$ **76.** $\dfrac{1,027,500}{250}$

77. *Temperature Change* The temperature at 6 A.M. was −10° F. Determine the temperature at noon if it had increased by 22° F.

78. *Balance in an Account* During a given month you made withdrawals of $350 and $500 from your savings account. You also made one deposit of $450. The balance was $2,750 at the beginning of the month. Find the balance at the end of the month. (Disregard any interest that may have been earned.)

79. *Flying Altitude* A plane flying at an altitude of 31,000 feet is instructed to descend to an altitude of 24,000 feet. How many feet must the aircraft descend?

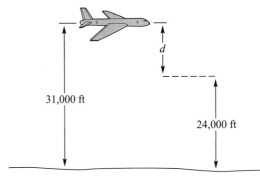

80. *Profit and Loss* A company showed a loss of $650,000 during the first six months of a given year. The company ended the year with an overall profit of $362,000. What was the profit during the last six months of the year?

81. *Savings Plan* Suppose you decide to save $50 per month for 10 years. Determine the total amount that you would save during that period of time.

82. *Total Cost* Computer printer ribbons cost $8 per ribbon. There are 10 ribbons per box, and you order six boxes. What is the total cost of the order?

83. *Stock Values* On Monday you purchased $800 worth of stock. The value of the stock during the remainder of the week is shown in the accompanying **bar chart**. Use the chart to complete the following table showing the daily gains and losses during the week.

Figure for 83

Day	Daily Gain or Loss
Tues	
Wed	
Thur	
Fri	

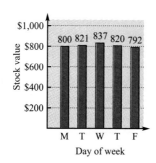

84. *College Enrollment* The total student enrollment at a college is shown in the accompanying bar chart. Complete the following table showing the annual enrollment gain or loss during the indicated five years. Assume the enrollment for 1987 was 3,560.

Year	Yearly Gain or Loss
1988	
1989	
1990	
1991	
1992	

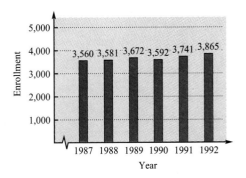

85. If a negative number is used as a factor 25 times, what is the sign of the product?

86. If a negative number is used as a factor 16 times, what is the sign of the product?

Operations on the Set of Rational Numbers

Introduction   •   Addition and Subtraction of Rational Numbers   •
Multiplication and Division of Rational Numbers   •   Decimal Fractions   •   Applications

Introduction

Remember that a rational number is a real number that can be written as the ratio of two integers, where the top number is called the **numerator** and the bottom number the **denominator**. Some examples are

$$\frac{3}{8}, \quad \frac{12}{5}, \quad \frac{1}{7}, \quad \text{and} \quad \frac{3}{1}.$$

Note that writing the fraction $\frac{3}{1}$ is simply another way of writing the integer 3.

Suppose a pizza is cut into 12 equal-sized pieces, as shown in Figure 1.8. If you eat 3 pieces and your friend eats 4 pieces, what portion of the pizza has been eaten? Because 7 of 12 pieces were eaten, we can reason that $\frac{7}{12}$ of the pizza was eaten. In other words, to find the portion that was eaten, we simply add the fractions $\frac{3}{12}$ and $\frac{4}{12}$ to get

$$\frac{3}{12} + \frac{4}{12} = \frac{3+4}{12} = \frac{7}{12}.$$

Now suppose that you ate $\frac{1}{4}$ of the pizza and your friend ate $\frac{1}{3}$ of it. In this case, what portion of the pizza was eaten? This seems to be a more difficult problem. But why? Isn't it because in the first instance, the fractions $\frac{3}{12}$ and $\frac{4}{12}$ are *alike* (they have *like* denominators), whereas in the second instance, the fractions $\frac{1}{4}$ and $\frac{1}{3}$ are *not alike* (they have *unlike* denominators)?

FIGURE 1.8

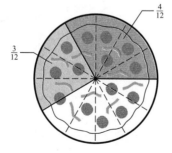

Addition and Subtraction of Rational Numbers

To add two fractions with unlike denominators, we first rewrite the two fractions so that they have like (or common) denominators. Once this is done, we can add the fractions by simply adding their numerators and writing the sum over the common denominator.

Let's see how this works with $\frac{1}{4}$ and $\frac{1}{3}$. First, we try to find the smallest positive integer, called the **least common multiple**, that is a multiple of both 3 and 4. In this case, the least common multiple is 12. We then use this number as our common denominator for rewriting the given fractions.

To write the fraction $\frac{1}{3}$ so that it has a denominator of 12, we multiply both the numerator and the denominator by 4. (This is equivalent to multiplying the fraction by 1.)

$$\frac{1}{3} = \frac{1(4)}{3(4)} = \frac{4}{12}$$

Similarly, to write the fraction $\frac{1}{4}$ so that it has a denominator of 12, we multiply its numerator and denominator by 3.

$$\frac{1}{4} = \frac{1(3)}{4(3)} = \frac{3}{12}$$

Since both fractions now have like denominators, we can add them as follows.

$$\frac{1}{4} + \frac{1}{3} = \frac{3}{12} + \frac{4}{12} = \frac{3+4}{12} = \frac{7}{12}$$

We summarize the rules for adding fractions as follows.

Addition of Fractions

1. To add two fractions *with like denominators*, add their numerators and write the sum over the like (or common) denominator.

2. To add two fractions *with unlike denominators*, rewrite both fractions so that they have like denominators. Then use the rule for adding fractions with like denominators.

NOTE: Adding fractions with unlike denominators is an example of a basic problem-solving strategy that is used in mathematics—rewriting a given problem in a simpler or more familiar form.

EXAMPLE 1   ■   Adding Fractions with Like Denominators

Find the following sum.

$$\frac{3}{7} + \frac{5}{7}$$

Solution

These two fractions have like denominators. Therefore, we have

$$\frac{3}{7} + \frac{5}{7} = \frac{8}{7}.$$ ■

EXAMPLE 2   ■   Adding Fractions with Unlike Denominators

Find the following sum.

$$\frac{3}{8} + \frac{-5}{12}$$

Solution

These two fractions have unlike denominators. In order to add them, we first find a common denominator. In this case, 24 is a common multiple of both 8 and 12. Thus, we add the two fractions as follows.

$$\frac{3}{8} + \frac{-5}{12} = \frac{3(3)}{8(3)} + \frac{(-5)(2)}{12(2)}$$ Find common denominator

$$= \frac{9}{24} + \frac{-10}{24}$$ Like fractions

$$= \frac{9 + (-10)}{24}$$ Add numerators

$$= \frac{-1}{24}$$

$$= -\frac{1}{24}$$ ■

In Example 2 note that the fraction $\frac{-1}{24}$ can be written as $-\frac{1}{24}$. Be sure you see that the following statements are true.

$$\frac{-1}{3} = \frac{1}{-3} = -\frac{1}{3}$$ *Unlike* signs yield a negative fraction

$$\frac{-2}{-5} = \frac{2}{5}$$ *Like* signs yield a positive fraction

We summarize these rules of signs for fractions as follows.

Rules of Signs for Fractions

1. If the numerator and denominator of a fraction have *like* signs, the value of the fraction is *positive*.

2. If the numerator and denominator of a fraction have *unlike* signs, the value of the fraction is *negative*.

EXAMPLE 3   ■   Adding a Fraction and an Integer

Find the following sum.

$$\frac{1}{4} + 2$$

Solution

Considering 2 to have a denominator of 1, we use the rule for adding fractions with unlike denominators.

$$\frac{1}{4} + \frac{2}{1} = \frac{1}{4} + \frac{2(4)}{1(4)} \qquad \text{Find common denominator}$$

$$= \frac{1}{4} + \frac{8}{4} \qquad \text{Like fractions}$$

$$= \frac{1 + 8}{4} \qquad \text{Add numerators}$$

$$= \frac{9}{4}$$

EXAMPLE 4 ■ Adding a Mixed Number and a Fraction

Find the following sum.

$$1\frac{4}{5} + \frac{11}{15}$$

Solution

To begin, we rewrite the **mixed number** $1\frac{4}{5}$ as a fraction.

$$1\frac{4}{5} = 1 + \frac{4}{5} = \frac{5}{5} + \frac{4}{5} = \frac{9}{5}$$

Then we add the two fractions as follows.

$$1\frac{4}{5} + \frac{11}{15} = \frac{9}{5} + \frac{11}{15} = \frac{9(3)}{5(3)} + \frac{11}{15} = \frac{27}{15} + \frac{11}{15} = \frac{38}{15}$$

■

NOTE: In Example 4, a common shortcut for writing $1\frac{4}{5}$ as $\frac{9}{5}$ is to multiply 1 by 5, add the result to 4, and then divide by 5, as follows.

$$1\frac{4}{5} = \frac{1(5) + 4}{5} = \frac{9}{5}$$

A fraction is said to be in **reduced form** if its numerator and denominator have no common factors (other than 1). For instance, the fraction $\frac{2}{3}$ is in reduced form because 2 and 3 have no common factors other than 1. On the other hand, the fraction $\frac{8}{12}$ is not in reduced form because 8 and 12 have a common factor of 4. When we rewrite a fraction so that it is in reduced form, we say we are **reducing** or **simplifying** the fraction. We do this by dividing the numerator and denominator by all common factors. For instance, we reduce the fraction $\frac{8}{12}$ as follows.

$$\frac{8}{12} = \frac{2(\overset{1}{\cancel{4}})}{3(\underset{1}{\cancel{4}})} = \frac{2(1)}{3(1)} = \frac{2}{3} \qquad \text{Reduce factors}$$

When you obtain a fraction as an answer to a problem, it is usually a good idea to check to see if the fraction can be reduced. Although it is not a hard and fast rule, we normally consider the reduced fraction $\frac{2}{3}$ to be a better form for an answer than the fraction $\frac{8}{12}$.

To subtract fractions, we use the strategy we used for integers. We simply add the opposite fraction. Note how this is done in Examples 5 and 6.

EXAMPLE 5 ■ **Subtracting Fractions**

Perform the following subtractions.

(a) $\dfrac{3}{7} - \dfrac{5}{7}$ (b) $\dfrac{7}{9} - \dfrac{11}{12}$

Solution

(a) $\dfrac{3}{7} - \dfrac{5}{7} = \dfrac{3}{7} + \dfrac{-5}{7}$ To subtract, add the opposite

$= \dfrac{3 + (-5)}{7}$

$= \dfrac{-2}{7}$

$= -\dfrac{2}{7}$

(b) $\dfrac{7}{9} - \dfrac{11}{12} = \dfrac{7(4)}{9(4)} + \dfrac{-11(3)}{12(3)}$ Find common denominator

$= \dfrac{28}{36} + \dfrac{-33}{36}$ Like fractions

$= \dfrac{28 + (-33)}{36}$ Add numerators

$= \dfrac{-5}{36}$

$= -\dfrac{5}{36}$ ■

EXAMPLE 6 ■ **Subtracting Fractions**

Perform the following subtractions.

(a) $\dfrac{5}{16} - \left(-\dfrac{7}{30}\right)$ (b) $3\frac{5}{8} - 1\frac{7}{10}$

Solution

(a) $\dfrac{5}{16} - \left(-\dfrac{7}{30}\right) = \dfrac{5(15)}{16(15)} + \dfrac{7(8)}{30(8)}$ Find common denominator

$= \dfrac{75}{240} + \dfrac{56}{240}$ Like fractions

$= \dfrac{75 + 56}{240}$ Add numerators

$= \dfrac{131}{240}$

(b) As we become more familiar with a mathematical process, our skills increase and we tend to perform some of the steps mentally (without writing out all of the details). For instance, a quick solution to the problem $3\frac{5}{8} - 1\frac{7}{10}$ might involve writing only the following steps.

$$3\frac{5}{8} - 1\frac{7}{10} = \frac{29}{8} - \frac{17}{10} = \frac{29(5)}{8(5)} - \frac{17(4)}{10(4)} = \frac{145}{40} - \frac{68}{40} = \frac{77}{40}$$ ■

EXAMPLE 7   ■   **Combining Three or More Fractions**

Evaluate the following.

$$\frac{5}{6} - \frac{7}{15} + \frac{3}{10} - 1$$

Solution

The least common denominator of 6, 15, and 10 is 30. Therefore, we can rewrite the given expression as

$$\frac{5}{6} - \frac{7}{15} + \frac{3}{10} - 1 = \frac{5(5)}{6(5)} + \frac{(-7)(2)}{15(2)} + \frac{3(3)}{10(3)} + \frac{(-1)(30)}{30}$$

$$= \frac{25}{30} + \frac{-14}{30} + \frac{9}{30} + \frac{-30}{30}$$

$$= \frac{25 - 14 + 9 - 30}{30}$$

$$= \frac{-10}{30}$$

$$= -\frac{1(\cancel{10})}{3(\cancel{10})}$$ Reduce fraction

$$= -\frac{1}{3}.$$ ■

In Example 7, the cancel marks are a shorthand notation for the following steps.

$$-\frac{1(10)}{3(10)} = -\frac{1}{3} \cdot \frac{10}{10} = -\frac{1}{3} \cdot 1 = -\frac{1}{3}$$

Multiplication and Division of Rational Numbers

To find the product of two rational numbers in fraction form, we simply multiply numerators and multiply denominators. Note how it's done in the next example.

Multiplication of Fractions	To multiply two fractions, multiply the two numerators to form the numerator of the product, and multiply the two denominators to form the denominator of the product.

EXAMPLE 8 ■ Multiplying Fractions

Multiply the following and give the answers in reduced form.

(a) $\dfrac{5}{8} \cdot \dfrac{3}{2}$

(b) $\left(-\dfrac{7}{9}\right)\left(-\dfrac{5}{21}\right)$

(c) $\dfrac{8}{15}\left(-\dfrac{6}{10}\right)$

(d) $\left(3\tfrac{1}{5}\right)\left(-\dfrac{7}{6}\right)\left(\dfrac{5}{3}\right)$

Solution

(a) $\dfrac{5}{8} \cdot \dfrac{3}{2} = \dfrac{5 \cdot 3}{8 \cdot 2} = \dfrac{15}{16}$ Multiply numerators and denominators

(b) $\left(-\dfrac{7}{9}\right)\left(-\dfrac{5}{21}\right) = \dfrac{7}{9} \cdot \dfrac{5}{21} = \dfrac{7(5)}{9(21)} = \dfrac{7(5)}{9(3)(7)} = \dfrac{5}{27}$ Multiply and reduce

(c) $\dfrac{8}{15}\left(-\dfrac{6}{10}\right) = -\dfrac{8(6)}{(15)(10)} = -\dfrac{(8)(2)(3)}{(3)(5)(2)(5)} = -\dfrac{8}{25}$ Multiply and reduce

(d) $\left(3\tfrac{1}{5}\right)\left(-\dfrac{7}{6}\right)\left(\dfrac{5}{3}\right) = \left(\dfrac{16}{5}\right)\left(-\dfrac{7}{6}\right)\left(\dfrac{5}{3}\right)$

$= -\dfrac{(8)(2)(7)(5)}{(5)(3)(2)(3)}$ Multiply

$= -\dfrac{56}{9}$ Reduce ■

The **reciprocal** of a number is the number by which it must be multiplied to obtain 1. For instance, the reciprocal of 3 is $\tfrac{1}{3}$ because $3\left(\tfrac{1}{3}\right) = 1$. Similarly, the reciprocal of $-\tfrac{2}{3}$ is $-\tfrac{3}{2}$ because $\left(-\tfrac{2}{3}\right)\left(-\tfrac{3}{2}\right) = 1$.

To divide two fractions, we multiply the first fraction by the reciprocal of the second fraction. Another way of saying this is "invert the divisor and multiply." The next two examples should make this clear.

EXAMPLE 9   ■   Dividing Fractions

Perform the following divisions and write the answers in reduced form.

(a) $\dfrac{5}{8} \div \dfrac{20}{12}$ (b) $\dfrac{6}{13} \div \left(-\dfrac{9}{26}\right)$

Solution

(a) $\dfrac{5}{8} \div \dfrac{20}{12} = \dfrac{5}{8} \cdot \dfrac{12}{20} = \dfrac{(5)(12)}{(8)(20)}$ Invert divisor and multiply

$= \dfrac{(5)(3)(4)}{(8)(4)(5)} = \dfrac{3}{8}$ Reduce

(b) $\dfrac{6}{13} \div \left(-\dfrac{9}{26}\right) = \dfrac{6}{13} \cdot \left(-\dfrac{26}{9}\right)$ Invert divisor and multiply

$= -\dfrac{(6)(26)}{(13)(9)}$

$= -\dfrac{(2)(3)(2)(13)}{(13)(3)(3)} = -\dfrac{4}{3}$ Reduce   ■

In Example 9(a), note that the expression $\frac{5}{8} \div \frac{20}{12}$ can also be written in the **compound fraction** form

$$\dfrac{5/8}{20/12}.$$

The same rule applies—invert the divisor (the bottom fraction) and multiply to obtain $\frac{5}{8} \cdot \frac{12}{20}$.

EXAMPLE 10   ■   Dividing Fractions

Perform the following divisions and write the answers in reduced form.

(a) $-\dfrac{8}{3} \div 6$ (b) $3\dfrac{9}{13} \div 1\dfrac{5}{8}$

Solution

(a) $-\dfrac{8}{3} \div 6 = -\dfrac{8}{3} \div \dfrac{6}{1}$

$= -\dfrac{8}{3} \cdot \dfrac{1}{6} = -\dfrac{(8)(1)}{(3)(6)}$ Invert divisor and multiply

$= -\dfrac{(2)(4)}{(3)(2)(3)}$ Reduce

$= -\dfrac{4}{9}$ Reduced form

(b) $3\frac{9}{13} \div 1\frac{5}{8} = \frac{48}{13} \div \frac{13}{8}$

$\qquad = \frac{48}{13} \cdot \frac{8}{13}$ Invert divisor and multiply

$\qquad = \frac{(48)(8)}{(13)(13)}$

$\qquad = \frac{384}{169}$ ■

Decimal Fractions

Rational numbers (integers and fractions) can be represented as **terminating** or **repeating decimals**. Such forms are called **decimal fractions**. Here are some examples.

Terminating Decimals	*Repeating Decimals*
$\frac{1}{4} = 0.25$	$\frac{1}{6} = 0.1666\ldots$ or $0.1\overline{6}$
$\frac{3}{8} = 0.375$	$\frac{1}{3} = 0.333\ldots$ or $0.\overline{3}$
$\frac{2}{10} = 0.2$	$\frac{1}{12} = 0.0833\ldots$ or $0.08\overline{3}$
$\frac{5}{16} = 0.3125$	$\frac{8}{33} = 0.2424\ldots$ or $0.\overline{24}$

Note that the *bar* notation is used to indicate the *repeated* digit (or digits) in the decimal notation. We can obtain the decimal representation of any fraction by long division. For instance, the decimal representation for $\frac{5}{12}$ is $0.41\overline{6}$, as can be seen from the following long division problem.

$$
\begin{array}{r}
0.4166\ldots = 0.41\overline{6} \\
12\overline{)5.000} \\
\underline{48} \\
20 \\
\underline{12} \\
80 \\
\underline{72} \\
80 \\
\end{array}
$$

For calculations involving decimal fractions such as $0.41666\ldots$, we must **round the decimal**. For instance, rounded to two decimal places, the number $0.41666\ldots$ is 0.42. Similarly, rounded to three decimal places, the number $0.41666\ldots$ is 0.417. When rounding decimals, we use the following rules.

Rounding a Decimal	1. Determine the number of digits of accuracy you wish to keep. The digit in the last position you keep is called the **rounding digit**, and the digit in the first position you discard is called the **decision digit**.
	2. If the decision digit is 5 or greater, round up by adding 1 to the rounding digit.
	3. If the decision digit is 4 or less, round down by leaving the rounding digit unchanged.

The rules for operating with decimal fractions are like those for integers with appropriate adjustments in the locations of the decimal points. These rules are illustrated in the next two examples.

EXAMPLE 11 ■ Operations with Decimal Fractions

Perform the indicated operations.

(a) $0.583 + 1.06 + 2.9104$ (b) $(-3.57)(0.032)$

Solution

(a) To add decimals, align the decimal points and proceed as in integer addition.

$$
\begin{array}{r}
^{1\ \ 1}\\
0.583\\
1.06\\
+\ 2.9104\\
\hline
4.5534
\end{array}
$$

(b) To multiply decimals, use integer multiplication and then place the decimal point (in the product) so that the number of decimal places equals the sum of the decimal places in the two factors.

$$
\begin{array}{r}
-3.57\\
\times\quad 0.032\\
\hline
714\\
1071\\
\hline
-0.11424
\end{array}
$$

       Two decimal places
       Three decimal places

       Five decimal places ■

EXAMPLE 12 ■ Dividing Decimal Fractions

Perform the indicated division and round your answer to two decimal places.

$1.483 \div 0.56$

Solution

To divide decimals, convert the divisor to an integer by moving its decimal point to the right. Move the decimal point in the dividend an equal number of places to the right.

$$0.56\overline{)1.483}$$ Relocate decimal points to obtain integer divisor

Place the decimal point in the quotient directly above the new decimal point in the dividend and then divide as with integers.

$$
\begin{array}{r}
2.648 \\
56\overline{)148.3} \\
\underline{112} \\
363 \\
\underline{336} \\
270 \\
\underline{224} \\
460 \\
\underline{448} \\
\end{array}
$$

Rounded to two decimal places, the answer is 2.65. This answer can be written as

$$\frac{1.483}{0.56} \approx 2.65$$

where the symbol $\approx$ means **approximately equal to**.   ■

Applications

The following example is similar to the stock investment problem given in Section 1.2. This time, however, the gains and losses are given in fraction form.

EXAMPLE 13   ■   An Application: Stock Purchases

Suppose on Monday you bought 50 shares of stock at $\$48\frac{1}{2}$ per share. During the week the stocks rose and fell, as shown in Table 1.2.

TABLE 1.2

Tuesday	Wednesday	Thursday	Friday
Up $\frac{3}{8}$	Up $1\frac{3}{4}$	Down $\frac{1}{2}$	Up $2\frac{7}{8}$

(a) What was the value of the stock at the close on Tuesday?

(b) What was the value of the stock at the close on Wednesday?

(c) What was the value of the stock at the end of the week?

(d) What would the total gain have been if Friday's gain had occurred on each of the four days?

(e) What was the average daily gain (loss) for the four days recorded?

Solution

(a) Since the original value of the stock was $50\left(48\frac{1}{2}\right) = \$2,425$, and each of the 50 shares gained $\$\frac{3}{8}$ by the close on Tuesday, the total value of your stock at the close on Tuesday was

$$2,425 + 50\left(\frac{3}{8}\right) = 2,425 + 18.75 = \$2,443.75.$$

(b) Using the result of part (a), the value at the close on Wednesday was

$$2,443.75 + 50\left(1\frac{3}{4}\right) = 2,443.75 + 87.5 = \$2,531.25.$$

(c) The value of the stock at the end of the week was

$$2,425 + 50\left(\frac{3}{8}\right) + 50\left(1\frac{3}{4}\right) + 50\left(-\frac{1}{2}\right) + 50\left(2\frac{7}{8}\right) = \$2,650.00.$$

(d) The gain on Friday was $2\frac{7}{8}$ per share. If this gain had occurred each day, the total gain would have been

$$4(50)\left(2\frac{7}{8}\right) = \$575.00.$$

(e) The total gain for the week was $2,650 - 2,425 = \$225$. Thus, the average daily gain for four days was

$$\text{Average} = \frac{225}{4} = \$56.25. \qquad \blacksquare$$

DISCUSSION PROBLEM   ■   Calculator Techniques

When using a calculator to perform operations with decimals, you should try to get in the habit of rounding your answers *only* after all the calculations are done. If you round the answer at a preliminary stage, you can introduce unnecessary round-off error. To see how this can occur, use your calculator to perform the following division in two ways. (Round your answer to two decimal places.)

(a) $\dfrac{(1.05)(2.34)}{0.57} = \dfrac{2.457}{0.57} \approx 4.31$

(b) $\dfrac{(1.05)(2.34)}{0.57} \approx \dfrac{2.46}{0.57} \approx 4.32$

Note that in version (b) of this problem, the preliminary result of 2.457 was rounded to 2.46 before dividing by 0.57. Which of these two techniques produced the more accurate answer?   ■

Warm-Up   *The following warm-up exercises involve skills that were covered in earlier sections. You will use these skills in the exercise set for this section.*

In Exercises 1–10, perform the indicated operations.

1. $10 + (-10)$
2. $3 \cdot 1$

3. $5 + |-3|$
4. $-4 - (-10)$

5. $-6 \cdot 2$
6. $5 + (4 \cdot 3)$

7. $\dfrac{10 + 8}{9}$
8. $\dfrac{|15 - 25|}{5}$

9. $\dfrac{(-3) \cdot (-4)}{2}$
10. $-\dfrac{(-2)(5)}{10}$

1.3   EXERCISES

In Exercises 1–4, find the sum of the two fractions indicated by the shaded regions of the given figure. (Assume each figure is divided in regions of equal areas.)

1. 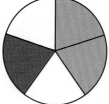   **2.** 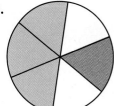   **3.**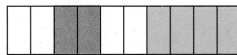

4.

In Exercises 5–10, write the given fraction in reduced form.

5. $\dfrac{2}{8}$
6. $\dfrac{21}{28}$
7. $\dfrac{60}{192}$
8. $\dfrac{45}{225}$

9. $\dfrac{36}{24}$
10. $\dfrac{72}{180}$

In Exercises 11–32, perform the given operations and write the answer in reduced form.

11. $\dfrac{7}{15} + \dfrac{2}{15}$
12. $\dfrac{13}{35} + \dfrac{5}{35}$
13. $\dfrac{9}{16} - \dfrac{3}{16}$
14. $\dfrac{15}{32} - \dfrac{7}{32}$

15. $\dfrac{9}{11} - \left(-\dfrac{5}{11}\right)$
16. $\dfrac{5}{6} - \left(-\dfrac{13}{6}\right)$
17. $\dfrac{3}{4} - \dfrac{5}{4}$
18. $\dfrac{3}{8} - \left|-\dfrac{5}{8}\right|$

19. $\frac{1}{2} + \frac{1}{3}$ **20.** $\frac{3}{5} + \frac{1}{2}$ **21.** $\frac{3}{16} + \frac{3}{8}$ **22.** $\frac{2}{3} + \frac{4}{9}$

23. $\left|-\frac{1}{8}\right| - \frac{1}{6}$ **24.** $\frac{13}{8} - \frac{3}{4}$ **25.** $1\frac{3}{16} - 2\frac{1}{4}$ **26.** $5\frac{7}{8} - 2\frac{1}{2}$

27. $-5\frac{2}{3} - 4\frac{5}{12}$ **28.** $-2\frac{3}{4} - 3\frac{1}{5}$ **29.** $4 - \frac{8}{3}$ **30.** $\frac{17}{25} + 2$

31. $1 + \frac{2}{3} - \frac{5}{6}$ **32.** $2 - \frac{15}{16} - \frac{7}{8} + |-5|$

In Exercises 33–48, perform the given operations and write the answer in reduced form.

33. $\frac{1}{2} \times \frac{3}{4}$ **34.** $-\frac{2}{3} \times \frac{5}{7}$ **35.** $\frac{2}{3}\left(-\frac{9}{16}\right)$ **36.** $\left(-\frac{3}{4}\right)\left(-\frac{4}{9}\right)$

37. $\left(-\frac{7}{16}\right)\left(-\frac{12}{5}\right)$ **38.** $\left(\frac{5}{3}\right)\left(-\frac{3}{5}\right)$ **39.** $\left(-\frac{3}{2}\right)\left(-\frac{15}{16}\right)\left(\frac{12}{25}\right)$ **40.** $\left(\frac{1}{2}\right)\left(-\frac{4}{15}\right)\left(-\frac{5}{24}\right)$

41. $\frac{3}{8} \div \frac{3}{4}$ **42.** $\frac{5}{16} \div \frac{25}{8}$ **43.** $-\frac{5}{12} \div \frac{45}{32}$ **44.** $\left(-\frac{16}{21}\right) \div \left(-\frac{12}{27}\right)$

45. $\frac{15/8}{3/8}$ **46.** $\frac{11/12}{-5/6}$ **47.** $\frac{-5}{15/16}$ **48.** $\frac{-35/12}{-14}$

In Exercises 49–58, write each rational number in decimal form. (Use the bar notation for repeating digits.)

49. $\frac{3}{4}$ **50.** $\frac{5}{8}$ **51.** $\frac{9}{16}$ **52.** $\frac{7}{20}$

53. $\frac{2}{3}$ **54.** $\frac{5}{6}$ **55.** $\frac{7}{12}$ **56.** $\frac{8}{15}$

57. $\frac{5}{11}$ **58.** $\frac{5}{21}$

In Exercises 59–68, perform the specified operations. (Round your answer to two decimal places.)

59. $1.21 + 4.06 - 3.00$ **60.** $-3.4 + 1.062 - 5.13$

61. $-0.0005 - 2.01 + 0.111$ **62.** $132.1 + (-25.45)$

63. $(-6.3)(9.05)$ **64.** $(-0.05)(-85.95)$

65. $4.69 \div 0.12$ **66.** $1.062 \div (-2.1)$

67. $\frac{-15.1}{-9.6}$ **68.** $\frac{5.42}{4.02}$

In Exercises 69 and 70, determine the unknown fractional part of the given pie chart.

69.

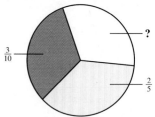

70.

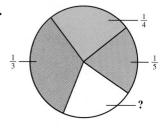

71. *U.S. Blood Types* The percentages of the U.S. population with different blood types are shown in the accompanying pie chart. (*Source:* U.S. Department of Health and Human Services.) What percentage of the population has a blood type of O⁺?

U.S. Blood Types

72. *U.S. Electricity Sources* The percentages of U.S. electricity sources are shown in the accompanying pie chart. (*Source:* Department of Energy.) What percentage of U.S. electricity is generated by nuclear power?

Figure for 72

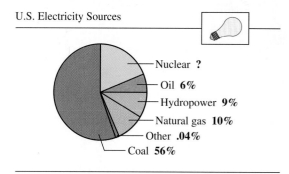

U.S. Electricity Sources

Nuclear **?**
Oil **6%**
Hydropower **9%**
Natural gas **10%**
Other **.04%**
Coal **56%**

73. *Stock Purchase* Two hundred shares of stock are purchased at a price of $23\frac{5}{8}$ per share and 300 shares at $86\frac{1}{4}$ per share. Find the total price of the purchased stock.

74. *Telephone Charge* A telephone company charges $1.16 for the first minute and $0.85 for each additional minute. Find the cost of a seven-minute phone call.

75. *Gasoline Price* The price per gallon of regular unleaded gasoline at three service stations is $1.159, $1.269, and $1.179, respectively. Find the average price per gallon.

76. *Grocery Purchase* Suppose you go to the grocery store and purchase two gallons of milk at $2.23 per gallon and three loaves of bread at $1.23 per loaf. If you give the cashier $20, how much change will you receive?

77. *Annual Fuel Cost* The sticker on a new car lists the fuel efficiency as 22.3 miles per gallon. Estimate the annual fuel cost if the car is to be driven approximately 12,000 miles per year and the fuel cost is $1.159 per gallon.

78. *Cost Per Pound* A can of food weighing 2.5 pounds was purchased for $4.95. Find the cost per pound.

In Exercises 79 and 80, determine whether the statement is true or false.

79. The reciprocal of every nonzero integer is an integer.

80. The reciprocal of every nonzero rational number is a rational number.

Exponents, Order of Operations, and Calculators
Positive Integer Exponents   •   Order of Operations   •   Calculators

Positive Integer Exponents

In Section 1.2 we saw that multiplication by a positive integer can be described as repeated addition.

Repeated Addition	*Multiplication*
$7 + 7 + 7 + 7$	4×7

4 terms of 7

In a similar way, *repeated multiplication* can be written in what is called **exponential form**.

Repeated Multiplication	*Exponential Form*
$7 \cdot 7 \cdot 7 \cdot 7$	7^4

4 factors of 7

In the exponential form 7^4, 7 is called the **base**, and it specifies the repeated factor; 4 is called the **exponent**, and it indicates how many times the base occurs as a factor. When we write the exponential form 7^4, we say that we are raising 7 to the **fourth power**. When a number is raised to the first power, we usually do not write the exponent 1. For instance, we usually write 5^1 simply as 5. Exponential forms are read as follows.

> 7^2 is read as "seven to the second power" or as "seven squared."
>
> 4^3 is read as "four to the third power" or as "four cubed."
>
> $(-2)^5$ is read as "negative two to the fifth power."

It is important to recognize how exponential forms like $(-2)^4$ and -2^4 differ. Note that

$$(-2)^4 = (-2)(-2)(-2)(-2) = 16$$ The negative sign is part of the base

whereas

$$-2^4 = -(2 \cdot 2 \cdot 2 \cdot 2) = -16.$$ The negative sign is *not* part of the base

EXAMPLE 1   ■   Evaluating Exponential Expressions

(a) $2^5 = 2 \cdot 2 \cdot 2 \cdot 2 \cdot 2 = 32$

(b) $5^2 = 5 \cdot 5 = 25$

(c) $(-3)^3 = (-3)(-3)(-3) = -27$ *Odd* number of negative factors yields a *negative* product

(d) $(-3)^4 = (-3)(-3)(-3)(-3) = 81$ *Even* number of negative factors yields a *positive* product

(e) $-3^4 = -(3 \cdot 3 \cdot 3 \cdot 3) = -81$ ■

EXAMPLE 2 ■   Evaluating Exponential Expressions

(a) $1^6 = 1 \cdot 1 \cdot 1 \cdot 1 \cdot 1 \cdot 1 = 1$ 1 to any positive power is 1

(b) $-1^6 = -(1^6) = -1$

(c) $0^5 = 0 \cdot 0 \cdot 0 \cdot 0 \cdot 0 = 0$ 0 to any positive power is 0

(d) $\left(\dfrac{2}{3}\right)^4 = \dfrac{2}{3} \cdot \dfrac{2}{3} \cdot \dfrac{2}{3} \cdot \dfrac{2}{3} = \dfrac{2 \cdot 2 \cdot 2 \cdot 2}{3 \cdot 3 \cdot 3 \cdot 3} = \dfrac{2^4}{3^4} = \dfrac{16}{81}$

(e) $\left(-\dfrac{1}{5}\right)^3 = \left(-\dfrac{1}{5}\right)\left(-\dfrac{1}{5}\right)\left(-\dfrac{1}{5}\right) = \dfrac{(-1)(-1)(-1)}{(5)(5)(5)} = \dfrac{(-1)^3}{5^3} = -\dfrac{1}{125}$ ■

EXAMPLE 3 ■   An Application: Transporting Capacity

A van can transport a load of cases of motor oil that is six cases high, six cases wide, and six cases long. If each case contains six quarts of motor oil, how many quarts can the truck transport?

FIGURE 1.9

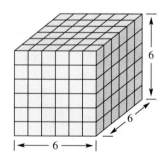

Solution

A picture is useful in this situation. We can see from Figure 1.9 that 6 occurs as a factor four times. That is,

$$\underbrace{6}_{\text{(length)}} \cdot \underbrace{6}_{\text{(width)}} \cdot \underbrace{6}_{\text{(height)}} \cdot \underbrace{6}_{\text{(\# per case)}} = 6^4 = 1{,}296.$$

Thus, the van can transport 1,296 quarts of oil.   ■

Order of Operations

Up to this point, we have discussed five operations of arithmetic—addition, subtraction, multiplication, division, and exponentiation (repeated multiplication). When we use more than one operation in a given problem, we face the question of which operation to do first. For example, without further guidelines, we could evaluate $4 + 3 \cdot 5$ in two ways. To avoid confusion, we use **symbols of grouping** such as parentheses ( ) or brackets [ ]. For instance, the parentheses in $(4 + 3) \cdot 5$ indicate that we first add 4 and 3, then multiply the result by 5 to obtain

$$(4 + 3) \cdot 5 = 7 \cdot 5 = 35.$$

On the other hand, the parentheses in $4 + (3 \cdot 5)$ make it clear that we should multiply first, then add to obtain

$$4 + (3 \cdot 5) = 4 + 15 = 19.$$

Sometimes symbols of groupings are *nested*, one within another. In such cases we work from the innermost grouping outward.

$$5[8 + (4 - 1)] = 5[8 + 3] \qquad \text{Subtract inside the parentheses}$$
$$= 5[11] \qquad \text{Add inside the brackets}$$
$$= 55 \qquad \text{Multiply}$$

When evaluating expressions with several arithmetic operations, we adhere to the following priority order of operations.

Order of Operations
1. Perform operations inside *symbols of groupings*, starting with the innermost symbol.

2. Evaluate all *exponential* expressions.

3. Perform all *multiplications* and *divisions* from left to right.

4. Perform all *additions* and *subtractions* from left to right.

EXAMPLE 4 ■ Order of Operations

(a) $7 - [(5 \cdot 3) + 2^3] = 7 - [15 + 2^3]$ Multiply inside the parentheses

$$= 7 - [15 + 8] \qquad \text{Evaluate the exponential expression}$$
$$= 7 - 23 \qquad \text{Add inside the brackets}$$
$$= -16 \qquad \text{Subtract}$$

(b) $[36 \div (3^2 \cdot 2)] - 6 = [36 \div (9 \cdot 2)] - 6$ Evaluate the exponential expression

$$= [36 \div 18] - 6 \qquad \text{Multiply inside the parentheses}$$
$$= 2 - 6 \qquad \text{Divide inside the brackets}$$
$$= -4 \qquad \text{Subtract} \qquad ■$$

EXAMPLE 5 ■ Order of Operations

(a) $36 - [3^2 \cdot (2 \div 6)] = 36 - \left[3^2 \cdot \dfrac{1}{3} \right]$ Divide inside the parentheses

$$= 36 - \left[9 \cdot \dfrac{1}{3} \right] \qquad \text{Evaluate the exponential expression}$$
$$= 36 - 3 \qquad \text{Multiply inside the brackets}$$
$$= 33 \qquad \text{Subtract}$$

(b) $(36 \div 3^2) + 2(-6) = (36 \div 9) + 2(-6)$ Evaluate the exponential expression

$$= 4 + (-12) \qquad \text{Divide and multiply}$$
$$= -8 \qquad \text{Add} \qquad ■$$

EXAMPLE 6   ■   Order of Operations

(a) $10 - 2[8 + (5 - 7)] = 10 - 2[8 + (-2)]$ Subtract inside the parentheses

$= 10 - 2[6]$ Add inside the brackets

$= 10 - 12$ Multiply

$= -2$ Subtract

(b) $10 \div (2[8 - (5 - 7)^2]) = 10 \div (2[8 - (-2)^2])$ Subtract inside the parentheses

$= 10 \div (2[8 - 4])$ Evaluate the exponential expression

$= 10 \div (2[4])$ Subtract inside the brackets

$= 10 \div 8$ Multiply inside the parentheses

$= \dfrac{5}{4}$ Divide and simplify

■

MATH MATTERS

History of Numbers

Today we take numbers for granted. We use numbers to count, to measure, to order things, and to calculate.

Early people knew little about numbers. They could see that an antelope had four legs, but if they saw a group of several antelope, they wouldn't have been able to tell others how many were in the group. We know that primitive people were aware of numbers because in some prehistoric caves there are

Lascaux cave, France.

pictures of animals with lines or dots scratched beside the pictures—perhaps indicating the number of animals killed in a hunt.

The fact that early people couldn't count might not have been important to tribes that relied primarily on hunting. But, as people began to farm, counting became more important. Farmers had to keep track of the number of animals they owned. To do this, people invented a way of writing numbers. By 2800 B.C. both the Sumerians and the Egyptians had devised ways of writing numbers. Moreover, as people traveled from one civilization to another, they borrowed writing techniques. For instance, the Romans borrowed their method of writing numbers from the Greeks.

Our present system of writing numbers is a combination of systems used by many ancient civilizations, primarily Babylonian, Egyptian, Arabian, Greek, and Indian.

EXAMPLE 7   ■   Order of Operations

$$6 + \frac{15}{3^2 - 4} - (-5) = 6 + \frac{15}{9 - 4} - (-5) \qquad \text{Evaluate exponential expression}$$

$$= 6 + \frac{15}{5} - (-5) \qquad \text{Subtract in denominator}$$

$$= 6 + 3 - (-5) \qquad \text{Divide}$$

$$= 9 + 5$$

$$= 14$$

Note how the fraction bar in this problem acts like a symbol of grouping, requiring the subtraction in the denominator *before* dividing by the denominator.   ■

Calculators

This text includes several examples and exercises that use a calculator. As we encounter each new calculator application, we will give general instructions for using a calculator. These instructions, however, may not agree precisely with the steps required by *your* calculator, so be sure you are familiar with the use of the keys on your own calculator.

   For each of the calculator examples in the text, we will give two possible keystroke sequences: one for a standard *scientific* calculator, and one for a *graphing* calculator.*

   Try the following evaluations on your own calculator to see whether the indicated keystrokes give the same display.

EXAMPLE 8   ■   Evaluating Expressions on a Calculator

(a) To evaluate the expression $-4 - 5$, use the following keystrokes.

Keystrokes	*Display*	
4 +/− − 5 =	−9	Scientific
(−) 4 − 5 ENTER	−9	Graphing

(b) To evaluate the expression $-3^2 + 4$, use the following keystrokes.

Keystrokes	*Display*	
3 x^2 +/− + 4 =	−5	Scientific
(−) 3 x^2 + 4 ENTER	−5	Graphing

(c) To evaluate the expression $5/(4 + 3 \cdot 2)$, use the following keystrokes.

Keystrokes	*Display*	
5 ÷ (4 + 3 × 2) =	0.5	Scientific
5 ÷ (4 + 3 × 2) ENTER	0.5	Graphing ■

---

*The graphing calculator keystrokes given in this text correspond to the *TI-81* graphing calculator from *Texas Instruments*. For other graphing calculators, the keystrokes may differ.

DISCUSSION PROBLEM   ■   A Computer Program

The formula

$$A = 1{,}000\left(1 + \frac{0.08}{N}\right)^{5N}$$

gives the balance A in an account in which \$1,000 is deposited for five years. The annual percentage rate is 8% and the interest is compounded N times per year. Use the following BASIC program to find the balance if the interest is compounded (a) annually, $N = 1$; (b) quarterly, $N = 4$; and (c) monthly, $N = 12$. From these results what can you conclude about the relationship between the amount A and the number of times per year the interest is compounded?

BASIC Program:

```
10 FOR I=1 TO 3
20 READ N
30 PRINT "N=";N;
40 LET A=1000*(1+0.8/N)^(5*N)
50 PRINT USING "$##,###.##";A
60 NEXT I
70 END
80 DATA 1, 4, 12
```

Warm-Up

The following warm-up exercises involve skills that were covered in earlier sections. You will use these skills in the exercise set for this section.

In Exercises 1–10, perform the indicated operations and write the answer in reduced form.

1. $16(-4)$ **2.** $-25(-3)$ **3.** $\dfrac{-49}{7}$ **4.** $-\dfrac{0}{32}$

5. $\dfrac{5}{16} - \dfrac{3}{16}$ **6.** $-2 + \dfrac{3}{2}$ **7.** $\left(-\dfrac{4}{3}\right)\left(-\dfrac{9}{16}\right)$ **8.** $\dfrac{7}{8} \div \dfrac{3}{16}$

9. $\dfrac{3/4}{-5/8}$ **10.** $\dfrac{\dfrac{1}{3} + \dfrac{5}{6}}{\dfrac{5}{12}}$

■ **1.4   EXERCISES**

In Exercises 1–4, rewrite the given quantity using exponents.

1. $2 \cdot 2 \cdot 2 \cdot 2 \cdot 2$ **2.** $(-5)(-5)(-5)(-5)$

3. $\left(-\dfrac{1}{4}\right)\left(-\dfrac{1}{4}\right)\left(-\dfrac{1}{4}\right)$ **4.** $(1.6)(1.6)(1.6)(1.6)(1.6)$

In Exercises 5–8, rewrite the given quantity as a repeated multiplication. (Do not evaluate the expression.)

5. $(-3)^6$

6. $\left(\dfrac{3}{8}\right)^5$

7. $(9.8)^3$

8. $(0.01)^8$

In Exercises 9–20, evaluate the given exponential expression.

9. 3^2

10. 4^3

11. $(-5)^3$

12. 3^4

13. -5^3

14. $-(-3)^2$

15. $\dfrac{1}{4^3}$

16. $\left(\dfrac{1}{4}\right)^3$

17. $\left(\dfrac{2}{3}\right)^3$

18. $\left(\dfrac{4}{5}\right)^3$

19. $(-1.2)^3$

20. $(1.5)^4$

In Exercises 21–30, evaluate the given expression, if possible. If it is not possible, state the reason.

21. $\dfrac{1-3^2}{-2}$

22. $\dfrac{3^2+4^2}{5}$

23. $\dfrac{5^2+12^2}{13}$

24. $\dfrac{4^2-2^3}{4}$

25. 2.1×10^2

26. 4.85×10^4

27. $\dfrac{8.4}{10^3}$

28. $\dfrac{6.23}{10^2}$

29. $\dfrac{3^2+1}{0}$

30. $\dfrac{0}{5^2+1}$

In Exercises 31–34, determine whether the statement is true or false.

31. $(-2)^3$ is negative.

32. $(-2)^4$ is positive.

33. -2^2 is positive.

34. -2^3 is negative.

In Exercises 35–62, evaluate the given expression. Write fractional answers in reduced form.

35. $16 + (3 \cdot 4)$

36. $25 - (32 \div 4)$

37. $4 - 6 + 10$

38. $(45 \div 10) \cdot 2$

39. $(16 - 5) \div (3 - 5)$

40. $(10 - 16) \cdot (20 - 26)$

41. $5 + (2^2 \cdot 3)$

42. $181 - (13 \cdot 3^2)$

43. $(-6)^2 - (5^2 \cdot 4) + 12$

44. $(-3)^3 + (12 \div 2^2)$

45. $(-4)^2 - (3 \cdot 2^4)$

46. $(3 \cdot 4^2) - 32$

47. $\left(3 \cdot \dfrac{5}{9}\right) + 1 - \dfrac{1}{3}$

48. $\dfrac{2}{3}\left(\dfrac{3}{4}\right)^2 + 2$

49. $5 - (8 - 15)$

50. $(12 + 2) \div 2$

51. $6(8 - 5)$

52. $4\left(-\dfrac{2}{3} + \dfrac{4}{3}\right)$

53. $-[2 - (6 + 5)]$

54. $[360 - (8 + 12)] \div 10$

55. $\dfrac{(3 \cdot 6) - (4 \cdot 6)}{5 + 1}$

56. $\dfrac{3 + [15 \div (-3)]}{16}$

57. $\dfrac{\dfrac{7}{3}\left(\dfrac{2}{3}\right)}{\dfrac{28}{15}}$

58. $\dfrac{3}{8}\left(\dfrac{1}{5}\right) \div \dfrac{25}{32}$

59. $\frac{3}{2}\left(\frac{2}{3} + \frac{1}{6}\right)$ **60.** $18\left(\frac{1}{2} + \frac{2}{3}\right)$ **61.** $\frac{3}{5}\left(\frac{3}{4} - \frac{2}{3}\right)$ **62.** $\frac{7}{25}\left(\frac{7}{16} - \frac{1}{8}\right)$

In Exercises 63–66, use a calculator to perform the specified calculations. (Round your answer to two decimal places.)

63. $(3.4)^2 - 6(1.2)^3$ **64.** $300\left(1 + \frac{0.1}{12}\right)^{24}$

65. $1,000 \div \left(1 + \frac{0.09}{4}\right)^8$ **66.** $\frac{1.32 + 4(3.68)}{1.5}$

In Exercises 67–72, show why the two quantities are not equal. (The symbol $\neq$ means **is not equal to**.)

67. $4 \cdot 10^2 \neq 4 \cdot 10 \cdot 4 \cdot 10$ **68.** $3(4 \cdot 5) \neq 3 \cdot 4 \cdot 3 \cdot 5$

69. $-3^2 \neq (-3)(-3)$ **70.** $4 \cdot 6^2 \neq 24^2$

71. $4 - (6 - 2) \neq 4 - 6 - 2$ **72.** $\frac{8 - 6}{2} \neq 4 - 6$

In Exercises 73–76, place the correct inequality symbol ($<$ or $>$) between the given numbers.

73. 2^2 ▨ 2^4 **74.** $(-3)^2$ ▨ $(-3)^3$ **75.** $\frac{3}{4}$ ▨ $\left(\frac{3}{4}\right)^2$ **76.** $\left(\frac{2}{3}\right)^3$ ▨ $\left(\frac{2}{3}\right)^2$

In Exercises 77 and 78, find the total area of the figure. (*Hint:* The area of a rectangle is the product of its length and width.)

77.

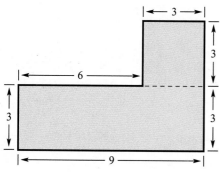

78.

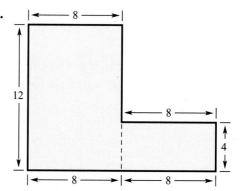

79. *Volume of a Cube* The volume of a cube with each edge of length 7 inches is given by $V = 7 \cdot 7 \cdot 7$. Write the volume using exponential notation. What is the unit measure for the volume V?

80. *Balance in an Account* $3,000 is deposited in an account earning 8% compounded yearly. Calculate the amount after four years by the given equation.

$$A = 3,000(1.08)^4$$

81. *Company Expenses* The portions of total expenses for a company are shown in the accompanying pie chart. What portion of the total expenses is spent on utilities? If the total expenses are $450,000, how much is spent on utilities?

Company Expenses

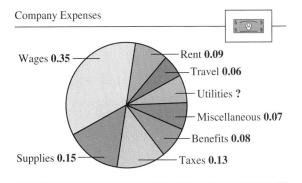

Wages **0.35**
Rent **0.09**
Travel **0.06**
Utilities **?**
Miscellaneous **0.07**
Benefits **0.08**
Supplies **0.15**
Taxes **0.13**

82. *Payroll Deductions* The portions of payroll deductions for an employee are shown in the accompanying pie chart. What portion of the gross amount is represented by the net amount (the take-home pay)? If the gross amount is $1,800, what is the net amount?

Payroll Deductions

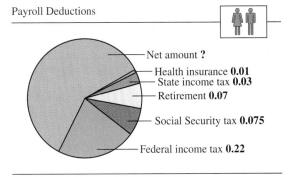

Net amount **?**
Health insurance **0.01**
State income tax **0.03**
Retirement **0.07**
Social Security tax **0.075**
Federal income tax **0.22**

Rules of Arithmetic and an Introduction to Algebra
The Arithmetic-Algebra Connection   •   What Is Algebra?   •   Definitions and Rules of Arithmetic   •
Properties of Real Numbers

The Arithmetic-Algebra Connection

Perhaps you have noticed that, so far, we have avoided the use of symbols* (letters like a, b, c, x, y, and z) to describe the definitions, rules, and operations of arithmetic. For example, we said in Section 1.2, "To subtract one integer from another, add the opposite of the number subtracted to the other integer." A specific illustration of this definition was

$$7 - 5 = 7 + (-5).$$

In the language of algebra, we can use letters to describe subtraction.

For any two integers a and b, $a - b = a + (-b)$.

---

*In algebra, we use a variety of letters to represent numbers. The most commonly used letters are a, b, c, x, y, and z. However, any other letter may be used. For instance, we might represent an integer by n, the cost by C, or two similar real numbers by x_1 and x_2. In the last case, the symbols x_1 and x_2 are read as "x sub one" and "x sub two."

Do you see how this statement describes subtraction as "adding the opposite?" This arithmetic-algebra connection can be illustrated in the following way.

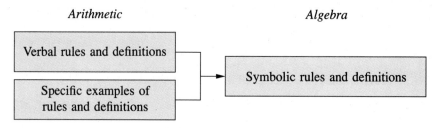

Arithmetic   *Algebra*

Here are three illustrations of this connection. The first is called the **Commutative Property of Addition**, and it tells us that two real numbers can be added in either order. The second illustration is a formal definition of absolute value, and the third is a formal definition of division of rational numbers.

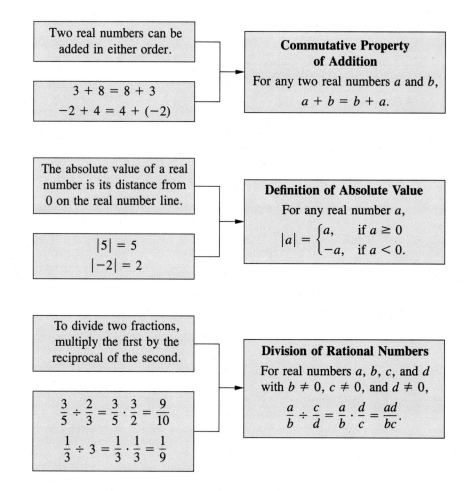

NOTE: The product a and b can be written as simply ab. Thus, $a \cdot b = ab = a(b) = (a)(b)$ all mean the same thing.

In the next example, we ask you to write an appropriate symbolic statement that generalizes an arithmetic rule.

EXAMPLE 1 ■   Generalizing a Rule of Arithmetic

Write an algebraic description of the following arithmetic rule: "The product of two real numbers with unlike signs is negative."

Solution

To guarantee that two letters represent numbers of unlike signs, we can choose two positive numbers and place a negative sign by just one of them. For instance, we could choose 3 and 7 and write $(-3) \cdot 7 = 3 \cdot (-7) = -21$. This leads to the following algebraic version of the rule.

If a and b are positive real numbers, then

$$\underbrace{(-a)(b)}_{\text{Unlike signs}} = \underbrace{(a)(-b)}_{\text{Unlike signs}} = \underbrace{-(a \cdot b)}_{\text{Negative product}}.$$

■

What Is Algebra?

Much of beginning algebra consists of learning to recognize and apply the symbolic versions of the rules of arithmetic. But what else is included in algebra? There is general agreement within the mathematical community* that algebra consists of the following.

1. Symbolic representations and applications of the rules of arithmetic.

2. Rewriting (reducing, simplifying, factoring) algebraic expressions into equivalent forms.

3. Creating and solving equations.

4. Studying relationships among variables by the use of functions and graphs.

These four components of algebra tend to nest within each other, as shown in Figure 1.10.

*"The Ideas of Algebra," Chap. 2 in *Yearbook of the National Council of Teachers of Mathematics* (Reston, VA: National Council of Teachers of Mathematics, 1988).

FIGURE 1.10

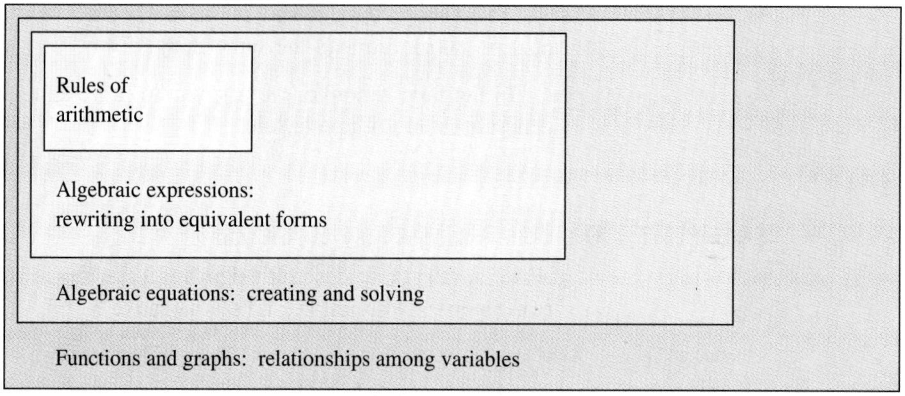

The next example uses the **Right Distributive Property** to illustrate some of the components of algebra. The Right Distributive Property states that for any real numbers a, b, and c

$$(a + b)c = ac + bc.$$ Right Distributive Property

NOTE: A statement of equality like $(a + b)c = ac + bc$ works both ways, from right to left as well as from left to right.

EXAMPLE 2 ■ Application of the Distributive Property

Arithmetic Version

(a) Evaluate the arithmetic expression $3 \cdot 7 + 9 \cdot 7$.

$$3 \cdot 7 + 9 \cdot 7 = (3 + 9)7$$ Distributive Property
$$= (12)7$$ Add
$$= 84$$ Multiply

Algebra Versions

(b) Simplify the algebraic expression $3x + 9x$.

$$3x + 9x = (3 + 9)x$$ Distributive Property
$$= 12x$$ Add

(c) Solve the algebraic equation $3x + 9x = 12$.

$$3x + 9x = 12$$ Given
$$(3 + 9)x = 12$$ Distributive Property
$$12x = 12$$ Add
$$x = 1$$ Divide by 12   ■

In algebra it is important to understand the difference between simplifying (rewriting) an algebraic *expression*, and solving an algebraic *equation*. Though it may be oversimplification, the following distinction is generally valid.

A mathematical expression *has no equal sign,* whereas a mathematical equation *must have an equal sign.*

When we use an equal sign to *rewrite* an expression, we are merely indicating the *equivalence* of the new expression and the previous one. In fact, the equal sign in each of the following statements does not mean we have an equation to solve; rather, it means that the expression to the right of the equal sign can be used as a *replacement* for the one on the left and vice versa.

Original Expression   *Equivalent Expression*

$$a + b = b + a$$
$$a - b = a + (-b)$$
$$(a + b)c = ac + bc$$

Definitions and Rules of Arithmetic

For convenience we list the algebraic versions of the definitions and rules of arithmetic that we have discussed so far. In each case we include a specific example for clarification.

Definitions

Let a, b, c, and d be real numbers.

Definition   *Example*

1. Subtraction:

$$a - b = a + (-b)$$ $$5 - 7 = 5 + (-7)$$

2. Multiplication: (a is a positive integer)

$$a \cdot b = \underbrace{b + b + \cdots + b}_{a \text{ terms}}$$ $$3 \cdot 5 = 5 + 5 + 5$$

3. Division: ($b \neq 0$)

$$a \div b = c, \text{ if and only if } a = c \cdot b$$ $$12 \div 4 = 3 \text{ because } 12 = 3 \cdot 4$$

4. Less than:
   $a < b$ if there is a positive real number $-2 < 1$ because $-2 + 3 = 1$
   c, such that $a + c = b$.

5. Absolute value: $|a| = \begin{cases} a, & \text{if } a \geq 0 \\ -a, & \text{if } a < 0 \end{cases}$ $$|-3| = -(-3) = 3$$

Rules of Arithmetic   Let a, b, c, and d be real numbers.

Rule   *Example*

1. Addition of fractions:

$$\frac{a}{b} + \frac{c}{d} = \frac{ad + bc}{bd}, \quad b \neq 0, d \neq 0$$ $$\frac{1}{3} + \frac{2}{7} = \frac{1 \cdot 7 + 3 \cdot 2}{3 \cdot 7} = \frac{13}{21}$$

2. Subtraction of fractions:

$$\frac{a}{b} - \frac{c}{d} = \frac{ad - bc}{bd}, \quad b \neq 0, d \neq 0$$ $$\frac{1}{3} - \frac{2}{7} = \frac{1 \cdot 7 - 3 \cdot 2}{3 \cdot 7} = \frac{1}{21}$$

3. Multiplication of fractions:

$$\frac{a}{b} \cdot \frac{c}{d} = \frac{a \cdot c}{b \cdot d}, \quad b \neq 0, d \neq 0$$ $$\frac{1}{3} \cdot \frac{2}{7} = \frac{1 \cdot 2}{3 \cdot 7} = \frac{2}{21}$$

4. Division of fractions:

$$\frac{a}{b} \div \frac{c}{d} = \frac{a}{b} \cdot \frac{d}{c}, \quad b \neq 0, d \neq 0, c \neq 0$$ $$\frac{1}{3} \div \frac{2}{7} = \frac{1}{3} \cdot \frac{7}{2} = \frac{7}{6}$$

5. Rule of signs for fractions:

$$\frac{a}{b} = \frac{-a}{-b} \quad \text{and} \quad \frac{-a}{b} = \frac{a}{-b} = -\frac{a}{b}$$ $$\frac{12}{4} = \frac{-12}{-4},$$ $$\frac{-12}{4} = \frac{12}{-4} = -\frac{12}{4}$$

6. Exponentiation: (n is a positive integer)

$$a^n = \underbrace{a \cdot a \cdot a \cdots a}_{n \text{ factors}}$$ $$4^5 = 4 \cdot 4 \cdot 4 \cdot 4 \cdot 4$$

7. Equivalent fractions:

$$\frac{a}{b} = \frac{c}{d}, \text{ if and only if } ad = bc;$$ $$\frac{1}{4} = \frac{3}{12} \text{ because}$$

$$b \neq 0, d \neq 0$$ $$1 \cdot 12 = 3 \cdot 4$$

EXAMPLE 3   ■   Using Definitions and Rules of Arithmetic

Complete the following statements using the specified definition or rule.

(a) Definition of multiplication:

$6 + 6 + 6 + 6 =$ ▨

(b) Exponentiation:

$6^4 =$ ▨

(c) Addition of fractions:

$\dfrac{3}{5} + \dfrac{1}{7} =$ ▨

(d) Multiplication of fractions:

$\dfrac{2 \cdot 6}{3 \cdot 5} =$ ▨

Solution

(a) By definition, multiplication is described as repeated addition. Therefore, we can write the following.

$$6 + 6 + 6 + 6 = 4 \cdot 6$$

(b) By definition, exponentiation is described as repeated multiplication. Therefore, we can write the following.

$$6^4 = 6 \cdot 6 \cdot 6 \cdot 6$$

(c) By definition of the sum of two fractions, we can write the following.

$$\frac{3}{5} + \frac{1}{7} = \frac{3 \cdot 7 + 5 \cdot 1}{5 \cdot 7}$$

(d) By definition of the product of two fractions, we can write the following.

$$\frac{2 \cdot 6}{3 \cdot 5} = \frac{2}{3} \cdot \frac{6}{5}$$ ■

Properties of Real Numbers

We are now ready to give the symbolic (or algebraic) versions of properties that we know are true about operations with real numbers. We refer to these properties as **properties of real numbers**. For each property we give a verbal description and an illustrative example. Keep in mind that for all the properties listed, the letters a, b, and c represent real numbers.

Properties of Real Numbers   Let a, b, and c be real numbers. Then the following properties are true.

Property	Verbal Description
Commutative Property of Addition	Two real numbers can be added in either order.
$a + b = b + a$	Example: $3 + 5 = 5 + 3$
Commutative Property of Multiplication	Two real numbers can be multiplied in either order.
$ab = ba$	Example: $4 \cdot (-7) = -7 \cdot 4$
Associative Property of Addition	When adding three real numbers, it makes no difference which two are added first.
$(a + b) + c = a + (b + c)$	Example: $(2 + 6) + 5 = 2 + (6 + 5)$
Associative Property of Multiplication	When multiplying three real numbers, it makes no difference which two are multiplied first.
$(ab)c = a(bc)$	Example: $(3 \cdot 5) \cdot 2 = 3 \cdot (5 \cdot 2)$
Distributive Property	Multiplication distributes over addition.
$a(b + c) = ab + ac$ $(b + c)a = ba + ca$	Examples: $3(8 + 5) = 3 \cdot 8 + 3 \cdot 5$ $(8 + 5)3 = 8 \cdot 3 + 5 \cdot 3$
Additive Identity Property	The sum of zero and a real number equals the number itself.
$a + 0 = 0 + a = a$	Example: $3 + 0 = 0 + 3 = 3$
Multiplicative Identity Property	The product of one and a real number equals the number itself.
$a \cdot 1 = 1 \cdot a = a$	Example: $4 \cdot 1 = 1 \cdot 4 = 4$
Additive Inverse Property	The sum of a real number and its opposite is zero.
$a + (-a) = 0$	Example: $3 + (-3) = 0$
Multiplicative Inverse Property	The product of a nonzero real number and its reciprocal is one.
$a \cdot \dfrac{1}{a} = 1, \quad a \neq 0$	Example: $8 \cdot \dfrac{1}{8} = 1$

NOTE: Why are subtraction and division not listed in the preceding collection? It is because they fail to possess many of these properties. For instance, subtraction and division are not commutative. To see this, consider $7 - 5 \neq 5 - 7$ and $12 \div 4 \neq 4 \div 12$. Similarly, the examples $9 - (5 - 3) \neq (9 - 5) - 3$ and $12 \div (4 \div 2) \neq (12 \div 4) \div 2$ illustrate the fact that subtraction and division are not associative.

EXAMPLE 4   ■   Identifying Properties of Real Numbers

Name the property of real numbers that justifies the given statement.

(a) $3(a + 2) = 3 \cdot a + 3 \cdot 2$ (b) $5 \cdot \dfrac{1}{5} = 1$

(c) $7 + (5 + b) = (7 + 5) + b$ (d) $(b + 3) + 0 = b + 3$

Solution

(a) This statement is justified by the Distributive Property.

(b) This statement is justified by the Multiplicative Inverse Property.

(c) This statement is justified by the Associative Property of Addition.

(d) This statement is justified by the Additive Identity Property.   ■

EXAMPLE 5   ■   Using the Properties of Real Numbers

Complete the following statements using the specified property of real numbers.

(a) Multiplicative Identity Property: (b) Associative Property of Addition:

  $(3b)1 = $ ▢ $(c + 2) + 7 = $ ▢

(c) Additive Inverse Property: (d) Distributive Property:

  $0 = 3a + $ ▢ $3 \cdot a + 3 \cdot 4 = $ ▢

Solution

(a) By the Multiplicative Identity Property, we can write

   $(3b)1 = 3b.$

(b) By the Associative Property of Addition, we can write

   $(c + 2) + 7 = c + (2 + 7).$

(c) By the Additive Inverse Property, we can write

   $0 = 3a + (-3a).$

(d) By the Distributive Property, we can write

   $3 \cdot a + 3 \cdot 4 = 3(a + 4).$ ■

DISCUSSION PROBLEM   ■   You Be the Instructor

Suppose you are teaching an algebra class and one of your students hands in the following problem. What is the error in this work?

   $-3(x + 2) = \cancel{-3x + 3(2)}$

     $= \cancel{-3x + 6}$

What would you say to your class to help them avoid this type of error?   ■

Warm-Up

The following warm-up exercises involve skills that were covered in earlier sections. You will use these skills in the exercise set for this section.

In Exercises 1–10, evaluate the given expression.

1. $3^2 - (-4)$

2. $(-5)^2 + 3$

3. 9.3×10^6

4. $6.6 \div 10^3$

5. $\dfrac{|12 - 4^2|}{2}$

6. $\dfrac{-|7 + 3^2|}{4}$

7. $3 + 2(6 + 10)$

8. $-50 - 4(3 - 8)$

9. $(-4)^2 - (30 \div 5)$

10. $(8 \cdot 9) + (-4)^3$

1.5   EXERCISES

In Exercises 1–20, state the property of real numbers that justifies the given statement.

1. $6(-3) = -3(6)$

2. $16 + 10 = 10 + 16$

3. $x + 10 = 10 + x$

4. $8x = x(8)$

5. $0 + 15 = 15$

6. $1 \cdot 4 = 4$

7. $(10 + 3) + 2 = 10 + (3 + 2)$

8. $25 + (-25) = 0$

9. $-16 + 16 = 0$

10. $(2 \cdot 3)4 = 2(3 \cdot 4)$

11. $4(3 \cdot 10) = (4 \cdot 3)10$

12. $(32 + 8) + 5 = 32 + (8 + 5)$

13. $7\left(\dfrac{1}{7}\right) = 1$

14. $14 + (-14) = 0$

15. $6(3 + x) = 6 \cdot 3 + 6x$

16. $(14 + 2)3 = 14 \cdot 3 + 2 \cdot 3$

17. $(4 + x)(2 - x) = 4(2 - x) + x(2 - x)$

18. $\dfrac{1}{a}(3 + y) = \dfrac{1}{a}(3) + \dfrac{1}{a}(y)$

19. $x + (y + 3) = (x + y) + 3$

20. $[(x + y)u]v = (x + y)(uv)$

In Exercises 21–30, complete the given statement using the specified property of real numbers.

21. Distributive Property: $6(x + 2) = $

22. Commutative Property of Multiplication: $10(-3) = $

23. Commutative Property of Addition: $y + 5 = $

24. Distributive Property: $x(4 - x) = $

25. Associative Property of Multiplication: $(6x)y = $

26. Associative Property of Addition: $(10 + x) + 2y =$

27. Associative Property of Addition: $3x + (2y + 5) =$

28. Associative Property of Multiplication: $12(3 \cdot 4) =$

29. Commutative Property of Addition: $5(u + v) =$

30. Commutative Property of Multiplication: $5(u + v) =$

In Exercises 31–38, give (a) the additive inverse, and (b) the multiplicative inverse of the given quantity. (Assume that the given quantity is not zero.)

31. 50 **32.** 12 **33.** $2x$ **34.** $5y$

35. $x + y$ **36.** $x - 6$ **37.** ab **38.** uv

In Exercises 39–46, rewrite the given expression using the Associative Property of Addition or the Associative Property of Multiplication.

39. $10 + (8 + 2)$ **40.** $16 + (4 + 3)$ **41.** $(x + 3) + 2$ **42.** $(z + 5) + 15$

43. $(2 \cdot 3)4$ **44.** $3(5 \cdot 10)$ **45.** $2(3y)$ **46.** $10(6x)$

In Exercises 47–54, rewrite the given expression using the Distributive Property, and simplify the answer.

47. $3(6 + 10)$ **48.** $4(8 + 3)$ **49.** $3(2x + 4)$ **50.** $10(15 + 3t)$

51. $\frac{2}{3}(9z + 24)$ **52.** $\frac{3}{5}(10y + 45)$ **53.** $x(3 + x)$ **54.** $y(4 + u)$

In Exercises 55 and 56, identify the property of real numbers used to justify each step in rewriting the given expression.

55. $3 + 10(x + 1) = 3 + 10x + 10$
$$= 3 + 10 + 10x$$
$$= (3 + 10) + 10x$$
$$= 13 + 10x$$

56. $4(x + 2) = 4(2 + x)$
$$= 8 + 4x$$

In Exercises 57–60, state why the two quantities are not equal.

57. $5(x + 3) \neq 5x + 3$ **58.** $7(x + 2) \neq 7x + 2$ **59.** $\frac{8}{0} \neq 0$ **60.** $5\left(\frac{1}{5}\right) \neq 0$

In Exercises 61 and 62, determine whether the order in which the two statements are performed is "commutative." That is, do you obtain the same result regardless of which statement is performed first?

61. (a) "Drain the used oil from the engine."
(b) "Fill the crankcase with five quarts of new oil."

62. (a) "Weed the flower beds."
(b) "Mow the lawn."

CHAPTER 1   SUMMARY

As you review and prepare for a test on this chapter, first try to obtain a global view of what was discussed. Then review the specific skills needed in each category.

Arithmetic Operations with Real Numbers

- *Order* real numbers—Be able to compare two real numbers using inequality symbols.

  Section 1.1, Exercises 1–20, Exercises 39–50

- *Evaluate* sums, differences, products, and quotients.

  Section 1.2, Exercises 1–76; Section 1.3, Exercises 1–58; Section 1.4, Exercises 35–62

- *Evaluate* exponential expressions.

  Section 1.4, Exercises 1–30

- *Evaluate* absolute values.

  Section 1.1, Exercises 29–54

- *Create* arithmetic expressions from verbal statements.

  Section 1.2, Exercises 77–86; Section 1.3, Exercises 71–78

- *Evaluate* arithmetic expressions on a calculator.

  Section 1.4, Exercises 63–66

The Arithmetic–Algebra Connection

- *Identify* the rule demonstrated by a given statement equation.

  Section 1.5, Exercises 1–20

- *Rewrite and/or simplify* a numerical expression using the rules of arithmetic and properties of real numbers.

  Section 1.5, Exercises 1–14

- *Rewrite and/or simplify* an algebraic expression using the rules of arithmetic and properties of real numbers.

  Section 1.5, Exercises 21–54

Chapter 1 Review Exercises

In Exercises 1–6, show each real number as a point on the real number line and place the correct inequality symbol ($<$ or $>$) between the real numbers.

1. $-\dfrac{1}{10}$ ▨ 4 **2.** $\dfrac{25}{3}$ ▨ $\dfrac{5}{3}$ **3.** -3 ▨ -7 **4.** $-\dfrac{8}{3}$ ▨ $-\dfrac{2}{3}$

5. 10.6 ▨ -3.5 **6.** -3 ▨ -4

In Exercises 7–10, evaluate the given expression.

7. $|-8.5|$ **8.** $|3.4|$ **9.** $-|-8.5|$ **10.** $|-9.6|$

In Exercises 11–14, place the correct symbol ($<$, $>$, or $=$) between the two real numbers.

11. $|-84|$ ▨ $|84|$ **12.** $|-10|$ ▨ $|4|$ **13.** $\left|\dfrac{3}{10}\right|$ ▨ $-\left|\dfrac{4}{5}\right|$ **14.** $|2.3|$ ▨ $-|2.3|$

In Exercises 15–74, evaluate the given expression, if possible. (Write fractional answers in reduced form.)

15. $32 + 68$ **16.** $14 + 54$ **17.** $16 + (-5)$ **18.** $-125 + 30$

19. $350 - 125 + 15$ **20.** $35 - 25 - 10$ **21.** $-114 + 76 - 230$ **22.** $-448 - 322 + 100$

23. $|-86| - |124|$ **24.** $67 + |-53|$ **25.** 15×3 **26.** -22×4

27. $-300(-5)$ **28.** $18(3{,}200)$ **29.** $131(-6)(3)$ **30.** $(-46)(-5)(-2)$

31. $\dfrac{-162}{9}$ **32.** $\dfrac{-52}{-4}$ **33.** $815 \div 0$ **34.** $-48 \div 6$

35. $\dfrac{78 - |-78|}{5}$ **36.** $\dfrac{144}{2 \cdot 3 \cdot 3}$ **37.** $\dfrac{54 - 4 \cdot 3}{6}$ **38.** $\dfrac{3 \cdot 5 + 125}{10}$

39. $\dfrac{3}{25} + \dfrac{7}{25}$ **40.** $\dfrac{9}{64} + \dfrac{7}{64}$ **41.** $\dfrac{27}{16} - \dfrac{15}{16}$ **42.** $-\dfrac{5}{12} + \dfrac{1}{12}$

43. $-\dfrac{5}{9} + \dfrac{2}{3}$ **44.** $\dfrac{7}{15} - \dfrac{2}{25}$ **45.** $\dfrac{25}{32} + \dfrac{7}{24}$ **46.** $-\dfrac{7}{8} - \dfrac{11}{12}$

47. $5\dfrac{3}{4} - 3\dfrac{5}{8}$ **48.** $-3\dfrac{7}{10} + 1\dfrac{1}{20}$ **49.** $\dfrac{5}{8} \cdot \dfrac{-2}{15}$ **50.** $\dfrac{3}{32} \cdot \dfrac{32}{3}$

51. $35\left(\dfrac{1}{35}\right)$ **52.** $-\dfrac{5}{12}\left(-\dfrac{4}{25}\right)$ **53.** $\dfrac{5}{14} \div \dfrac{15}{28}$ **54.** $-\dfrac{7}{10} \div \dfrac{4}{15}$

55. $\dfrac{-3/4}{-7/8}$

56. $\dfrac{15/32}{-5}$

57. $\dfrac{\frac{3}{2} - \frac{1}{4}}{\frac{3}{8}}$

58. $\dfrac{-\frac{3}{5} - \frac{1}{2}}{\frac{9}{10}}$

59. $3.2 - 1.5 + 11.4$

60. $-25 + 16.5 + 3.75$

61. $\dfrac{5.25}{0.25}$

62. $(5.2)(16.8)$

63. 7^3

64. $(-5)^2$

65. $(-7)^3$

66. $-(-2)^4$

67. $\left(\dfrac{3}{5}\right)^4$

68. $\dfrac{2}{6^3}$

69. $240 - (4^2 \cdot 5)$

70. $5^2 - (625 \cdot 5^2)$

71. $3^2(10 - 2^2)$

72. $-5(16 - 5^2)$

73. $\left(\dfrac{3}{4}\right)\left(\dfrac{5}{6}\right) + 4$

74. $75 - (24 \div 2^3)$

In Exercises 75–84, state the property of real numbers that justifies the given statement.

75. $-10 + 6 = 6 + (-10)$

76. $123 + (-123) = 0$

77. $9 \cdot \dfrac{1}{9} = 1$

78. $8(7 + 5) = 8 \cdot 7 + 8 \cdot 5$

79. $-2(7 + x) = -2 \cdot 7 + (-2)x$

80. $14(3) = 3(14)$

81. $4 + (3 + x) = 4 + 3 + x$

82. $5(3x) = (5 \cdot 3)x$

83. $17 \cdot 1 = 17$

84. $r + (2s + 3) = (r + 2s) + 3$

85. *Telephone Charge* A telephone call costs $0.64 for the first minute plus $0.72 for each additional minute. Find the cost of a five-minute call.

86. *Total Charge* To purchase a product, suppose you make a down payment of $75 plus nine monthly payments of $25 each. What is the total amount paid for the product?

87. *Cost Per Ounce* A container of food weighing 22 ounces is purchased for $1.43. Find the cost per ounce.

88. *Volume of Water in Hot Tub* The volume of water in a hot tub is given by $V = 6^2 \cdot 3$ (see figure). How many cubic feet of water will the hot tub hold? Find the total weight of the water in the tub. (Use the fact that one cubic foot of water weighs 62.4 pounds.)

89. *Calculator Experiment* Enter any number between 0 and 1 in a calculator. Square the number, and then square the result. Continue this process. What number does the calculator display seem to be approaching?

90. *Calculator Experiment* Use a calculator to calculate 15^4 in two ways.

15	y^x	4	$=$		
15	x^2	x^2			} Scientific
15	$\wedge$	4	ENTER		
15	x^2	x^2	ENTER		} Graphing

Why do these two methods give the same result?

Figure for 88

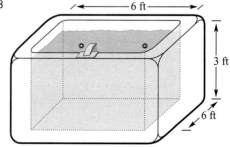

6 ft

3 ft

6 ft

CHAPTER 1 TEST

Take this test as you would take a test in class. After you are done, check your work with the answers given in the back of the book.

1. Place the correct symbol ($<$ or $>$) between the real numbers $-\frac{3}{5}$ and $-|-2|$.

2. Evaluate: $16 + (-20)$

3. Evaluate: $-50 - (-60)$

4. Evaluate: $-5(32)$

5. Evaluate: $\frac{72}{-9}$

6. Evaluate: $\frac{5}{6} - \frac{1}{8}$

7. Evaluate: $\left(-\frac{3}{4}\right)\left(-\frac{6}{15}\right)$

8. Evaluate: $\frac{7}{16} \div \frac{21}{28}$

9. Evaluate: $\frac{-8.1}{0.3}$

10. Evaluate: $(-4)^3$

11. Evaluate: $-\left(\frac{2}{3}\right)^2$

12. Evaluate: $35 - (50 \div 5^2)$

13. Evaluate: $\frac{1}{4}\left(\frac{3}{5} - \frac{1}{10}\right)$

14. State the property of real numbers demonstrated by $3(4 + 6) = 3 \cdot 4 + 3 \cdot 6$.

15. State the property of real numbers demonstrated by $5 \cdot \frac{1}{5} = 1$.

16. State the property of real numbers demonstrated by $3 + (4 + 8) = (3 + 4) + 8$.

17. It is necessary to cut a 90-foot rope into six pieces of equal length. What is the length of each piece?

18. A **cord** of wood is a pile 4 feet long, 4 feet wide, and 8 feet high. The volume of a rectangular solid is its length times its width times its height. Find the number of cubic feet in a cord of wood.

19. Show the fraction $\frac{2}{3}$ graphically in the accompanying figure. Divide the whole into the correct number of equal-sized parts and shade the appropriate number of these parts.

CHAPTER TWO

Fundamentals of Algebra

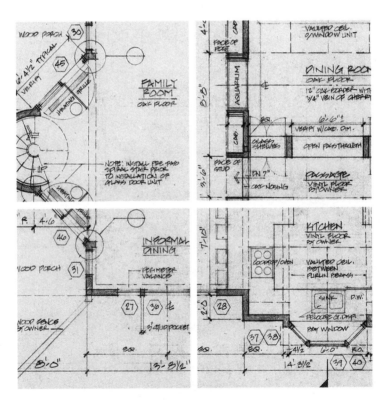

Chapter Overview

In this chapter we show how to work with algebraic expressions by extending the rules of arithmetic to form rules of algebra. We then show how these rules of algebra can be used to rewrite algebraic expressions in simpler, more usable forms. This is one of the three major goals of algebra. In the last part of the chapter we explore some strategies for translating verbal expressions into algebraic forms.

Algebraic Expressions and Properties of Exponents

Introduction • Algebraic Expressions • Positive Integer Exponents • Properties of Exponents

Introduction

One of the distinguishing characteristics of algebra is its use of symbols (usually a combination of numbers and letters) to represent a quantity whose numerical value is unknown. Let's look at a simple example.

Suppose you accept a part-time job for $6 an hour. The job offer states that you will be expected to work between 15 and 30 hours a week. Because you don't know how many hours you will work during a given week, your total income for a week is unknown. Moreover, your income would likely *vary* from week to week. By representing the variable quantity (the number of hours worked) by the letter x, we can represent the weekly income by the following **algebraic expression**.

Number of hours
worked in week
$$6\ \ x$$
$6 per hour

In the product $6x$, the number 6 is a *constant* and the letter x is a *variable*.

Algebraic Expressions

We define an algebraic expression as follows.

Algebraic Expression	A collection of letters (called **variables**) and real numbers (called **constants**) combined by using addition, subtraction, multiplication, or division is called an **algebraic expression**.

Some examples of algebraic expressions are

$$3x + y, \quad -5a^3, \quad 2W - 7, \quad \frac{x}{y + 3}, \quad \text{and} \quad x^2 - 4x + 5.$$

The **terms** of an algebraic expression are those parts that are separated by *addition*. For example, the expression $x^2 - 4x + 5$ has three terms: x^2, $-4x$, and 5. Note that $-4x$, rather than $4x$, is a term of $x^2 - 4x + 5$ because

$$x^2 - 4x + 5 = x^2 + (-4x) + 5.$$

For variable terms like x^2 and $-4x$, the numerical factor is called the **coefficient** of the term. Here, the coefficient of x^2 is 1 and the coefficient of $-4x$ is -4.

EXAMPLE 1   ■   Identifying the Terms of an Algebraic Expression

Identify the terms of the following algebraic expressions.

(a) $3x - \dfrac{1}{2}$ (b) $2y - 5x - 7$

(c) $5(x - 3) + 3x - 4$ (d) $4 - 6x + \dfrac{x + 9}{3}$

Solution

Algebraic Expression	*Terms*
(a) $3x - \dfrac{1}{2}$	$3x, \ -\dfrac{1}{2}$
(b) $2y - 5x - 7$	$2y, \ -5x, \ -7$
(c) $5(x - 3) + 3x - 4$	$5(x - 3), \ 3x, \ -4$
(d) $4 - 6x + \dfrac{x + 9}{3}$	$4, \ -6x, \ \dfrac{x + 9}{3}$

■

EXAMPLE 2   ■   Identifying Coefficients

Identify the coefficient of each of the following terms.

(a) x^3 (b) $\dfrac{2x}{3}$ (c) $3(2 - x)$

Solution

Term	*Coefficient*
(a) x^3	1 because $x^3 = 1 \cdot x^3$
(b) $\dfrac{2x}{3}$	$\dfrac{2}{3}$ because $\dfrac{2x}{3} = \dfrac{2}{3}(x)$
(c) $3(2 - x)$	3

■

Positive Integer Exponents

We know from Section 1.4 that exponents can be used to denote repeated multiplication. For example, 7^4 represents the product obtained by repeatedly multiplying 7 by itself four times.

$$\overset{\text{Exponent}}{\underset{\text{Base}}{7^4}} = \underbrace{7 \cdot 7 \cdot 7 \cdot 7}_{\text{4 factors}}$$

In general, for any positive integer n and any real number a, we said that

$$a^n = \underbrace{a \cdot a \cdot a \cdots a}_{n \text{ factors}}.$$

This rule applies to factors that are variables as well as to factors that are algebraic expressions.

Definition of Exponential Form

Let n be a positive integer and let a be a real number, a variable, or an algebraic expression.

$$a^n = \underbrace{a \cdot a \cdot a \cdots a}_{n \text{ factors}}$$

In this definition remember that the letter a can be a number, a variable, or an algebraic expression. It may be helpful to think of a as a box into which we can place any algebraic expression. Here are some examples.

$$3^4 = 3 \cdot 3 \cdot 3 \cdot 3 \qquad\qquad \text{Where } a = 3$$
$$x^4 = x \cdot x \cdot x \cdot x \qquad\qquad \text{Where } a = x$$
$$(-2x)^4 = (-2x)(-2x)(-2x)(-2x) \qquad \text{Where } a = -2x$$
$$(y + 2)^4 = (y + 2)(y + 2)(y + 2)(y + 2) \qquad \text{Where } a = y + 2$$
$$(5m^2)^4 = (5m^2)(5m^2)(5m^2)(5m^2) \qquad \text{Where } a = 5m^2$$

EXAMPLE 3 ■ **Writing Products in Exponential Form**

Write the following products in exponential form.

(a) $3 \cdot x \cdot x \cdot x$ (b) $3x \cdot 3x \cdot 3x$ (c) $xxyyy$

(d) $2 \cdot 2 \cdot 2 \cdot aab$ (e) $xzzyzy$

Solution

(a) $3 \cdot \underbrace{x \cdot x \cdot x}_{3 \text{ factors}} = 3 \cdot x^3 = 3x^3$ 3 is *not* a factor in the base

(b) $\underbrace{3x \cdot 3x \cdot 3x}_{3 \text{ factors}} = (3x)^3$ 3 *is* a factor in the base

(c) $xxyyy = x^2y^3$

(d) $2 \cdot 2 \cdot 2 \cdot aab = 2^3a^2b$

(e) $xzzyzy = xyyzzz = xy^2z^3$ Commutative Property of Multiplication ■

EXAMPLE 4 ■ **Writing Exponential Forms as Products**

Write each expression as a product of factors.

(a) $(5x)^2y^3$ (b) $5x^2y^3$ (c) $-x^4z$ (d) $(-x)^4z$

Solution

(a) $(5x)^2y^3 = (5x)(5x)yyy = 5 \cdot 5xxyyy$

(b) $5x^2y^3 = 5xxyyy$

(c) $-x^4z = -(x^4)z = -xxxxz$

(d) $(-x)^4z = (-x)(-x)(-x)(-x)z = xxxxz$ ∎

Properties of Exponents

When multiplying two exponential expressions that have the *same* base, we add exponents. To see why this is true, consider the product

$a^3 \cdot a^2.$

Since the first expression represents $a \cdot a \cdot a$ and the second represents $a \cdot a$, it follows that the product of the two expressions represents $a \cdot a \cdot a \cdot a \cdot a$, as follows.

$$a^3 \cdot a^2 = \underbrace{(a \cdot a \cdot a)}_{\substack{Three \\ factors}} \cdot \underbrace{(a \cdot a)}_{\substack{Two \\ factors}} = \underbrace{(a \cdot a \cdot a \cdot a \cdot a)}_{\substack{Five \\ factors}} = a^{3+2} = a^5$$

We summarize this property of exponents as follows.

Multiplying Exponential Forms with the Same Base

Let m and n be positive integers, and let a be a real number, a variable, or an algebraic expression.

$$a^m \cdot a^n = a^{m+n}$$

This rule extends to three or more factors of a raised to positive integer powers. For example,

$a^m \cdot a^n \cdot a^k = a^{m+n+k}.$

In the next example we show how this rule can be used to simplify products involving exponential forms.

EXAMPLE 5 ∎ **Simplifying Products Involving Exponential Forms**

Simplify the following expressions.

(a) b^4b^2b (b) $3^2x^3 \cdot x$ (c) $(-9x^2)(-3x^5)$

Solution

(a) $b^4b^2b = b^{4+2+1} = b^7$

(b) $3^2x^3 \cdot x = (3^2)(x^3 \cdot x) = 9x^4$

(c) $(-9x^2)(-3x^5) = (-9)(-3)(x^2 \cdot x^5) = 27x^7$ ∎

EXAMPLE 6 ■ **Simplifying Products Involving Exponential Forms**

Simplify the following expressions.

(a) $(-3x^2y)(5xy)(2y^2)$ (b) $(-2x^4)^3$ (c) $2xy^3(3x^2y)^2$

Solution

(a) $(-3x^2y)(5xy)(2y^2) = (-3)(5)(2)(x^2 \cdot x)(y \cdot y \cdot y^2) = -30x^3y^4$

(b) $(-2x^4)^3 = (-2x^4)(-2x^4)(-2x^4) = (-2)(-2)(-2)(x^4)(x^4)(x^4) = -8x^{12}$

(c) $2xy^3(3x^2y)^2 = 2xy^3(3x^2y)(3x^2y) = (2 \cdot 3 \cdot 3)(x \cdot x^2 \cdot x^2)(y^3 \cdot y \cdot y) = 18x^5y^5$ ■

Be sure you see the difference between the expressions

$$x^3 \cdot x^4 \quad \text{and} \quad x^3 + x^4.$$

The first is a product of exponential forms, whereas the second is a sum of exponential forms. The rule for multiplying exponential forms having the same base can be applied to the first expression, but *not* to the second expression.

EXAMPLE 7 ■ **Simplifying Expressions with More than One Term**

Simplify the following expressions.

(a) $(3x)^2 - 5x^3$ (b) $(-x^2y)^3 + 2(x^2y)^2$

Solution

(a) $(3x)^2 - 5x^3 = (3x)(3x) - 5x^3$

$$= (3 \cdot 3)(x \cdot x) - 5x^3$$

$$= 9x^2 - 5x^3$$

(b) $(-x^2y)^3 + 2(x^2y)^2 = (-x^2y)(-x^2y)(-x^2y) + 2(x^2y)(x^2y)$

$$= (-1)(x^2y)(-1)(x^2y)(-1)(x^2y) + 2(x^2y)(x^2y)$$

$$= (-1)(-1)(-1) \cdot x^2 \cdot x^2 \cdot x^2 \cdot y \cdot y \cdot y + 2 \cdot x^2 \cdot x^2 \cdot y \cdot y$$

$$= (-1)x^6y^3 + 2x^4y^2$$

$$= -x^6y^3 + 2x^4y^2$$ ■

We have already discussed one rule of exponents: $a^m \cdot a^n = a^{m+n}$. There are two other rules of exponents that you need to know. The first deals with an exponential expression that is itself raised to a power. For instance, how would you evaluate the expression $(a^m)^n$? As an example, let's try writing out the repeated multiplication for $(a^2)^3$.

$$(a^2)^3 = (a^2)(a^2)(a^2) = a^{2+2+2} = a^6$$

From this result, it appears that the rule is

$$(a^m)^n = a^{mn}.$$

The second deals with a product that is raised to a power. For instance, how would you evaluate $(ab)^m$? Let's look at a simple example, say $(ab)^3$. By writing out the repeated multiplication and applying the Commutative Property of Multiplication we obtain

$$(ab)^3 = (ab)(ab)(ab) = (a \cdot a \cdot a)(b \cdot b \cdot b) = a^3 b^3.$$

Thus, it appears that the general rule is

$$(ab)^m = a^m b^m.$$

We summarize the three rules of exponents we have discussed thus far in the following list.

Rules of Exponents

Let m and n be positive integers, and let a and b be real numbers, variables, or algebraic expressions. Then the following properties are true.

1. $a^m \cdot a^n = a^{m+n}$ 2. $(a^m)^n = a^{m \cdot n}$ 3. $(ab)^m = a^m b^m$

EXAMPLE 8 ■ Applying the Rules of Exponents

Use the rules of exponents to rewrite the following expressions.
(a) $(2^3)^4$ (b) $(y^2)^3$ (c) $[(x + 2)^3]^3$
(d) $(3x)^3$ (e) $(-x)^4$ (f) $(2x^2)^3$

Solution

(a) $(2^3)^4 = 2^{3 \cdot 4} = 2^{12} = 4{,}096$ (b) $(y^2)^3 = y^{2 \cdot 3} = y^6$
(c) $[(x + 2)^3]^3 = (x + 2)^{3 \cdot 3} = (x + 2)^9$ (d) $(3x)^3 = 3^3 \cdot x^3 = 27x^3$
(e) $(-x)^4 = (-1)^4 x^4 = x^4$ (f) $(2x^2)^3 = 2^3 (x^2)^3 = 2^3 x^{2 \cdot 3} = 8x^6$

■

EXAMPLE 9 ■ Applying the Rules of Exponents

Use the rules of exponents to rewrite the following expressions.
(a) $(x^2)^3 (2x)^2$ (b) $(-3x)^3 (2x^2)2 + 3x^3$ (c) $(xy^2)^2 (x^2 y)^3$

Solution

(a) $(x^2)^3 (2x)^2 = x^6 \cdot 2^2 \cdot x^2 = 4x^8$
(b) $(-3x)^3 (2x^2)2 + 3x^3 = (-3)^3 \cdot x^3 \cdot 2^2 \cdot (x^2)^2 + 3x^3$
$$= -27 \cdot 4 \cdot x^3 \cdot x^4 + 3x^3$$
$$= -108x^7 + 3x^3$$
(c) $(xy^2)^2 (x^2 y)^3 = x^2 \cdot (y^2)^2 \cdot (x^2)^3 \cdot y^3$
$$= x^2 \cdot y^4 \cdot x^6 \cdot y^3$$
$$= x^8 y^7$$

■

DISCUSSION PROBLEM ■ A Mathematical Riddle

Suppose you were asked to write the largest number that can be written using only three digits. What would the answer be? The number 999 seems to be the obvious answer. However, if we allow the digits to be exponents, then we can obtain numbers that are much larger than 999. For instance, consider the numbers

$$99^9 \approx 913{,}517{,}000{,}000{,}000{,}000$$

and

$$9^{99} \approx \underbrace{29{,}513{,}000{,}000{,}\dots{,}000}_{90 \text{ zeros}}.$$

Can you think of a number that can be written using only three nines that is *much larger* than either of these two numbers? ■

Warm-Up

The following warm-up exercises involve skills that were covered in earlier sections. You will use these skills in the exercise set for this section.

In Exercises 1–6, evaluate the given expression.

1. $2(4 - 3^2)$ **2.** $(3 \div 9) - (12 \div 6)$

3. $(36 \div 3^2) + 4^2$ **4.** $(4 \cdot 2^3) + 10$

5. $\dfrac{120}{2^3 + 4^2}$ **6.** $-2[3 - (12 - 3)]$

In Exercises 7–10, state the property of real numbers that justifies the given statement.

7. $-5(3 + 6) = (-5)3 + (-5)6$ **8.** $2(-3) = -3(2)$

9. $4 + (13 - 5) = (4 + 13) - 5$ **10.** $15\left(\dfrac{1}{15}\right) = 1$

2.1 EXERCISES

In Exercises 1–10, list the terms of the given algebraic expression.

1. $3x^2 + 5$ **2.** $5 - 3t^2 - 4t^3$ **3.** $\dfrac{5}{3} - 3y^3$ **4.** $6x - \dfrac{2}{3}$

5. $2x - 3y + 1$ **6.** $x^2 + 18xy + y^2$ **7.** $3(x + 5) + 10$ **8.** $7 - 2(x - 5)$

9. $4x + \dfrac{x + 2}{3}$ **10.** $x^2 + \dfrac{3x + 1}{x - 1} + 4$

In Exercises 11–20, find the coefficient of the given term.

11. $-6x^2$ 　　　　　　**12.** $25y^5$ 　　　　　　**13.** $\frac{1}{2}y$ 　　　　　　**14.** $-\frac{3}{8}t^2$

15. $\frac{3x}{4}$ 　　　　　　**16.** $-\frac{7y}{12}$ 　　　　　　**17.** $-\frac{3}{2}x$ 　　　　　　**18.** $\frac{1}{8}(x + y)$

19. $-\frac{1}{3}(y - 6)$ 　　　　**20.** $5\left(x^2 + \frac{1}{6}\right)$

In Exercises 21–30, rewrite the given product in exponential form.

21. $2 \cdot u \cdot u \cdot u \cdot u$ 　　　　　　　　　**22.** $\frac{1}{3}x \cdot x \cdot x \cdot x \cdot x$

23. $2u \cdot 2u \cdot 2u \cdot 2u$ 　　　　　　　　**24.** $\frac{1}{3}x \cdot \frac{1}{3}x \cdot \frac{1}{3}x \cdot \frac{1}{3}x \cdot \frac{1}{3}x$

25. $a \cdot a \cdot a \cdot b \cdot b$ 　　　　　　　　**26.** $y \cdot y \cdot z \cdot z \cdot z \cdot z$

27. $4 \cdot x \cdot x \cdot y \cdot x \cdot y$ 　　　　　　　**28.** $u \cdot 7 \cdot v \cdot v \cdot 7 \cdot u$

29. $3 \cdot (x - y) \cdot (x - y) \cdot 3 \cdot 3$ 　　　　**30.** $(u - v) \cdot (u - v) \cdot 8 \cdot 8 \cdot 8 \cdot (u - v)$

In Exercises 31–44, write the given expression as a product of factors.

31. 2^2x^4 　　　　**32.** 5^3x^2 　　　　**33.** $4y^2z^3$ 　　　　**34.** $3uv^4$

35. $(a^2)^3$ 　　　　**36.** $(v^3)^2$ 　　　　**37.** $5x^3 \cdot x^4$ 　　　　**38.** $a^2y^2 \cdot y^3$

39. $(ab)^3$ 　　　　**40.** $2(xz)^4$ 　　　　**41.** $(-2y)^3$ 　　　　**42.** $(-3z)^4$

43. $(x + y)^2$ 　　　**44.** $(s - t)^5$

In Exercises 45–60, simplify the given expression.

45. $u^2 \cdot u^4$ 　　　　**46.** $5z^3 \cdot z$ 　　　　**47.** $5x \cdot (x^6)$ 　　　　**48.** $(-6x^2)x^4$

49. $(-2x^2)(4x)$ 　　　**50.** $(-xz)(-2y^2z)$ 　　　**51.** $2b^4(-ab)(3b^2)$ 　　　**52.** $(4xy)(-3x^2)(-2y^3)$

53. $(x - 2y)(x - 2y)^3$ 　　**54.** $10(x - 3)^2(x - 3)^5$ 　　**55.** $(-2x)^3(3x^2)^2 + 5x^2$ 　　**56.** $(2y)^3 - 3y$

57. $-2(3x)^2(3x) + 5x^2$ 　　**58.** $(x^3)^2(x^2)^4 + 7x^3$ 　　**59.** $(a^2b)^3(ab^2)^4$ 　　**60.** $(-2st^3)^5(s^2t)^4$

In Exercises 61–64, determine whether the two expressions are equal.

61. $x^5 \cdot x^3, \quad x^{15}$ 　　　　　　　**62.** $(-2x)^4, \quad -2x^4$

63. $-3x^3, \quad -27x^3$ 　　　　　　**64.** $(xy)^2, \quad x^2y^2$

65. *Balance in an Account* The balance in an account that has an annual percentage rate of r, compounded quarterly for one year, is given by

$$P\left(1 + \frac{r}{4}\right)\left(1 + \frac{r}{4}\right)\left(1 + \frac{r}{4}\right)\left(1 + \frac{r}{4}\right).$$

Simplify this expression.

66. *Moment of Inertia* The moment of inertia of a solid is given by

$$\frac{1}{2}m(2a)^2(2L).$$

Simplify this expression.

67. What power of 10 is 10,000?

68. What power of 10 is 1,000,000?

In Exercises 69 and 70, perform the indicated operations and simplify.

69. $8 \cdot 10^3 + 3 \cdot 10^2 + 9 \cdot 10^1$

70. $3 \cdot 10^6 + 5 \cdot 10^4 + 7 \cdot 10^2$

71. *Reasonable Wages* Suppose you accept a job for 25 days with the following pay schedule. On the first day the wages are 2¢, the second day 4¢, the third day 8¢, and so on. Complete the following table, which gives the wages for selected days in the 25-day assignment. Would you accept these wages?

t	1	5	10	20	25
2^t					

72. *Area and Volume* The two accompanying figures are a square and a cube, respectively. Each has an edge of length x. Use exponential notation to find an expression that represents the area of the square, and an expression that represents the volume of the cube. (*Hint:* The area of a rectangle is given by the product of its length and width. The volume of a rectangular box is given by the product of its length, width, and height.)

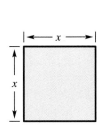
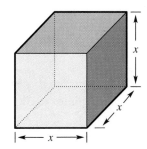

73. *Volume of a Safe* A fireproof safe has a cubical shape (see figure). Use the formula for the volume of a cube from Exercise 72 to find the volume of the storage space in the safe.

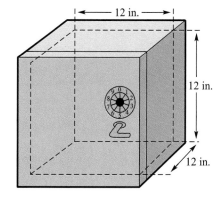

Basic Rules of Algebra

Basic Rules of Algebra • *Combining Like Terms*

Basic Rules of Algebra

The properties of real numbers that we discussed in Section 1.5 are often used to rewrite algebraic expressions. The following list is the same as that given in Section 1.5, except that the examples involve algebraic expressions.

Basic Rules of Algebra

Let a, b, and c represent real numbers, variables, or algebraic expressions. Then the following properties are true.

Property	*Example*
Commutative Property of Addition:	
$a + b = b + a$	$3x + x^2 = x^2 + 3x$
Commutative Property of Multiplication:	
$ab = ba$	$(5 + x)x^3 = x^3(5 + x)$
Associative Property of Addition:	
$(a + b) + c = a + (b + c)$	$(2x + 7) + x^2 = 2x + (7 + x^2)$
Associative Property of Multiplication:	
$(ab)c = a(bc)$	$(2x \cdot 5y) \cdot 7 = 2x \cdot (5y \cdot 7)$
Distributive Property:	
$a(b + c) = ab + ac$	$4x(7 + 3x) = 4x \cdot 7 + 4x \cdot 3x$
$(a + b)c = ac + bc$	$(2y + 5)y = 2y \cdot y + 5 \cdot y$
Additive Identity Property:	
$a + 0 = a$	$3y^2 + 0 = 3y^2$
Multiplicative Identity Property:	
$a \cdot 1 = 1 \cdot a = a$	$(-2x^3) \cdot 1 = 1 \cdot (-2x^3) = -2x^3$
Additive Inverse Property:	
$a + (-a) = 0$	$3y^2 + (-3y^2) = 0$
Multiplicative Inverse Property:	
$a \cdot \dfrac{1}{a} = 1, \quad a \neq 0$	$(x^2 + 2) \cdot \dfrac{1}{x^2 + 2} = 1$

NOTE: Since subtraction is defined as "adding the opposite," the Distributive Properties are also true for subtraction. That is,

$$a(b - c) = ab - ac \quad \text{and} \quad (a - b)c = ac - bc.$$

The next three examples illustrate the use of a variety of the basic rules of algebra.

EXAMPLE 1 ■ Identifying the Basic Rules of Algebra

Identify the rule of algebra illustrated in each of the following.

(a) $(3x^2)7 = 7(3x^2)$ (b) $8x \cdot \dfrac{1}{8x} = 1, \quad x \neq 0$

(c) $(4x^2 + x) - (4x^2 + x) = 0$ (d) $(2 + x^2) + 2x^2 = 2 + (x^2 + 2x^2)$

(e) $(y - 5)2 + (y - 5)y = (y - 5)(2 + y)$

Solution

(a) This equation illustrates the Commutative Property of Multiplication. In other words, we obtain the same result whether we multiply $3x^2$ by 7 or multiply 7 by $3x^2$.

(b) This equation illustrates the Multiplicative Inverse Property. Note that it is important that x be a nonzero number. If x were allowed to be zero, we would be in trouble because the reciprocal of zero is undefined.

(c) This equation illustrates the Additive Inverse Property. In terms of subtraction, this property simply states that when any expression is subtracted from itself the result is zero.

(d) This equation illustrates the Associative Property of Addition. In other words, to form the sum $2 + x^2 + 2x^2$, it doesn't matter whether 2 and x^2 are added first or x^2 and $2x^2$ are added first.

(e) This equation illustrates the Distributive Property.

$$ab + ac = a(b + c) \qquad \text{Distributive Property}$$
$$(y - 5)2 + (y - 5)y = (y - 5)(2 + y)$$

Note in this case that $a = y - 5$, $b = 2$, and $c = y$. ■

EXAMPLE 2 ■ Applying the Basic Rules of Algebra

Use the given rule to complete the following statements.

(a) Additive Identity Property: $(x - 2) + = x - 2$

(b) Commutative Property of Multiplication: $5(y + 6) = $

(c) Commutative Property of Addition: $5(y + 6) = $

(d) Distributive Property: $5(y + 6) = $

(e) Associative Property of Addition: $(x^2 + 3) + 7 = $

Solution

(a) $(x - 2) + 0 = x - 2$ (b) $5(y + 6) = (y + 6)5$

(c) $5(y + 6) = 5(6 + y)$ (d) $5(y + 6) = 5y + 5(6)$

(e) $(x^2 + 3) + 7 = x^2 + (3 + 7)$ ∎

In Example 3 we illustrate some common uses of the Distributive Property. Study this example carefully. Such uses of the Distributive Property are very important in algebra.

EXAMPLE 3   ∎   Using the Distributive Property

Expand the following expressions using the Distributive Property.

(a) $2(7 - x)$ (b) $(10 - 2y)3$ (c) $2x(x + 4)$ (d) $-3(1 - 2y + x)$

Solution

(a) $2(7 - x) = 2 \cdot 7 - 2 \cdot x = 14 - 2x$

(b) $(10 - 2y)3 = 10(3) - 2y(3) = 30 - 6y$

(c) $2x(x + 4) = 2x(x) + 2x(4) = 2x^2 + 8x$

(d) $-3(1 - 2y + x) = (-3)(1) - (-3)(2y) + (-3)(x) = -3 + 6y - 3x$ ∎

NOTE: When we apply the Distributive Property as demonstrated in Example 3, we say that we are **expanding** an algebraic expression. For instance, we can expand $3(x + y)$ by writing the expression as $3x + 3y$.

Combining Like Terms

Two or more terms of an algebraic expression can be combined only if they are *like terms*.

Definition of Like Terms   In an algebraic expression, two terms are said to be **like terms** if they are both constant terms or if they have the same variable factor(s).

The terms $5x$ and $-3x$ are like terms because they have the same variable factor. Similarly, $3x^2y$, $-x^2y$, and $\frac{1}{3}(x^2y)$ are like terms. Note that $3x^2y$ and $-x^2y^2$ are not like terms because their variable factors are different.

EXAMPLE 4   ∎   Identifying Like Terms

List the like terms in each of the following algebraic expressions.

(a) $3x + 3y + 5$ (b) $5xy + 1 - xy$

(c) $12 - x^2 + 3x - 5$ (d) $7x - 3 - 2x + 5$

Solution

(a) The expression $3x + 3y + 5$ has no like terms.

(b) In the expression $5xy + 1 - xy$, the terms $5xy$ and $-xy$ are like terms.

(c) In the expression $12 - x^2 + 3x - 5$, the only like terms are the constant terms 12 and -5.

(d) In the expression $7x - 3 - 2x + 5$, the terms $7x$ and $-2x$ are like terms, and the constant terms -3 and 5 are like terms.   ■

To combine like terms in an algebraic expression, we simply add (or subtract) their respective coefficients and attach the common variable factor. This is actually an application of the Distributive Property, as shown in Example 5.

EXAMPLE 5   ■   **Using the Distributive Law to Combine Like Terms**
Simplify the following expressions by combining like terms.

(a) $5x + 2x - 4$ (b) $-5 + 8 + 7y - 5y$ (c) $2y - 3x - 4x$

Solution

(a) $5x + 2x - 4 = (5 + 2)x - 4$ Distributive Property

$\qquad\qquad\quad = 7x - 4$ Simplest form

(b) $-5 + 8 + 7y - 5y = (-5 + 8) + (7 - 5)y$ Distributive Property

$\qquad\qquad\qquad\qquad = 3 + 2y$ Simplest form

(c) $2y - 3x - 4x = 2y - (3 + 4)x$ Distributive Property

$\qquad\qquad\quad\; = 2y - 7x$ Simplest form   ■

Often, we need to use other rules of algebra before we can apply the Distributive Property to combine like terms. This is illustrated in the next example.

EXAMPLE 6   ■   **Using Rules of Algebra to Combine Like Terms**
Simplify the following expressions by combining like terms.

(a) $7x + 3y - 4x$ (b) $12a - 5 - 3a + 7$ (c) $y - 4x - 7y + 9y$

Solution

(a) $7x + 3y - 4x = 3y + 7x - 4x$ Commutative Property

$\qquad\qquad\quad = 3y + (7x - 4x)$ Associative Property

$\qquad\qquad\quad = 3y + (7 - 4)x$ Distributive Property

$\qquad\qquad\quad = 3y + 3x$ Simplest form

(b) $12a - 5 - 3a + 7 = 12a - 3a - 5 + 7$ Commutative Property

$\qquad\qquad\qquad\quad = (12a - 3a) + (-5 + 7)$ Associative Property

$\qquad\qquad\qquad\quad = (12 - 3)a + (-5 + 7)$ Distributive Property

$\qquad\qquad\qquad\quad = 9a + 2$ Simplest form

(c) $y - 4x - 7y + 9y = -4x + (y - 7y + 9y)$ Collect like terms

$= -4x + (1 - 7 + 9)y$ Distributive Property

$= -4x + 3y$ Simplest form ∎

As you gain experience with the rules of algebra, you may want to combine some of the steps in your work. For instance, you might feel comfortable listing only the following steps to solve part (b) of Example 6.

$12a - 5 - 3a + 7 = (12a - 3a) + (-5 + 7)$ Collect like terms

$= 9a + 2$ Simplify

DISCUSSION PROBLEM ▪ **Expressing Algebraic Rules in Words**

This section lists ten basic rules of algebra. There are two commutative properties, two associative properties, two identity properties, two inverse properties, and two distributive properties. State the algebraic version of each of these properties from memory. Then write a sentence that describes each property in words. ▪

Warm-Up

The following warm-up exercises involve skills that were covered in earlier sections. You will use these skills in the exercise set for this section.

In Exercises 1 and 2, rewrite the given expression in exponential form.

1. $3z \cdot 3z \cdot 3z \cdot 3z$ **2.** $x \cdot 8 \cdot y \cdot y \cdot 8 \cdot 8 \cdot x \cdot x$

In Exercises 3 and 4, rewrite the given expression as a product of factors.

3. $3^2x^4y^3$ **4.** $5(uv)^4$

In Exercises 5–10, simplify the given expression.

5. $v^2 \cdot v^3$ **6.** $(u^3)^2$

7. $(-2x)^2x^4$ **8.** $-y^2(-2y)^3$

9. $5z^3(z^2)^2$ **10.** $(a + 3)^2(a + 3)^5$

2.2	**EXERCISES**

In Exercises 1–12, identify the rule (or rules) of algebra illustrated by the given equation.

1. $x + 2y = 2y + x$

2. $-10(xy^2) = (-10x)y^2$

3. $(x^2 + y^2) \cdot 1 = x^2 + y^2$

4. $rt + 0 = rt$

5. $(3x + 2y) + z = 3x + (2y + z)$

6. $2zy = 2yz$

7. $16xy \cdot \dfrac{1}{16xy} = 1, \quad xy \neq 0$

8. $(5m + 3) - (5m + 3) = 0$

9. $x(y + z) = xy + xz$

10. $(x + y) \cdot \dfrac{1}{x + y} = 1, \quad x + y \neq 0$

11. $x^2 + (y^2 - y^2) = x^2$

12. $(x + 2)(x + y) = x(x + y) + 2(x + y)$

In Exercises 13–20, use the indicated property to complete the given statement.

13. Additive Inverse Property: $(x + 1) - $ ▭ $= 0$

14. Associative Property of Multiplication: $(-5r)s = -5($ ▭ $)$

15. Commutative Property of Multiplication: $v(2) = $ ▭

16. Additive Identity Property: $(4x - 3y) + $ ▭ $= 4x - 3y$

17. Distributive Property: $(t + 5)(t - 2) = t($ ▭ $) + 5($ ▭ $)$

18. Additive Inverse Property: $(2z - 3) + $ ▭ $= 0$

19. Multiplicative Inverse Property: $5x($ ▭ $) = 1, \quad x \neq 0$

20. Multiplicative Identity Property: $(s - 5)($ ▭ $) = s - 5$

In Exercises 21–30, expand the expression by using the Distributive Property.

21. $-5(2x - y)$

22. $\dfrac{1}{8}(16 + 8z)$

23. $(x + 2)(3)$

24. $(4 - t)(-6)$

25. $x(x + xy + y^2)$

26. $r(r - t + s)$

27. $3(x^2y + x)$

28. $4(2y^2 - y)$

29. $(-1)(u - v)$

30. $(-1)(x^2 - 2xy + y^2)$

In Exercises 31–36, list the terms of the given expression and the coefficient of each term.

31. $6x^2 - 3xy + y^2$

32. $-4xy + 2xz - yz$

33. $4x^3 - 3x^2 + 2x$

34. $-x - 5x^3 + x^4$

35. $0.12x + 0.36x^2y - 1.40y$

36. $\dfrac{1}{3}m^2 + \dfrac{5}{6}n^2$

In Exercises 37–40, list the like terms of the given expression.

37. $16t^3 + 4 - 5 + 3t^3$

38. $-\frac{1}{4}x^2 - 3x + \frac{3}{4}x^2 + x$

39. $6x^2y + 2xy - 4x^2y$

40. $a^2 + 5ab^2 - 3b^2 + 7a^2b - ab^2 + a^2$

In Exercises 41–50, simplify the given expression by combining like terms.

41. $3y - 5y$ **42.** $-16x + 25x$ **43.** $x + 5y - 3x - y$ **44.** $7s + 3 - 3s - 4$

45. $x^2 - 2xy + 4 + xy$ **46.** $5z - 5 + 10z$

47. $3\left(\dfrac{1}{x}\right) - \dfrac{1}{x} + 8$ **48.** $16\left(\dfrac{a}{b}\right) - 6\left(\dfrac{a}{b}\right) + \dfrac{3}{2} - \dfrac{1}{2}$

49. $3x^2 - x^2y + 4xy^2 + 3x^2y - xy^2 + y^2$ **50.** $z^3 + 2z^2 + z + z^2 + 2z + 1$

In Exercises 51 and 52, state why the two algebraic expressions are not like terms.

51. $\dfrac{1}{2}x^2y, \quad -\dfrac{5}{2}xy^2$

52. $-16x^2y^3, \quad 7x^2y$

In Exercises 53–56, determine whether the given expressions are equal.

53. $3(x - 4), \quad 3x - 4$ **54.** $-3(x - 4), \quad -3x - 12$ **55.** $6x - 4x, \quad 2x$ **56.** $12y^2 + 3y^2, \quad 36y^2$

In Exercises 57–62, use the Distributive Property to perform the required arithmetic *mentally*. For example, suppose you work in an industry where the wage is \$14 per hour and "time and a half" for overtime. Thus, your hourly wage for overtime is $14(1.5) = 14\left(1 + \frac{1}{2}\right) = 14 + 7 = \$21$.

57. $16(1.5) = 16\left(1 + \frac{1}{2}\right)$ **58.** $15(1.5) = 15\left(1 + \frac{1}{2}\right)$ **59.** $8(52) = 8(50 + 2)$

60. $6(29) = 6(30 - 1)$ **61.** $5(7.98) = 5(8 - 0.02)$ **62.** $12(11.95) = 12(12 - 0.05)$

63. *Area of a Rectangle* The accompanying figure shows two adjoining rectangles. Demonstrate the Distributive Property by filling in the blanks to express the combined area of the two rectangles in two ways.

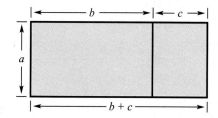

64. *Area of a Rectangle* The accompanying figure shows two adjoining rectangles. Demonstrate the subtraction version of the Distributive Property by filling in the blanks to express the area on the left in two ways.

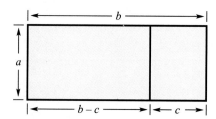

Rewriting and Evaluating Algebraic Expressions
Simplifying Algebraic Expressions • *Symbols of Grouping* • *Evaluating Algebraic Expressions*

Simplifying Algebraic Expressions

Simplifying an algebraic expression by rewriting it into a more usable form is one of the three most frequently used skills in algebra. The other two—solving an equation and sketching the graph of an equation—will be studied in later chapters of this text.

To "simplify an algebraic expression" generally means to remove symbols of grouping and combine like terms. For instance, the expression $x + (3 + x)$ can be simplified as $2x + 3$.

We learned to combine like terms in the last section. In this section we focus on removing symbols of grouping. To remove symbols of grouping you must be familiar with the use of the properties of multiplication and exponents. We review some of these properties in the next two examples.

EXAMPLE 1 ■ Simplifying Algebraic Expressions

Simplify the following expressions.

(a) $-3(-5x)$ (b) $7(-x)$

Solution

(a) $-3(-5x) = (-3)(-5)x$ Associative Property

$\qquad\qquad\quad = 15x$ Simplest form

(b) $7(-x) = 7(-1)(x)$ Coefficient of $-x$ is -1

$\qquad\quad = -7x$ Simplest form   ■

EXAMPLE 2 ■ Simplifying Algebraic Expressions

(a) $\dfrac{5x}{3} \cdot \dfrac{3}{5} = \left(\dfrac{5}{3} \cdot x\right) \cdot \dfrac{3}{5}$ Coefficient of $5x/3$ is $5/3$

$\qquad\quad = \left(\dfrac{3}{5} \cdot \dfrac{5}{3}\right) \cdot x$ Commutative and Associative Properties

$\qquad\quad = 1 \cdot x$ Multiplicative Inverse

$\qquad\quad = x$ Multiplicative Identity

(b) $x^2(-2x^3) = (-2)(x^2 \cdot x^3)$ Commutative and Associative Properties

$\qquad\qquad\quad = -2x^{2+3}$ Property of Exponents

$\qquad\qquad\quad = -2x^5$ ■

MATH MATTERS

Find the Day of the Week for Any Date

Many people are interested in finding the day of the week on which they were born. There is a way to do this (without looking at old calendars) by using the following number values.

To use these values to find the day of the week for any date between January 1, 1901, and December 31, 1999, use the following steps. (This example uses August 16, 1967.)

Month	Value	Day	Value
Jan	0	Sun	0
Feb	3	Mon	1
Mar	3	Tues	2
Apr	6	Wed	3
May	1	Thu	4
Jun	4	Fri	5
Jul	6	Sat	6
Aug	2		
Sep	5		
Oct	0		
Nov	3		
Dec	5		

Step	*Example*	
1. Determine the number of leap days between January 1, 1901 and date. (Divide year of century by 4; ignore remainder.*)	16	67 ÷ 4
2. Add this to the day of the month,	16	
3. the value of the month, and	2	August
4. the year of the century.	+ 67	1967
	101	Total
5. Divide total by 7; keep remainder.	101 ÷ 7 = 14	Remainder = 3
6. Use the remainder to determine the day of the week.	3	Wednesday

Thus, August 16, 1967, occurred on a Wednesday.

The constellation Leo represents the Zodiac sign for those born between July 23 and August 22.

*Do not count the leap day if the date is earlier than March 1 on a leap year. For instance, there were only four leap days between January 1, 1901, and February 28, 1920.

Symbols of Grouping

The main tool for removing symbols of grouping is the Distributive Property. For nested symbols of grouping like

$$5x - 2x[3 + 2(x - 7)]$$

we first remove the innermost grouping symbol and combine like terms. The process is then repeated for subsequent grouping symbols, as follows.

$5x - 2x[3 + 2(x - 7)] = 5x - 2x[3 + 2x - 14]$	Remove innermost parentheses
$= 5x - 2x[2x - 11]$	Combine like terms in brackets
$= 5x - 2x(2x) - 2x(-11)$	Remove brackets
$= 5x - 4x^2 + 22x$	Multiply
$= 27x - 4x^2$	Combine like terms

Note that this answer could also have been written as

$$-4x^2 + 27x.$$

EXAMPLE 3   ■   **Removing Symbols of Grouping**

Simplify the following expressions.

(a) $5x + (x - 7)2$ (b) $-2(4x - 1) + 3x$ (c) $3(y - 5) - (2y - 7)$

Solution

(a) $5x + (x - 7)2 = 5x + 2x - 14$	Distributive Property
$= 7x - 14$	Combine like terms
(b) $-2(4x - 1) + 3x = -8x + 2 + 3x$	Distributive Property
$= -8x + 3x + 2$	Commutative Property
$= -5x + 2$	Combine like terms
(c) $3(y - 5) - (2y - 7) = 3y - 15 - (2y - 7)$	Distributive Property
$= 3y - 15 - 2y + 7$	Remove parentheses by changing signs of terms inside
$= (3y - 2y) + (-15 + 7)$	Collect like terms
$= y - 8$	Combine like terms ■

NOTE: If a parenthetical expression is preceded by a *plus* sign, the parentheses can be removed without changing the signs of the terms inside, as follows.

$$3y + (-2y + 7) = 3y - 2y + 7$$

If a parenthetical expression is preceded by a *minus* sign, the parentheses can be removed by changing the sign of each term inside, as follows.

$$3y - (2y - 7) = 3y - 2y + 7$$

EXAMPLE 4   ■   Removing Nested Symbols of Grouping

(a) $5x - 2[4x + 3(x - 1)] = 5x - 2[4x + 3x - 3]$ Remove innermost parentheses

$\qquad\qquad\qquad\qquad\quad = 5x - 2[7x - 3]$ Combine like terms in brackets

$\qquad\qquad\qquad\qquad\quad = 5x - 14x + 6$ Remove brackets

$\qquad\qquad\qquad\qquad\quad = -9x + 6$ Combine like terms

(b) $-7y + 3[2y - (3 - 2y)] - 5y + 4$

$\qquad = -7y + 3[2y - 3 + 2y] - 5y + 4$ Remove innermost parentheses

$\qquad = -7y + 3[4y - 3] - 5y + 4$ Combine like terms in brackets

$\qquad = -7y + 12y - 9 - 5y + 4$ Remove brackets

$\qquad = (-7y + 12y - 5y) + (-9 + 4)$ Collect like terms

$\qquad = -5$ Combine like terms   ■

EXAMPLE 5   ■   Simplifying Algebraic Expressions

(a) $(-3x)(5x^4) + 7x^5 = (-3)(5)x \cdot x^4 + 7x^5$ Commutative and Associative Properties

$\qquad\qquad\qquad\qquad = -15x^5 + 7x^5$ Property of Exponents

$\qquad\qquad\qquad\qquad = -8x^5$ Combine like terms

(b) $2x(x + 3y) + 4(5 - xy) = 2x^2 + 6xy + 20 - 4xy$

$\qquad\qquad\qquad\qquad\qquad = 2x^2 + 6xy - 4xy + 20$

$\qquad\qquad\qquad\qquad\qquad = 2x^2 + 2xy + 20$ ■

In the next example, we show how the Distributive Property can be used with fractional expressions.

EXAMPLE 6   ■   Simplifying Fractional Expressions Using the Distributive Property

(a) $\dfrac{3x}{5} - \dfrac{x}{5} = \dfrac{3}{5}x - \dfrac{1}{5}x$ Write with fractional coefficients

$\qquad\qquad = \left(\dfrac{3}{5} - \dfrac{1}{5}\right)x$ Distributive Property

$\qquad\qquad = \dfrac{2}{5}x$ Simplest form

(b) $\dfrac{x}{4} + \dfrac{2x}{7} = \left(\dfrac{1}{4} + \dfrac{2}{7}\right)x$ Write with fractional coefficients and distribute

$\qquad\qquad = \left(\dfrac{1(7)}{4(7)} + \dfrac{2(4)}{7(4)}\right)x$ Common denominator

$\qquad\qquad = \dfrac{15}{28}x$ Simplest form

(c) $\dfrac{3x}{10} - \dfrac{4x}{15} = \left(\dfrac{3}{10} - \dfrac{4}{15}\right)x$ 　　　Write with fractional coefficients and distribute

$\qquad = \left(\dfrac{3(3)}{10(3)} - \dfrac{4(2)}{15(2)}\right)x$ 　　　Common denominator

$\qquad = \dfrac{1}{30}x$ 　　　Simplest form ■

Evaluating Algebraic Expressions

In applications of algebra, we are often required to **evaluate** an algebraic expression. This means we are to find the *value* of an expression when its variables are replaced by real numbers. For instance, when $x = 2$, the value of the expression $2x + 3$ is as follows.

$2x + 3$ 　　　Expression

↓ 　　　Replace x by 2

$2(2) + 3 = 7$ 　　　Value of expression

When finding the value of an algebraic expression, be sure to replace every occurrence of the specified variable with the appropriate real number. For instance, when $x = -2$, the value of $x^2 - x + 3$ is

$(-2)^2 - (-2) + 3 = 4 + 2 + 3 = 9.$

EXAMPLE 7 ■ **Evaluating Algebraic Expressions**

Evaluate the following expressions when $x = -3$ and $y = 5$.

(a) $-x$ 　　(b) $3x + 2y$ 　　(c) $y - 2(x + y)$ 　　(d) $y^2 - 3y - 6$

Solution

(a) When $x = -3$, the value of $-x$ is

$-x = -(-3) = 3.$

(b) When $x = -3$ and $y = 5$, the value of $3x + 2y$ is

$3x + 2y = 3(-3) + 2(5) = -9 + 10 = 1.$

(c) When $x = -3$ and $y = 5$, the value of $y - 2(x + y)$ is

$y - 2(x + y) = 5 - 2((-3) + 5) = 5 - 2(2) = 5 - 4 = 1.$

(d) When $y = 5$, the value of $y^2 - 3y - 6$ is

$y^2 - 3y - 6 = 5^2 - 3(5) - 6 = 25 - 15 - 6 = 4.$ ■

Note in parts (a) and (c) of Example 7 that it is a good idea to use parentheses when using a negative number as a replacement for a variable. This is further demonstrated in Example 8.

EXAMPLE 8 ■   **Evaluating Algebraic Expressions**

Evaluate the following expressions when $x = 4$ and $y = -6$.

(a) y^2 (b) $-y^2$ (c) $y - x$ (d) $|y - x|$ (e) $|x - y|$

Solution

(a) When $y = -6$, the value of the expression y^2 is

$$y^2 = (-6)^2 = 36.$$

(b) When $y = -6$, the value of the expression $-y^2$ is

$$-y^2 = -(y^2) = -(-6)^2 = -36.$$

(c) When $x = 4$ and $y = -6$, the value of the expression $y - x$ is

$$y - x = (-6) - 4 = -6 - 4 = -10.$$

(d) When $x = 4$ and $y = -6$, the value of the expression $|y - x|$ is

$$|y - x| = |-6 - 4| = |-10| = 10.$$

(e) When $x = 4$ and $y = -6$, the value of the expression $|x - y|$ is

$$|x - y| = |4 - (-6)| = |4 + 6| = |10| = 10.$$

  ■

EXAMPLE 9 ■   **Evaluating Algebraic Expressions**

Evaluate the following expressions when $x = 0$, $y = -2$, and $z = 3$.

(a) $3x - 2yz + 7xz$ (b) $\dfrac{y + 2z}{5y - xz}$ (c) $(y + 2z)(z - 3y)$

Solution

(a) When $x = 0$, $y = -2$, and $z = 3$, the value of the given expression is

$$3x - 2yz + 7xz = 3(0) - 2(-2)(3) + 7(0)(3) = 0 + 12 + 0 = 12.$$

(b) When $x = 0$, $y = -2$, and $z = 3$, the value of the given expression is

$$\frac{y + 2z}{5y - xz} = \frac{-2 + 2(3)}{5(-2) - (0)(3)} = \frac{-2 + 6}{-10 - 0} = \frac{4}{-10} = -\frac{2}{5}.$$

(c) When $x = 0$, $y = -2$, and $z = 3$, the value of the given expression is

$$\begin{aligned}
(y + 2z)(z - 3y) &= ((-2) + 2(3))(3 - 3(-2)) \\
&= (-2 + 6)(3 + 6) \\
&= 4(9) \\
&= 36.
\end{aligned}$$

  ■

On occasion we may need to evaluate an algebraic expression for *several* values of x. In such cases you should consider simplifying the expression as much as possible *before* substituting for x. For instance, the evaluation of

$$5x - 2[4x + 3(x - 1)]$$

when $x = 0$, when $x = 2$, and when $x = -5$ could be done more efficiently by first simplifying the expression to the form $-9x + 6$ [see Example 4(a)] and then substituting for x to obtain the following.

When $x = 0$: $-9x + 6 = -9(0) + 6 = 6$

When $x = 2$: $-9x + 6 = -9(2) + 6 = -12$

When $x = -5$: $-9x + 6 = -9(-5) + 6 = 51$

Try evaluating the original expression for the same three values of x to see how much longer it takes.

DISCUSSION PROBLEM ■ **You Be the Instructor**

Suppose you are teaching an algebra class and one of your students hands in the following problem. What is the error in this work?

$$2[3x - 2(x - 3)] = 2[3x - 2x + 3]$$
$$= 2[x + 3]$$
$$= 2x + 6$$

What would you say to your class to help them avoid this type of error? ■

Warm-Up

The following warm-up exercises involve skills that were covered in earlier sections. You will use these skills in the exercise set for this section.

In Exercises 1–6, use the Distributive Property to expand the given expression.

1. $-4(2x - 5)$ **2.** $10(3t - 4)$

3. $x(-2xy + y^3)$ **4.** $-z(xz - 2y^2)$

5. $-\dfrac{3}{4}(12 - 8x)$ **6.** $\dfrac{2}{3}(-15x - 12z)$

In Exercises 7–10, simplify the given expression by combining like terms.

7. $4s - 6t + 7s + t$ **8.** $2x^2 - 4 + 5 - 3x^2$

9. $\dfrac{5x}{3} - \dfrac{2x}{3} - 4$ **10.** $3x^2y + xy - xy^2 - 6xy$

2.3 EXERCISES

In Exercises 1–10, use the properties of multiplication and the properties of exponents to simplify the given expression.

1. $-2(6x)$

2. $-4(-3y)$

3. $(-2x)(-3x)$

4. $(-5z)(2z^2)$

5. $\dfrac{5x}{8} \cdot \dfrac{16}{5}$

6. $\left(\dfrac{4x}{3}\right)\left(\dfrac{3x}{2}\right)$

7. $(-1.5x^2)(4x^3)$

8. $(-32t)(-0.5t^2)$

9. $(12xy^2)(-2x^3y^2)$

10. $(7r^2s^3)(3rs)$

In Exercises 11–24, simplify the given expression by removing symbols of grouping and combining like terms.

11. $2(x - 2) + 7$

12. $6(2s - 1) + s + 4$

13. $m - 3(m - 5) + 25$

14. $5l - 6(3l - 5l)$

15. $-6(1 - 2x) + 10(5 - x)$

16. $3(r - 2s) - 5(3r - 5s)$

17. $\dfrac{2}{3}(12x + 15) + 16$

18. $\dfrac{3}{8}(4 - y) - \dfrac{5}{2} + 10$

19. $3 - 2[6 + (4 - x)]$

20. $10x + 5[6 - (2x + 3)]$

21. $4x^2 + x(5 - x)$

22. $-z(z - 2) + 3z^2 + 5$

23. $-3t(4 - t) + t(t + 1)$

24. $4y[5 - (y + 1)] + 3y(y + 1)$

In Exercises 25–32, use the Distributive Property to simplify the given expression.

25. $\dfrac{2x}{3} - \dfrac{x}{3}$

26. $\dfrac{7y}{8} - \dfrac{3y}{8}$

27. $\dfrac{4z}{5} + \dfrac{3z}{5}$

28. $\dfrac{5t}{12} + \dfrac{7t}{12}$

29. $\dfrac{x}{3} - \dfrac{5x}{4}$

30. $\dfrac{5x}{7} + \dfrac{2x}{3}$

31. $\dfrac{3x}{10} - \dfrac{x}{10} + \dfrac{4x}{5}$

32. $\dfrac{3z}{4} - \dfrac{z}{2} - \dfrac{z}{3}$

In Exercises 33–48, evaluate the algebraic expression for the specified values of the variables. (If the evaluation is not possible, state the reason.)

Expression	Values	
33. $2x - 1$	(a) $x = \dfrac{1}{2}$	(b) $x = 4$
34. $-\lvert x \rvert$	(a) $x = \dfrac{4}{3}$	(b) $x = -\dfrac{4}{3}$
35. $2x^2 + 4x - 5$	(a) $x = -2$	(b) $x = 3$
36. $64 - 16t^2$	(a) $t = 2$	(b) $t = 3$
37. $3x - 2y$	(a) $x = 4, y = 3$	(b) $x = \dfrac{2}{3}, y = 1$
38. $10u - 3v$	(a) $u = 3, v = 10$	(b) $u = -2, v = -7$
39. $x - 3(x - y)$	(a) $x = 3, y = 3$	(b) $x = 4, y = -4$

40. $-3x + 2(x + y)$ (a) $x = -2$, $y = 2$ (b) $x = 0$, $y = 5$

41. $b^2 - 4ac$ (a) $a = 2$, $b = -3$, $c = -1$ (b) $a = -4$, $b = 6$, $c = -2$

42. $(2a - b) - (-a - 3b)$ (a) $a = \dfrac{3}{2}$, $b = 3$ (b) $a = -2$, $b = 4$

43. $\dfrac{x - 2y}{x + 2y}$ (a) $x = 4$, $y = 2$ (b) $x = 4$, $y = -2$

44. $\dfrac{-y}{x^2 + y^2}$ (a) $x = 0$, $y = 5$ (b) $x = 1$, $y = -3$

45. *Area of a Triangle:* $\dfrac{1}{2}bh$ (a) $b = 3$, $h = 5$ (b) $b = 2$, $h = 10$

46. *Volume of a Rectangular Box:* lwh   (a) $l = 4$, $w = 2$, $h = 9$ (b) $l = 10$, $w = 5$, $h = 20$

47. *Distance traveled:* rt   (a) $r = 50$, $t = 3.5$ (b) $r = 35$, $t = 4$

48. *Simple interest:* Prt   (a) $P = 1{,}000$, $r = 0.08$, $t = 3$ (b) $P = 500$, $r = 0.07$, $t = 5$

49. (a) Complete the following table by evaluating the expression $3x - 2$.

x	-1	0	1	2	3	4
$3x - 2$						

(b) From the table in part (a), determine the increase in the value of the expression for each one-unit increase in x.

(c) Use the results of parts (a) and (b) to determine the increase in the algebraic expression $\frac{2}{3}x + 4$ for each one-unit increase in x.

50. (a) Complete the following table by evaluating the expression $2x + 5y$ for the indicated values of x and y.

y \ x	-1	0	1	2	3
-1					
0					
1					
2					
3					

(b) Determine the increase in the value of the algebraic expression for each one-unit increase in x. Does the answer depend on the value of y?

51. *Area of Floor Space* The basic floor plan for a one-story house is shown in the accompanying figure.

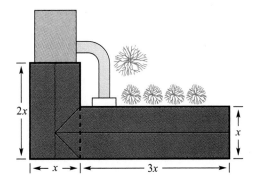

(a) Write a simplified algebraic expression for the amount of floor space in the house.

(b) Determine the amount of floor space if $x = 15$ feet. What units of measure are used for the amount of floor space?

52. *Area of a Trapezoid* The area of a trapezoid with parallel bases of length b_1 and b_2 and height h is given by

$$\frac{h}{2}(b_1 + b_2),$$

as shown in the figure.

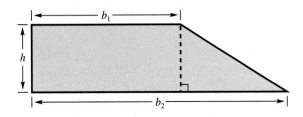

(a) Show that the area can also be expressed as

$$b_1 h + \frac{1}{2}(b_2 - b_1)h$$

and give a geometric explanation for the area represented by each term of this expression.

(b) Find the area of a trapezoid with $b_1 = 7$, $b_2 = 12$, and $h = 3$.

Area of a Trapezoid In Exercises 53 and 54, use the formula for the area of a trapezoid, $\frac{1}{2}h(b_1 + b_2)$, to find the area of the given trapezoid.

53.

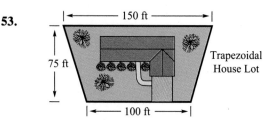

54.

Trapezoidal Floor Tile

SECTION
2.4

Translating Expressions: Verbal to Algebraic
Introduction • Translating Phrases • Hidden Operations

Introduction

In the first three sections of this chapter, we focused on rewriting and simplifying algebraic expressions. In this section we look at ways to *construct* algebraic expressions.

Let's take another look at the example discussed at the beginning of this chapter. In that example you were paid $6 an hour and your weekly pay was represented by

$$\boxed{\text{Pay per hour}} \cdot \boxed{\text{Number of hours}} = \boxed{6 \text{ dollars}} \cdot \boxed{x \text{ hours}} = 6x.$$

Note the hidden operation of multiplication in this expression. Nowhere in the verbal problem does it say we are to multiply 6 times x. It is *implied* in the problem—we just use common sense to come up with the product. This is often the case when using algebra to solve real-life problems.

EXAMPLE 1   ■   Constructing an Algebraic Expression

A person is paid 3¢ for each aluminum soda can, and 2¢ for each steel soda can collected. Write an algebraic expression that represents the total weekly income for this recycling activity.

Solution

Before writing an algebraic expression for the weekly income, it is helpful to construct an informal verbal model. For instance, the following verbal model could be used.

| Pay per can | . | Number of *aluminum* cans | + | Pay per can | . | Number of *steel* cans |

Note that the word *and* in the problem statement indicates addition. Because both the number of aluminum cans and the number of steel cans can vary from week to week, we use the two variables x and y, respectively, and write the following algebraic expression.

| 3 cents | · | x cans | + | 2 cents | · | y cans | $= 3x + 2y$

■

EXAMPLE 2   ■   Constructing an Algebraic Expression

A person is paid 3¢ for each aluminum soda can collected, 2¢ for each steel soda can collected, and $45 a week for collecting other kinds of trash in the city park. Write an expression that represents the total weekly income.

Solution

When creating a model involving two different units, like cents and dollars, we must convert to a single unit. In Example 1 we used cents as the unit, so let's use dollars here. A verbal model for the weekly income could be as follows.

| Pay per can | . | Number of aluminum cans | + | Pay per can | . | Number of steel cans | + | 45 dollars |

$= $ | $0.03 | · | x cans | + | $0.02 | · | y cans | + | $45 |

$= 0.03x + 0.02y + 45$

Remember that this expression represents the weekly pay in *dollars*. Try writing an algebraic expression that represents the weekly pay in cents.

■

Translating Phrases

When you are translating verbal sentences and phrases into algebraic expressions, it is sometimes helpful to watch for key words and phrases that indicate the four different operations of arithmetic. The following list gives several examples.

Translating Phrases into Algebraic Expressions	Key Words and Phrases	Verbal Description	Algebraic Expression
	Addition:		
	Sum, plus, greater, increased by, more than, exceeds, total of	The sum of 6 and x	$6 + x$
		Eight more than y	$y + 8$
	Subtraction:		
	Difference, minus, less, decreased by, subtracted from, reduced by, the remainder	Five decreased by a	$5 - a$
		Four less than z	$z - 4$
	Multiplication:		
	Product, multiplied by, twice, times, percent of	Five times x	$5x$
	Division:		
	Quotient, divided by, ratio, per	The ratio of x to 3	$\dfrac{x}{3}$

This symbol always before the word to is always the numerator

EXAMPLE 3 ■ **Translating Verbal Phrases Containing a Specified Variable**

Translate each of the following into an algebraic expression.

(a) Three more than five times x

(b) Three less than the product of five and m

(c) y decreased by the sum of ten and x^2

Solution

(a) Three more than five times x

$$5x + 3$$ *Think:* 3 added to what?

(b) Three less than the product of five and m

$$5m - 3$$ *Think:* 3 subtracted from what?

(c) y decreased by the sum of ten and x^2

$$y - (10 + x^2)$$ *Think:* What is subtracted from y?   ■

EXAMPLE 4 ■ **Translating Verbal Phrases Containing a Specified Variable**

Translate each of the following into an algebraic expression.

(a) The sum of seven and six times x

(b) Six times the sum of x and seven

(c) The product of four and x, all divided by three

Solution

(a) The sum of seven and six times x

$7 + 6x$ *Think:* Add 7 and what?

(b) Six times the sum of x and seven

$6(x + 7)$ *Think:* 6 multiplied by what?

(c) The product of four and x, all divided by three

$\dfrac{4x}{3}$ *Think:* What is divided by 3?   ■

In most applications of algebra, the variables are not specified and it is your task to assign variables to the *appropriate* quantities. Though similar to the problems in Examples 3 and 4, the problems in the next example may seem more difficult because variables have not been assigned to the unknown quantities.

EXAMPLE 5 ■   **Translating Verbal Phrases Having No Specified Variable**

Translate each of the following into a variable expression.

$3 + x$

$5 - 3x$

(a) The sum of three and a number

(b) Five decreased by the product of three and a number

$y + 2x$ ⟶ (c) A number increased by twice another number

$\dfrac{x + 3}{12}$ ⟶ (d) The sum of a number and three, divided by twelve

Solution

In all cases we will let x be a number and y be another number.

(a) The sum of three and a number

$3 + x$ *Think:* 3 added to what?

(b) Five decreased by the product of three and a number

$5 - 3x$ *Think:* What is subtracted from 5?

(c) A number increased by twice another number

$x + 2y$ *Think:* A number added to what?

(d) The sum of a number and three, divided by twelve

$\dfrac{x + 3}{12}$ *Think:* What is divided by 12?   ■

A good way to learn algebra is to do it forwards and backwards. In the next example, we translate algebraic expressions into verbal form. Keep in mind that other key words could be used to describe the operations in each expression. In this example we try to use key words or phrases that keep the verbal expression clear and concise.

EXAMPLE 6 ■ Translating Algebraic Expressions into Verbal Form

Without using a variable, write a verbal description for each of the following.

(a) $7x - 12$ (b) $7(x - 12)$ (c) $5 + \dfrac{x}{2}$ (d) $\dfrac{5 + x}{2}$

Solution

(a) *Algebraic expression:* $7x - 12$
Primary operation: Subtraction
Terms: $7x$ and 12
Verbal description: Twelve less than the product of seven and a number

(b) *Algebraic expression:* $7(x - 12)$
Primary operation: Multiplication
Factors: 7 and $(x - 12)$
Verbal description: Twelve is subtracted from a number and the result is multiplied by seven.

(c) *Algebraic expression:* $5 + \dfrac{x}{2}$
Primary operation: Addition
Terms: 5 and $\dfrac{x}{2}$
Verbal description: Five added to the quotient of a number and two

(d) *Algebraic expression:* $\dfrac{5 + x}{2}$
Primary operation: Division
Numerator, denominator: Numerator is $5 + x$; denominator is 2
Verbal description: The sum of five and a number, all divided by two ■

Hidden Operations

Most real-life problems do not contain verbal expressions that clearly identify the arithmetic operations involved. We need to rely on common sense, past experience, and the physical nature of the problem in order to identify the operations hidden in the problem statement. Watch for *hidden products* in the next two examples.

EXAMPLE 7 ■ Discovering Hidden Products

(a) A cash register contains x quarters. Write an expression for this amount of money in dollars.

(b) A cash register contains n nickels and d dimes. Write an expression for this amount of money in cents.

(c) Write an expression showing how far a person can ride a bicycle in t hours if the person travels at a constant rate of 15 miles per hour.

Solution

(a) The amount of money is a product.

Verbal model: | Value of coin | · | Number of coins |

Labels: Value of coin = 0.25 (in dollars)
              Number of coins = x

Algebraic expression: 0.25x (in dollars)

(b) The amount of money is a sum of products.

Verbal model: | Value of nickel | · | Number of nickels | + | Value of dime | · | Number of dimes |

Labels: Value of nickel = 5 (in cents)
              Number of nickels = n
              Value of dime = 10 (in cents)
              Number of dimes = d

Algebraic expression: $5n + 10d$ (in cents)

(c) The distance traveled is a product.

Verbal model: | Rate of travel | · | Time traveled |

Labels: Rate of travel = 15 (miles per hour)
              Time traveled = t (in hours)

Algebraic expression: 15t (miles)   ■

NOTE: In Example 7(c) the final answer is listed in terms of miles. This makes sense in the following way.

$$15 \frac{\text{miles}}{\text{hour}} \cdot t \text{ hours}$$

Note that the hours "cancel," leaving the answer in terms of miles. This technique, called *unit analysis*, can be very helpful in determining the final unit of measure.

EXAMPLE 8   ■   Discovering Hidden Operations

(a) A person adds k liters of a fluid containing 45% antifreeze to a car radiator. Write an expression that indicates how much antifreeze was added.

(b) A consumer received a 30% discount on an item priced at d dollars. Write an expression, in dollars, for the discount.

(c) A person paid x dollars plus a 6% sales tax for an automobile. Write an expression for the total cost of the automobile.

Solution

(a) The amount of antifreeze is a product.

Verbal model: $\boxed{\text{Percent of antifreeze}} \cdot \boxed{\text{Number of liters}}$

Labels: Percent of antifreeze = 0.45 (in decimal form)
Number of liters = k

Algebraic expression: $0.45k$ (liters)

(b) The amount of the discount is a product.

Verbal model: $\boxed{\text{Percent of discount}} \cdot \boxed{\text{Original price}}$

Labels: Percent of discount = 0.3 (in decimal form)
Original price = d (in dollars)

Algebraic expression: $0.3d$ (in dollars)

Percentages means rates

(c) The total cost is a sum.

Verbal model: $\boxed{\text{Cost of automobile}} + \boxed{\text{Percent of sales tax}} \cdot \boxed{\text{Cost of auto}}$

Labels: Percent of sales tax = 0.06 (in decimal form)
Cost of automobile = x (in dollars)

Algebraic expression: $x + 0.06x = (1 + 0.06)x = 1.06x$ ∎

Hidden operations are often involved when assigning a variable name (label) to two unknown quantities. For example, suppose two numbers add up to 17 and one of the numbers is assigned the variable x. What expression can we use to represent the second number? Let's try a specific case first, then apply it to a general case.

Specific case: If the first number is 12, then the second number is
$$17 - 12 = 5.$$
General case: If the first number is x, then the second number is $17 - x$.

The strategy of using a *specific* case to help determine the general case is often useful in applications of algebra. Observe the use of this strategy in the next example.

EXAMPLE 9 ■ Using a Specific Case to Find a General Case

In each of the following, use the given variable to label the unknown quantity.

(a) A person's weekly salary is d dollars. What is the annual salary for this person?

(b) A person's annual salary is y dollars. What is the monthly salary for this person?

Solution

(a) There are 52 weeks in a year.

Specific case: If the weekly salary is $200, then the annual salary is $52 \cdot 200$ dollars.

General case: If the weekly salary is d dollars, then the annual salary is $52 \cdot d$ or $52d$ dollars.

(b) There are 12 months in a year.

Specific case: If the annual salary is $24,000, then the monthly salary (in dollars) is $24,000 \div 12$.

General case: If the annual salary is y, then the monthly salary (in dollars) is $y \div 12$ or $y/12$. ∎

EXAMPLE 10 ■ **Using a Specific Case to Find a General Case**

In each of the following, use the given variables to label the unknown quantity.

(a) One person is k inches shorter than another person. If the first person is 60 inches tall, how tall is the second person?

(b) A consumer buys g gallons of gasoline for a total of d dollars. What is the price per gallon?

Solution

(a) The first person is k inches shorter than the second person.

Specific case: If the first person is 10 inches shorter than the second person, then the second person is $60 + 10$ inches tall.

General case: If the first person is k inches shorter than the second person, then the second person is $60 + k$ inches tall.

(b) To obtain the price per gallon, we divide the total price by the total number of gallons.

Specific case: If the total price is $11.50 and the total number of gallons is 10, then the price per gallon is $11.50 \div 10$ dollars per gallon.

General case: If the total price is d dollars and the total number of gallons is g, then the price per gallon is $d \div g$ or d/g dollars per gallon. ∎

In mathematics it is useful to know how to represent certain types of integers algebraically. For instance, consider the set $\{2, 4, 6, 8, \ldots\}$ of even integers. What algebraic symbol could we use to denote an even integer? Since every even integer has 2 as a factor,

$$2 = 2 \cdot 1, \qquad 4 = 2 \cdot 2, \qquad 6 = 2 \cdot 3, \qquad 8 = 2 \cdot 4, \qquad \ldots,$$

it follows that any integer n multiplied by 2 must be an *even* integer. Moreover, if $2n$ is even, then $2n - 1$ and $2n + 1$ must be *odd* integers. For example, choose $n = 5$. Then $2n = 2 \cdot 5 = 10$ is even, whereas $2n - 1 = 10 - 1 = 9$ and $2n + 1 = 10 + 1 = 11$ are both odd.

Two integers are called **consecutive integers** if they differ by 1. Hence, for any integer n, its next two larger consecutive integers are $n + 1$ and $(n + 1) + 1$ or $n + 2$. Thus, we can denote three consecutive integers by n, $n + 1$, and $n + 2$. In summary, we have the following.

Expressions for Special Types of Integers

Let n be an integer. Then the following expressions can be used to denote even integers, odd integers, and consecutive integers.

1. $2n$ denotes an *even* integer.

2. $2n - 1$ and $2n + 1$ denote *odd* integers.

3. The set $\{n, n + 1, n + 2\}$ denotes three *consecutive* integers.

DISCUSSION PROBLEM ■ **Problems with Insufficient Information**

Most of the verbal problems you encounter in a mathematics text have precisely the right amount of information necessary to solve the problem. In real life, however, we often encounter problems for which we must obtain more information before we can solve the problem. What additional information would you need in order to solve the following problem?

During a given week, a person worked 46 hours for the same employer. The regular hourly rate is $8. Write an expression for the person's gross pay during the week, including any pay received for overtime. ■

Warm-Up

The following warm-up exercises involve skills that were covered in earlier sections. You will use these skills in the exercise set for this section.

In Exercises 1–4, simplify the given expression.

1. $-3(3x - 2y) + 5y$ **2.** $3v - (4 - 5v)$

3. $-y^2(y^2 + 4) + 6y^2$ **4.** $5t(2 - t) + t^2$

In Exercises 5–10, evaluate the algebraic expression for the specified values of the variables. (If not possible, state the reason.)

5. $x^2 - y^2$, $x = 4$, $y = 3$ **6.** $4s + st$, $s = 3$, $t = -4$

7. $\dfrac{x}{x^2 + y^2}$, $x = 0$, $y = 3$ **8.** $\dfrac{z^2 + 2}{x^2 - 1}$, $x = 1$, $z = 1$

9. $\dfrac{a}{1 - r}$, $a = 2$, $r = \dfrac{1}{2}$ **10.** $2l + 2w$, $l = 3$, $w = 1.5$

2.4   EXERCISES

In Exercises 1–20, translate the phrase into an algebraic expression. (Let x represent the arbitrary real number.)

1. A number is increased by 5.

2. A number is decreased by 7.

3. Six less than a number

4. Ten more than a number

5. Twice a number

6. A number is divided by 100.

7. One-fourth of a number

8. Twenty-five percent of a number

9. A number is divided by 3.

10. Three-tenths of a number

11. Eight plus the product of 5 and a number

12. A number is divided by 5, and the quotient is decreased by 15.

13. A number is tripled, and the product is increased by 5.

14. A number is increased by 5, and the sum is tripled.

15. Ten times the sum of a number and 4

16. Seven more than 5 times a number

17. The absolute value of the result obtained by subtracting a number from 16

18. The absolute value of the product of 4 times a number, subtracted from 95

19. The square of a number, increased by 1

20. Twice the cube of a number, increased by 4 times the number

In Exercises 21–30, write a verbal description of the algebraic expression. Use words only; do not use the variable. (There is more than one correct answer.)

21. $x - 10$

22. $x + 9$

23. $3x + 2$

24. $4 - 7x$

25. $3(2 + x)$

26. $10(t - 6)$

27. $\dfrac{t + 1}{2}$

28. $\dfrac{1}{2} - \dfrac{t}{5}$

29. $x^2 + 5$

30. $x(x + 1)$

In Exercises 31–36, translate the phrase into an algebraic expression. Then simplify the expression.

31. The sum of x and 3 is multiplied by x.

32. The sum of 25 and x is added to x.

33. Nine is subtracted from x and the result is multiplied by 3.

34. The sum of 4 and x is added to the sum of x and -8.

35. The square of x is added to the sum of x and 1.

36. Eight times the sum of x and 24, all divided by 5

In Exercises 37–52, construct an algebraic expression that represents the indicated quantity.

37. *Total Amount of Money* A person has n dimes. Write an algebraic expression that represents the total amount of money (in dollars).

38. *Total Amount of Money* A person has m dimes and n quarters. Write an algebraic expression that represents the total amount of money (in dollars).

39. *Sales Tax* The sales tax on a purchase of L dollars is 6%. Write an algebraic expression that represents the total amount of sales tax.

40. *Income Tax* The state income tax on a gross income of I dollars is 2.2%. Write an algebraic expression that represents the total amount of income tax.

41. *Travel Time* A truck travels 100 miles at an average speed of r miles per hour (see figure). Write an algebraic expression that represents the total travel time.

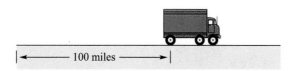

|←———— 100 miles ————→|

42. *Distance Traveled* A plane travels at the rate of r miles per hour for three hours. Write an algebraic expression that represents the total distance traveled by the plane.

43. *Camping Fee* A family stays in a campground that charges $15 for the two parents, plus $2 for each of the n children. Write an algebraic expression that represents the total camping fee.

44. *Hourly Wage* The hourly wage for an employee is $6.50 per hour plus 75 cents for each of q units produced during the hour. Write an algebraic expression that represents the total hourly earnings for the employee.

45. *Computer Screen* A computer screen has sides of length s inches (see figure). Write an algebraic expression that represents the area of the computer screen. What are the units of measure for the area of the screen?

46. *Area of a Rectangle* A rectangle has sides of length $3w$ and w. Write an algebraic expression that represents the area of the rectangle.

47. *Perimeter of a Picture* A rectangular picture frame has sides of length $1.5w$ and w (see figure). Write an algebraic expression that represents the perimeter of the picture frame.

48. *Perimeter of a Square* A square has sides of length s. Write an algebraic expression that represents the perimeter of the square.

49. Write an algebraic expression that represents the sum of three consecutive integers, the first of which is n.

Figure for 45   Figure for 47

50. Write an algebraic expression that represents the sum of two consecutive even integers, the first of which is $2n$.

51. Write an algebraic expression that represents the sum of two consecutive odd integers, the first of which is $2n + 1$.

52. Write an algebraic expression that represents the sum of the squares of two consecutive even integers, the first of which is $2n$.

53. (a) Complete the following table in which the third row is the difference between consecutive entries of the second row.

n	0	1	2	3	4	5
$3n + 2$						
Differences						

(b) Using the results of the table, determine the third row difference that would result in a similar table if the algebraic expression were $an + b$.

54. Find a and b so that the expression $an + b$ would yield the following table.

n	0	1	2	3	4	5
$an + b$	4	9	14	19	24	29

CHAPTER 2   SUMMARY

As you review and prepare for a test on this chapter, first try to obtain a global view of what was discussed. Then review the specific skills needed in each category.

The Basic Rules of Algebra

■ *Simplify* an exponential expression by using the properties of real numbers to remove parentheses and write the result with positive exponents.

Section 2.1, Exercises 45–60; Section 2.3, Exercises 1–10

■ *Identify* the basic rules of algebra demonstrated by a given equation.

Section 2.2, Exercises 1–12

■ *Add or subtract* algebraic expressions by using the basic rules of algebra with emphasis on the Distributive Property.

Section 2.2, Exercises 21–30, Exercises 41–62; Section 2.3, Exercises 25–32

■ *Simplify* an algebraic expression by using the basic rules of algebra to remove symbols of grouping.

Section 2.3, Exercises 11–24

Evaluate and Create Algebraic Expressions

■ *Evaluate* an algebraic expression.

Section 2.3, Exercises 33–54

■ *Translate* a verbal statement into algebraic form.

Section 2.4, Exercises 1–20

■ *Translate* an algebraic expression into verbal form.

Section 2.4, Exercises 21–30

■ *Create* an algebraic expression from real-life verbal statements.

Section 2.4, Exercises 37–52

Chapter 2 Review Exercises

In Exercises 1–4, list the terms of the given algebraic expression and find the coefficient of each term.

1. $4x - \frac{1}{2}x^3$

2. $5x^2 - 3x$

3. $y^2 - 10yz + \frac{2}{3}z^2$

4. $-x^3 + y^3$

In Exercises 5–8, rewrite the given product in exponential form.

5. $5z \cdot 5z \cdot 5z$

6. $\frac{3}{8}y \cdot \frac{3}{8}y \cdot \frac{3}{8}y \cdot \frac{3}{8}y$

7. $a(b - c) \cdot a(b - c)$

8. $3 \cdot (y - x) \cdot (y - x) \cdot 3 \cdot 3$

In Exercises 9–16, simplify the given expression.

9. $(x^3)^2$

10. $y^2 \cdot y^4 \cdot y$

11. $t^4(-2t^2)$

12. $(-3u^2)^2(3u^2)$

13. $(xy)(-5x^2y^3)$

14. $5v^2(3uv)(-2uv^2)$

15. $(-2y^2)^3(8y)$

16. $2(x - y)^4(x - y)^2$

In Exercises 17–22, identify the rule of algebra illustrated by the given equation.

17. $(x - y)(2) = 2(x - y)$

18. $(a + b) + 0 = a + b$

19. $2x + (3y + z) = (2x + 3y) + z$

20. $x(y + z) = xy + xz$

21. $xy \cdot \frac{1}{xy} = 1, \quad xy \neq 0$

22. $u(vw) = (uv)w$

In Exercises 23–30, use the Distributive Property to expand the given expression.

23. $4(x + 3y)$

24. $\frac{3}{4}(8s - 12t)$

25. $-5(2u - 3v)$

26. $-3(-2x - 8y)$

27. $x(8x + 5y)$

28. $-u(3u - 10v)$

29. $-(-a + 3b)$

30. $(7 - 2j)(-6)$

In Exercises 31–38, simplify the given expression by combining like terms.

31. $3a - 5a$

32. $6c - 2c$

33. $3p - 4q + q + 8p$

34. $10x - 4y - 25x + 6y$

35. $x^2 + 3xy - xy + 4$

36. $uv^2 + 10 - 2uv^2 + 2$

37. $5(x - y) + 3xy - 2(x - y)$

38. $y^3 + 2y^2 + 2y^3 - 3y^2 + 1$

In Exercises 39–46, simplify the given algebraic expression.

39. $5(u - 4) + 10$

40. $16 - 3(v + 2)$

41. $3s - (r - 2s)$

42. $50x - (30x + 100)$

43. $-3(1 - 10z) + 2(1 - 10z)$

44. $8(15 - 3y) - 5(15 - 3y)$

45. $2[x + 2(y - x)]$

46. $2t[4 - (3 - t)] + 5t$

In Exercises 47–50, evaluate the given expression for the specified values of the variables.
(If not possible, state the reason.)

Expression	*Values*	
47. $x^2 - 2x + 5$	(a) $x = 0$	(b) $x = 2$
48. $x^3 - 8$	(a) $x = 2$	(b) $x = 4$
49. $x^2 - x(y + 1)$	(a) $x = 2,\ y = -1$	(b) $x = 1,\ y = 2$
50. $\dfrac{x + 5}{y}$	(a) $x = -5,\ y = 3$	(b) $x = 2,\ y = 0$

In Exercises 51–60, translate the phrase into an algebraic expression. (Let x represent the arbitrary real number.)

51. Two-thirds of a number, plus 5

52. One hundred, decreased by 5 times a number

53. Ten less than twice a number

54. The ratio of a number to 10

55. Fifty, increased by the product of 7 and a number

56. Seventy-five, decreased by the quotient of a number and 2

57. The sum of a number and 10 is divided by 8.

58. The product of 15 and a number, decreased by 2

59. The sum of the square of a number and 64

60. The absolute value of the sum of a number and -10

In Exercises 61–64, write a verbal description of the algebraic expression without using the variable. (There is more than one correct answer.)

61. $x + 3$

62. $3x - 2$

63. $\dfrac{y - 2}{3}$

64. $4(x + 5)$

65. *Income Tax* The income tax rate on a taxable income of I dollars is 28%. Write an algebraic expression that represents the total amount of income tax.

66. *Total Amount of Money* A person has n nickels and q quarters. Write an algebraic expression that represents the total amount of money.

67. *Area* The front of a built-in refrigerator has a width of w feet and a height that is 3 feet greater than the width (see figure). Write an algebraic expression that represents the area of the front of the refrigerator.

68. *Distance Traveled* A car travels for 10 hours at an average speed of s miles per hour. Write an algebraic expression that represents the total distance traveled.

Figure for 67

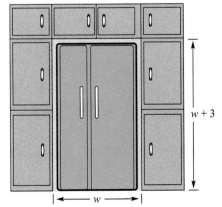

69. *Sum* Write an algebraic expression that represents the sum of three consecutive odd integers, the first of which is $2n - 1$.

70. *Rental Income* Write an expression that represents the rent for n months if the monthly rent is $625.

71. *Area* The face of a tape deck has the dimensions shown in the accompanying figure. Find an algebraic expression that represents the area of the face of the tape deck. (*Hint:* The area is given by the difference of the areas of two rectangles.)

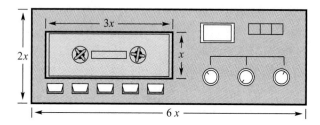

72. Perform the indicated operations and simplify.

$$7 \cdot 10^4 + 2 \cdot 10^3 + 8 \cdot 10^1$$

73. (a) Complete the following table in which the third and fourth rows are the differences between consecutive entries of the preceding rows.

n	0	1	2	3	4	5
$n^2 + 3n + 2$						
Differences						
Differences						

(b) Describe the patterns for the third and fourth rows of the table.

CHAPTER 2 TEST

Take this test as you would take a test in class. After you are done, check your work with the answers given in the back of the book.

1. List the terms and the numerical coefficients of the algebraic expression $2x^2 - 7xy + 3y^3$.

2. Rewrite the product $x \cdot (x + y) \cdot x \cdot (x + y) \cdot x$ in exponential form.

3. Simplify: $(c^2)^4$

4. Simplify: $t^4 \cdot t^3 \cdot t$

5. Simplify: $-5uv(2u^3)$

6. Identify the rule of algebra demonstrated by $(5x)y = 5(xy)$.

7. Identify the rule of algebra demonstrated by $2 + (x - y) = (x - y) + 2$.

8. Identify the rule of algebra demonstrated by $7xy - 7xy = 0$.

9. Expand by using the Distributive Property: $3(x + 8)$

10. Expand by using the Distributive Property: $-y(3 - 2y)$

11. Simplify: $3b - 2a + a - 10b$

12. Simplify: $15(u - v) - 7(u - v)$

13. Simplify: $3z - (4 - z)$

14. Simplify: $2[10 - (t + 1)]$

15. Evaluate the expression $x^3 - 2$ when $x = 3$.

16. Evaluate the expression $r^2 + 4(s + 2)$ when $r = 5$ and $s = -12$.

17. Evaluate (if possible) the expression $\dfrac{a + 2b}{3a - b}$ when $a = 2$ and $b = 6$.

18. Translate the phrase, "one-fifth of a number, increased by two," into an algebraic expression, letting n represent the arbitrary real number.

19. Write an algebraic expression for the length of a rectangle if the length is four units less than twice the width. Let w represent the width.

20. Write an algebraic expression for the income from a concert if the prices of the tickets for adults and children were $3 and $2, respectively. Let n represent the number of adults in attendance and m represent the number of children.

CHAPTER THREE

Solving Equations

Chapter Overview

So far, we have discussed algebra as an extension of the rules of arithmetic and as a means of constructing, rewriting, and simplifying algebraic expressions. In this chapter we introduce the third major component of algebra—constructing, rewriting, and solving algebraic equations. We will limit our discussion to solving linear equations. This type of equation has a wide variety of real-life applications.

Introduction to Equations
Introduction • Equations • Forming Equivalent Equations • Constructing Equations

Introduction

In Chapter 2 we used the rules of algebra to construct, rewrite, and simplify algebraic *expressions*. In this chapter we use the same rules plus some properties of equality to construct, rewrite, and solve algebraic *equations*. Figure 3.1 illustrates the parts of algebra that you will have studied by the time you finish with Chapter 3.

FIGURE 3.1

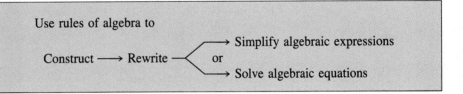

Equations

An **equation** is a statement that two mathematical expressions are equal. Some examples of equations are

$$x = 3, \qquad 5x - 2 = 8, \qquad 3x - 12 = 3(x - 4), \quad \text{and} \quad x^2 - 9 = 0.$$

To **solve** an equation involving x means to find all values of x for which the equation is true. Such values are called **solutions**, and we say that the solutions **satisfy** the equation. For instance, 3 is a solution of $x = 3$ because $3 = 3$ is a true statement. Similarly, 2 is a solution of $5x - 2 = 8$ because $5(2) - 2 = 8$ is a true statement.

The **solution set** of an equation is the set of all real numbers that are solutions of the equation. Sometimes an equation will have the set of all real numbers as its solution set. Such an equation is called an **identity**. For instance, the equation

$$3x - 12 = 3(x - 4) \qquad\qquad \text{Identity}$$

is an identity because the equation is true for all real values of x. Try values like 0, 1, -2, and 5 in this equation to see that each one is a solution.

An equation whose solution set is not the entire set of real numbers is called a **conditional equation**. For instance, the equation

$$x^2 - 9 = 0 \qquad\qquad \text{Conditional equation}$$

is a conditional equation because it has only two solutions, 3 and -3.

In Examples 1 and 2 we show how to **check** whether a given value of x is a solution of an equation.

EXAMPLE 1 ■ Checking a Solution of an Equation

Determine whether -2 is a solution of the equation

$$x^2 - 5 = 4x + 7.$$

Solution

To check whether -2 is a solution of this equation, we replace the variable x by the number -2. This must be done for all occurrences of x on *both sides* of the equation. After replacing, we simplify both sides of the equation. If both sides turn out to be the same number, then -2 is a solution, and we say that the solution "checks." If the two sides turn out to be different numbers, then -2 is not a solution.

$$x^2 - 5 = 4x + 7 \qquad \text{Given equation}$$
$$(-2)^2 - 5 \overset{?}{=} 4(-2) + 7 \qquad \text{Replace } x \text{ by } -2$$
$$4 - 5 \overset{?}{=} -8 + 7 \qquad \text{Simplify}$$
$$-1 = -1 \qquad -2 \text{ is a solution}$$

Since both sides of the equation turn out to be the same number, we conclude that -2 is a solution of the given equation. ■

NOTE: When checking a solution, we like to write a question mark over the equal sign to indicate that we are not sure of the validity of the equation.

Just because you have found one solution of an equation, you should not conclude that you have found all the solutions. For instance, try checking that 6 is also a solution of the equation given in Example 1.

EXAMPLE 2 ■ A Trial Solution that Does not Check

Determine whether 2 is a solution of the equation

$$x^2 - 5 = 4x + 7.$$

Solution

$$x^2 - 5 = 4x + 7 \qquad \text{Given equation}$$
$$(2)^2 - 5 \overset{?}{=} 4(2) + 7 \qquad \text{Replace } x \text{ by } 2$$
$$4 - 5 \overset{?}{=} 8 + 7 \qquad \text{Simplify}$$
$$-1 \neq 15 \qquad 2 \text{ is not a solution}$$

Since the two sides of the equation turn out to be different after the replacement of x by 2, we conclude that 2 is not a solution of the given equation. ■

It is important to understand the following distinction between an algebraic expression and an algebraic equation.

1. An algebraic *expression* contains no equal sign and cannot be solved. We often try to simplify an expression by rewriting it. We then use an equal sign to show that the rewritten expression is equivalent to the original expression. For instance, we might rewrite the expression $4(x - 1)$ as $4x - 4$ and indicate the equivalence of the two expressions by writing

$$4(x - 1) = 4x - 4.$$

The resulting equation is actually an identity. In other words, the expression on the left yields the same value as the expression on the right when x is replaced by any real number.

2. An algebraic *equation* has an equal sign. An equation may be either an identity or a conditional equation. We do not try to solve an equation that is an identity because (by definition) the solution set consists of all real numbers. We do, however, try to solve conditional equations. In doing so we try to find all real numbers that are solutions of the conditional equation.

Forming Equivalent Equations

It is helpful to think of an equation as having two sides that are in balance. Consequently, when we try to solve an equation, we must be careful to maintain that balance by performing the same operation on both sides.

Two equations that have the same set of solutions are called **equivalent**. For instance, the equations $x = 3$ and $x - 3 = 0$ are equivalent because both have only one solution—the number 3. When any one of the four operations in the following list is applied to an equation, the resulting equation is equivalent to the original equation.

Forming Equivalent Equations	A given equation can be transformed into an *equivalent equation* by one or more of the following steps.	*Given Equation*	*Equivalent Equation*
	1. Remove symbols of grouping, combine like terms, or reduce fractions on one or both sides of the equation.	$3x - x = 8$	$2x = 8$
	2. Add (or subtract) the same quantity to *both* sides of the equation.	$x - 3 = 5$	$x = 8$
	3. Multiply (or divide) *both* sides of the equation by the same *nonzero* quantity.	$3x = 9$	$x = 3$
	4. Interchange the two sides of the equation.	$7 = x$	$x = 7$

The second and third operations in this list can be used to eliminate terms or factors in an equation. For example, to solve the equation $x - 5 = 1$, we need to eliminate the term -5 on the left side. This is accomplished by adding its opposite, 5, to both sides.

$x - 5 = 1$	Given equation
$x - 5 + 5 = 1 + 5$	Add 5 to both sides
$x + 0 = 6$	Combine like terms
$x = 6$	Solution

All four of the equations listed above are equivalent, and we call them the **steps** of the solution.

EXAMPLE 3 ■ Operations Used to Solve Equations

(a) To eliminate the -2 on the left side of the following equation, we add 2 to both sides of the equation.

$x - 2 = 5$	Given equation
$x - 2 + 2 = 5 + 2$	Add 2 to both sides
$x = 7$	Solution

(b) To eliminate the $2x$ on the right side of the following equation, we subtract $2x$ from both sides of the equation.

$3x = 2x + 4$	Given equation
$3x - 2x = 2x - 2x + 4$	Subtract $2x$ from both sides
$x = 4$	Solution

(c) To eliminate the 5 in the denominator on the left side of the equation, we multiply both sides of the equation by 5.

$\dfrac{x}{5} = -2$	Given equation
$\dfrac{x}{5}(5) = -2(5)$	Multiply both sides by 5
$x = -10$	Solution

(d) To eliminate the factor of 4 on the left side of the equation, we divide both sides of the equation by 4.

$4x = 9$	Given equation
$\dfrac{4x}{4} = \dfrac{9}{4}$	Divide both sides by 4
$x = \dfrac{9}{4}$	Solution

■

Note in Example 3(b) that we *subtracted* $2x$ from both sides of the equation to get rid of $2x$ on the right side. We could just as easily have *added* $-2x$ to both sides. Both techniques are legitimate—which one you choose to use is a matter of personal preference.

Constructing Equations

As a motivation for learning to solve equations in the next section, we now look at some suggestions for *constructing* equations that arise from real-life problems.

When trying to use algebra to solve a real-life problem, it is helpful to look for one or more implied equalities in the verbal description. These equalities might be stated explicitly, but often you will have to use some prior knowledge or experience to determine the equalities.

Sometimes it is helpful to use two stages in constructing equations that represent real-life problems. In the first stage we translate the verbal description into a *verbal model*. In the second stage we translate the verbal model into a *mathematical model* or *algebraic equation*.

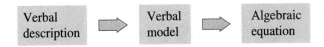

Here are a couple of examples of verbal models.

1. The sale price of a basketball is $18. If the sale price is $7 less than the original price, what is the original price?

   Verbal model: Sale price = Original price − Discount

   $18 = Original price − $7

2. The original price of a basketball is $25. If the discount is $7, what is the sale price?

   Verbal model: Sale price = Original price − Discount

   Sale price = $25 − $7

To obtain an algebraic equation from a verbal model, we use the strategies discussed in Section 2.4.

EXAMPLE 4 ■ Using Verbal Models to Construct Equations

Write an algebraic equation for the following problem. The total income that an employee received in 1992 was $21,550. Of that, $750 represented a bonus given at the end of the year. How much was the employee paid each week? Assume that each weekly paycheck contained the same amount, and that the year consisted of 52 weeks.

Solution

Verbal model:

| Income for year | = | 52 times weekly pay | + | Bonus |

Labels: Income for year = $21,550
Weekly pay = x (in dollars)
Bonus = $750

Algebraic equation: $21,550 = 52x + 750$ ■

In Example 4 note that *both* sides of the equation $21,550 = 52x + 750$ represent dollar amounts. When you construct an equation, be sure to check that both sides of the equation represent the *same* unit of measure.

EXAMPLE 5 ■ Using Verbal Models to Construct Equations

Write an algebraic equation for the following problem. Tickets for a concert were $15 for each floor seat and $10 for each stadium seat. There were 800 seats on the main floor, and these were sold out. If the total revenue from ticket sales was $52,000, how many stadium seats were sold?

Solution

Verbal model:

| Total revenue | = | Revenue from floor seats | + | Revenue from stadium seats |

Labels: Total revenue = $52,000
Price per floor seat = $15
Number of floor seats = 800
Price per stadium seat = $10
Number of stadium seats = x

Algebraic equation: $52,000 = 15(800) + 10x$

Note that both sides of the equation represent dollar amounts. ■

In the next section we will learn how to solve the equations constructed in Examples 4 and 5.

DISCUSSION PROBLEM ■ Problems with Red Herrings

When constructing an equation to represent a word problem, we are occasionally given too much information. The unnecessary information in a word problem is sometimes called a "red herring." Find the red herring in the following problem.

A customer purchases a sweater and a pair of jeans for $51, plus sales tax. The sales tax on the purchase is 6%. If the sweater was half the price of the jeans, how much sales tax was charged? ■

Warm-Up

The following warm-up exercises involve skills that were covered in earlier sections. You will use these skills in the exercise set for this section.

In Exercises 1–4, simplify the algebraic expression.

1. $(-2y^2)^3$

2. $(3a^2)(4ab)$

3. $3x - 2(x - 5)$

4. $x(x - 3) - (x^2 + 6x)$

In Exercises 5 and 6, evaluate the expression for the specified value of the variable.

Expression	*Values*

5. $x^2 - 2x + 1$ (a) $x = 1$ (b) $x = 3$

6. $\dfrac{1}{x^2 + 1}$ (a) $x = 0$ (b) $x = 2$

In Exercises 7–10, translate the phrase into an algebraic expression. (Let x represent the arbitrary number.)

7. Four more than twice a number

8. A number is decreased by 10, and the result is doubled.

9. A number is decreased by five, and the result is squared.

10. A number is increased by 25, and the sum is halved.

3.1 EXERCISES

In Exercises 1–14, determine whether the given value of x is a solution to the equation.

Equation	*Values*		*Equation*	*Values*	
1. $2x - 6 = 0$	(a) $x = 3$	(b) $x = 1$	**2.** $5x - 25 = 0$	(a) $x = 10$	(b) $x = 5$
3. $2x + 4 = 2$	(a) $x = 0$	(b) $x = -1$	**4.** $3x + 10 = 4$	(a) $x = -2$	(b) $x = 2$
5. $x + 5 = 2x$	(a) $x = -1$	(b) $x = 5$	**6.** $2x - 3 = 5x$	(a) $x = 0$	(b) $x = -1$
7. $x + 3 = 2(x - 4)$	(a) $x = 11$	(b) $x = -5$	**8.** $5x - 1 = 3(x + 5)$	(a) $x = 8$	(b) $x = -2$
9. $2x + 10 = 7(x + 1)$	(a) $x = \frac{3}{5}$	(b) $x = \frac{2}{3}$	**10.** $9x + 6 = 9 - x$	(a) $x = -\frac{3}{4}$	(b) $x = \frac{3}{10}$
11. $x^2 - 4 = x + 2$	(a) $x = 3$	(b) $x = -2$	**12.** $x^2 = 2(4 - x)$	(a) $x = 2$	(b) $x = -4$
13. $\frac{2}{x} - \frac{1}{x} = 1$	(a) $x = 1$	(b) $x = \frac{1}{3}$	**14.** $\frac{5}{x - 1} + \frac{1}{x} = 5$	(a) $x = 3$	(b) $x = \frac{1}{6}$

In Exercises 15–20, describe each step of the given solution.

15.
$$5x + 12 = 22$$
$$5x + 12 - 12 = 22 - 12$$
$$5x = 10$$
$$\frac{5x}{5} = \frac{10}{5}$$
$$x = 2$$

16.
$$14 - 3x = 5$$
$$14 - 3x - 14 = 5 - 14$$
$$-3x = -9$$
$$\frac{-3x}{-3} = \frac{-9}{-3}$$
$$x = 3$$

17.
$$\frac{2}{3}x = 12$$
$$\frac{3}{2}\left(\frac{2}{3}x\right) = \frac{3}{2}(12)$$
$$x = 18$$

18.
$$\frac{4}{5}x = -28$$
$$\frac{5}{4}\left(\frac{4}{5}x\right) = \frac{5}{4}(-28)$$
$$x = -35$$

19.
$$2(x - 1) = x + 3$$
$$2x - 2 = x + 3$$
$$-x + 2x - 2 = -x + x + 3$$
$$x - 2 = 3$$
$$x - 2 + 2 = 3 + 2$$
$$x = 5$$

20.
$$x + 6 = -6(4 - x)$$
$$x + 6 = -24 + 6x$$
$$-x + x + 6 = -x - 24 + 6x$$
$$6 = 5x - 24$$
$$6 + 24 = 5x - 24 + 24$$
$$30 = 5x$$
$$\frac{30}{5} = \frac{5x}{5}$$
$$6 = x$$

In Exercises 21–42, construct an equation for the given word problem. Do *not* solve the equation.

21. *Test Score* After your instructor added six points to each student's test score, your score was 94. What was your original score?

22. *Rainfall* With the 1.2-inch rainfall today, the total for the month is 4.5 inches. How much had been recorded for the month before today's rainfall?

23. *Computer Purchase* You presently have $3,650 saved for the purchase of a new computer that will cost $4,532. How much more must you save?

24. *List Price* The sale price of a coat is $225.98. If the discount is $64, what is the list price?

25. The sum of a number and 12 is 45. What is the number?

26. The sum of three times a number and 4 is 16. What is the number?

27. Four times the sum of a number and 6 is 100. What is the number?

28. Find a number such that six times the number, subtracted from 120, is 96.

29. Find a number such that 15 less than twice the number equals the number divided by 3.

30. Determine a number such that when the sum of the number and 8 is divided by 4, the result is 32.

31. The sum of three consecutive even integers is 18. Find the first number.

32. The sum of two consecutive odd integers is 100. Find the first number.

33. *Dimensions of a Mirror* The width of a rectangular mirror is one-third its length, as shown in the figure. The perimeter of the mirror is 96 inches. What are the dimensions of the mirror?

$|\!\!\leftarrow\! \frac{1}{3}l \!\rightarrow\!|$

l

34. *Height of a Box* A rectangular toolbox has a base that is 2 feet by 3 feet, as shown in the figure. The volume of the box is 24 cubic feet. What is the height of the box?

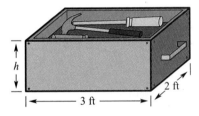

h

$|\!\!\leftarrow\!\!\!-\!\!\!-\!\!\! 3\ \text{ft} \!\!\!-\!\!\!-\!\!\!\rightarrow\!|$ 2 ft

35. *Average Speed* After traveling for three hours, you are still 25 miles from completing a 160-mile trip (see figure). What was your average speed during the first three hours?

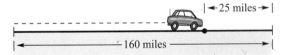

$|\!\!\leftarrow\! 25\ \text{miles} \!\rightarrow\!|$

$|\!\!\leftarrow\!\!\!-\!\!\!-\!\!\! 160\ \text{miles} \!\!\!-\!\!\!-\!\!\!\rightarrow\!|$

36. *Average Speed* A group of students plan to take two cars to a soccer game. The first car leaves on time, travels at an average speed of 45 miles per hour, and arrives at the destination in three hours. What average speed must be obtained by the students in the second car if they leave one-half hour after the first car and arrive at the game at the same time as the students in the first car?

37. *Price of a Product* The price of a product increased by 25% during the past year. It is now selling for $375. What was the price one year ago?

38. *Dow-Jones Average* The Dow-Jones Average fell by 5% during a week and was 2,695 at the close of the market on Friday. What was the average at the close of the market on the previous Friday?

39. *Annual Depreciation* A corporation buys equipment for an initial purchase price of $750,000. It is estimated that the useful life of the equipment will be three years and at that time its value will be $75,000. The total depreciation is divided equally among the three years. Determine the amount of depreciation declared each year.

40. *Car Payments* Suppose you made 48 monthly payments of $158 each to buy a used car. The total amount financed was $6,000. Find the amount of interest that you paid.

41. *Average Speed* After traveling for four hours, you are still 24 miles from completing a 200-mile trip. If it requires one-half hour to travel the last 24 miles, find the average speed during the first four hours of the trip.

42. *Price of a Product* The price of a product increased by 12% during the past year. The price of the product was $9,850 two years ago and $10,120 one year ago. What is its current price?

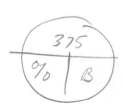

Solving Linear Equations

Introduction • *Linear Equations* • *Solving a Linear Equation in Standard Form* •
Solving a Linear Equation in Nonstandard Form • *Applications*

Introduction

This is an exciting and crucial stage in the study of algebra. In Chapters 1 and 2 you
were introduced to the rules of algebra, and you learned to use these rules to rewrite
and simplify algebraic expressions. In Sections 2.4 and 3.1 you gained experience in
translating verbal expressions and problems into algebraic forms. You are now ready
to use these skills and experiences to *solve equations*.

Linear Equations

The most common type of equation involving one variable is a **linear equation**.

Definition of Linear Equation	A **linear equation** involving one variable x is an equation that can be written in the standard form $$ax + b = c$$ where a, b, and c are real numbers with $a \neq 0$.

A linear equation involving one variable is also called a **first-degree equation**
because its variable has an (implied) exponent of one. Some examples of linear equations in standard form are

$$2x = 3, \quad x - 7 = 5, \quad 4x + 6 = 0, \quad \text{and} \quad \frac{x}{2} - 1 = \frac{5}{3}.$$

Remember that to *solve* an equation involving x means that we are to find the
values of x that satisfy the equation. For a linear equation $ax + b = c$, the goal is to
isolate x by rewriting the equation in the form

$$x = \text{(a number)}.$$

This rewriting is accomplished by using the techniques discussed in Section 3.1.
Beginning with the given equation, we write a sequence of equivalent equations, each
having the same solution as the original equation. As mentioned in the previous section,
each equivalent equation is called a **step** of the solution.

EXAMPLE 1 ■ **Solving an Equation**

Solve the following linear equation.

$$3x - 2 = 10$$

Solution

$3x - 2 = 10$	Given equation
$3x - 2 + 2 = 10 + 2$	Add 2 to both sides
$3x = 12$	Combine like terms
$\dfrac{3x}{3} = \dfrac{12}{3}$	Divide both sides by 3
$x = 4$	Isolate x

Thus, it appears that the solution of the given equation is 4. We can check this solution as follows.

Check

$3x - 2 = 10$	Given equation
$3(4) - 2 \overset{?}{=} 10$	Replace x by 4
$12 - 2 \overset{?}{=} 10$	
$10 = 10$	4 is a solution

Therefore, the solution is 4. ■

In Example 1 be sure you see that solving an equation has two basic stages. The first stage is to *find* the solution (or solutions). The second stage is to *check* that each solution you find actually satisfies the original equation. You can improve your accuracy in algebra by developing the habit of checking each solution.

A question you might be asking at this point is this: "We know that 4 is a solution of the equation given in Example 1. But how can we be sure that the equation does not have other solutions?" The answer is that a linear equation always has exactly one solution. We can prove this in the following way. Let $ax + b = c$ represent an arbitrary linear equation, where $a \neq 0$.

$ax + b = c$	Given equation
$ax + b - b = c - b$	Subtract b from both sides
$ax = c - b$	Combine like terms
$\dfrac{ax}{a} = \dfrac{c - b}{a}$	Divide both sides by a
$x = \dfrac{c - b}{a}$	Isolate x

Because a, b, and c are constant, the last equation has only one solution: $x = (c - b)/a$. Moreover, because the last equation is equivalent to the given equation, we can also conclude that the given equation has only one solution.

Solution of a Linear Equation	A linear equation of the form

A linear equation of the form

$$ax + b = c, \quad a \neq 0$$

has exactly one solution.

Remember that this result applies only to linear equations. Other types of equations may have two or more solutions. For instance, the nonlinear equation $x^2 = 4$ has two solutions: 2 and -2.

Solving a Linear Equation in Standard Form

In Example 1 note the steps that were used to isolate the variable x. A good question at this point would be: "How do I know which step to use *first* to isolate x?"

The answer is that you need practice. By solving many linear equations, you will find that your skill will improve. To begin, it may help to compare solving a linear equation in standard form to unwrapping a stick of gum. First you remove the outer wrapping, then you remove the foil wrap to isolate the stick of gum. In the equation $3x - 2 = 10$, think of x as the stick of gum, as shown in Figure 3.2. First you remove the outer paper, the term -2, which is further from x than is the factor 3. Then you remove the foil, the factor 3, which is the number closest to x. Thus, we have isolated (unwrapped) x and have exposed the solution to be 4. With this analogy in mind, go back and review the steps in Example 1.

FIGURE 3.2

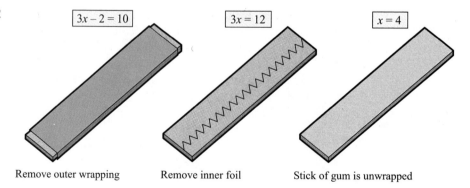

Remove outer wrapping   Remove inner foil   Stick of gum is unwrapped

In the preceding analogy, note that the term 10, on the right side of the equation, does not influence the "unwrapping order." Also keep in mind that we remove factors and terms by using the *inverse* operation. We remove a sum by subtracting, a difference by adding, a product by dividing, and a quotient by multiplying.

EXAMPLE 2   ■   Solving a Linear Equation in Standard Form

Solve the following linear equation.

$$2x + 7 = 3$$

Solution

$$2x + 7 = 3 \qquad \text{Given equation}$$
$$2x + 7 - 7 = 3 - 7 \qquad \text{Subtract 7 from both sides}$$
$$2x = -4 \qquad \text{Combine like terms}$$
$$\frac{2x}{2} = -\frac{4}{2} \qquad \text{Divide both sides by 2}$$
$$x = -2 \qquad \text{Isolate } x$$

Check

$$2x + 7 = 3 \qquad \text{Given equation}$$
$$2(-2) + 7 \overset{?}{=} 3 \qquad \text{Replace } x \text{ by } -2$$
$$-4 + 7 \overset{?}{=} 3$$
$$3 = 3 \qquad -2 \text{ is a solution}$$

Thus, the solution is -2.   ■

EXAMPLE 3   ■   Solving a Linear Equation in Standard Form

Solve the following linear equation.

$$5x = 9$$

Solution

In the equation $5x = 9$, note that there is no *term* that we need to get rid of on the left side of the equation, and so we go immediately to removing the coefficient of x.

$$5x = 9 \qquad \text{Given equation}$$
$$\frac{5x}{5} = \frac{9}{5} \qquad \text{Divide both sides by 5}$$
$$x = \frac{9}{5} \qquad \text{Isolate } x$$

Check

$$5x = 9 \qquad \text{Given equation}$$
$$5\left(\frac{9}{5}\right) \overset{?}{=} 9 \qquad \text{Replace } x \text{ by } \tfrac{9}{5}$$
$$9 = 9 \qquad \tfrac{9}{5} \text{ is a solution}$$

Thus, the solution is $\frac{9}{5}$.   ■

EXAMPLE 4   ■   **Solving a Linear Equation in Standard Form**

Solve the following linear equation.

$$\frac{x}{3} - 2 = -4$$

Solution

$\dfrac{x}{3} - 2 = -4$	Given equation
$\dfrac{x}{3} - 2 + 2 = -4 + 2$	Add 2 to both sides
$\dfrac{x}{3} = -2$	Combine like terms
$3\left(\dfrac{x}{3}\right) = 3(-2)$	Multiply both sides by 3
$x = -6$	Solution is -6

Thus, the solution is -6. Check this solution in the original equation.   ■

As you gain experience in solving linear equations, you will probably find that you can perform some of the solution steps in your head. For instance, you might solve the equation given in Example 4 by writing only the following steps.

$\dfrac{x}{3} - 2 = -4$	Given equation
$\dfrac{x}{3} = -2$	Add 2 to both sides
$x = -6$	Multiply both sides by 3

Remember, however, that you should not skip the final step—checking your solution.

Solving a Linear Equation in Nonstandard Form

The definition of a linear equation contains the phrase "that can be written" in the standard form $ax + b = c$. This suggests that some linear equations may come in nonstandard or disguised form.

A common nonstandard form of linear equations is one in which the variable terms are not combined into one term. Some examples are

$$3x + 8 - 5x = 4, \qquad 7x = 4x, \quad \text{and} \quad x + 2 = 2x - 6.$$

In such cases we rewrite the equation in standard form. Note how this is done in the next two examples.

EXAMPLE 5 ■ Solving a Linear Equation in Nonstandard Form

Solve the following linear equation.

$$3x + 8 - 5x = 4$$

Solution

$3x + 8 - 5x = 4$	Given equation
$3x - 5x + 8 = 4$	Regroup terms
$-2x + 8 = 4$	Standard form
$-2x + 8 - 8 = 4 - 8$	Subtract 8 from both sides
$-2x = -4$	Combine like terms
$\dfrac{-2x}{-2} = \dfrac{-4}{-2}$	Divide both sides by -2
$x = 2$	Solution is 2

Check

$3x + 8 - 5x = 4$	Given equation
$3(2) + 8 - 5(2) \stackrel{?}{=} 4$	Replace x by 2
$6 + 8 - 10 \stackrel{?}{=} 4$	
$4 = 4$	2 is a solution

Thus, the solution is 2. ■

EXAMPLE 6 ■ Solving a Linear Equation in Nonstandard Form

Solve the following linear equation.

$$x + 2 = 2x - 6$$

Solution

$x + 2 = 2x - 6$	Given equation
$x - 2x + 2 = 2x - 2x - 6$	Subtract $2x$ from both sides
$-x + 2 = -6$	Combine like terms
$-x + 2 - 2 = -6 - 2$	Subtract 2 from both sides
$-x = -8$	Combine like terms
$(-1)(-x) = (-1)(-8)$	Multiply both sides by -1
$x = 8$	Solution is 8

Thus, the solution is 8. Check this solution in the original equation. ■

We mentioned that every linear equation in *standard form* has exactly one solution. When an equation is written in *nonstandard form*, we cannot be sure that it has a solution. For instance, the equation

$$2x + 3 = 2(x + 4)$$

has no solution. To see this, we try to write the equation in standard form, as follows.

$$2x + 3 = 2(x + 4)$$
$$2x + 3 = 2x + 8$$
$$2x - 2x + 3 = 2x - 2x + 8$$
$$3 = 8$$

Because there are no values of x that make the last equation true, we can conclude that the original equation has no solution. Watch out for this type of equation in the exercise set.

Applications

In the next example we look back at one of the real-life problems that we introduced in Section 3.1.

EXAMPLE 7   ■   **An Application: Ticket Sales**

Tickets for a concert were $15 for each floor seat and $10 for each stadium seat. There were 800 seats on the main floor, and these were sold out. If the total revenue from ticket sales was $52,000, how many stadium seats were sold?

Solution

In Example 5 in the previous section, we developed the following model.

Verbal model: Total revenue = Revenue from floor seats + Revenue from stadium seats

Labels:
Total revenue = $52,000
Price per floor seat = $15
Number of floor seats = 800
Price per stadium seat = $10
Number of stadium seats = x

Algebraic equation: $52,000 = 15(800) + 10x$

We now solve this linear equation for x. To begin, we interchange the left and right sides of the equation to obtain

$$15(800) + 10x = 52,000.$$

This step is not really necessary, but we like to start this way because it is customary to have the term involving x on the left side of the equation.

$$15(800) + 10x = 52,000 \qquad \text{Given equation}$$
$$12,000 + 10x = 52,000$$
$$12,000 - 12,000 + 10x = 52,000 - 12,000 \qquad \text{Subtract 12,000 from both sides}$$
$$10x = 40,000 \qquad \text{Combine like terms}$$
$$\frac{10x}{10} = \frac{40,000}{10} \qquad \text{Divide both sides by 10}$$
$$x = 4,000 \qquad \text{Solution is 4,000}$$

Thus, there were 4,000 stadium seats sold. To check this solution we should go back to the original statement of the problem. In this case there were 4,000 stadium seats sold and 800 seats sold on the main floor. Thus, the total revenue is

$$\underbrace{10(4,000)}_{\text{Stadium seats}} + \underbrace{15(800)}_{\text{Floor seats}} = 40,000 + 12,000 = \$52,000,$$

which agrees with the original problem statement.

DISCUSSION PROBLEM   ■   Checking the Sensibility of a Solution

When you are solving a word problem, be sure to ask yourself whether your solution makes sense. Write a short paragraph explaining why the following answers don't make sense.

(a) A problem asks you to find the number of square feet in a suburban home. The answer you obtain is 26,780 square feet.

(b) A problem asks you to find the weight of a suitcase. The answer you obtain is 54 ounces.

(c) A problem asks you to find the net weight of a box of frozen food. The answer you obtain is 5.45 liters.

Warm-Up

The following warm-up exercises involve skills that were covered in earlier sections. You will use these skills in the exercise set for this section.

In Exercises 1–6, perform the indicated operation.

1. $2(-3) + 9$ **2.** $-10(6 - 2)$ **3.** $4 - \dfrac{5}{2}$

4. $\dfrac{|18 - 25|}{6}$ **5.** $\left(-\dfrac{7}{12}\right)\left(\dfrac{3}{28}\right)$ **6.** $\dfrac{4}{3} \div \dfrac{5}{6}$

In Exercises 7–10, determine whether the given value of the variable is a solution of the equation.

7. $6x - 5 = 0, \quad x = \dfrac{5}{6}$ **8.** $4 - 3x = 0, \quad x = \dfrac{4}{3}$

9. $3(x - 4) = x, \quad x = 6$ **10.** $x + 6 = 2(3x + 1), \quad x = -3$

3.2   EXERCISES

In Exercises 1–4, verify that the solution of the given equation is also the solution of each equation shown in the solution process.

1. $5x + 15 = 0$

$5x + 15 - 15 = 0 - 15$

$5x = -15$

$\dfrac{5x}{5} = \dfrac{-15}{5}$

$x = -3$

2. $7x - 14 = 0$

$7x - 14 + 14 = 0 + 14$

$7x = 14$

$\dfrac{7x}{7} = \dfrac{14}{7}$

$x = 2$

3. $-2x + 5 = 13$

$-2x + 5 - 5 = 13 - 5$

$-2x = 8$

$\dfrac{-2x}{-2} = \dfrac{8}{-2}$

$x = -4$

4. $22 - 3x = 10$

$22 - 3x - 22 = 10 - 22$

$-3x = -12$

$\dfrac{-3x}{-3} = \dfrac{-12}{-3}$

$x = 4$

In Exercises 5–8, solve the linear equation and state the algebraic property used in each step of the solution process. (Use Example 1 as a model.)

5. $8x - 2 = 20$

6. $-7x + 24 = 3$

7. $10 - 4x = -6$

8. $6x + 1 = -11$

In Exercises 9–50, solve the given equation and check your solution. (Some of the equations have no solution.)

9. $4x = 12$

10. $6x = 18$

11. $-5x = 30$

12. $-14x = 42$

13. $12x = 18$

14. $9x = -21$

15. $6x - 4 = 0$

16. $8z + 10 = 0$

17. $25x - 4 = 46$

18. $15x - 18 = 12$

19. $3y - 2 = 2y$

20. $24 - 5x = x$

21. $4 - 7x = 5x$

22. $2s - 13 = 28s$

23. $4 - 5t = 16 + t$

24. $3x + 4 = x + 10$

25. $15x - 3 = 15 - 3x$

26. $2x - 5 = 7x + 10$

27. $-3t = 0$

28. $4z + 2 = 4z$

29. $-3t + 5 = -3t$

30. $4z - 8 = 2$

31. $2x + 4 = -3x + 6$

32. $4y + 4 = -y + 5$

33. $2x = -3x$

34. $2x = 3x - 3$

35. $2x - 5 + 10x = 3$

36. $-4x + 10 + 10x = 4$

37. $\dfrac{x}{3} = 10$

38. $-\dfrac{x}{2} = 3$

39. $x - \dfrac{1}{3} = \dfrac{4}{3}$

40. $2x + \dfrac{5}{2} = \dfrac{9}{2}$

41. $3x - \dfrac{1}{4} = \dfrac{3}{4}$

42. $2x - \dfrac{3}{8} = \dfrac{5}{8}$

43. $t - \dfrac{1}{3} = \dfrac{1}{2}$

44. $z + \dfrac{2}{5} = -\dfrac{3}{10}$

45. $2s + \dfrac{3}{2} = 2s + 2$

46. $14 - 5s = -2 + 5s$

47. $\dfrac{1}{5}x + 1 = \dfrac{3}{10}x - 4$

48. $\dfrac{1}{8}x + \dfrac{1}{2} = \dfrac{9}{4}$

49. $0.2x + 5 = 6$

50. $4 - 0.3x = 1$

In Exercises 51–54, solve the given equation and round your answer to two decimal places. (A calculator may be helpful.)

51. $0.234x + 1 = 2.805$ **52.** $275x - 3130 = 512$ **53.** $\dfrac{x}{3.155} = 2.850$ **54.** $\dfrac{2x}{3.7} = \dfrac{3}{4}$

In Exercises 55–66, solve the given word problem.

55. Find a number such that the sum of that number and 45 is 75.

56. The sum of four times a number and 16 is 100. Find the number.

57. The sum of two consecutive odd integers is 72. Find the two integers.

58. The sum of three consecutive even integers is 192. Find the three integers.

59. *Dimensions of a Tennis Court* The length of a rectangular tennis court is six feet more than twice the width (see figure). Find the width of the court if the length is 78 feet.

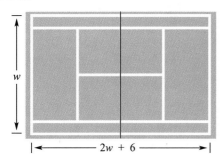

w

$\longleftarrow 2w + 6 \longrightarrow$

60. *Dimensions of a Sign* A sign has the shape of an equilateral triangle (see figure). The perimeter of the sign is 225 centimeters. Find the length of the sides of the sign. (An equilateral triangle is one whose sides have the same length.)

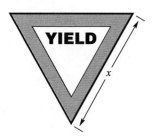

YIELD

x

61. *Cost of Housing* Suppose you budget 30% of your annual after-tax income for housing. If your after-tax income is $32,500, what amount can you spend on housing?

62. *Retirement Plan* Suppose you budget $7\frac{1}{2}$% of your gross income for an individual retirement plan. If your annual gross income is $29,800, how much would you put in your retirement plan each year?

63. *Original Price* A coat sells for $250 during a 20% storewide clearance sale. What was the original price of the coat?

64. *Buy Now or Wait?* A sales representative indicates that if a customer waits another month for a new car that currently costs $17,800, the price will increase by 6%. However, the customer will pay an interest penalty of $450 for the early withdrawal of a certificate of deposit if the car is purchased now. Determine whether the customer should buy now or wait another month.

65. *Height of a Fountain* Water is forced out of a fountain with an initial velocity of 28 feet per second (see figure). The velocity of the water stream is then given by $v = -32t + 28$, where t is the time in seconds. The height of the water is given by $h = -16t^2 + 28t$. What is the maximum height of the fountain? (*Hint:* Find the time when the velocity is zero, and then find the height for that time.)

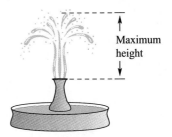

Maximum height

66. *Hours of Labor* The bill (including parts and labor) for the repair of an automobile was $357. The cost for parts was $285. How many hours were spent repairing the car if the cost of labor was $32 per hour?

67. *Dimensions of a Rectangle* The length of a rectangle is t times its width (see figure). Thus, the perimeter P is given by $P = 2w + 2(tw)$ where w is the width of the rectangle. If the perimeter of the rectangle is 1,200 meters, complete the following table giving the length, width, and area of the rectangle for the specified values of t.

68. Use the table in Exercise 67 to draw a conclusion concerning the area of a rectangle of given perimeter as its length increases relative to its width.

Figure for 67

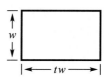

t	1	1.5	2	3	4	5
Width						
Length						
Area						

SECTION
3.3

More About Solving Linear Equations

Introduction   •   Linear Equations Containing Symbols of Grouping   •   Linear Equations Involving Fractions

Introduction

In Section 3.2 we looked at techniques for solving linear equations in standard form such as

$$3x - 2 = 10 \quad \text{and} \quad \frac{x}{3} - 2 = -4.$$

We also looked at techniques for solving some simpler linear equations in nonstandard form such as

$$3x + 8 - 5x = 4 \quad \text{and} \quad x + 2 = 2x - 6.$$

In this section we continue our study of linear equations by looking at more complicated nonstandard forms.

Linear Equations Containing Symbols of Grouping

To solve a linear equation that contains symbols of grouping, we first remove the symbols of grouping from each side. We then combine like terms and proceed to solve the resulting linear equation the usual way. Study the next two examples carefully, paying attention to the mental steps that can usually be performed without writing them down.

EXAMPLE 1 ■ **Solving a Linear Equation Involving Parentheses**

Solve the following linear equation.

$$4(x - 3) = 8$$

Solution

$4(x - 3) = 8$	Given equation
$4 \cdot x - 4 \cdot 3 = 8$	Distributive Property (mental step)
$4x - 12 = 8$	
$4x - 12 + 12 = 8 + 12$	Add 12 to both sides (mental step)
$4x = 20$	Collect like terms
$\dfrac{4x}{4} = \dfrac{20}{4}$	Divide both sides by 4 (mental step)
$x = 5$	Solution is 5

Check

$4(x - 3) = 8$	Given equation
$4(5 - 3) \overset{?}{=} 8$	Replace x by 5
$4(2) \overset{?}{=} 8$	
$8 = 8$	Solution checks

Thus, the solution is 5. ■

EXAMPLE 2 ■ **Solving a Linear Equation Involving Parentheses**

Solve the following linear equation.

$$3(2x - 1) + x = 11$$

Solution

$3(2x - 1) + x = 11$	Given equation
$3 \cdot 2x - 3 \cdot 1 + x = 11$	Distributive Property (mental step)
$6x - 3 + x = 11$	
$6x + x - 3 = 11$	Collect like terms (mental step)
$7x - 3 = 11$	Combine like terms
$7x - 3 + 3 = 11 + 3$	Add 3 to both sides (mental step)
$7x = 14$	Combine like terms
$\dfrac{7x}{7} = \dfrac{14}{7}$	Divide both sides by 7 (mental step)
$x = 2$	Solution is 2

Thus, the solution is 2. Check this solution in the original equation. ■

In the examples that follow, we will often omit some of the mental steps illustrated in Examples 1 and 2. When reading the examples, however, be sure that you can fill in any of the missing steps.

EXAMPLE 3   ■   **Solving a Linear Equation Involving Parentheses**
Solve the following linear equation.

$$5(x + 2) = 2(x - 1)$$

Solution

$5(x + 2) = 2(x - 1)$	Given equation
$5x + 10 = 2x - 2$	Distributive Property
$5x - 2x + 10 = -2$	Subtract $2x$ from both sides
$3x + 10 = -2$	Combine like terms
$3x = -2 - 10$	Subtract 10 from both sides
$3x = -12$	Combine like terms
$x = -4$	Divide both sides by 3

Thus, the solution is -4. Check this solution in the original equation.   ■

EXAMPLE 4   ■   **Solving a Linear Equation Involving Parentheses**
Solve the following linear equation.

$$2(x - 7) - 3(x + 4) = 4 - (5x - 2)$$

Solution

$2(x - 7) - 3(x + 4) = 4 - (5x - 2)$	Given equation
$2x - 14 - 3x - 12 = 4 - 5x + 2$	Distributive Property
$-x - 26 = -5x + 6$	Combine like terms
$-x + 5x - 26 = 6$	Add $5x$ to both sides
$4x - 26 = 6$	Combine like terms
$4x = 32$	Add 26 to both sides
$x = 8$	Divide both sides by 4

Thus, the solution is 8. Check this solution in the original equation.   ■

The linear equation in the next example involves both brackets and parentheses. Watch out for nested symbols of grouping such as these. We suggest removing the innermost symbols of grouping first.

EXAMPLE 5   ■   **A Linear Equation Involving Nested Symbols of Grouping**

Solve the following linear equation.

$$5x - 2[4x + 3(x - 1)] = 8 - 3x$$

Solution

$5x - 2[4x + 3(x - 1)] = 8 - 3x$	Given equation
$5x - 2[4x + 3x - 3] = 8 - 3x$	Distributive Property applied to innermost parentheses
$5x - 2[7x - 3] = 8 - 3x$	Combine like terms inside brackets
$5x - 14x + 6 = 8 - 3x$	Distributive Property
$-9x + 6 = 8 - 3x$	Combine like terms
$-9x + 3x + 6 = 8$	Add $3x$ to both sides
$-6x = 2$	Combine like terms and subtract 6 from both sides
$x = \dfrac{2}{-6}$	Divide both sides by -6
$x = -\dfrac{1}{3}$	Reduce fraction

Thus, the solution is $-\frac{1}{3}$. Check this solution in the original equation.   ■

MATH MATTERS

Divisibility Tests

There are several tests that you can perform in your head to determine whether a number is divisible by another number. Here are some of the better known divisibility tests.

Divisor	Test
2	Does the number end with an even digit (0, 2, 4, 6, 8)?
3	Is the sum of the digits of the number divisible by 3?
4	Are the last two digits of the number divisible by 4?
5	Is the last digit 0 or 5?
6	Is the number divisible by both 2 and 3?
8	Are the last three digits divisible by 8?
9	Is the sum of the digits of the number divisible by 9?

Here are two examples of the use of these tests. The number 4,617 *is* divisible by 9 because $4 + 6 + 1 + 7 = 18$ is divisible by 9. However, the number 3,827 *is not* divisible by 9 because $3 + 8 + 2 + 7 = 20$ is not divisible by 9.

Linear Equations Involving Fractions

We can solve a linear equation such as $2x - \frac{3}{4} = 1$ by adding $\frac{3}{4}$ to both sides and then dividing by 2 as follows.

$$2x - \frac{3}{4} + \frac{3}{4} = 1 + \frac{3}{4}$$

$$2x = \frac{7}{4}$$

$$x = \frac{7}{8}$$

For linear equations that contain more than one fraction, however, it is usually better to clear the equation of fractions by multiplying both sides of the equation by the least common multiple of the denominators of all the fractions. For example, the equation

$$\frac{3x}{2} - \frac{1}{3} = \frac{x}{4} + 2$$

can be cleared of fractions by multiplying both sides by 12, the least common multiple of 2, 3, and 4. (The least common multiple of the numbers 2, 3, and 4 is also called the **least common denominator** of the corresponding fractions having denominators 2, 3, and 4.) Notice how this is done in the next example.

EXAMPLE 6 ■ **Solving a Linear Equation Involving Fractions**

Solve the linear equation $\dfrac{3x}{2} - \dfrac{1}{3} = \dfrac{x}{4} + 2$.

Solution

$$\frac{3x}{2} - \frac{1}{3} = \frac{x}{4} + 2 \qquad\qquad \text{Given equation}$$

$$12\left(\frac{3x}{2} - \frac{1}{3}\right) = 12\left(\frac{x}{4} + 2\right) \qquad\qquad \text{Multiply both sides by 12}$$

$$12 \cdot \frac{3x}{2} - 12 \cdot \frac{1}{3} = 12 \cdot \frac{x}{4} + 12 \cdot 2 \qquad\qquad \text{Distributive Property}$$

$$18x - 4 = 3x + 24 \qquad\qquad \text{Clear fractions}$$

$$15x - 4 = 24 \qquad\qquad \text{Subtract } 3x \text{ from both sides}$$

$$15x = 28 \qquad\qquad \text{Add 4 to both sides}$$

$$x = \frac{28}{15} \qquad\qquad \text{Divide both sides by 15}$$

Thus, the solution is $\frac{28}{15}$. Check this solution in the original equation. ■

EXAMPLE 7 ■ **Solving a Linear Equation Involving Fractions**

Solve the linear equation $\dfrac{x}{3} + \dfrac{3x}{4} = 13$.

Solution

$$\frac{x}{3} + \frac{3x}{4} = 13 \qquad \text{Given equation}$$

$$12\left(\frac{x}{3}\right) + 12\left(\frac{3x}{4}\right) = 12(13) \qquad \text{Multiply both sides by 12}$$

$$4x + 9x = 156$$

$$13x = 156 \qquad \text{Combine like terms}$$

$$x = 12 \qquad \text{Divide both sides by 13}$$

Thus, the solution is 12. Check this solution in the original equation.   ∎

EXAMPLE 8 ■ **Solving a Linear Equation Involving Fractions**

Solve the linear equation $\dfrac{x}{5} - 7 = \dfrac{3x}{2} + \dfrac{4}{5}$.

Solution

$$\frac{x}{5} - 7 = \frac{3x}{2} + \frac{4}{5} \qquad \text{Given equation}$$

$$10\left(\frac{x}{5} - 7\right) = 10\left(\frac{3x}{2} + \frac{4}{5}\right) \qquad \text{Multiply both sides by 10}$$

$$2x - 70 = 15x + 8 \qquad \text{Clear fractions}$$

$$-13x - 70 = 8 \qquad \text{Subtract } 15x \text{ from both sides}$$

$$-13x = 78 \qquad \text{Add 70 to both sides}$$

$$x = -6 \qquad \text{Divide both sides by } -13$$

Thus, the solution is -6. Check this solution in the original equation.   ∎

EXAMPLE 9 ■ **Solving a Linear Equation Involving Fractions**

Solve the linear equation $\dfrac{x - 2}{4} + \dfrac{2x + 1}{6} = \dfrac{17}{12}$.

Solution

$$\frac{x - 2}{4} + \frac{2x + 1}{6} = \frac{17}{12} \qquad \text{Given equation}$$

$$12 \cdot \frac{x - 2}{4} + 12 \cdot \frac{2x + 1}{6} = 12 \cdot \frac{17}{12} \qquad \text{Multiply both sides by 12}$$

$$3(x - 2) + 2(2x + 1) = 17 \qquad \text{Clear fractions}$$

$$3x - 6 + 4x + 2 = 17 \qquad \text{Distributive Property}$$

$$7x - 4 = 17 \qquad \text{Combine like terms}$$

$$7x = 21 \qquad \text{Add 4 to both sides}$$

$$x = 3 \qquad \text{Divide both sides by 7}$$

Thus, the solution is 3. Check this solution in the original equation.   ∎

A common type of linear equation is one that equates two fractions. To solve such equations, we consider the fractions to be **equivalent** and use **cross-multiplication**. That is, if

$$\frac{a}{b} = \frac{c}{d}, \quad \text{then} \quad a \cdot d = b \cdot c.$$

Note how cross-multiplication is used in the next example.

EXAMPLE 10   ■   Using Cross-Multiplication to Solve a Linear Equation

Use cross-multiplication to solve the linear equation $\dfrac{x + 2}{3} = \dfrac{8}{5}$.

Solution

$$\frac{x + 2}{3} = \frac{8}{5} \qquad \text{Given equation}$$

$$5(x + 2) = 3(8) \qquad \text{Cross-multiply}$$

$$5x + 10 = 24 \qquad \text{Distributive Property}$$

$$5x = 14 \qquad \text{Subtract 10 from both sides}$$

$$x = \frac{14}{5} \qquad \text{Divide both sides by 5}$$

Check

$$\frac{x + 2}{3} = \frac{8}{5} \qquad \text{Given equation}$$

$$\frac{\frac{14}{5} + 2}{3} \stackrel{?}{=} \frac{8}{5} \qquad \text{Replace } x \text{ by } \frac{14}{5}$$

$$\frac{\frac{14}{5} + \frac{10}{5}}{3} \stackrel{?}{=} \frac{8}{5}$$

$$\frac{\frac{24}{5}}{3} \stackrel{?}{=} \frac{8}{5}$$

$$\frac{24}{5}\left(\frac{1}{3}\right) \stackrel{?}{=} \frac{8}{5}$$

$$\frac{8}{5} = \frac{8}{5} \qquad \text{Solution checks}$$

Thus, the solution is $\frac{14}{5}$.

■

More extensive applications of cross-multiplication will be discussed when we study ratios and proportions in Section 4.1.

DISCUSSION PROBLEM ■ You Be the Instructor

Suppose you are teaching an algebra class and one of your students hands in the following problem. Is the answer that the student obtained correct? Is the work correct? Would you give partial credit for getting the correct answer using *incorrect* steps?

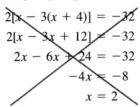

$$2[x - 3(x + 4)] = -32$$
$$2[x - 3x + 12] = -32$$
$$2x - 6x + 24 = -32$$
$$-4x = -8$$
$$x = 2$$

■

Warm-Up

The following warm-up exercises involve skills that were covered in earlier sections. You will use these skills in the exercise set for this section.

In Exercises 1–10, solve the linear equation and check your solution.

1. $-t = 6$

2. $8 - z = 3$

3. $50 - z = 15$

4. $x - 6 = 3x + 10$

5. $2x - 5 = x + 9$

6. $6x + 8 = 8 - 2x$

7. $2x + \dfrac{3}{2} = \dfrac{3}{2}$

8. $-10x + \dfrac{2}{3} = \dfrac{7}{3} - 5x$

9. $\dfrac{x}{6} + \dfrac{x}{3} = 1$

10. $\dfrac{x}{5} + \dfrac{1}{5} = \dfrac{7}{10}$

3.3   EXERCISES

In Exercises 1–34, solve the equation and check your solution. (Some of the equations have no solution.)

1. $2(x - 3) = 4$

2. $4(x + 1) = 24$

3. $4 - (z + 6) = 8$

4. $25 - (y + 3) = 15$

5. $3 - (2x - 4) = 3$

6. $16 - (3x - 10) = 5$

7. $-3(t + 5) = 0$

8. $4(z - 2) = 0$

9. $-3(t + 5) = 6$

10. $4(z - 2) = 32$

11. $3(x + 4) = 10(x + 4)$

12. $-8(x - 6) = 3(x - 6)$

13. $7x - 2(x - 2) = 12$

14. $15(x + 1) - 8x = 29$

15. $6 = 3(y + 1) - 4(1 - y)$

16. $100 = 4(y - 6) - (y - 1)$

17. $2(x + 5) - 7 = 3(x - 2)$

18. $6[x - (2x + 3)] = 8 - 5x$

19. $\dfrac{x}{2} = \dfrac{3}{2}$

20. $\dfrac{z}{3} = -\dfrac{5}{3}$

21. $\dfrac{y}{5} = -\dfrac{3}{10}$

22. $\dfrac{t}{4} = \dfrac{3}{8}$

23. $\dfrac{6x}{25} = \dfrac{3}{5}$

24. $-\dfrac{8x}{9} = \dfrac{2}{3}$

25. $\dfrac{t + 4}{6} = \dfrac{2}{3}$

26. $\dfrac{x - 6}{10} = \dfrac{3}{5}$

27. $\dfrac{5x}{4} + \dfrac{1}{2} = 0$

28. $\dfrac{y}{4} - \dfrac{5}{8} = 2$

29. $\dfrac{x}{5} - \dfrac{x}{2} = 1$

30. $\dfrac{x}{3} + \dfrac{x}{4} = 1$

31. $\dfrac{2}{3}(z + 5) - \dfrac{1}{4}(z + 24) = 0$

32. $\dfrac{3x}{2} + \dfrac{1}{4}(x - 2) = 10$

33. $\dfrac{100 - 4u}{3} = \dfrac{5u + 6}{4} + 6$

34. $\dfrac{8 - 3x}{2} - 4 = \dfrac{x}{6}$

In Exercises 35–40, solve the equation by first cross-multiplying to obtain a linear equation.

35. $\dfrac{5x - 4}{5x + 4} = \dfrac{2}{3}$

36. $\dfrac{10x + 3}{5x + 6} = \dfrac{1}{2}$

37. $\dfrac{1}{x} = 7$

38. $\dfrac{5}{x + 1} = 1$

39. $6 = \dfrac{5}{x}$

40. $\dfrac{14}{x} = 7$

41. *Time to Complete Task* Two people can complete 80% of a task in t hours, where t must satisfy the equation

$$\frac{t}{10} + \frac{t}{15} = 0.8.$$

Solve this equation for t.

42. *Time to Complete Task* The time to complete 100% of the task in Exercise 41 is given by the equation

$$\frac{t}{10} + \frac{t}{15} = 1.$$

Solve this equation for t.

43. *Course Grade* To get an A in a course a student must have an average of at least 90 on four tests of 100 points each. A certain student scores 87, 92, and 84 on the first three tests. What must the student score on the fourth exam to earn a 90% average for the course?

44. *Course Grade* To get an A in a course a student must have an average of at least 90 on four tests of 100 points each. A certain student scores 87, 92, and 84 on the first three tests. The fourth test is weighted so that it counts twice as much as each of the first three tests. What must the student score on the fourth exam to earn a 90% average for the course?

In Exercises 45–48, solve the following equation for x.

$$p_1 x + p_2(a - x) = p_3 a$$

(*Note:* The symbols p_1, p_2, and p_3 use **subscripts** to denote different variables. The symbol p_1 is read "p sub one," the symbol p_2 is read "p sub two," and the symbol p_3 is read "p sub three.")

45. *Mixture Problem* Determine the number of quarts of a 10% solution that must be mixed with a 30% solution to obtain 100 quarts of a 25% solution. ($p_1 = 0.1$, $p_2 = 0.3$, $p_3 = 0.25$, and $a = 100$.)

46. *Mixture Problem* Determine the number of gallons of a 25% solution that must be mixed with a 50% solution to obtain five gallons of a 30% solution. ($p_1 = 0.25$, $p_2 = 0.5$, $p_3 = 0.3$, and $a = 5$.)

47. *Mixture Problem* An eight-quart automobile cooling system is filled with coolant that is 40% antifreeze. Determine the amount that must be withdrawn and replaced with pure antifreeze so that the eight quarts of coolant will be 50% antifreeze. ($p_1 = 1$, $p_2 = 0.4$, $p_3 = 0.5$, and $a = 8$.)

48. *Mixture Problem* A grocer mixes two kinds of nuts costing $2.49 per pound and $3.89 per pound to make 100 pounds of a mixture costing $3.19 per pound. How many pounds of the nuts costing $2.49 per pound must be put into the mixture? ($p_1 = 2.49$, $p_2 = 3.89$, $p_3 = 3.19$, and $a = 100$.)

In Exercises 49 and 50, use the following equation.

$$W_1 x = W_2(a - x)$$

49. *Balancing a Seesaw* The board on a seesaw is 10 feet long. Find the position of the fulcrum so that the seesaw will balance for two children weighing 60 pounds and 90 pounds, as shown in the figure. ($W_1 = 90$, $W_2 = 60$, and $a = 10$.)

Figure for 49

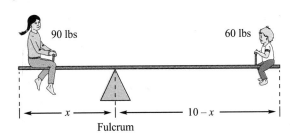

Fulcrum

50. *Raising a Weight* The fulcrum of a six-foot-long lever is six inches from a weight, as shown in the figure. Find the maximum weight that a 190-pound person can lift using this lever. ($W_1 = 190$, $x = 5\frac{1}{2}$, and $a = 6$.)

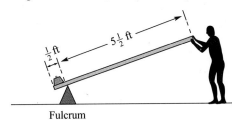

Fulcrum

SECTION

3.4

Inequalities

Inequalities and Their Graphs • *Properties of Inequalities* • *Solving a Linear Inequality* • *Solving a Double Linear Inequality* • *Applications*

Inequalities and Their Graphs

Simple inequalities were introduced in Section 1.1 to *order* the real numbers. There, we compared two real numbers using the inequality symbols $<$, $\leq$, $>$, and $\geq$. For instance, because -5 lies to the left of 3 on the number line, we say that -5 is less than 3 and write $-5 < 3$. In this section we study **algebraic inequalities**, which are inequalities that contain one or more variable terms. Some examples are

$$x \leq 3, \qquad x \geq -2, \qquad x - 5 < 2, \quad \text{and} \quad 5x - 7 < 3x + 9.$$

Each of these inequalities is a **linear inequality** in the variable x because the (implied) exponent of x is 1. (The inequality $x^2 < 2$ is *not* linear in the variable x because x has an exponent of 2.)

As with an equation, we **solve an inequality** in the variable x by finding all values of x for which the inequality is true. Such values are called **solutions** and we say that the solutions **satisfy** the inequality. The **solution set** of an inequality is the set of all real numbers that are solutions of the inequality.

Often, the solution set of an inequality will consist of infinitely many real numbers. To get a visual image of the solution set, it is helpful to sketch its **graph** on the real number line. For instance, the graph of the solution set of $x < 2$ consists of all points on the real number line that are to the left of 2. Example 1 lists some inequalities and the corresponding graphs of their solution sets.

EXAMPLE 1   ■   Graphs of Inequalities

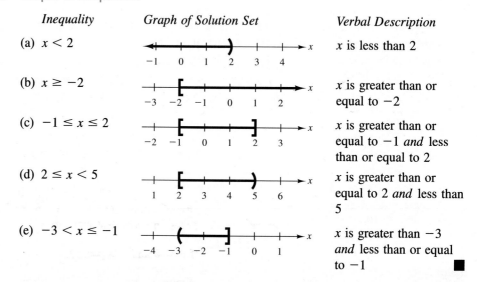

Inequality	Graph of Solution Set	Verbal Description
(a) $x < 2$		x is less than 2
(b) $x \geq -2$		x is greater than or equal to -2
(c) $-1 \leq x \leq 2$		x is greater than or equal to -1 *and* less than or equal to 2
(d) $2 \leq x < 5$		x is greater than or equal to 2 *and* less than 5
(e) $-3 < x \leq -1$		x is greater than -3 *and* less than or equal to -1

Note that in the graphs we use a parenthesis to denote the inequality symbols $<$ and $>$, and a bracket to denote the inequality symbols $\leq$ and $\geq$.

Properties of Inequalities

The procedures for solving linear inequalities in one variable are much like those for solving linear equations. To isolate the variable, we make use of **properties of inequalities**. These properties are similar to the properties of equality, but there are two important exceptions. When both sides of an inequality are multiplied or divided by a negative number, the direction of the inequality symbol must be reversed. Here is an example.

$$-2 < 5 \qquad \text{Given inequality}$$
$$(-3)(-2) > (-3)(5) \qquad \text{Multiply both sides by } -3 \text{ and reverse the inequality}$$
$$6 > -15$$

Two inequalities that have the same solution set are called **equivalent**. The following list describes operations that can be used to create equivalent inequalities.

Properties of Inequalities

Let a, b, and c be real numbers, variables, or algebraic expressions. Then the following properties are true.

Property	Verbal and Algebraic Descriptions
Addition Property	Adding the same quantity to both sides of an inequality produces an equivalent inequality. If $a < b$, then $a + c < b + c$.
Subtraction Property	Subtracting the same quantity from both sides of an inequality produces an equivalent inequality. If $a < b$, then $a - c < b - c$.
Multiplication Properties	Multiplying both sides of an inequality by a *positive* quantity produces an equivalent inequality. If $a < b$ and c is positive, then $ac < bc$. Multiplying both sides of an inequality by a *negative* quantity produces an equivalent inequality in which the inequality symbol is reversed. If $a < b$ and c is negative, then $ac > bc$.
Division Properties	Dividing both sides of an inequality by a *positive* quantity produces an equivalent inequality. If $a < b$ and c is positive, then $\dfrac{a}{c} < \dfrac{b}{c}$. Dividing both sides of an inequality by a *negative* quantity produces an equivalent inequality in which the inequality symbol is reversed. If $a < b$ and c is negative, then $\dfrac{a}{c} > \dfrac{b}{c}$.
Transitive Property	Consider three quantities for which the first quantity is less than the second, and the second is less than the third. It follows that the first quantity must be less than the third quantity. If $a < b$ and $b < c$, then $a < c$.

Each of the above properties is true if the symbol $<$ is replaced by $\leq$ and the symbol $>$ is replaced by $\geq$.

Solving a Linear Inequality

In the next several examples we apply the properties of inequalities to solve linear inequalities. As you read through each example, pay special attention to the steps in which the inequality symbol is reversed. Remember that when you multiply or divide an inequality by a negative number, you must reverse the inequality symbol.

EXAMPLE 2 ■ Solving a Linear Inequality

Solve the following linear inequality and sketch the graph of its solution set.

$$x + 5 < 8$$

Solution

$x + 5 < 8$	Given inequality
$x + 5 - 5 < 8 - 5$	Subtract 5 from both sides
$x < 3$	Solution set

Thus, the solution set consists of all real numbers that are less than 3. The graph of this solution set is shown in Figure 3.3.

FIGURE 3.3

■

Checking the solution set of an inequality is not as simple as checking the solutions of an equation. (There are usually too many x-values to substitute back into the original inequality.) We can, however, get an indication of the validity of a solution set by substituting a few convenient values of x. For instance, in Example 2 we found the solution of $x + 5 < 8$ to be $x < 3$. Try checking that $x = 0$ satisfies the original inequality, whereas $x = 4$ does not.

EXAMPLE 3 ■ Solving a Linear Inequality

Solve the following linear inequality and sketch the graph of its solution set.

$$3y - 1 \leq -7$$

Solution

$3y - 1 \leq -7$	Given inequality
$3y - 1 + 1 \leq -7 + 1$	Add 1 to both sides
$3y \leq -6$	Combine like terms
$\dfrac{3y}{3} \leq \dfrac{-6}{3}$	Divide both sides by (positive) 3
$y \leq -2$	Solution set

Thus, the solution set consists of all real numbers that are less than or equal to -2. The graph of this solution set is shown in Figure 3.4.

FIGURE 3.4

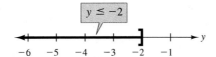

■

EXAMPLE 4   ■   Solving a Linear Inequality

Solve the following linear inequality and sketch the graph of its solution set.

$$12 - 2x > 10$$

Solution

$12 - 2x > 10$	Given inequality
$12 - 12 - 2x > 10 - 12$	Subtract 12 from both sides
$-2x > -2$	Combine like terms
$\dfrac{-2x}{-2} < \dfrac{-2}{-2}$	Divide both sides by -2 and reverse inequality symbol
$x < 1$	Solution set

Thus, the solution set consists of all real numbers that are less than 1. The graph of this solution set is shown in Figure 3.5.

FIGURE 3.5

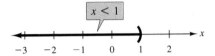

In the next two examples, we solve inequalities in which the variable occurs more than once.

EXAMPLE 5   ■   Solving a Linear Inequality

Solve the following linear inequality and sketch the graph of its solution set.

$$5x - 7 > 3x + 9$$

Solution

$5x - 7 > 3x + 9$	Given inequality
$5x > 3x + 16$	Add 7 to both sides
$5x - 3x > 16$	Subtract $3x$ from both sides
$2x > 16$	Combine like terms
$x > 8$	Divide both sides by 2

Thus, the solution set consists of all real numbers that are greater than 8, as shown in Figure 3.6.

FIGURE 3.6

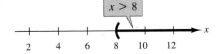

EXAMPLE 6 ■   Solving a Linear Inequality

Solve the following linear inequality and sketch the graph of its solution set.

$$1 - \frac{3x}{2} \geq x - 4$$

Solution

$$1 - \frac{3x}{2} \geq x - 4 \qquad \text{Given inequality}$$

$$2 - 3x \geq 2x - 8 \qquad \text{Multiply both sides by 2}$$

$$-3x \geq 2x - 10 \qquad \text{Subtract 2 from both sides}$$

$$-5x \geq -10 \qquad \text{Subtract } 2x \text{ from both sides}$$

$$x \leq 2 \qquad \begin{array}{l}\text{Divide both sides by } -5 \\ \text{and reverse the inequality}\end{array}$$

Thus, the solution set consists of all real numbers that are less than or equal to 2, as shown in Figure 3.7.

FIGURE 3.7

Solving a Double Linear Inequality

Sometimes it is convenient to write two inequalities as a **double inequality**. For instance, we can write the two inequalities $-3 \leq 6x - 1$ and $6x - 1 < 3$ more simply as

$$-3 \leq 6x - 1 < 3.$$

This form allows us to solve the two given inequalities together, as demonstrated in Example 7.

EXAMPLE 7 ■   Solving a Double Inequality

Solve the inequality $-3 \leq 6x - 1 < 3$.

Solution

$$-3 \leq 6x - 1 < 3 \qquad \text{Given inequality}$$

$$-3 + 1 \leq 6x - 1 + 1 < 3 + 1 \qquad \text{Add 1 to all three parts}$$

$$-2 \leq 6x < 4 \qquad \text{Combine like terms}$$

$$\frac{-2}{6} \leq \frac{6x}{6} < \frac{4}{6} \qquad \text{Divide each part by 6}$$

$$-\frac{1}{3} \leq x < \frac{2}{3} \qquad \text{Solution set}$$

Thus, the solution set consists of all real numbers that are greater than or equal to $-\frac{1}{3}$ and less than $\frac{2}{3}$, as shown in Figure 3.8.

FIGURE 3.8

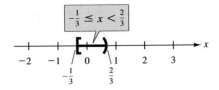

The double inequality in Example 7 could have been solved in two parts as follows.

$$-3 \le 6x - 1 \quad \text{and} \quad 6x - 1 < 3$$
$$-2 \le 6x \quad\quad \text{and} \quad\quad 6x < 4$$
$$-\frac{1}{3} \le x \quad\quad \text{and} \quad\quad x < \frac{2}{3}$$

The solution set consists of all real numbers that satisfy *both* inequalities. In other words, the solution set is the set of all values of x for which $-\frac{1}{3} \le x < \frac{2}{3}$.

Applications

Before looking at some applications, we give one example of translating verbal statements into inequalities.

EXAMPLE 8 ■ Translating Verbal Statements

Verbal Statement	*Inequality*
(a) x is at most 2.	$x \le 2$
(b) x is no more than 2.	$x \le 2$
(c) x is at least 2.	$x \ge 2$
(d) x is more than 2.	$x > 2$
(e) x is less than 2.	$x < 2$

We conclude this section with two applications that involve inequalities.

EXAMPLE 9   ■   An Application: Car Rental

A subcompact car can be rented from Company A for $190 per week with no extra charge for mileage. A similar car can be rented from Company B for $100 per week, plus 20¢ for each mile driven. How many miles must you drive in a week to make the rental fee for Company A less than that for Company B?

Solution

Verbal model: Weekly cost from A $<$ Weekly cost from B

Labels: Number of miles driven in one week $= m$
Weekly cost from A $= 190$ (in dollars)
Weekly cost from B $= 100 + 0.2m$ (in dollars)

Inequality: $190 < 100 + 0.2m$

$90 < 0.2m$

$450 < m$

Note that the inequality $450 < m$ is equivalent to writing $m > 450$. Thus, the car from Company A is cheaper if you plan to drive more than 450 miles in a week. ■

EXAMPLE 10   ■   An Application: Course Grade

Suppose you are taking a college course in which the grade is based on six 100-point exams. To earn an A in the course, you must have a total of at least 90% of the points. On the first five exams, your scores were 85, 92, 88, 96, and 87. How many points do you have to obtain on the sixth test in order to earn an A in the course?

Solution

Verbal model: Total points $\geq$ 90% of 600

Labels: Score for sixth exam $= x$
Total points $= (85 + 92 + 88 + 96 + 87) + x$

Inequality: $(85 + 92 + 88 + 96 + 87) + x \geq 0.9(600)$

$448 + x \geq 540$

$x \geq 540 - 448$

$x \geq 92$

Thus, you must get at least 92 points on the sixth exam to earn an A in the course.

■

DISCUSSION PROBLEM   ■   **Properties of Inequalities**

After listing the properties of inequalities, we mentioned that each of the properties is true if the symbol $<$ is replaced by $\leq$ and the symbol $>$ is replaced by $\geq$. Rewrite each of the properties of inequalities using the symbols $\leq$ and $\geq$.   ■

Warm-Up

The following warm-up exercises involve skills that were covered in earlier sections. You will use these skills in the exercise set for this section.

In Exercises 1–4, place the correct inequality symbol ($<$ or $>$) between the two real numbers.

1. $-\dfrac{1}{2}$ ▓▓▓ -7

2. $-\dfrac{1}{3}$ ▓▓▓ $-\dfrac{1}{6}$

3. -2 ▓▓▓ -3

4. -6 ▓▓▓ $-\dfrac{13}{2}$

In Exercises 5–10, solve the equation.

5. $-2n = 5$

6. $16 + 2l = 64$

7. $\dfrac{9 + x}{3} = 15$

8. $20 - \dfrac{9}{x} = 2$

9. $4 - 3(1 - x) = 7$

10. $6(t - 6) = 0$

3.4   **EXERCISES**

In Exercises 1–4, determine whether the given value of x satisfies the inequality.

Inequality   *Values*

1. $5x - 12 > 0$ (a) $x = 3$ (b) $x = -3$ (c) $x = \dfrac{5}{2}$ (d) $x = \dfrac{3}{2}$

2. $x + 1 < \dfrac{2x}{3}$ (a) $x = 0$ (b) $x = 4$ (c) $x = -4$ (d) $x = -3$

3. $0 < \dfrac{x - 2}{4} < 2$ (a) $x = 4$ (b) $x = 10$ (c) $x = 0$ (d) $x = \dfrac{7}{2}$

4. $-1 < \dfrac{3 - x}{2} \leq 1$ (a) $x = 0$ (b) $x = 3$ (c) $x = 1$ (d) $x = 5$

In Exercises 5–10, match the given inequality with its graph. [The graphs are labeled (a), (b), (c), (d), (e), and (f).]

5. $x < 4$

(a)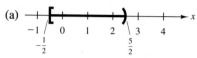

6. $x \geq 4$

(b)

7. $\dfrac{3}{2} < x$

(c)

8. $-\dfrac{1}{2} < x \leq \dfrac{5}{2}$

(d)

9. $\dfrac{3}{2} < x < 4$

(e)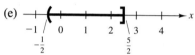

10. $-\dfrac{1}{2} \leq x < \dfrac{5}{2}$

(f)

In Exercises 11–50, solve the given inequality and sketch the graph of its solution set.

11. $t - 3 \geq 2$ **12.** $t + 1 < 6$ **13.** $x + 4 \leq 6$ **14.** $z - 2 > 0$

15. $4x < 12$ **16.** $2x > 3$ **17.** $-10x < 40$ **18.** $-6x > 18$

19. $-3n > -9$ **20.** $-7n < -21$ **21.** $\dfrac{2}{3}x \leq 12$ **22.** $\dfrac{5}{8}x \geq 10$

23. $-\dfrac{3}{4}x > -3$ **24.** $-\dfrac{1}{6}x < -2$ **25.** $2x - 5 > 7$ **26.** $3x + 2 \leq 14$

27. $5 - x \leq 1$ **28.** $3 - y \geq -3$ **29.** $4 - 2x < 3$ **30.** $14 - 3x > 5$

31. $2x - 5 > -x + 6$ **32.** $25x + 4 \leq 10x + 19$

33. $6 < 3(y + 1) - 4(1 - y)$ **34.** $6[x - (2x + 3)] < 8 - 5x$

35. $-2(z + 1) \geq 3(z + 1)$ **36.** $8(t - 3) < 4(t - 3)$

37. $10(1 - y) < 3 - 2y$ **38.** $6(3 - z) \geq 5(3 + z)$

39. $\dfrac{5x}{4} + \dfrac{1}{2} > 0$ **40.** $\dfrac{y}{4} - \dfrac{5}{8} < 2$ **41.** $\dfrac{x}{5} - \dfrac{x}{2} \leq 1$ **42.** $\dfrac{x}{3} + \dfrac{x}{4} \geq 1$

43. $1 < 2x + 3 < 9$ **44.** $-8 \leq 1 - 3(x - 2) < 13$

45. $-4 < \dfrac{2x - 3}{3} < 4$ **46.** $0 \leq \dfrac{x + 3}{2} < 5$ **47.** $6 > \dfrac{x - 2}{-3} > -2$ **48.** $-2 < \dfrac{x - 4}{-2} \leq 3$

49. $\dfrac{3}{4} > x + 1 > \dfrac{1}{4}$ **50.** $-1 < -\dfrac{x}{3} < 1$

In Exercises 51–56, use inequality notation to denote the given statement.

51. x is nonnegative.   **52.** P is no more than 2.   **53.** y is more than -6.   **54.** t is less than 8.

55. z is at least 3.   **56.** x is a real number greater than or equal to -2 and less than 5.

In Exercises 57–60, write a verbal statement describing the set of real numbers satisfying the given inequality.

57. $x \leq 10$ **58.** $z > 8$ **59.** $-\dfrac{3}{2} < y \leq 5$ **60.** $-3 \leq t < 3.8$

61. *Planet Distances* Mars is further from the Sun than Venus, and Venus is further from the Sun than Mercury (see figure). What can be said about the relationship between Mars's and Mercury's distances from the Sun?

Sun   Mercury   Venus   Earth   Mars

62. *Budget for Trip* Suppose you have $2,500 budgeted for a trip. The transportation for the trip will cost $900. To stay within your budget, all other costs must be no more than what amount?

63. *Annual Operating Cost* A utility company has a fleet of vans. The annual operating cost C (in dollars) per van is

$$C = 0.32m + 2{,}300$$

where m is the number of miles traveled by a van in a year. What number of miles will yield an annual operating cost that is less than $10,000?

64. *Profit* The revenue for selling x units of a product is

$$R = 115.95x.$$

The cost of producing x units is

$$C = 95x + 750.$$

In order to obtain a profit, the revenue must be greater than the cost. For what values of x will this product produce a profit?

65. *Telephone Cost* The cost of a long-distance telephone call is $0.46 for the first minute and $0.31 for each additional minute. The total cost of the call cannot exceed $4. Find the interval of time that is available for the call.

66. *Sides of a Triangle* The lengths of the sides of the triangle in the accompanying figure are a, b, and c. Find an inequality that relates $a + b$ and c.

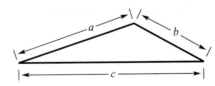

67. *Comparing Distances* Suppose you live three miles from college and your parents live two miles from you (see figure). Let d represent the distance between your parents' house and the college. Write an inequality involving d.

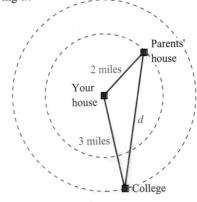

CHAPTER 3   SUMMARY

As you review and prepare for a test on this chapter, first try to obtain a global view of what was discussed. Then review the specific skills needed in each category.

Rewrite and Solve Linear Equations

- *Check* whether a specified value of x is a solution of a given equation.

  Section 3.1, Exercises 1–14

- *Identify* the properties used to form equivalent equations.

  Section 3.1, Exercises 15–20;
  Section 3.2, Exercises 5–8

- *Solve and check* a linear equation given in standard form or non-standard form.

  Section 3.1, Exercises 9–50

- *Solve and check* a linear equation that contains symbols of grouping.

  Section 3.3, Exercises 1–18

- *Solve and check* a linear equation that contains a fraction.

  Section 3.3, Exercises 19–40

- *Create, solve, and check* a linear equation from a verbal statement.

  Section 3.2, Exercises 55–56;
  Section 3.3, Exercises 41–50;
  Section 3.4, Exercises 51–54

Rewrite and Solve Linear Inequalities

- *Check* whether a specified value of x satisfies a given inequality.

  Section 3.4, Exercises 1–4

- *Solve and check* a linear inequality.

  Section 3.4, Exercises 11–50

- *Sketch* the graph of the solution set of a linear inequality.

  Section 3.4, Exercises 11–50

- *Translate* a verbal phrase into inequality form and conversely.

  Section 3.4, Exercises 55–60

- *Create, solve, and check* a linear inequality from verbal statements.

  Section 3.4, Exercises 61–67

Chapter 3 Review Exercises

In Exercises 1–10, determine whether the given value of x is a solution of the equation.

Equation	*Values*	
1. $5x + 6 = 36$	(a) $x = 3$	(b) $x = 6$
2. $17 - 3x = 8$	(a) $x = 3$	(b) $x = -3$
3. $3x - 12 = x$	(a) $x = -1$	(b) $x = 6$
4. $8x + 24 = 2x$	(a) $x = 0$	(b) $x = -4$
5. $4(2 - x) = 3(2 + x)$	(a) $x = \dfrac{2}{7}$	(b) $x = -\dfrac{2}{3}$
6. $5x + 2 = 3(x + 10)$	(a) $x = 14$	(b) $x = -10$
7. $\dfrac{4}{x} - \dfrac{2}{x} = 5$	(a) $x = -1$	(b) $x = \dfrac{2}{5}$
8. $\dfrac{x}{3} + \dfrac{x}{6} = 1$	(a) $x = \dfrac{2}{9}$	(b) $x = -\dfrac{2}{9}$
9. $x(x - 7) = -12$	(a) $x = 3$	(b) $x = 4$
10. $x(x + 1) = 2$	(a) $x = 1$	(b) $x = -2$

In Exercises 11–30, solve the given equation and check your solution.

11. $10x = 50$

12. $-3x = 21$

13. $8x + 7 = 39$

14. $12x - 5 = 43$

15. $24 - 7x = 3$

16. $13 + 6x = 61$

17. $15x - 4 = 16$

18. $3x - 8 = 2$

19. $3x - 2(x + 5) = 10$

20. $4x + 2(7 - x) = 5$

21. $2x + 3 = 5x - 2$

22. $8(x - 2) = 3(x + 2)$

23. $\dfrac{x}{5} = 4$

24. $-\dfrac{x}{14} = \dfrac{1}{2}$

25. $\dfrac{2}{3}x - \dfrac{1}{6} = \dfrac{9}{2}$

26. $\dfrac{1}{8}x + \dfrac{3}{4} = \dfrac{5}{2}$

27. $\dfrac{3}{x} = 9$

28. $\dfrac{4 - x}{x} = 7$

29. $6 + \dfrac{5}{x} = 1$

30. $\dfrac{x}{3} + \dfrac{x}{5} = 1$

In Exercises 31–34, solve the given equation and round your answer to two decimal places. (A calculator may be helpful.)

31. $516x - 875 = 3{,}250$

32. $2.825x + 3.125 = 12.5$

33. $\dfrac{x}{4.625} = 48.5$

34. $5x + \dfrac{1}{4.5} = 18.125$

In Exercises 35–50, solve the given inequality and sketch the graph of its solution.

35. $x + 5 \geq 7$

36. $x - 2 \leq 1$

37. $3x - 8 < 1$

38. $4x + 3 > 15$

39. $-11x \leq -22$

40. $-7x \geq 21$

41. $\frac{4}{5}x > 8$

42. $\frac{2}{3}n < -4$

43. $14 - \frac{1}{2}t < 12$

44. $32 + \frac{7}{8}k > 11$

45. $3(1 - y) \geq 2(4 + y)$

46. $4 - 3y \leq 8(10 - y)$

47. $-2 < \frac{x}{3} \leq 2$

48. $-5 \leq 3 - 4x < 5$

49. $3 > \frac{x + 1}{-2} > 0$

50. $5 \geq \frac{x - 3}{3} > 2$

In Exercises 51–54, write an equation that represents the given statement. (Identify the letters you choose as labels.)

51. The sum of a number and its reciprocal is $\frac{37}{6}$.

52. *Distance* An automobile travels 135 miles in t hours with an average speed of 45 miles per hour (see figure).

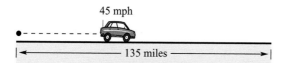

45 mph

135 miles

53. *Area* The area of the shaded region in the accompanying figure is 24 square inches. (The area of a triangle is $A = \frac{1}{2}bh$, where b is the base and h is the height.)

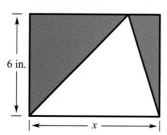

6 in.

x

54. *Perimeter* The perimeter of the face of a rectangular traffic light is 72 inches (see figure).

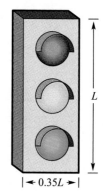

L

$0.35L$

In Exercises 55–60, write an inequality that represents the given statement.

55. z is at least 10.

56. x is nonnegative.

57. y is more than 8 but less than 12.

58. The area A is no more than 100 square feet.

59. The volume V is less than 12 cubic feet.

60. The perimeter P is at least 24 inches.

CHAPTER 3 TEST

Take this test as you would take a test in class. After you are done, check your work with the answers given in the back of the book.

1. Determine whether the given value of x is a solution of the equation $6(3 - x) - 5(2x - 1) = 7$.

   (a) $x = -2$ (b) $x = 1$

2. Solve: $4x - 3 = 18$

3. Solve: $10 - (2 - x) = 2x + 1$

4. Solve: $\dfrac{5x}{4} = \dfrac{5}{2} + x$

5. Solve: $\dfrac{7}{x} - 5 = 2$

6. Solve using a calculator and round the solution to two decimal places:
   $4.08(x + 10) = 9.50(x - 2)$

7. Solve and sketch the solution: $x + 3 \le 7$

8. Solve and sketch the solution: $-\dfrac{2x}{3} > 4$

9. Solve and sketch the solution: $-3 < 2x - 1 \le 3$

10. Solve and sketch the solution: $2 \ge \dfrac{3 - x}{2} > -1$

11. Use inequality notation to denote the phrase, "y is no more than 10."

12. Use inequality notation to denote the phrase, "t is at least 4."

13. When the sum of a number and 6 is divided by 8 the result is 7. Find the number.

14. The price of a product increased by 20% during the past year. It is now selling for $240. What was the price one year ago?

15. The bill (including parts and labor) for the repair of a home appliance was $134. The cost for parts was $62. How many hours were spent repairing the appliance if the cost of labor was $18 per half hour?

CHAPTER FOUR

Applications of Linear Equations

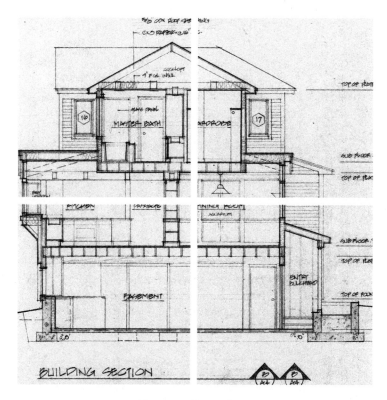

Chapter Overview

Linear equations have many different uses, both in problems from everyday life and in more technical problems encountered in business and science. In this chapter we will study a wide variety of uses of linear equations. The chapter is lengthy and you may find it difficult, but mastery of the skills discussed in this chapter is an important part of success in algebra.

Ratio and Proportion
Setting Up Ratios   •   Unit Prices   •   Solving Proportions   •   The Consumer Price Index

Setting Up Ratios

A comparison of one number to another by division is called a **ratio**. For example, in a class of 29 students made up of 16 women and 13 men, the ratio of women to men is 16 to 13 or $\frac{16}{13}$. Some other ratios for this class are as follows.

Men to women: $\dfrac{13}{16}$

Men to students: $\dfrac{13}{29}$

Students to women: $\dfrac{29}{16}$

Note the order implied by a ratio. The ratio of a to b means a/b, whereas the ratio of b to a means b/a.

Definition of Ratio

The **ratio** of the real number a to the real number b is given by

$$\frac{a}{b}.$$

The ratio of a to b is sometimes written as $a : b$.

EXAMPLE 1 ■ Finding Ratios

(a) The ratio of 7 to 5 is given by

$$\frac{7}{5}.$$

(b) The ratio of 12 to 8 is given by

$$\frac{12}{8} = \frac{3}{2}.$$

Note that the fraction $\frac{12}{8}$ can be written in reduced form as $\frac{3}{2}$. In this text we will follow the practice of writing ratios in reduced form whenever possible.

(c) The ratio of $3\frac{1}{2}$ to $5\frac{1}{4}$ is given by

$$\frac{3\frac{1}{2}}{5\frac{1}{4}} = \frac{\frac{7}{2}}{\frac{21}{4}} = \frac{7}{2} \cdot \frac{4}{21} = \frac{2}{3}.$$

■

Real-world applications of ratios are numerous. They are used to describe opinion surveys (for/against), populations (male/female, unemployed/employed, predator/prey), mixtures (oil/gasoline, water/alcohol, active/inert), unit prices (dollars/ounce), and so on.

When comparing two *measurements* by a ratio, we prefer to use the same unit of measurement in both the numerator and the denominator. For example, to find the ratio of 4 feet to 8 inches, we could convert 4 feet to 48 inches (by multiplying by 12) to obtain

$$\frac{4 \text{ feet}}{8 \text{ inches}} = \frac{48 \text{ inches}}{8 \text{ inches}} = \frac{48}{8} = \frac{6}{1}.$$

Or we could convert 8 inches to $\frac{8}{12}$ feet (by dividing by 12) to obtain

$$\frac{4 \text{ feet}}{8 \text{ inches}} = \frac{4 \text{ feet}}{\frac{8}{12} \text{ feet}} = 4 \cdot \frac{12}{8} = \frac{6}{1}.$$

The point is that if you use different units of measurement in the numerator and denominator, then you *must* include the units in the ratio. If you use the same units of measurement in the numerator and denominator, then it is not necessary to write the units.

EXAMPLE 2 ■ **Comparing Measurements**

Find a ratio to compare the relative sizes of the following. Use the same unit of measurement in the numerator and denominator.

(a) 5 gallons to 7 gallons   (b) 3 meters to 40 centimeters

(c) 200 cents to 3 dollars   (d) 30 months to $1\frac{1}{2}$ years

Solution

(a) Since the units of measurement are the same, the ratio is $\frac{5}{7}$.

(b) Since the units of measurement are different, we begin by converting meters to centimeters *or* centimeters to meters. Here, it is easier to convert meters to centimeters by multiplying by 100. (Note that 1 meter is equivalent to 100 centimeters.)

$$\frac{3 \text{ meters}}{40 \text{ centimeters}} = \frac{3(100) \text{ centimeters}}{40 \text{ centimeters}} = \frac{300}{40} = \frac{15}{2}$$

Therefore, the ratio is 15 to 2.

(c) Since 200 cents is the same as 2 dollars, the ratio is

$$\frac{200 \text{ cents}}{3 \text{ dollars}} = \frac{2 \text{ dollars}}{3 \text{ dollars}} = \frac{2}{3}.$$

Note that if we had converted 3 dollars to 300 cents, we would have obtained the same ratio since

$$\frac{200 \text{ cents}}{3 \text{ dollars}} = \frac{200 \text{ cents}}{300 \text{ cents}} = \frac{200}{300} = \frac{2}{3}.$$

(d) Since $1\frac{1}{2}$ years $= 18$ months, the ratio is

$$\frac{30 \text{ months}}{1\frac{1}{2} \text{ years}} = \frac{30 \text{ months}}{18 \text{ months}} = \frac{30}{18} = \frac{5}{3}.$$ ■

Table 4.1 gives some common conversion factors that you will need in order to solve the exercises in this section.

TABLE 4.1   Conversion Factors

Length:	Weight:	Volume:	Time:
1 foot = 12 inches	1 pound = 16 ounces	1 pint = 16 fluid ounces	1 minute = 60 seconds
1 yard = 3 feet	1 ton = 2,000 pounds	1 quart = 2 pints	1 hour = 60 minutes
1 mile = 5,280 feet	1 kilogram = 1,000 grams	1 gallon = 4 quarts	
1 meter = 100 centimeters		1 liter = 1,000 milliliters	
1 kilometer = 1,000 meters			

Unit Prices

As a consumer, it is important for you to be able to determine the **unit prices** of items you buy in order to make the best use of your money. The **unit price** of an item is given by the ratio of the total price to the total units.

$$\frac{\text{Unit}}{\text{price}} = \frac{\text{Total price}}{\text{Total units}}$$

To state unit prices, we usually use the word *per*. For instance, the unit price for a particular brand of coffee might be 4.69 dollars *per* pound, or $4.69 per pound.

EXAMPLE 3   ■   Finding Unit Prices

Find the unit price (in dollars per ounce) for each of the following items.

(a) A 12-ounce jar of strawberry jam for 99¢

(b) A 4.5-ounce tube of toothpaste for $1.89

(c) A 5-pound, 4-ounce box of detergent for $4.62

Solution

(a) *Verbal model:* $\boxed{\text{Unit price}} = \boxed{\dfrac{\text{Total price}}{\text{Total units}}}$

   Unit price: $\dfrac{99 \text{ cents}}{12 \text{ ounces}} = \dfrac{0.99 \text{ dollars}}{12 \text{ ounces}} = \$0.0825 \text{ per ounce}$

(b) *Verbal model:* $\boxed{\text{Unit price}} = \boxed{\dfrac{\text{Total price}}{\text{Total units}}}$

   Unit price: $\dfrac{\$1.89}{4.5 \text{ ounces}} = \$0.42 \text{ per ounce}$

(c) Since the weight of this item is given in pounds *and* ounces, we first convert to ounces by using the conversion factor 1 pound = 16 ounces.

$$\begin{aligned}
\text{Total units} &= 5 \text{ lb} + 4 \text{ oz} \\
&= 5(16 \text{ oz}) + 4 \text{ oz} \\
&= 80 \text{ oz} + 4 \text{ oz} \\
&= 84 \text{ oz}
\end{aligned}$$

The unit price is determined as follows.

Verbal model: $\boxed{\text{Unit price}} = \boxed{\dfrac{\text{Total price}}{\text{Total units}}}$

Unit price: $\dfrac{\$4.62}{84 \text{ ounces}} = \$0.055 \text{ per ounce}$ ∎

EXAMPLE 4 ■ Comparing Unit Prices

Which has the smaller unit price: a 12-ounce box of breakfast cereal for $2.69 or a 16-ounce box of the same cereal for $3.49?

Solution

The unit price for the smaller box is

$$\text{Unit price} = \frac{\text{Total price}}{\text{Total units}} = \frac{\$2.69}{12 \text{ ounces}} \approx \$0.224 \text{ per ounce.}$$

The unit price for the larger box is

$$\text{Unit price} = \frac{\text{Total price}}{\text{Total units}} = \frac{\$3.49}{16 \text{ ounces}} \approx \$0.218 \text{ per ounce.}$$

Thus, the larger box has a slightly smaller unit price. ∎

Solving Proportions

A **proportion** is a statement that equates two ratios. For example, if the ratio of a to b is the same as the ratio of c to d, we can write the proportion as

$$\frac{a}{b} = \frac{c}{d}.$$

In typical applications we know values for three of the letters (quantities) and we are required to find the value of the fourth. To solve such a fractional equation, we can use the *cross-multiplication* procedure introduced in Section 3.3.

If $\dfrac{a}{b} = \dfrac{c}{d}$, then $ad = bc$.

The quantities a and d are called the **extremes** of the proportion, whereas b and c are called the **means** of the proportion.

EXAMPLE 5 ■ Solving Proportions

Solve the following proportions for x.   (a) $\dfrac{50}{x} = \dfrac{2}{28}$ (b) $\dfrac{x}{3} = \dfrac{10}{6}$

Solution

(a)
$$\frac{50}{x} = \frac{2}{28} \qquad \text{Given}$$
$$50(28) = 2x \qquad \text{Cross multiply}$$
$$\frac{1,400}{2} = x \qquad \text{Divide both sides by 2}$$
$$700 = x \qquad \text{Solution}$$

Thus, the ratio of 50 to 700 is the same as the ratio of 2 to 28.

(b)
$$\frac{x}{3} = \frac{10}{6} \qquad \text{Given}$$
$$x = \frac{30}{6} \qquad \text{Multiply both sides by 3}$$
$$x = 5 \qquad \text{Solution}$$

Thus, the ratio of 5 to 3 is the same as the ratio of 10 to 6. ■

By setting up a proportion so that the unknown always occurs in the numerator, we can save a step or two in our solution process. For instance, by writing the proportion

$$\frac{50}{x} = \frac{2}{28} \quad \text{as} \quad \frac{x}{50} = \frac{28}{2}$$

we can solve for x in *one step* by multiplying both sides by 50 to get

$$x = 50 \cdot \frac{28}{2} = 700.$$

Keep this in mind as you study the following examples and do the exercises for this section.

EXAMPLE 6 ■ **Solving a Proportion**

The ratio of a certain number to 6 is the same as the ratio of 2 to 5. What is the number?

Solution

We let x represent the unknown number and write the following proportion.

$$\frac{x}{6} = \frac{2}{5}$$

We can solve this equation for x by multiplying both sides of the equation by 6 to obtain

$$x = 6 \cdot \frac{2}{5} = \frac{12}{5}.$$

Thus, the number is $\frac{12}{5}$.   ■

EXAMPLE 7 ■ **An Application: Gasoline Cost**

Suppose you are driving from New York to Arizona, a trip of 2,750 miles. You begin the trip with a full tank of gas. After traveling 416 miles, you refill the tank for $24. How much should you plan to spend on gasoline for the entire trip? (Assume the driving conditions are similar throughout the trip.)

Solution

Verbal model: $\dfrac{\text{Cost of long trip}}{\text{Miles for long trip}} = \dfrac{\text{Cost of short trip}}{\text{Miles for short trip}}$

Labels: Cost of short trip = \$24
            Miles for short trip = 416
            Cost of long trip = x (dollars)
            Miles for long trip = 2,750

Proportion: $\dfrac{x}{2,750} = \dfrac{24}{416}$

$$x = (2,750)\left(\frac{24}{416}\right)$$

$$x \approx \$158.65$$

Thus, you should plan to spend \$158.65 (or approximately \$160.00) for gasoline on the trip. Check this solution in the original statement of the problem.   ■

EXAMPLE 8   ■   An Application: Amount of Fertilizer

The recommended application for a particular type of lawn fertilizer is one 50-pound bag for 575 square feet. How many bags of this type of fertilizer would be required to fertilize 2,850 square feet of lawn?

Solution

Verbal model: $\dfrac{\text{Pounds for lawn}}{\text{Square feet of lawn}} = \dfrac{\text{Pounds per bag}}{\text{Square feet per bag}}$

Labels: Pounds for lawn = x
Square feet of lawn = 2,850
Pounds per bag = 50
Square feet per bag = 575

Proportion: $\dfrac{x}{2,850} = \dfrac{50}{575}$

$$x = (2,850)\left(\dfrac{50}{575}\right)$$

$$x \approx 247.8$$

Since the fertilizer comes in 50-pound bags, the lawn will require five bags (or 250 pounds). Check this solution in the original statement of the problem.   ■

EXAMPLE 9   ■   Using a Proportion to Calculate Taxes

You have just moved into a new house valued at $110,000. If your next-door neighbor pays $1,150 in real estate taxes each year on a house valued at $89,000, how much a year should you expect to pay in real estate taxes? (Assume that the rate is the same.)

Solution

Verbal model: $\dfrac{\text{Your taxes}}{\text{Your house value}} = \dfrac{\text{Neighbor's taxes}}{\text{Neighbor's house value}}$

Labels: Your real estate taxes = x
Your house value = $110,000
Neighbor's real estate taxes = $1,150
Neighbor's house value = $89,000

Proportion: $\dfrac{x}{110,000} = \dfrac{1,150}{89,000}$

$$x = 110,000 \cdot \dfrac{1,150}{89,000}$$

$$x \approx \$1,421$$

Thus, you should expect your real estate taxes to be about $1,421. Check this solution in the original statement of the problem.   ■

The Consumer Price Index

The rate of inflation is important to all of us. Simply stated, **inflation** is an economic condition in which the price of a fixed amount of goods or services increases. (Thus, a fixed amount of money buys less in a given year than in previous years.) The most widely used measurement of inflation in the United States is the **Consumer Price Index** (CPI), often called the **Cost-of-Living Index**. Table 4.2 shows the "All Items" or general index for the years 1950 to 1989. (*Source:* U.S. Bureau of Labor and Statistics.)

TABLE 4.2   **The Consumer Price Index**

Year	CPI	Year	CPI	Year	CPI	Year	CPI
1950	72.1	1960	88.7	1970	116.3	1980	246.8
1951	77.8	1961	89.6	1971	121.3	1981	272.4
1952	79.5	1962	90.6	1972	125.3	1982	289.1
1953	80.1	1963	91.7	1973	133.1	1983	298.4
1954	80.5	1964	92.9	1974	147.7	1984	311.1
1955	80.2	1965	94.5	1975	161.2	1985	322.2
1956	81.4	1966	97.2	1976	170.5	1986	328.4
1957	84.3	1967	100.0	1977	181.5	1987	345.7
1958	86.6	1968	104.2	1978	195.4	1988	358.1
1959	87.3	1969	109.8	1979	217.4	1989	375.4

To determine (from the CPI) the change in the buying power of a dollar from one year to another, we make use of the following proportion.

$$\frac{\text{Price year } n}{\text{Index year } n} = \frac{\text{Price year } m}{\text{Index year } m}$$

As before, three of these quantities will be known, and we are to solve for the fourth.

EXAMPLE 10 ■ Using the Consumer Price Index

Suppose you purchased a piece of jewelry in 1980 for $750. Using the Consumer Price Index, what would you expect the replacement value of the jewelry to have been in 1988?

Solution

Verbal model: $\dfrac{\text{Price in 1988}}{\text{Index in 1988}} = \dfrac{\text{Price in 1980}}{\text{Index in 1980}}$

Labels: Price in 1988 = x (dollars)
Index in 1988 = 358.1
Price in 1980 = $750
Index in 1980 = 246.8

Proportion: $\dfrac{x}{358.1} = \dfrac{750}{246.8}$

$$x = 358.1 \cdot \dfrac{750}{246.8}$$

$$x \approx \$1{,}088$$

Thus, you should expect the replacement value of the jewelry to have been approximately \$1,088 in 1988. Check this solution in the original statement of the problem. ■

DISCUSSION PROBLEM ■   The Value of π

One of the best known ratios in mathematics is denoted by the Greek letter π, pronounced "pie." This number represents the ratio of the circumference of *any* circle to its diameter. To estimate the value of π, try the following experiment. Measure the diameter of a circular cylinder (such as a soda can), and then measure the circumference of the cylinder, as shown in Figure 4.1. Then use your calculator to approximate the value of π as follows.

$$\pi = \dfrac{\text{Circumference}}{\text{Diameter}}$$

FIGURE 4.1

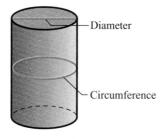

Diameter

Circumference

■

In Exercises 1–4, write the fraction in reduced form.

1. $\dfrac{15}{25}$ **2.** $\dfrac{36}{42}$ **3.** $\dfrac{33}{123}$ **4.** $\dfrac{28}{42}$

In Exercises 5–10, solve the given equation.

5. $\dfrac{x}{3} = \dfrac{4}{9}$ **6.** $\dfrac{t}{16} = \dfrac{1}{4}$ **7.** $\dfrac{n}{5} = \dfrac{8}{25}$ **8.** $\dfrac{x}{2} = \dfrac{15}{4}$

9. $\dfrac{6}{x} = \dfrac{3}{5}$ **10.** $\dfrac{5}{8} = \dfrac{4}{x}$

4.1 EXERCISES

In Exercises 1–8, express the ratio as a fraction in reduced form. (Use the same units of measure for both quantities.)

1. Thirty-six inches to 24 inches

2. Twenty-four pounds to 30 pounds

3. One quart to one gallon

4. Three inches to two feet

5. Seventy-five centimeters to two meters

6. Sixty milliliters to one liter

7. Ninety minutes to two hours

8. Three thousand pounds to five tons

In Exercises 9–14, express each statement as a ratio in reduced form. (Use the same units of measure for both quantities.)

9. *Study Hours* Suppose you study six hours per day and are in class three hours per day. Find the ratio of the number of study hours to class hours.

10. *Income Tax* You have $10 of state income tax withheld from your paycheck per week when your gross pay is $500. Find the ratio of tax to gross pay.

11. *Price-Earnings Ratio* The ratio of the price of a stock to its earnings is called the **price-earnings ratio**. A certain stock sells for $78 per share and earns $6.50 per share. What is the price-earnings ratio of this stock?

12. *Compression Ratio* The **compression ratio** of an engine is the ratio of the expanded volume of gas in one of its cylinders to the compressed volume of gas in the cylinder (see figure). A cylinder in a certain engine has an expanded volume of 345 cubic centimeters and a compressed volume of 17.25 cubic centimeters. What is the compression ratio of this engine?

Figure for 12

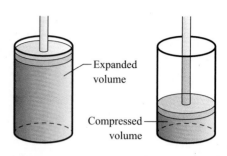

Expanded volume

Compressed volume

13. *Gear Ratio* The **gear ratio** of two gears is the number of teeth in one gear to the number of teeth in a second gear. One gear in a gear box has 45 teeth, and the other has 30 teeth (see figure). Find the gear ratio of the larger gear to the smaller gear.

30 teeth   45 teeth

14. *Specific Gravity* The **specific gravity** of a substance is the ratio of its weight to the weight of the same volume of water. Find the specific gravity of kerosene if kerosene weighs 0.82 grams per cubic centimeter and water weighs 1 gram per cubic centimeter.

In Exercises 15–18, find the unit price (in dollars per ounce) of the product.

15. A 20-ounce can of pineapple for 89¢

16. An 18-ounce box of cereal for $1.99

17. A 1-pound, 4-ounce loaf of bread for $1.29

18. A 1-pound package of cheese for $2.89

In Exercises 19 and 20, determine which product has the smaller unit price.

19. A $27\frac{1}{2}$-ounce can of spaghetti sauce for \$1.29, or a 32-ounce jar for \$1.45

20. A 16-ounce package of margarine for \$1.29, or a 3-pound tub for \$3.29

In Exercises 21–36, solve the given proportion.

21. $\dfrac{3}{5} = \dfrac{y}{20}$

22. $\dfrac{x}{9} = \dfrac{5}{18}$

23. $\dfrac{t}{4} = \dfrac{25}{2}$

24. $\dfrac{y}{25} = \dfrac{12}{10}$

25. $\dfrac{x}{5} = \dfrac{2}{3}$

26. $\dfrac{z}{35} = \dfrac{5}{14}$

27. $\dfrac{3}{8} = \dfrac{6}{t}$

28. $\dfrac{12}{7} = \dfrac{6}{x}$

29. $\dfrac{x+1}{5} = \dfrac{3}{10}$

30. $\dfrac{z-3}{8} = \dfrac{3}{16}$

31. $\dfrac{x+6}{3} = \dfrac{x-5}{2}$

32. $\dfrac{x-2}{4} = \dfrac{x+10}{10}$

33. $\dfrac{8}{x+2} = \dfrac{3}{x-1}$

34. $\dfrac{5}{x-4} = \dfrac{6}{x}$

35. $\dfrac{0.5}{0.8} = \dfrac{n}{0.3}$

36. $\dfrac{4.5}{2} = \dfrac{0.5}{t}$

37. *Fuel Efficiency* A car uses 20 gallons of gasoline for a trip of 360 miles. How many gallons would be used on a trip of 400 miles? (Assume there is no change in the fuel efficiency.)

38. *Amount of Fuel* A tractor requires 4 gallons of diesel fuel to plow for 90 minutes. How many gallons of fuel would be required to plow for 8 hours?

39. *Building Material* One hundred cement blocks are needed to build a 16-foot wall. How many blocks are needed to build a 40-foot wall?

40. *Force on a Spring* A force of 50 pounds stretches a spring 4 inches. How many pounds of force are required to stretch the spring 6 inches?

41. *Amount of Gasoline* The gasoline to oil ratio for a two-cycle engine is 30 to 1. How much gasoline is required to produce a mixture that contains one-half pint of oil?

42. *Enlarging a Recipe* Two cups of flour are required to make one batch of cookies. How many cups are required for $2\frac{1}{2}$ batches?

43. *Real Estate Taxes* The tax on a property with an assessed value of \$65,000 is \$825. Find the tax on a property with an assessed value of \$90,000.

44. *Real Estate Taxes* The tax on a property with an assessed value of \$65,000 is \$1,100. Find the tax on a property with an assessed value of \$90,000.

45. *Polling Results* In a public opinion poll, 624 people from a sample of 1,100 indicated that they would vote for a specific candidate. Assuming this poll to be a correct indicator of the electorate, how many votes can the candidate expect to receive from 40,000 votes cast?

46. *Defective Units* A quality control engineer for a certain buyer found 2 defective units in a sample of 50. At this rate, what is the expected number of defective units in a shipment of 10,000 units?

47. *Map Scale* The scale represents 100 miles on the accompanying map. Approximate the distance between Philadelphia and Pittsburgh.

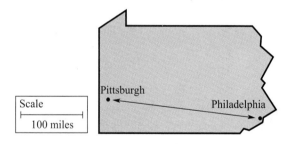

48. *Map Scale* If $1\frac{1}{2}$ inches represents 40 miles on a map, approximate the distance between two cities that are 4 inches apart on the map.

49. *Pumping Time* A pump can fill a 750-gallon tank in 35 minutes. How long will it take to fill a 1,000-gallon tank with this pump?

50. *Pounds of Sand* The ratio of cement to sand in an 80-pound bag of dry mix is 1 to 4. Find the number of pounds of sand in the bag. (*Note:* Dry mix is composed of only two ingredients—cement and sand.)

Similar Triangles In Exercises 51–54, solve for the length *x* of the side of a triangle. (Assume that the two triangles are similar, and use the fact that corresponding sides of similar triangles are proportional.)

51.

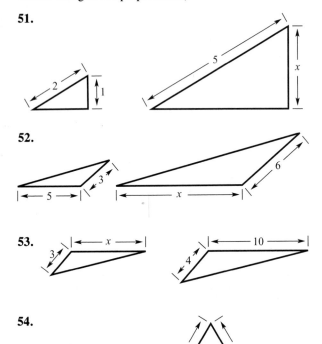

52.

53.

54.

55. *Shadow Length* Find the length of the shadow of a person who is 6 feet tall and is standing 10 feet from a streetlight that has a height of 15 feet. (See figure.) (*Note:* The two triangles in the figure are similar and hence, their corresponding sides are proportional.)

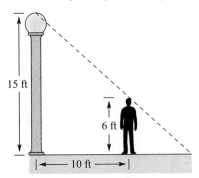

56. *Shadow Length* A man who is 6 feet tall walks directly toward the tip of the shadow of a tree. When the man is 100 feet from the tree, he starts forming his own shadow beyond the shadow of the tree. The length of the shadow of the tree beyond this point is 8 feet. Find the height of the tree. (See note in Exercise 55.)

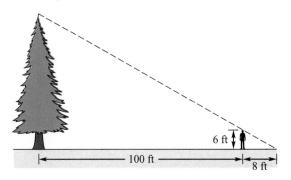

Consumer Price Index In Exercises 57–60, estimate the price of the given item for the specified year by using the Consumer Price Index found in Table 4.2.

57. The 1988 price of a lawn tractor that cost $2,875 in 1978

58. The 1988 price of a watch that cost $58 in 1973

59. The 1960 price of a gallon of milk that cost $2.07 in 1988

60. The 1950 price of a coat that cost $225 in 1985

61. *Area of a Circle* Find the ratio of the area of the larger pizza to the smaller pizza in the accompanying figure. (*Note:* The area of a circle is $A = \pi r^2$. The Greek letter π is used in mathematics to represent the ratio of the circumference of a circle to its diameter. The value of π is approximately 3.14.)

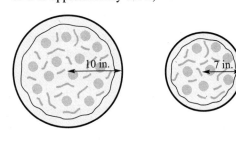

Percents and the Percent Equation

Percents • The Percent Equation • Applications

Percents

Numbers that describe rates, increases, decreases, and discounts are often given as percents, denoted by the symbol %. **Percent** means *per hundred*, or *parts of 100*. (The Latin word for 100 is *centum*.) For example, 30% means 30 parts of 100, which is equivalent to the fraction $\frac{30}{100}$ or $\frac{3}{10}$.

In applications involving percents, you must usually convert the percent number to decimal (or fraction) form before performing any arithmetic operations. Consequently, you need to be able to convert from percents to decimals (or fractions) and conversely. This can be accomplished by using the following relationship.

$$\boxed{\text{Decimal or fraction}} \cdot \boxed{100\%} = \boxed{\text{Percent}}$$

For example, the decimal 0.38 corresponds to the percent

$$0.38(100\%) = 38\%.$$

EXAMPLE 1   ■   **Converting Decimals and Fractions to Percents**

Convert the following numbers to percents.

(a) $\frac{3}{5}$ (b) 1.2

Solution

(a) *Verbal model:* $\boxed{\text{Fraction}} \cdot \boxed{100\%} = \boxed{\text{Percent}}$

   Equation: $\frac{3}{5}(100\%) = \frac{300}{5}\% = 60\%$

   Thus, the fraction $\frac{3}{5}$ corresponds to 60%.

(b) *Verbal model:* $\boxed{\text{Decimal}} \cdot \boxed{100\%} = \boxed{\text{Percent}}$

   Equation: $(1.2)(100\%) = 120\%$

   Thus, the decimal 1.2 corresponds to 120%.   ■

Note in Example 1(b) that it is possible to have percents that are larger than 100%.

EXAMPLE 2 ■ Converting Percents to Decimals and Fractions

(a) Convert 3.5% to a decimal.

(b) Convert 55% to a fraction.

Solution

(a) *Verbal model:* | Decimal | · | 100% | = | Percent |

Label: x = decimal

Equation: $x(100\%) = 3.5\%$

$$x = \frac{3.5\%}{100\%}$$

$$x = 0.035$$

Thus, 3.5% corresponds to the decimal 0.035.

(b) *Verbal model:* | Fraction | · | 100% | = | Percent |

Label: x = fraction

Equation: $x(100\%) = 55\%$

$$x = \frac{55\%}{100\%}$$

$$x = \frac{11}{20}$$

Thus, 55% corresponds to the fraction $\frac{11}{20}$. ■

Some percents occur so commonly that it is helpful to memorize them. Table 4.3 shows the decimal and fraction conversions for several common percents.

TABLE 4.3 **Common Percent Conversions**

Percent	10%	$12\frac{1}{2}\%$	20%	25%	$33\frac{1}{3}\%$	50%	$66\frac{2}{3}\%$	75%
Decimal	0.1	0.125	0.2	0.25	0.33 . . .	0.5	0.66 . . .	0.75
Fraction	$\frac{1}{10}$	$\frac{1}{8}$	$\frac{1}{5}$	$\frac{1}{4}$	$\frac{1}{3}$	$\frac{1}{2}$	$\frac{2}{3}$	$\frac{3}{4}$

The Percent Equation

The primary use of percents is to compare two numbers. For example, we say 2 is 50% of 4, 5 is 25% of 20, and 150% of 10 is 15. The following model is helpful.

Verbal model: a = p percent of b

Labels: b = base number
                   p = percent (in decimal form)
                   a = number being compared to b

Percent equation: $a = p \cdot b$

Remember to convert p to a decimal value before multiplying by b.

EXAMPLE 3 ■   **Solving Percent Equations**

Solve the following problems.

(a) What is 30% of 70?

(b) Fourteen is 25% of what?

(c) One hundred thirty-five is what percent of 27?

Solution

(a) *Verbal model:*   What number =   30% of 70

    Label: a = unknown number

    Percent equation: $a = (0.3)(70) = 21$

Therefore, 21 is 30% of 70.

(b) *Verbal model:*   14 =   25% of what number

    Label: b = unknown number

    Percent equation: $14 = 0.25b$

$$\frac{14}{0.25} = b$$

$$56 = b$$

Therefore, 14 is 25% of 56.

(c) *Verbal model:*   135 =   What percent of 27

    Label: p = unknown percent (in decimal form)

    Percent equation: $135 = p(27)$

$$\frac{135}{27} = p$$

$$5 = p$$

Therefore, 135 is 500% of 27.   ■

EXAMPLE 4   ■   Solving Percent Equations

Solve the following problems.

(a) $761.25 is $14\frac{1}{2}\%$ of what?   (b) 19 is what percent of 95?

Solution

(a) *Verbal model:* $761.25 = $14\frac{1}{2}\%$ of what number

    Label: b = unknown number (in dollars)

    Percent equation: $761.25 = 0.145b$

    $$\frac{761.25}{0.145} = b$$

    $$5{,}250 = b$$

Therefore, $761.25 is $14\frac{1}{2}\%$ of $5,250. Check this by multiplying 0.145 by 5,250 to obtain 761.25.

(b) *Verbal model:*   19 = What percent of 95

    Label: p = unknown percent (in decimal form)

    Percent equation: $19 = p(95)$

    $$\frac{19}{95} = p$$

    $$0.2 = p$$

Therefore, 19 is 20% of 95.   ■

Applications

In most real-life applications, the base number b and the number a are much more disguised than they are in Examples 3 and 4. It sometimes helps to think of a as a "new" amount and b as the "original" amount. Try using these suggestions in the examples and exercises that follow.

EXAMPLE 5   ■   An Application: Real Estate Commission

A real estate agency receives a commission of $5,167.50 for the sale of a $79,500 house. What percent commission is this?

Solution

Verbal model: Commission = Percent of sale price

Labels: a = commission = $5,167.50
                   p = unknown percent (in decimal form)
                   b = sale price = $79,500

Equation: $5{,}167.50 = p \cdot (79{,}500)$

$$\frac{5{,}167.50}{79{,}500} = p$$

$$0.065 = p$$

Therefore, the real estate agency receives a commission of 6.5%. Check this solution in the original statement of the problem.   ■

EXAMPLE 6   ■   **An Application: Cost-of-Living Raise**

A union negotiates for a cost-of-living raise of 7%. Find the amount of the raise for a union member whose salary is $17,240. What is the new salary for this person?

Solution

Verbal model: Raise = Percent of salary

Labels: a = raise (dollars)
p = 0.07 (percent in decimal form)
b = salary = $17,240

Equation: $a = p \cdot b$
$a = 0.07(17{,}240) = 1{,}206.80$

Therefore, the raise amounts to $1,206.80 and the new salary for the person is

New salary = $17{,}240.00 + 1{,}206.80 = \$18{,}446.80$.

Check this solution in the original statement of the problem.   ■

EXAMPLE 7   ■   **An Application: Course Grade**

Suppose you missed getting an A in your chemistry course by only three points. If your point total for the course is 402, how many points were possible in the course? (Assume that you needed 90% of the course total for an A.)

Solution

Verbal model: Your points + 3 = 90% of total points

Labels: a = your points + 3 = 402 + 3 = 405
p = 0.9 (percent in decimal form)
b = total points for course

Equation: $a = p \cdot b$

$405 = 0.9b$

$$\frac{405}{0.9} = b$$

$450 = b$

Therefore, there were 450 total points for the course. Check this solution in the original statement of the problem.   ■

EXAMPLE 8   ■   An Application: Seating Capacity

The seating capacity of the university football stadium was increased from 68,000 to 78,500.

(a) What percent increase in seating capacity does this represent?

(b) The old capacity is what percent of the new capacity?

Solution

(a) *Verbal model:* | Increase | = | Percent of original capacity |

   Labels: a = amount of increase = 78,500 − 68,000 = 10,500
   p = unknown percent (in decimal form)
   b = original capacity = 68,000

   Equation: $a = p \cdot b$
   $$10,500 = p \cdot 68,000$$
   $$\frac{10,500}{68,000} = p$$
   $$0.154 \approx p$$

Therefore, the increase in seating capacity is approximately 15.4%.

(b) *Verbal model:* | Old capacity | = | What percent of new capacity |

   Labels: a = old capacity = 68,000
   p = unknown percent (in decimal form)
   b = new capacity = 78,500

   Equation: $a = p \cdot b$
   $$68,000 = p(78,500)$$
   $$\frac{68,000}{78,500} = p$$
   $$0.866 \approx p$$

Therefore, the old capacity is approximately 86.6% of the new capacity.   ■

Starting with Section 3.1, we have been demonstrating the problem-solving style used in this text. As a summary of what we have been doing and for future reference in work with applied problems, we recommend the following approach to solving word problems.

Guidelines for Solving Word Problems	1. Search for the *hidden equality*—the two essential expressions said to be equal or known to be equal. (A sketch may be helpful.)
	2. Write a *verbal model* that equates these two essential expressions.
	3. Assign *labels* to fixed quantities and the variable quantities.
	4. Rewrite the verbal model as an *algebraic equation* using the assigned labels.
	5. *Solve* the algebraic equation.
	6. *Check* to see that your solution satisfies the word problem as stated.

DISCUSSION PROBLEM ■   The Latin Word *Centum*

The word *percent* comes from the Latin word *centum*, which means one hundred. How many other English words can you think of that are related to *centum*? How is each of the words related to the number 100? ■

Warm-Up

The following warm-up exercises involve skills that were covered in earlier sections. You will use these skills in the exercise set for this section.

In Exercises 1–10, solve the given equation.

1. $-10x = 1,000$

2. $15x = 60$

3. $-\dfrac{x}{10} = 1,000$

4. $\dfrac{x}{15} = 60$

5. $-\dfrac{10}{x} = 1,000$

6. $\dfrac{15}{x} = 60$

7. $0.35x = 70$

8. $0.60x = 24$

9. $125(1 - r) = 100$

10. $3,050(1 - r) = 1,830$

4.2 EXERCISES

In Exercises 1–10, complete the table showing the equivalent forms of a percent: the number of parts out of 100, a decimal, and a fraction (in reduced form).

1.

Percent	Parts out of 100	Decimal	Fraction
40%			

2.

Percent	Parts out of 100	Decimal	Fraction
15%			

3.

Percent	Parts out of 100	Decimal	Fraction
7.5%			

4.

Percent	Parts out of 100	Decimal	Fraction
75%			

5.

Percent	Parts out of 100	Decimal	Fraction
	63		

6.

Percent	Parts out of 100	Decimal	Fraction
	10.5		

7.

Percent	Parts out of 100	Decimal	Fraction
		0.155	

8.

Percent	Parts out of 100	Decimal	Fraction
		0.8	

9.

Percent	Parts out of 100	Decimal	Fraction
			$\frac{3}{5}$

10.

Percent	Parts out of 100	Decimal	Fraction
			$\frac{9}{20}$

In Exercises 11–14, change the percent to a decimal.

11. 12.5% **12.** 95% **13.** 250% **14.** 0.3%

In Exercises 15–18, change the decimal to a percent.

15. 0.075 **16.** 0.57 **17.** 0.62 **18.** 1.75

In Exercises 19–22, change the fraction to a percent.

19. $\frac{4}{5}$ **20.** $\frac{5}{4}$ **21.** $\frac{7}{20}$ **22.** $\frac{2}{3}$

In Exercises 23–26, determine what percent of the entire figure is shaded.

23.

| | $\frac{1}{4}$ | $\frac{1}{4}$ | $\frac{1}{4}$ | $\frac{1}{4}$ |

24.

| | $\frac{1}{3}$ | $\frac{1}{3}$ | $\frac{1}{3}$ |

25.

150°

26.

60° 60° 60° 60° 60° 60°

27. What is 30% of 150?

28. What is 62% of 1,200?

29. What is 9.5% of 816?

30. What is 32% of 516?

31. What is $\frac{3}{4}$% of 56?

32. What is 0.2% of 100,000?

33. What is 200% of 88?

34. What is 325% of 450?

35. 43% of what number is 903?

36. 85% of what number is 425?

37. $12\frac{1}{2}$% of what number is 275?

38. $37\frac{1}{2}$% of what number is 813?

39. 450% of what number is 594?

40. 250% of what number is 210?

41. 0.6% of what number is 2.16?

42. 0.08% of what number is 51.2?

43. 576 is what percent of 800?

44. 1,950 is what percent of 5,000?

45. 45 is what percent of 360?

46. 148.8 is what percent of 960?

47. 38 is what percent of 5,700?

48. 22 is what percent of 800?

49. 1,000 is what percent of 200?

50. 110 is what percent of 110?

51. *Rent Payment* Suppose you spend 17% of your monthly income of $2,500 for rent. What is your monthly rent payment?

52. *College Freshmen* Thirty-five percent of the students enrolled in a college are freshmen. The enrollment of the college is 2,800. Find the number of freshmen.

53. *Snowfall* During a given winter, there were 120 inches of snow. Of that, 86 inches fell in December. What percent of the snow fell in December?

54. *Layoff* Because of slumping sales, a small company laid off 30 of its 153 employees. What percentage of the work force was laid off?

55. *Eligible Voters* The news media reported that 6,432 votes were cast in the last election and that this rep- resented 63% of the eligible voters of a district. How many eligible voters are in the district?

56. *Defective Parts* A quality control engineer reported that 2.5% of a sample of parts were defective. Find the size of the sample if the engineer detected two defective parts.

57. *Price* A new van costs approximately 110% of what it was three years ago. The current price is $22,850. What was the approximate price three years ago?

58. *Membership Drive* Because of a membership drive for a public television station, the current membership is 125% of what it was a year ago. The current number is 7,815. How many members did the station have last year?

59. *Course Grade* Suppose you missed getting a B in your mathematics course by 6 points. If your point total for the course is 394, how many points were possible in the course? (Assume that you needed 80% of the course total for a B.)

60. *Target Size* A circular target is attached to a rectangular board, as shown in the accompanying figure. The radius of the circle is $4\frac{1}{2}$ inches, and the dimensions of the board are 12 inches by 15 inches. What percentage of the board is covered by the target? (The area of a circle is $A = \pi r^2$, where r is the radius of the circle.)

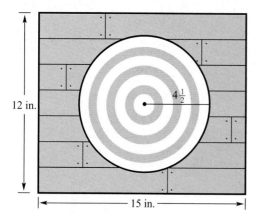

61. *Revenue* The total 1989 revenue for the Oakland A's baseball team is given in the accompanying pie chart. (*Source:* Oakland A's Baseball Team.) Find the percentage that each of the four sources of revenue is to the total revenue.

Revenue for the Oakland A's

Attendance **$21,340,000**

Advertising, royalties, promotions, and other **$6,095,000**

Concessions **$5,910,000**

TV and radio **$17,200,000**

62. *Expenses* The total 1989 expenses for the Oakland A's baseball team is given in the accompanying pie chart. (*Source:* Oakland A's Baseball Team.) Find the percentage that each of the seven categories of expense is to the total expense. Use the total revenue from Exercise 61 to find the profit the Oakland A's made in 1989.

Expenses for the Oakland A's

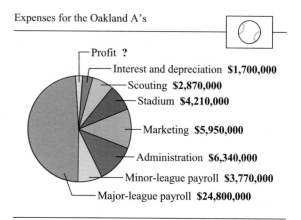

Profit **?**

Interest and depreciation **$1,700,000**

Scouting **$2,870,000**

Stadium **$4,210,000**

Marketing **$5,950,000**

Administration **$6,340,000**

Minor-league payroll **$3,770,000**

Major-league payroll **$24,800,000**

63. *Rearranging Furniture* Three hundred people were surveyed and reported that they rearrange furniture in their homes for several different reasons (see figure). How many people in the survey said that they rearranged furniture for each of the given reasons? (*Source:* Southwestern Bell.)

Why We Rearrange Furniture

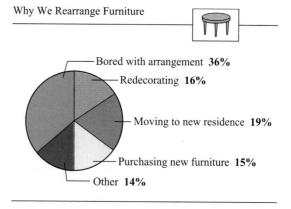

Bored with arrangement **36%**

Redecorating **16%**

Moving to new residence **19%**

Purchasing new furniture **15%**

Other **14%**

64. *Scientists* The accompanying table shows the number of women scientists in the United States in 1978 and 1988. (*Source:* U.S. Bureau of Labor and Statistics.)

(a) Use the table to find the total number of computer scientists (men and women) in 1988.

(b) Use the table to find the total number of physical scientists in 1978.

Women Scientists in the United States

	1978		1988	
Field	Number	%	Number	%
Computer Science	40,200	23%	218,700	31%
Biology	30,000	18%	89,200	30%
Mathematics	13,100	24%	44,900	27%
Physical Science	18,500	9%	46,500	15%
Environmental Science	7,200	10%	12,300	11%

Photo Enlarged In Exercises 65 and 66, Figure (a) was put into a photocopier and reduced or enlarged to produce Figure (b). Estimate the percentage of reduction or enlargement.

65. (a)   (b)

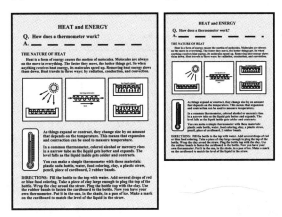

66. (a)   (b)

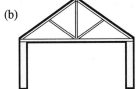

Business and Consumer Problems
Introduction • Markup and Discounts • Bills, Charges, and Payments • Wages, Salaries, and Commissions

Introduction

So far in the text our applications of linear equations have been limited to either proportion problems or problems that fit the percent equation model. In this section we expand our applications of linear equations to include common consumer problems involving sales tax, repair bills, discounts, markup, salaries, admission fees, car rentals, and more. Our goal is for you to gain confidence in setting up and solving such problems by discovering how much alike these problems are. Many of the applications in this section have the algebraic structure given in the following verbal model.

$$\boxed{\text{Total amount}} = \boxed{\text{Base amount}} + \boxed{\text{Percent}} \cdot \boxed{\text{Base amount}}$$

The variable term in the preceding sum is often a *hidden product* like those referred to in Section 2.4. As an example, consider the total cost of a taxable item with a list price of d dollars. It fits the following verbal model.

$$\boxed{\text{Cost of taxable item}} = \boxed{\text{List price}} + \boxed{\text{Tax}}$$

$$= \boxed{d \text{ dollars}} + \boxed{\text{Percent}} \cdot \boxed{d \text{ dollars}}$$

Markup and Discounts

You may have had the experience of buying an item at one store and later finding that you could have paid less for the same item at another store. The basic reason for this price difference is **markup**, which is the difference between the **cost** (the amount a retailer pays for the item) and the **price** (the amount for which the retailer sells the item to the consumer). A verbal model for this problem is as follows.

$$\boxed{\text{Selling price}} = \boxed{\text{Cost}} + \boxed{\text{Markup}}$$

In such a problem, the markup may be known or it may be expressed as a percent of the cost. This percent is called the **markup rate**.

$$\boxed{\text{Markup}} = \boxed{\text{Markup rate}} \cdot \boxed{\text{Cost}}$$

EXAMPLE 1 ■ Finding Selling Price and Cost

A sporting goods store uses a markup rate of 55% on all items.

(a) The cost of a golf bag is $35. What is the selling price of the bag?

(b) The selling price of a pair of ski boots is $98. What is the cost of the boots?

Solution

(a) *Verbal model:* $\boxed{\text{Selling price}} = \boxed{\text{Cost}} + \boxed{\text{Markup}}$

Labels: Selling price $= x$ (dollars)
Cost $= \$35$
Markup rate $= 0.55$ (percent in decimal form)
Markup $= (0.55)(35)$ (dollars)

Equation: $x = 35 + (0.55)(35)$
$= 35 + 19.25$
$= \$54.25$

Therefore, the selling price is $54.25. Check this solution in the original statement of the problem.

(b) *Verbal model:* $\boxed{\text{Selling price}} = \boxed{\text{Cost}} + \boxed{\text{Markup}}$

Labels: Selling price = $98
                    Cost = x (dollars)
                    Markup rate = 0.55 (percent in decimal form)
                    Markup = 0.55x (dollars)

Equation: $98 = x + 0.55x$

                    $98 = 1.55x$

                    $\dfrac{98}{1.55} = x$

                    $\$63.23 \approx x$

Therefore, the cost is $63.23. Check this solution in the original statement of the problem.   ∎

EXAMPLE 2   ■   Finding the Markup Rate

A shoe store sells a pair of shoes for $60. The cost of the shoes is $24. What is the markup rate?

Solution

Verbal model: | Price | = | Cost | + | Markup |

Labels: Selling price = $60
                    Cost = $24
                    Markup rate = p (percent in decimal form)
                    Markup = $p(24)$ (dollars)

Equation: $60 = 24 + p(24)$

                    $60 - 24 = (p)(24)$

                    $36 = (p)(24)$

                    $\dfrac{36}{24} = p$

                    $1.5 = p$

Thus, since $p = 1.5$, it follows that the markup rate is 150%. (Remember that when we convert the decimal form $p = 1.5$ to percent form, we obtain 150%.) Check this solution in the original statement of the problem.   ∎

As consumers, we all like a bargain! We often delay buying an item until it goes on sale—that is, until it is offered or sold at a discounted price. The model for this situation is

| Sale price | = | List price | − | Discount |

where the **discount** is given in dollars, and the **discount rate** is given as a percent of the list price.

| Discount | = | Discount rate | · | List price |

Keep these models in mind as you study the following examples.

EXAMPLE 3 ■ **Finding the Discount Rate**

During a midsummer sale, a lawn mower listed at $199.95 is on sale for $139.95. What is the discount rate?

Solution

Verbal model: $\boxed{\text{Discount}}$ = $\boxed{\text{Discount rate}}$ · $\boxed{\text{List price}}$

Labels: Discount = 199.95 − 139.95 = $60
List price = $199.95
Discount rate = p (percent in decimal form)

Equation: $60 = p(199.95)$

$$\frac{60}{199.95} = p$$

$$0.30 \approx p$$

Thus, the discount rate is 30%. Check this solution in the original statement of the problem. ■

EXAMPLE 4 ■ **Finding the Sale Price**

A drugstore advertises 40% off on all summer tanning products. A bottle of suntan oil lists for $3.49. What is the sale price?

Solution

Verbal model: $\boxed{\text{Sale price}}$ = $\boxed{\text{List price}}$ − $\boxed{\text{Discount}}$

Labels: List price = $3.49
Discount rate = 0.4 (percent in decimal form)
Discount = 0.4(3.49) (dollars)
Sale price = x (dollars)

Equation: $x = 3.49 - (0.4)(3.49)$

$$\approx 3.49 - 1.40$$

$$= \$2.09$$

Thus, the sale price is $2.09. Check this solution in the original statement of the problem. ■

Bills, Charges, and Payments

As consumers, we pay a wide variety of fees, bills, and charges that include descriptive statements such as "parts plus labor," "daily rate plus mileage," "food plus tips," or "down payment plus so much a month." Study the following examples to see how they fit the basic verbal model introduced in this section.

EXAMPLE 5 ■ **Finding the Hours of Labor**

An auto repair bill of $250 lists $110 for parts and the rest for labor. The cost of labor is $28 per hour. How many hours of labor did it take to repair the auto?

Solution

Verbal model: | Total bill | = | Price of parts | + | Price of labor |

Labels: Total bill = $250
Price of parts = $110
Number of hours of labor = x
Hourly rate for labor = 28 (dollars per hour)
Price of labor = $28x$ (dollars)

Equation: $250 = 110 + 28x$

$140 = 28x$

$\dfrac{140}{28} = x$

$5 = x$

Thus, it took five hours (of labor) to repair the auto. Check this solution in the original statement of the problem.   ■

Total amount = Base + Percent · Base
                    amount   amount

EXAMPLE 6 ■ **Finding the Tip Rate**

For dinner at a restaurant, a customer left $18.50 for a meal that cost $15.95. What percent is the tip?

Solution

Verbal model: | Tip | = | Percent | · | Price of meal |

Labels: Tip = $18.50 - 15.95 = \$2.55$
Percent = p (in decimal form)
Price of meal = $15.95

Equation: $2.55 = p(15.95)$

$\dfrac{2.55}{15.95} = p$

$0.16 \approx p$

Thus, the customer left a tip that amounted to approximately 16% of the price of the meal. Check this solution in the original statement of the problem.   ■

EXAMPLE 7   ■   Finding the Cost of a Telephone Call

Suppose you made a 12-minute call from Denver to Atlanta. The call cost $2.05 for the first three minutes and 34¢ for each additional minute.

(a) How much did the call cost?

(b) Suppose the evening rates are 35% less than the daytime rates. How much would it have cost to make the same call during the evening?

Solution

(a) *Verbal model:*   Total price = Cost of first 3 minutes + Cost of additional 9 minutes

   Labels: Total cost = c (dollars)
                 Cost of first 3 minutes = $2.05
                 Cost for additional 9 minutes = 9(0.34) (dollars)

   Equation: $c = 2.05 + (0.34)9$
                  $c = 2.05 + 3.06$
                  $c = \$5.11$

   Thus, the total cost of the call was $5.11. Check this solution in the original statement of the problem.

(b) *Verbal model:*   Evening cost = Daytime cost − Discount

   Labels: Evening cost = x (dollars)
                 Daytime cost = $5.11
                 Discount = (0.35)(5.11) (dollars)

   Equation: $x = 5.11 - (0.35)(5.11)$
                  $x \approx \$3.32$

   Thus, the call would have cost $3.32 had it been made in the evening. Check this solution in the original statement of the problem.   ■

EXAMPLE 8   ■   Finding the Monthly Payment

Including finance charges, the total purchase price of a large-screen television is $1,596. Of this amount, $120 is to be given as a down payment and the remainder is to be paid in 12 equal monthly payments. How much is each monthly payment?

Solution

Verbal model: Total price = Down payment + 12 · Monthly payment

Labels: Total price = $1,596
Down payment = $120
Monthly payment = x (dollars)

Equation: $1,596 = 120 + 12x$

$1,476 = 12x$

$\dfrac{1,476}{12} = x$

$\$123 = x$

Thus, each monthly payment is $123. Check this solution in the original statement of the problem. ■

Wages, Salaries, and Commissions

In the last two examples in this section, we look at applications involving wages, salaries, and commissions. Here again the applications fit the basic verbal model given at the beginning of this section.

EXAMPLE 9 ■ Finding Hours of Overtime

An employee's regular hourly rate is $9.50, and the overtime hourly rate is $14.25 (for each hour over 40 hours in a week). During a given week the employee earned $551. How many hours of overtime did the employee work?

Solution

Verbal model: | Total pay | = | Regular pay | + | Overtime pay |

Labels: Total pay = $551
Regular pay = (9.50)(40) (dollars)
Number of hours of overtime = x
Overtime pay = $14.25x$ (dollars)

Equation: $551 = (9.50)(40) + 14.25x$

$551 - 380 = 14.25x$

$171 = 14.25x$

$\dfrac{171}{14.25} = x$

$12 = x$

Therefore, the employee worked 12 hours of overtime during the week. Check this solution in the original statement of the problem. ■

EXAMPLE 10 ■   **Finding Commission Rate**

A sales representative receives a weekly salary of $250 plus a commission on the total weekly sales. For a given week, the sales representative's total income was $530, and the total sales were $8,000. Find the commission rate paid to the sales representative.

Solution

Verbal model: | Total income | = | Salary | + | Commission |
| --- | --- | --- | --- | --- |

Labels: Total income = $530
Salary = $250
Commission rate = p (percent in decimal form)
Commission = $p(8,000)$ (dollars)

Equation: $530 = 250 + p(8,000)$

$280 = 8,000p$

$$\frac{280}{8,000} = p$$

$0.035 = p$

Thus, the commission rate is 3.5%. Check this solution in the original statement of the problem.   ■

DISCUSSION PROBLEM ■   **Comparison of Markup Rates**

Suppose you buy an item that has a retail price of $10. The markup on the item is $5. The seller claims that the markup rate is only 50% (this is called the markup rate on sales price). You claim that the markup rate is 100% (this is called the markup rate on cost). Write a short paragraph describing how each of these two types of markup rates is calculated.   ■

Warm-Up

The following warm-up exercises involve skills that were covered in earlier sections. You will use these skills in the exercise set for this section.

In Exercises 1–4, solve the given equation.

1. $3x - 42 = 0$ **2.** $64 - 16x = 0$

3. $2 - 3x = 14 + x$ **4.** $7 + 5x = 7x - 1$

In Exercises 5–10, solve the given percent problem.

5. What is 62% of 25?   **6.** What is $\frac{1}{2}$% of 6,000?

7. 300 is what percent of 150?   **8.** 600 is what percent of 900?

9. 145.6 is 32% of what number?   **10.** 2 is 0.8% of what number?

4.3   EXERCISES*

In Exercises 1–10, find the missing quantities. (Assume the markup rate is a percentage of the cost.)

Merchandise	Cost	Selling Price	Markup	Markup Rate
1. Wristwatch	$26.97	$49.95		
2. Bicycle	$71.97	$119.95		
3. Sleeping bag		$125.98	$56.69	
4. Calculator		$224.87	$75.08	
5. Automobile		$15,900.00	$2,650.00	
6. Lawn mower		$350.00	$80.77	
7. Battery charger		$74.38		81.5%
8. Gas lantern		$69.99		55.5%
9. Camera	$107.97			85.2%
10. Refrigerator	$680.00			32%

In Exercises 11–20, find the missing quantities. (Assume the discount rate is a percentage of the list price.)

Merchandise	List Price	Sale Price	Discount	Discount Rate
11. Wheel alignment	$39.95	$29.95		
12. Car battery	$50.99	$45.99		
13. Chair	$189.99		$30.00	
14. Shirt	$18.95		$8.00	
15. Digital watch	$119.96			50%
16. Athletic shoes	$84.95			65%
17. Gallon of paint		$18.95		20%
18. Coat		$189.00		40%
19. Camcorder		$695.00	$300.00	
20. Ring		$259.97	$135.00	

Discount = Discount Rate · List Price (handwritten)

*In this section your answers may differ slightly from those listed in the back of the text due to round-off error.

21. *Labor Charge* An appliance repair shop charges $30 for a service call and one-half hour of service. For each additional half hour of labor there is a charge of $16. Find the length of a service call if the bill is $78.

22. *Labor Charge* An auto repair bill of $450 lists $178 for parts and the rest for labor. The cost of labor is $32 per hour. How many hours did it take to repair the car?

23. *Tip Rate* A customer left $20 for a meal that cost $16.95. What is the tip rate?

24. *Tip Rate* A customer left $65 for a meal that cost $56.52. What is the tip rate?

25. *Phone Charge* The weekday rate for a telephone call is $0.55 for the first minute plus $0.40 for each additional minute (see figure). Determine the length of a call that cost $2.95. What would a call of the same length have cost if it had been made during the weekend when there is a 60% discount?

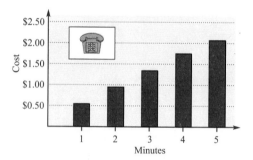

26. *Phone Charge* The weekday rate for a telephone call is $0.72 for the first minute plus $0.55 for each additional minute. Determine the length of a call that cost $5.12. What would a call of the same length have cost if it had been made during the evening when there is a 35% discount?

27. *Insurance Premium* The annual insurance premium for a policyholder is normally $739. However, after having an automobile accident, the policyholder was charged an additional 30%. What is the new annual premium?

28. *Insurance Premium* The annual insurance premium for a policyholder is normally $925. Find the annual premium if the policyholder must pay an additional 75% surcharge because of a traffic violation.

29. *Amount Financed* Suppose you buy a motorbike that costs $1,450 plus 6% sales tax. Find the amount of the sales tax and the total bill. How much of the total bill must you finance if you make a down payment of $500?

30. *Monthly Payment* Including finance charges, the total purchase price of a diamond ring is $1,050. Of this amount, $450 is given as a down payment and the remainder is paid in nine monthly payments. How much is each payment?

31. *Weekly Pay* The weekly salary of an employee is $300 plus a 5% commission on the employee's total sales. Find the pay for a week when sales amounted to $5,500.

32. *Weekly Pay* The weekly salary of an employee is $400 plus a 6% commission on the employee's total sales. Find the pay for a week when sales amounted to $6,000.

33. *Amount of Sales* The monthly salary of an employee is $1,000 plus a 7% commission on the total sales. How much must the employee sell in order to obtain a monthly salary of $3,500?

34. *Amount of Sales* Determine the sales of an employee who earned $2,100 as a sales commission. (Assume the sales commission rate is 7%.)

35. *Commission Rate* Determine the commission rate for an employee who earned $500 in commissions on sales of $4,000.

36. *Commission Rate* Determine the commission rate for an employee who earned $207 in commissions on sales of $2,300.

37. *Hours of Overtime* An employee is paid $11.25 per hour for the first 40 hours and $16 for each additional hour (see figure). During the first week on the job, the employee's gross pay was $622. How many hours of overtime did the employee work?

Figure for 37

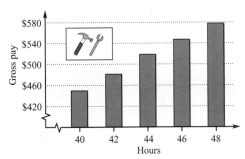

38. *Overtime Rate* Suppose you worked 48 hours during a given week (40 hours at the regular hourly rate and 8 hours at the overtime rate). Your gross pay for the week was $672. If your regular hourly rate is $12, how much is your overtime rate?

39. *Comparing Prices* A mail-order catalog lists automobile shock absorbers for $48.99 a pair, plus a shipping charge of $4.69. A local store has a special sale with 25% off a list price of $63.99 a pair. Which is the better bargain?

40. *Comparing Prices* A department store is offering a discount of 20% on a sewing machine with a list price of $239.95. A mail-order catalog has the same machine for $188.95 plus $4.32 for shipping. Which is the better bargain?

41. *Wholesale Cost* The list price of an automobile tire is $88. During a promotional sale the fourth tire is free with the purchase of three at the list price. Counting the free tire, the markup rate on cost is 10%. Find the cost of each tire.

42. *Price per Pound* The produce manager of a supermarket pays $27 for a 100-pound box of bananas. From past experience the manager estimates that 10% of the bananas will spoil before they are sold. At what price per pound should the bananas be sold to give the supermarket an average markup rate on cost of 30%?

SECTION 4.4	**Scientific Problems**
	Mixture Problems • Rate Problems • Other Applications

Mixture Problems

Many real-world problems involve combinations of two or more quantities that make up a new or different quantity. We call such problems **mixture problems**. They are usually composed of the sum of two or more "hidden products" that fit the following verbal model.

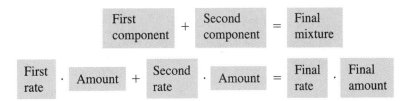

EXAMPLE 1 ■ **A Nut Mixture Problem**

A grocer wants to mix cashew nuts worth $7 per pound with 15 pounds of peanuts worth $2.50 per pound.

(a) To obtain a nut mixture worth $4 per pound, how many pounds of cashews are needed?

(b) How many pounds of mixed nuts will be produced for the grocer to sell?

Solution

(a) *Verbal model:* | Total cost of cashews | $+$ | Total cost of peanuts | $=$ | Total cost of mixed nuts |

    Labels: Cashews: cost per pound = $7.00, number of pounds = x
               Peanuts: cost per pound = $2.50, number of pounds = 15
               Mixed nuts: cost per pound = $4.00, number of pounds = $x + 15$

    Equation:

$$7(x) + 2.5(15) = 4(x + 15)$$
$$7x + 37.5 = 4x + 60$$
$$3x = 22.5$$
$$x = \frac{22.5}{3}$$
$$x = 7.5$$

Therefore, 7.5 pounds of cashews are needed.

(b) The final amount of mixed nuts for sale is

$$x + 15 = 7.5 + 15 = 22.5 \text{ pounds.}$$ ■

There are a couple of guidelines that you should remember when you are working mixture problems. First, when you set up your verbal model, be sure to check that you are working with the *same type of units* in each part of the model. For instance, in Example 1, note that each of the three parts of the verbal model is measuring cost. (If two parts were measuring cost and the other part was measuring pounds, you would know that the model was incorrect.)

Second, when you have the solution, go back to the original statement of the problem and check to see that the solution makes sense—both algebraically and from a commonsense point of view. For instance, in Example 1, you might perform a check as follows.

$$\overbrace{\left(\begin{array}{c}\$2.50 \\ \text{per} \\ \text{pound}\end{array}\right)\left(\begin{array}{c}15 \\ \text{pounds}\end{array}\right)}^{\text{Peanuts}} + \overbrace{\left(\begin{array}{c}\$7.00 \\ \text{per} \\ \text{pound}\end{array}\right)\left(\begin{array}{c}7.5 \\ \text{pounds}\end{array}\right)}^{\text{Cashews}} = \overbrace{\left(\begin{array}{c}\$4.00 \\ \text{per} \\ \text{pound}\end{array}\right)\left(\begin{array}{c}22.5 \\ \text{pounds}\end{array}\right)}^{\text{Mixed nuts}}$$

$$\$37.50 + \$52.50 = \$90.00$$

EXAMPLE 2 ■ A Coin Mixture Problem

A cash register contains $20.90 in dimes and quarters. If there are 119 coins in all, how many of each coin are in the cash register?

Solution

Verbal model: | Value of dimes | + | Value of quarters | = | Total value |

Labels: Mixed coins: total value = $20.90,   number of coins = 119
Dimes: value per coin = $0.10, number of coins = x
Quarters: value per coin = $0.25, number of coins = $119 - x$

Equation: $0.10(x) + 0.25(119 - x) = 20.90$

$$0.1x + 29.75 - 0.25x = 20.90$$

$$-0.15x = -8.85$$

$$x = \frac{-8.85}{-0.15} = 59 \text{ dimes}$$

$$119 - x = 60 \text{ quarters}$$

Therefore, the cash register has 59 dimes and 60 quarters.

Check

$$\underbrace{(0.10)(59)}_{\$5.90 \text{ in dimes}} + \underbrace{(0.25)(60)}_{\$15.00 \text{ in quarters}} = \$20.90 \quad \overbrace{}^{\substack{\text{Total for} \\ 119 \text{ coins}}}$$ ■

EXAMPLE 3 ■ A Solution Mixture Problem

A pharmacist needs to strengthen a 15% alcohol solution so that it contains 32% alcohol. How much pure alcohol should be added to 100 milliliters of the 15% solution? (See Figure 4.2.)

FIGURE 4.2

15% alcohol   100% alcohol   32% alcohol

Solution

Verbal model: | Amount of alcohol in original solution | + | Amount of alcohol in pure solution | = | Amount of alcohol in final solution |

Labels: Original solution: percent alcohol = 0.15, amount = 100 (ml)
Pure alcohol: percent alcohol = 1.00, amount = x (ml)
Final solution: percent alcohol = 0.32, amount = x + 100 (ml)

Equation: $0.15(100) + 1.00(x) = 0.32(100 + x)$

$$15 + x = 32 + 0.32x$$

$$0.68x = 17$$

$$x = \frac{17}{0.68} = 25 \text{ ml}$$

Thus, the pharmacist should add 25 milliliters of pure alcohol to the original solution. Check this solution in the original statement of the problem. ∎

.75(100) 15	15% .15	100ml
X	100%	10X
.32(100+X) 32+.32X	32%	100+X

$15 + x = 32 + .32x$
$-.32x -.32x$
$\dfrac{.68x}{.68} = \dfrac{17}{.68}$
$x = 25 \, ml$

multiply

EXAMPLE 4 ∎   **An Inventory Mixture Problem**

A department store has $30,000 of inventory in 12-inch and 19-inch color televisions. The profit on a 12-inch set is 22%, while the profit on a 19-inch set is 40%. If the profit on the entire stock is 35%, how much was invested (in inventory) in each type of television?

Rate

.22x	22% .22	X
.40(30,000-x) 12,000-.4x	40% .40	30,000-X
.35(30,000) 10,500	35% .35	$30,000

Solution

Verbal model:

$$\boxed{\begin{array}{c}\text{Profit for}\\\text{12-inch TVs}\end{array}} + \boxed{\begin{array}{c}\text{Profit for}\\\text{19-inch TVs}\end{array}} = \boxed{\begin{array}{c}\text{Total}\\\text{profit}\end{array}}$$

Labels: Total inventory: percent profit = 0.35, inventory = 30,000 (dollars)
12-inch TV: percent profit = 0.22, inventory = x (dollars)
19-inch TV: percent profit = 0.40, inventory = 30,000 − x (dollars)

Equation: $0.22(x) + 0.40(30,000 - x) = 0.35(30,000)$

$$0.22x + 12,000 - 0.4x = 10,500$$

$$-0.18x = -1,500$$

$$x = \frac{-1,500}{-0.18}$$

$$\approx \$8,333.33 \text{ (12-inch TVs)}$$

$$30,000 - x = 30,000 - 8,333.33$$

$$\approx \$21,666.67 \text{ (19-inch TVs)}$$

$.22 + 12,000 + .40x = 10,500$

$\begin{array}{r}.40\\-.22\\\hline.18\end{array}$

$-.18x + 12,000 = 10,500$
$ -12,000 \quad -12,000$
$\dfrac{-.18x}{-.18} = \dfrac{-1500}{-.18}$

$x = 8333.33$
$30,000 - x = 21,666.67$

Thus, the department store has invested approximately $8,333.33 in the 12-inch sets and approximately $21,666.67 in the 19-inch sets. Check this solution in the original statement of the problem. ∎

MATH MATTERS

Composition of the Earth's Atmosphere

The atmosphere of the Earth is a gaseous envelope of air that encloses our planet. It consists of colorless, odorless, and tasteless gases, and particles of dust. The composition of the air that we breathe is relatively constant throughout the world, although there are some slight variations depending on the zone and altitude. The eight basic gases that make up our atmosphere are as follows.

Gas	Percent
Nitrogen	78.08%
Oxygen	20.95%
Argon	0.93%
Carbon dioxide	0.03%
Neon	0.0018%
Helium	0.0005%
Krypton	0.0001%
Xenon	0.00001%

If you add these percentages, you will see that the total is not 100%. The remaining 0.00759% consists of small amounts of hydrocarbons, hydrogen, peroxide, sulfur, water vapor, and dust.

Rate Problems

Time-dependent problems involving distance are called **distance-rate problems**. They fit the following verbal model.

$$\text{Distance} \;=\; \text{Rate} \;\cdot\; \text{Time}$$

For instance, if you are traveling at a constant (or average) rate of 50 miles per hour for 45 minutes, then the total distance traveled is given by

$$\left(50 \,\frac{\text{miles}}{\text{hour}}\right) \cdot \left(\frac{45}{60}\,\text{hour}\right) = 37.5 \text{ miles.}$$

As with all problems involving applications, be sure to check that the units in the verbal model make sense. For instance, in this problem, the rate is given in *miles per hour*. Therefore, in order for the solution to be given in *miles*, we must convert the time (from minutes) to *hours*. In the model, we think of canceling the two "hours," as follows.

$$\left(50 \,\frac{\text{miles}}{\cancel{\text{hour}}}\right) \cdot \left(\frac{45}{60}\,\cancel{\text{hour}}\right) = 37.5 \text{ miles}$$

EXAMPLE 5 ■   **A Distance-Rate Problem**

Suppose you jog at an average rate of 8 kilometers per hour. How long will it take you to jog 14 kilometers?

Solution

Verbal model: Distance = Rate · Time

Labels: Distance = 14 (kilometers)
Rate = 8 (kilometers per hour)
Time = t (hours)

Equation: $14 = 8(t)$

$$\frac{14}{8} = t$$

$$1.75 = t$$

Thus, it would take you 1.75 hours (or one hour and 45 minutes). Check this solution in the original statement of the problem.   ■

EXAMPLE 6 ■   **A Distance-Rate Problem**

In a 10-kilometer race, one person runs at an average rate of 9 kilometers per hour and another runs at an average rate of 8.5 kilometers per hour. How far will the second runner be behind the first runner after 40 minutes?

Solution

Verbal model:

First runner's distance	−	Second runner's distance	=	Difference between distances

Labels: First runner's rate = 9 (km/hr)
Second runner's rate = 8.5 (km/hr)
Time (both runners) = 40 minutes = $\frac{2}{3}$ (hour)
Difference in distances = x (kilometers)

Equation: $9\left(\frac{2}{3}\right) - 8.5\left(\frac{2}{3}\right) = x$

$$\frac{18}{3} - \frac{17}{3} = x$$

$$\frac{1}{3} = x$$

Therefore, after 40 minutes the second runner will be $\frac{1}{3}$ kilometer behind the first runner. Check this solution in the original statement of the problem.   ■

NOTE: The distance-rate problem in Example 6 actually fits our mixture model because there are *two* different rates involved. Watch for such hidden mixture problems in the examples and exercises given in this section.

In **work-rate problems**, the work rate is the *reciprocal* of the time needed to do the entire job. For instance, if it takes 7 hours to complete a job, then the per-hour work rate is

$$\frac{1}{7} \text{ job per hour.}$$

Similarly, if it takes $4\frac{1}{2}$ minutes to complete a job, then the per-minute rate is

$$\frac{1}{4\frac{1}{2}} = \frac{1}{\frac{9}{2}} = \frac{2}{9} \text{ job per minute.}$$

EXAMPLE 7 ■ A Work-Rate Problem

Suppose it takes 3 hours and 20 minutes for you to type a research report. What percentage of the job will you have completed after 2 hours?

Solution

Verbal model: | Work done | = | Work rate | · | Time |

Labels: Work done = x (portion of total job)

Work rate = $\dfrac{1}{3 + \frac{1}{3}} = \dfrac{1}{\frac{10}{3}} = \dfrac{3}{10}$ job per hour

Time = 2 hours

Equation: $x = \left(\dfrac{3}{10}\right)(2)$

$x = \dfrac{6}{10}$

Therefore, $\frac{6}{10}$ of the job (or 60%) will be completed after 2 hours. Check this solution in the original statement of the problem. ■

EXAMPLE 8 ■ A Work-Rate Problem

Consider two machines in a paper manufacturing plant. Machine 1 can produce 2,000 pounds of paper in 3 hours. Machine 2 is newer and can produce 2,000 pounds of paper in $2\frac{1}{2}$ hours. How long will it take the two machines working together to produce 2,000 pounds of paper?

Solution

Verbal model: | Work done | = | Portion done by Machine 1 | + | Portion done by Machine 2 |

Labels: Both machines: work done = 1 complete job, time = t (hours)

Machine 1:   rate = $\frac{1}{3}$ job per hour,   time = t (hours)

Machine 2:   rate = $\frac{2}{5}$ job per hour,   time = t (hours)

Equation:

$$1 = \left(\frac{1}{3}\right)(t) + \left(\frac{2}{5}\right)(t)$$

$$1 = \left(\frac{1}{3} + \frac{2}{5}\right)(t)$$

$$1 = \left(\frac{11}{15}\right)(t)$$

$$\frac{15}{11} = t$$

Thus, it would take $\frac{15}{11}$ hours (or about 1.36 hours) for both machines to complete the job. Check this solution in the original statement of the problem. ■

Note in Example 8 that the "2,000 pounds" of paper was unnecessary information. We simply represented the 2,000 pounds as "one complete job." This unnecessary information was a red herring.

EXAMPLE 9   ■   A Fluid-Rate Problem

An above-ground swimming pool has a capacity of 15,600 gallons, as shown in Figure 4.3.

(a) If a primary drain pipe can empty the pool in $6\frac{1}{2}$ hours, at what rate (in gallons per minute) does the water flow through the drain pipe?

(b) A second drain pipe allows water to flow out at 25 gallons per minute. How long will it take to empty a full pool, using both drain pipes?

Solution

(a) *Verbal model:* | Volume | = | Rate of flow | · | Time |

Labels: Volume = 15,600 (gallons)
Rate of flow = r (gallons per minute)
Time = $6\frac{1}{2}$ hours

Equation: $15,600 = (r)\left(6\frac{1}{2}\right)$

$$\frac{15,600}{6\frac{1}{2}} = r$$

$$2,400 = r$$

Thus, the rate for the primary pipe is 2,400 gallons per hour. To obtain the rate in gallons per minute, we divide by 60 to obtain a rate of $\frac{2,400}{60} = 40$ gallons per minute. Check this solution in the original statement of the problem.

FIGURE 4.3

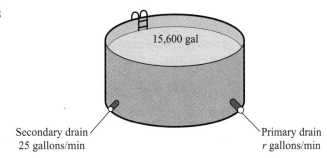

15,600 gal

Secondary drain
25 gallons/min

Primary drain
r gallons/min

(b) Using both drains, we have a mixture problem.

Verbal model: Volume = Volume through first drain + Volume through second drain

Labels: Total pool: volume = 15,600 (gallons)
First drain: rate = 40 (gallons per minute), time = t (minutes)
Second drain: rate = 25 (gallons per minute), time = t (minutes)

Equation: $15,600 = (40)(t) + (25)(t)$

$15,600 = 65t$

$\dfrac{15,600}{65} = t$

$240 = t$

Thus, it will take 240 minutes (or 4 hours) to empty the pool using both drains. Check this solution in the original statement of the problem. ■

Other Applications

In the remainder of this section, we will give examples of hidden mixture problems involving equal times, equal distances, equal rates, or equal ages.

EXAMPLE 10 ■ A Problem Involving Equal Distances

A cyclist leaves a town and travels at an average speed of 16 kilometers per hour. A second cyclist leaves 15 minutes later (from the same town) and travels at an average speed of 20 kilometers per hour, as shown in Figure 4.4.

(a) How long will it take the second cyclist to overtake the first one?

(b) How far will they have traveled when they meet?

FIGURE 4.4

20 km/hr   16 km/hr

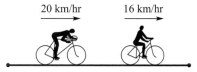

Solution

Even though we are asked to find the time involved, the essential nature of the problem is that the two cyclists traveled equal distances.

(a) *Verbal model:* | Distance for first cyclist | = | Distance for second cyclist |

Labels: Second cyclist: rate = 20 (km/hr), time = t (hours)
First cyclist: rate = 16 (km/hr), time = $t + \frac{1}{4}$ (hours)

Equation: $(16)\left(t + \dfrac{1}{4}\right) = (20)(t)$

$$16t + 4 = 20t$$
$$4 = 4t$$
$$1 = t$$

Thus, the second cyclist will overtake the first one in 1 hour. Check this solution in the original statement of the problem.

(b) *Verbal model:* | Distance | = | Rate | · | Time |

Labels: Distance (for second cyclist) = x (km)
Rate = 20 (km/hr)
Time = 1 (hour)

Equation: $x = 20(1)$
$x = 20$ km

Thus, the total distance traveled by the second cyclist (and also by the first) is 20 kilometers. Check this solution in the original statement of the problem. ■

EXAMPLE 11   ■   **A Problem Involving Equal Ages**

One bacterial culture is 18 days older than a second bacterial culture. In 10 days the first culture will be twice the age of the second. Find the present age of the two cultures.

Solution

Verbal model: | Age of first culture in 10 days | = 2 · | Age of second culture in 10 days |

Labels: Second culture: present age = x,   age in 10 days = $x + 10$
First culture: present age = $x + 18$, age in 10 days = $x + 28$

Equation:
$$x + 28 = 2(x + 10)$$
$$x + 28 = 2x + 20$$
$$8 = x$$

Thus, the second culture is presently 8 days old. Since the first culture is presently 18 days older, its present age must be 26 days. Check this solution in the original statement of the problem.   ∎

EXAMPLE 12 ∎   **A Consecutive Integer Problem**

Find three consecutive integers such that twice the sum of the first two is 8 more than three times the third integer.

Solution

Verbal model: $2 \cdot \left[\boxed{\text{First integer}} + \boxed{\text{Second integer}} \right] = 3 \cdot \boxed{\text{Third integer}} + \boxed{8}$

Labels:
First integer $= n$
Second integer $= n + 1$
Third integer $= n + 2$

Equation:
$$2[n + (n + 1)] = 8 + 3(n + 2)$$
$$2(2n + 1) = 8 + 3n + 6$$
$$4n + 2 = 3n + 14$$
$$n = 12$$
$$n + 1 = 13$$
$$n + 2 = 14$$

Thus, the three consecutive integers are 12, 13, and 14. Check this solution in the original statement of the problem.   ∎

DISCUSSION PROBLEM ∎   **You Be the Instructor**

Suppose you are teaching an algebra class and are making up a test that contains a mixture problem and a distance-rate problem. Your students have ten minutes to answer each question. Write two questions that you think would be reasonable to ask on the test.   ∎

The following warm-up exercises involve skills that were covered in earlier sections. You will use these skills in the exercise set for this section.

In Exercises 1–10, solve the given equation.

1. $14 - 2x = x + 2$

2. $2(x + 1) = 0$

3. $5[1 + 2(x + 3)] = 6 - 3(x - 1)$

4. $2 - 5(x - 1) = 2[x + 10(x - 1)]$

5. $\dfrac{x}{3} + 5 = 8$

6. $\dfrac{3x}{4} + \dfrac{1}{2} = 8$

7. $\dfrac{x}{3} + \dfrac{x}{2} = \dfrac{1}{3}$

8. $\dfrac{2}{x} + \dfrac{2}{5} = 1$

9. $\dfrac{3}{x} + \dfrac{4}{3} = 1$

10. $\dfrac{x}{x + 1} - \dfrac{1}{2} = \dfrac{4}{3}$

4.4   EXERCISES

1. *Number of Stamps* You have 100 stamps that have a total value of $18. Some of the stamps are worth 15¢ each and the others are worth 30¢ each. How many stamps of each type do you have?

2. *Number of Stamps* You have 100 stamps that have a total value of $25.20. Some of the stamps are worth 15¢ each and the others are worth 30¢ each. How many stamps of each type do you have?

3. *Number of Coins* A person has 20 coins in nickels and dimes with a combined value of $1.60. Determine the number of coins of each type.

4. *Number of Coins* A person has 50 coins in dimes and quarters with a combined value of $7.70. Determine the number of coins of each type.

5. *Poll Results* One thousand people were surveyed in an opinion poll. Candidates A and B received the same number of votes. Candidate C received twice as many votes as each of the other two candidates. How many votes did each candidate receive?

6. *Poll Results* One thousand people were surveyed in an opinion poll. The number of votes for candidates A, B, and C had ratios of 5 to 3 to 2, respectively. How many voted for each candidate?

7. *Ticket Sales* Ticket sales for a play total $1,700. The number of adult tickets is three times the number of children's tickets. The prices of the tickets for adults and children are $5 and $2, respectively. How many of each type of ticket were sold?

8. *Flower Order* A floral shop receives an order for flowers that totals $760. The prices per dozen for roses and carnations are $15 and $8, respectively. The order contains twice as many roses as carnations. How many of each type of flower are in the order?

9. *Feed Mixture* An agricultural corporation must purchase 100 tons of cattle feed. The feed is to be a mixture of oats, which costs $300 per ton, and corn, which costs $450 per ton. Complete the following table, where x is the number of tons of oats in the mixture.

Oats x	Corn $100 - x$	Price per ton of the mixture
0		
20		
40		
60		
80		
100		

(a) How does the increase in the number of tons of oats affect the number of tons of corn in the mixture?

(b) How does the increase in the number of tons of oats affect the price per ton of the mixture?

(c) If there were equal numbers of tons of oats and corn in the mixture, how would the price of the mixture be related to the price of each component?

10. *Metal Mixture* A metallurgist is making 5 ounces of an alloy out of metal A, which costs $52 per ounce, and metal B, which costs $16 per ounce. Complete the following table, where x is the number of ounces of metal A in the alloy.

Metal A x	Metal B $5 - x$	Price/ounce of the alloy
0		
1		
2		
3		
4		
5		

(a) How does the increase in the number of ounces of metal A in the alloy affect the number of ounces of metal B in the alloy?

(b) How does the increase in the number of ounces of metal A in the alloy affect the price of the alloy?

(c) If there were equal amounts of metal A and metal B in the alloy, how would the price of the alloy be related to the price of each of the components?

11. *Flower Mixture* A floral shop creates a mixed arrangement of roses, which cost $1.25 each, and carnations, which cost $0.75 each. An arrangement of one dozen flowers costs $11.00. Determine the number of roses per dozen flowers in an arrangement.

12. *Nut Mixture* A grocer mixes two kinds of nuts that cost $2.49 and $3.89 per pound to make 100 pounds of a mixture costing $3.47 per pound. How many pounds of each kind of nut were put into the mixture?

In Exercises 13–16, determine the number of units of alcohol concentrations of Solution 1 and Solution 2 needed to obtain the desired amount and alcohol concentration of the final solution.

	Concentration			Amount of
	Sol. 1	Sol. 2	Final	Final Solution
13.	10%	30%	25%	100 gallons
14.	25%	50%	30%	5 liters
15.	15%	45%	30%	10 quarts
16.	70%	90%	75%	25 gallons

17. *Antifreeze* The cooling system in a truck contains 4 gallons of coolant that is 30% antifreeze. How much must be withdrawn and replaced with 100% antifreeze in order to bring the coolant in the system to 50% antifreeze?

18. *Gas and Oil Mixture* Suppose you mixed gasoline and oil to make 2 gallons of fuel for your lawn mower. The mixture has 31 parts gasoline and 1 part oil. How much gasoline must be added to bring the mixture to 39 parts gasoline and 1 part oil?

In Exercises 19–24, determine the unknown distance, rate, or time.

	Distance, d	Rate, r	Time, t
19.		55 mph	3 hrs
20.		32 ft/sec	10 sec
21.	500 km	90 km/hr	
22.	128 ft	16 ft/sec	
23.	5,280 ft		$\frac{5}{2}$ sec
24.	432 mi		9 hr

25. *Space Shuttle Time* The speed of a space shuttle is 17,000 miles per hour (see figure). How long will it take the space shuttle to travel a distance of 3,000 miles?

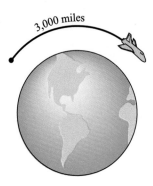

3,000 miles

26. *Speed of Light* The speed of light is 670,616,625 miles per hour, and the distance between the Earth and the Sun is 93,000,000 miles (see figure). How long does it take light from the Sun to reach the Earth?

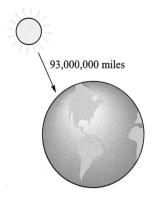

93,000,000 miles

27. *Distance* Two cars start at a given point and travel in the same direction at average speeds of 45 miles per hour and 52 miles per hour (see figure). How far apart will they be in 4 hours?

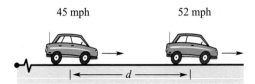

45 mph   52 mph

d

28. *Distance* Two planes leave an airport at approximately the same time and fly in opposite directions (see figure). Their speeds are 510 miles per hour and 600 miles per hour. How far apart will the planes be after $1\frac{1}{2}$ hours?

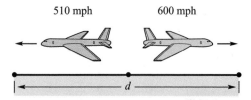

510 mph   600 mph

d

29. *Speed* Determine the average speed of an experimental plane that can travel a distance of 3,000 miles in 2.6 hours.

30. *Speed* Determine the average speed of an Olympic runner who completes the 10,000-meter race in 27 minutes and 45 seconds.

31. *Time* Two cars start at the same point and travel in the same direction at average speeds of 40 miles per hour and 55 miles per hour. How much time must elapse before the two cars are 5 miles apart?

32. *Time* Suppose on the first part of a 225-mile automobile trip you averaged 55 miles per hour. On the last part of the trip you only averaged 48 miles per hour because of increased traffic congestion. The total trip took 4 hours and 15 minutes. Find the amount of time at each speed.

33. *Speed* A driver of a truck traveled at an average speed of 52 miles per hour on a 200-mile trip to pick up a load of freight. On the return trip with the truck fully loaded, the average speed was 40 miles per hour. Find the average speed for the round-trip.

34. *Speed* A driver of a truck traveled at an average speed of 60 miles per hour on a 200-mile trip to pick up a load of freight. On the return trip with the truck fully loaded, the average speed was 40 miles per hour. Find the average speed for the round-trip.

35. *Time* A jogger leaves a given point on a fitness trail running at a rate of 4 miles per hour. Fifteen minutes later a second jogger leaves from the same location running at 5 miles per hour. How long will it take the second jogger to overtake the first and how far will each have run at that point?

36. *Time* A jogger leaves a given point on a fitness trail running at a rate of 4 miles per hour. Ten minutes later a second jogger leaves from the same location running at 5 miles per hour. How long will it take the second jogger to overtake the first and how far will each have run at that point?

37. *Work Rate* Suppose you can mow a large lawn in 2 hours using a riding mower, and in 3 hours using a push mower. Using both machines together, how long will it take you and a friend to mow the lawn?

38. *Work Rate* One person can complete a typing project in 6 hours and another can complete the same project in 8 hours. If they both work on the project, in how many hours can it be completed?

39. *Work Rate* One worker can complete a task in h hours while a second can complete the task in $\frac{3}{2}h$ hours.
   (a) Show that if both people work together the task can be completed in $t = \frac{3}{5}h$ hours.
   (b) Use the result of part (a) to complete the following table.

h	1	2	3	4	5
t					

40. *Work Rate* One worker can complete a task in h hours while a second can complete the task in $3h$ hours.
   (a) Show that by working together they can complete the task in $t = \frac{3}{4}h$ hours.
   (b) Use the result of part (a) to complete the following table.

h	1	2	3	4	5
t					

41. *Age Problem* Your age is twice that of one of your cousins. What is the age of your cousin if your combined age totals 39?

42. *Age Problem* Your age is three times that of one of your cousins. What is the age of your cousin if your combined age totals 32?

43. A mother was 30 years old when her son was born. How old will the son be when his age is one-third of his mother's age?

44. The difference in age between a father and daughter is 32 years. Determine the age of the father when his age is twice that of his daughter.

45. Find two consecutive integers whose sum is 525.

46. Find three consecutive integers whose sum is 804.

47. Find two consecutive even integers whose sum is 110.

48. Find two consecutive odd integers whose sum is 168.

49. Find three consecutive integers such that the sum of the first two is 13 less than three times the third integer.

50. Find three consecutive integers such that the sum of the first integer and twice the second integer is 12 more than twice the third integer.

51. One number is five times a second number. Find the numbers if their sum is 90.

52. One number is one-fifth that of a second number. Find the numbers if their sum is 144.

<table>
<tr><td>SECTION
4.5</td><td>### Formulas
Introduction • *Geometric Formulas* • *Business Formulas*</td></tr>
</table>

Introduction

In this section we show how to solve common types of geometric, scientific, and investment problems by making use of ready-made equations called **formulas**. Knowing formulas like those in the following lists will help you translate and solve a wide variety of real-world problems. (Note that $\pi \approx 3.14$.)

Common Formulas for Area and Perimeter

The following formulas give the area A and perimeter P (or circumference C) of the indicated figures.

Square

$A = s^2$

$P = 4s$

Rectangle

$A = lw$

$P = 2l + 2w$

Circle

$A = \pi r^2$

$C = 2\pi r$

Triangle

$A = \dfrac{1}{2}bh$

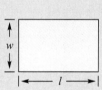
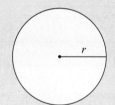
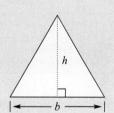

Common Formulas for Volume

The following formulas give the volume V of the indicated figures.

Cube

$V = s^3$

Rectangular Solid

$V = lwh$

Circular Cylinder

$V = \pi r^2 h$

Sphere

$V = \dfrac{4}{3}\pi r^3$

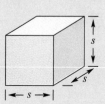
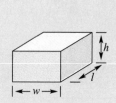
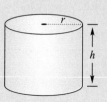
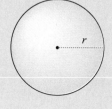

Miscellaneous Common Formulas	*Temperature:* $F = \dfrac{9}{5}C + 32$

F = degrees Fahrenheit C = degrees Celsius

Simple interest: $I = Prt$

I = interest P = principal

r = annual interest rate t = time in years

Compound interest: $A = P\left(1 + \dfrac{r}{n}\right)^{nt}$

A = balance P = principal

r = annual interest rate n = compoundings per year

t = time in years

Distance: $d = rt$

d = distance traveled r = rate

t = time

Geometric Formulas

It is helpful to be able to manipulate formulas—that is, to rewrite them in other forms. For instance, the formula for the perimeter of a rectangle can be rewritten or solved for w in the following manner.

$P = 2l + 2w$ Given formula

$P - 2l = 2w$ Subtract $2l$ from both sides

$\dfrac{P - 2l}{2} = w$ Divide both sides by 2

EXAMPLE 1   ■   Manipulating a Formula

Find a formula for the base of a triangle in terms of its area and height.

Solution

Common formula: $A = \dfrac{1}{2}bh$

Labels: A = area of triangle
b = length of base
h = height

Equation: $A = \dfrac{1}{2}bh$

$2A = bh$

$\dfrac{2A}{h} = b$ ■

EXAMPLE 2 ■ Using a Geometric Formula

A rectangular plot has an area of 100,000 square feet. The plot is 200 feet wide. How long is it?

Solution

In a problem like this, we find it helpful to begin by drawing a picture, as shown in Figure 4.5. In this picture, we label the width of the rectangle as $w = 200$ feet, and the unknown length as l.

FIGURE 4.5

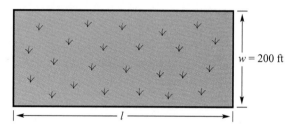

$w = 200$ ft

l

To solve for the unknown length, we use the following steps.

Common formula: $A = lw$

Labels: $A = 100{,}000$ (square feet)
$w = 200$ (feet)
$l = $ length (feet)

Equation: $A = lw$

$100{,}000 = l(200)$

$$\frac{100{,}000}{200} = l$$

$500 = l$

Thus, the length of the rectangular plot is 500 feet. (We can check this by multiplying 200 feet by 500 feet to obtain an area of 100,000 square feet.) ■

EXAMPLE 3 ■ Using Geometric Formulas

A rectangular box has a base that is 12 inches by 18 inches. The height of the box is 16 inches. (See Figure 4.6.)

(a) Find the perimeter of the base of the box.

(b) Find the area of the base of the box.

(c) Find the volume of the box.

FIGURE 4.6

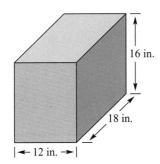

16 in.

18 in.

|← 12 in. →|

Solution

(a) *Common formula:* $P = 2l + 2w$

   Labels: P = perimeter of rectangular base (in inches)
   $w = 12$ (inches)
   $l = 18$ (inches)

   Equation: $P = 2l + 2w$
   $= 2(18) + 2(12)$
   $= 60$ inches

Thus, the perimeter of the base is 60 inches.

(b) *Common formula:* $A = lw$

   Labels: A = area of rectangular base (in square inches)
   $w = 12$ (inches)
   $l = 18$ (inches)

   Equation: $A = lw$
   $= (18)(12)$
   $= 216$ square inches

Thus, the area of the base is 216 square inches.

(c) *Common formula:* $V = lwh$

   Labels: V = volume of box (in cubic inches)
   $w = 12$ (inches)
   $l = 18$ (inches)
   $h = 16$ (inches)

   Equation: $V = lwh$
   $= (18)(12)(16)$
   $= 3{,}456$ cubic inches

Thus, the volume of the box is 3,456 cubic inches.   ■

   In Example 3(c) suppose you were asked to find the volume of the box in cubic feet (rather than in cubic inches). To do this, you could use the fact that one cubic foot is $(12)(12)(12) = 1{,}728$ cubic inches. This implies that the volume of the box is

$$\text{Volume} = \frac{3{,}456 \text{ cubic inches}}{1{,}728 \text{ cubic inches per cubic foot}} = 2 \text{ cubic feet.}$$

EXAMPLE 4 ■ Perimeter Problem

A rectangle is twice as long as it is wide, and its perimeter is 132 inches. Find the dimensions of the rectangle.

Solution

In this case, a picture is appropriate (see Figure 4.7).

FIGURE 4.7

Common formula: $P = 2l + 2w$

Labels: w = width (in inches)
                    l = length = $2w$ (in inches)
                    P = perimeter = 132 (inches)

Equation: $132 = 2(2w) + 2w$

                    $132 = 4w + 2w$

                    $132 = 6w$

                    $\dfrac{132}{6} = w$

                    $22 = w$

Therefore, the width of the rectangle is $w = 22$ inches. Since the length is twice the width, the length must be $l = 2w = 44$ inches. ■

Business Formulas

If a principal of P dollars is invested in an account paying an annual interest rate of r (in decimal form) for t years, then the interest (in dollars) is given by

$I = Prt.$ Simple interest

The balance A in the account is the sum of the interest and the original principal. That is,

$A = P + I = P + Prt = P(1 + rt).$

EXAMPLE 5 ■ Simple Interest

(a) Solve for r in $I = Prt$.   (b) Solve for P in $A = P + Prt$.

Solution

(a) $I = Prt$ Given
    $I = (Pt)r$ Group factors
    $\dfrac{I}{Pt} = r$ Divide by Pt

(b) $A = P + Prt$ Given
        $A = P(1 + rt)$ Distributive Property
    $\dfrac{A}{1 + rt} = P$ Divide by $1 + rt$ ■

EXAMPLE 6   ■   Simple Interest

An amount of $5,000 was deposited in an account paying simple interest. After six months, the account had earned $162.50 in interest. What is the annual interest rate for this account?

Solution

Common formula: $I = Prt$

Labels: I = interest = $162.50
P = principal = $5,000
t = time = $\frac{1}{2}$ (year)
r = annual interest rate (in decimal form)

Equation: $I = Prt$

$$162.50 = 5{,}000(r)\left(\frac{1}{2}\right)$$

$$\frac{2(162.50)}{5{,}000} = r$$

$$0.065 = r$$

Therefore, the annual interest rate is $r = 0.065$ (or 6.5%). Check this solution in the original statement of the problem.   ■

EXAMPLE 7   ■   Simple Interest

Suppose your parents asked you to review their income tax return. In the return you noticed that they earned interest of $13,500 from a savings account during the year. How much would you estimate they have in the savings account?

Solution

The answer, of course, depends on several different things. To find the exact answer, you would need to know the annual interest rate and the type of compounding. But just to obtain a rough estimate, let's assume that the savings account pays simple interest at 8%. We can then solve the problem as follows.

Common formula: $I = Prt$

Labels: P = unknown principal (in dollars)
r = 0.08 (annual interest rate in decimal form)
t = 1 (year)
I = $13,500

Equation: $I = Prt$

$$13{,}500 = P(0.08)(1)$$

$$\frac{13{,}500}{0.08} = P$$

$$\$168{,}750 = P$$

Thus, you can estimate that the savings account has a balance of approximately $168,750. (If the interest rate were as low as 6%, the balance would be $225,000, and if it were as high as 10%, the balance would be $135,000.) ■

EXAMPLE 8 ■ **A Simple Interest Mixture Problem**

A sum of $10,000 is invested in two different type of accounts. Part of the money is invested in an account paying $9\frac{1}{2}$% simple interest, and the remainder is invested in an account paying 11% simple interest. At the end of one year, the two accounts pay a total interest of $1,038.50. How much was invested in each account?

Solution

Simple interest problems are based on the formula $I = rP$, where I is the interest, r is the annual interest rate, and P is the principal.

Verbal model:
$\boxed{\begin{array}{c}\text{Interest on}\\\text{first account}\end{array}} + \boxed{\begin{array}{c}\text{Interest on}\\\text{second account}\end{array}} = \boxed{\begin{array}{c}\text{Total}\\\text{interest}\end{array}}$

Labels:
Both accounts: interest = $1,038.50, principal = $10,000
First account: interest rate = 0.095, principal = x
Second account: interest rate = 0.11,   principal = $10,000 - x$

Equation: $0.095(x) + 0.11(10,000 - x) = 1,038.50$

$$0.095x + 1,100 - 0.11x = 1,038.50$$

$$-0.015x = -61.50$$

$$x = \frac{-61.50}{-0.015} = \$4,100 \text{ at } 9\tfrac{1}{2}\%$$

$$10,000 - x = \$5,900 \text{ at } 11\%$$

Therefore, $4,100 was invested in the $9\frac{1}{2}$% account and $5,900 was invested in the 11% account. Check this solution in the original statement of the problem. ■

DISCUSSION PROBLEM ■ **Creating a Formula**

Suppose you have forgotten the formula that converts degrees Fahrenheit to degrees Celsius. However, you do remember the following facts.

(a) The formula has the form $F = aC + b$.

(b) When $F = 32°$, $C = 0°$.

(c) When $F = 212°$, $C = 100°$.

From this information, describe how you could create the formula. ■

Warm-Up

The following warm-up exercises involve skills that were covered in earlier sections. You will use these skills in the exercise set for this section.

In Exercises 1–4, simplify the algebraic expression.

1. $-5(-3x)$ **2.** $-2x(4x)$ **3.** $\dfrac{3x}{4} \cdot \dfrac{2x}{3}$ **4.** $4\left(\dfrac{x}{2}\right)$

In Exercises 5–8, remove the symbols of grouping and simplify by combining like terms.

5. $4(7x - y) - (y - 2x)$ **6.** $3s - 5(2s - 3)$

7. $\dfrac{3}{4}(12x - 8) + 10$ **8.** $2z[4 - (z + 1)] + 3z(z + 1)$

In Exercises 9 and 10, solve the given equation.

9. $\dfrac{x}{2} = \dfrac{x + 5}{4}$ **10.** $\dfrac{1 - z}{3} - \dfrac{z}{2} = 2$

4.5 EXERCISES

In Exercises 1–6, evaluate the formula for the given values of the variables. (List the *units* for each answer.)

1. *Distance-Rate-Time Formula:* $r = \dfrac{d}{t}$
$d = 100$ miles, $t = 2.5$ hours

2. *Distance-Rate-Time Formula:* $t = \dfrac{d}{r}$
$d = 100$ kilometers, $r = 60$ kilometers/hour

3. *Volume of a Circular Cylinder:* $V = \pi r^2 h$
$r = 2$ meters, $h = 3$ meters

4. *Volume of a Sphere:* $V = \dfrac{4}{3}\pi r^3$
$r = 6$ inches

5. *Simple Interest:* $A = P + Prt$
$P = \$1,000$, $r = 0.09$, $t = 5$

6. *Power* (in amps): $I = \dfrac{P}{V}$
$P = 1,500$ watts, $V = 110$ volts

In Exercises 7–26, solve for the specified variable.

7. Solve for h.
Area of a Triangle: $A = \dfrac{1}{2}bh$

8. Solve for r.
Circumference of a Circle: $C = 2\pi r$

9. Solve for R.
Ohm's Law: $E = IR$

10. Solve for l.
Perimeter of a Rectangle: $P = 2l + 2w$

11. Solve for l.
Volume of a Rectangular Solid: $V = lwh$

12. Solve for h.
Volume of a Circular Cylinder: $V = \pi r^2 h$

13. Solve for C.
Markup: $S = C + RC$

14. Solve for L.
Discount: $S = L - RL$

15. Solve for r.

  Simple Interest Balance: $A = P + Prt$

16. Solve for P.

  Compound Interest Balance: $A = P\left(1 + \dfrac{r}{n}\right)^{nt}$

17. Solve for b.

  Area of a Trapezoid: $A = \dfrac{1}{2}(a + b)h$

18. Solve for θ. (θ is the Greek letter Theta.)

  Area of a Sector of a Circle: $A = \dfrac{r^2\theta}{2}$

19. Solve for r.

  Volume of a Spherical Segment: $V = \dfrac{1}{3}\pi h^2(3r - h)$

20. Solve for b.

  Volume of an Oblate Spheroid: $V = \dfrac{4}{3}\pi a^2 b$

21. Solve for a.

  Free-Falling Body: $h = v_0 t + \dfrac{1}{2}at^2$

22. Solve for m_2.

  Newton's Law of Universal Gravitation: $F = \alpha\dfrac{m_1 m_2}{r^2}$

23. Solve for n.

  Arithmetic Progression: $L = a + (n - 1)d$

24. Solve for a.

  Arithmetic Progression: $S = \dfrac{n}{2}[2a + (n - 1)d]$

25. Solve for r.

  Geometric Progression: $S = \dfrac{rL - a}{r - 1}$

26. Solve for S_1.

  Prismoidal Formula: $V = \dfrac{1}{6}H(S_0 + 4S_1 + S_2)$

In Exercises 27–30, use the closed rectangular box shown in the accompanying figure to answer the questions.

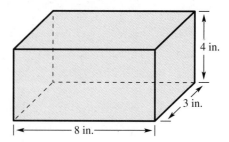

27. Find the area of the base.

28. Find the perimeter of the base.

29. Find the volume of the box.

30. Find the surface area of the box. (*Note:* This is the combined area of the six surfaces.)

31. *Floor Plan* The accompanying floor plan shows three square rooms. The perimeter of the bathroom is 32 feet. The perimeter of the kitchen is 80 square feet. Find the area of the living room.

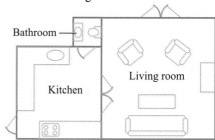

32. *Diameter of a Wheel* The circumference of a wheel is 30π inches (see figure). Find the diameter of the wheel.

In Exercises 33–42, assume that the interest is computed using the formula for *simple* interest.

33. Find the interest on a $1,000 bond paying an annual percentage rate of 9% for six years.

34. Find the interest on a $1,000 bond paying an annual percentage rate of 7.5% for 10 years.

35. Find the annual interest rate on a savings account that accumulated $110 in interest in one year on a principal of $1,000.

36. Find the annual interest rate on a certificate of deposit that accumulated $128.98 in interest in one year on a principal of $1,500.

37. Find the principal required to earn $51.75 in interest in three years, if the annual interest rate is $7\frac{1}{2}\%$.

38. Find the principal required to earn $408 in interest in four years, if the annual interest rate is $8\frac{1}{2}\%$.

39. Suppose you borrow $15,000 for one-half year. You agree to pay back the principal and the interest in one lump sum. The annual interest rate is 13%. What is your payment?

40. Suppose you borrow $15,000 for one-half year. You agree to pay back the principal and the interest in one lump sum. The annual interest rate is $11\frac{1}{2}\%$. What is your payment?

41. An inheritance of $25,000 is divided between two investments earning 8.5% and 10% simple interest. How much is in each investment if the total interest for one year is $2,350?

42. Six thousand dollars is divided between two investments earning 7% and 9% simple interest. How much is in each investment if the total interest for one year is $500?

CHAPTER 4   SUMMARY

As you review and prepare for a test on this chapter, first try to obtain a global view of what was discussed. Then review the specific skills needed in each category.

Creating Mathematical Models for Consumer Problems

- *Create* a ratio from a verbal statement.   Section 4.1, Exercises 1–14

- *Create* a ratio to determine a unit price.   Section 4.1, Exercises 15–18

- *Solve* a proportion.   Section 4.1, Exercises 21–36

- *Create and solve* a proportion from a verbal statement.   Section 4.1, Exercises 37–61

- *Convert* a fraction or decimal to a percent and vice versa.   Section 4.2, Exercises 1–22

- *Solve* the percent equation for any of its parts.   Section 4.2, Exercises 27–50

- *Create and solve* a percent equation from verbal statements.   Section 4.2, Exercises 51–61

- *Create* mathematical models for business and consumer problems.   Section 4.3, Exercises 1–42

Scientific Applications and Formulas

- *Create* a mathematical model for solving a mixture problem.   Section 4.4, Exercises 1–18

- *Create* a mathematical model for solving a rate problem.   Section 4.4, Exercises 19–40

- *Use* common geometric formulas as mathematical models.   Section 4.5, Exercises 1–12, Exercises 17–26

- *Use* common business formulas as mathematical models.   Section 4.5, Exercises 12–16, Exercises 33–42

Chapter 4 Review Exercises

In Exercises 1–4, express the ratio as a fraction in reduced form. (Use the same units of measure for both quantities.)

1. Eighteen inches to 4 yards

2. One pint to 2 gallons

3. Two hours to 90 minutes

4. Four meters to 150 centimeters

In Exercises 5–10, solve the given proportion.

5. $\dfrac{7}{16} = \dfrac{z}{8}$

6. $\dfrac{x}{12} = \dfrac{5}{4}$

7. $\dfrac{10}{a} = \dfrac{5}{2}$

8. $\dfrac{16}{5} = \dfrac{3}{r}$

9. $\dfrac{x+2}{4} = \dfrac{x-1}{3}$

10. $\dfrac{1}{x-4} = \dfrac{4}{9}$

In Exercises 11–14, use a proportion to solve the given problem.

11. *Real Estate Taxes* The tax on a property with an assessed value of \$75,000 is \$1,150. Find the tax on a property with an assessed value of \$110,000.

12. *Recipe Proportions* One and one-half cups of milk are needed to make one batch of pudding. How much is needed to make three batches?

13. *Map Distance* The scale represents 100 miles on the accompanying map. Use the map to approximate the distance between St. Petersburg and Tallahassee.

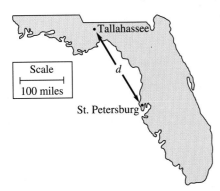

14. *Gasoline and Oil Ratio* The gasoline-to-oil ratio for a lawn mower is forty to one. Determine the amount of gasoline required to produce a mixture that contains one-half pint of oil.

Similar Triangles In Exercises 15 and 16, solve for the length x of the side of a triangle. (Assume that the two triangles are similar, and use the fact that corresponding sides of similar triangles are proportional.)

15.

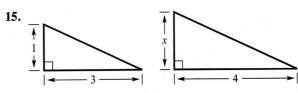

16.

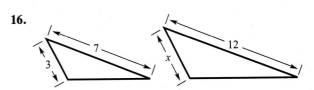

In Exercises 17 and 18, complete the table showing the equivalent forms of a percent: the number of parts out of 100, a decimal, and a fraction (in reduced form).

17.

Percent	Parts out of 100	Decimal	Fraction
35%			

18.

Percent	Parts out of 100	Decimal	Fraction
			$\frac{4}{5}$

In Exercises 19–24, solve the given percent problem.

19. What is 125% of 16?

20. What is 0.8% of 3,250?

21. 150 is $37\frac{1}{2}\%$ of what number?

22. 323 is 95% of what number?

23. 150 is what percent of 250?

24. 130.6 is what percent of 3,265?

25. *Revenue* The revenues for a corporation (in millions of dollars) in the years 1988 and 1989 were $4,521.40 and $4,679.00, respectively. Determine the percentage increase in revenue from 1988 to 1989.

26. *Price Increase* The manufacturer's suggested retail price for a certain car model is $18,459. Estimate the price of a comparably equipped car for the next model year. Assume that car prices will increase by $4\frac{1}{2}\%$.

27. *Price* A food processor that costs a retailer $85 is marked up by 35%. Find the price for the consumer.

28. *Markup Rate* A set of tools has a sale price of $199.98. The cost for the retailer is $155. Find the markup rate.

29. *Sale Price* While shopping for a new suit you find one you like with a list price of $279. The list price has been reduced by 35% to form the sale price. Find the sale price of the suit.

30. *Comparing Two Prices* A mail-order catalog lists the price of a piece of luggage at $89 plus $4 for shipping and handling. A local department store has the same luggage for $112.98. If the department store has a special 20% off sale on luggage, which is the better price?

31. *Tip Rate* A customer left $40 for a meal that cost $34.25. Determine the tip rate.

32. *Sales Commission* The weekly salary of an employee is $250 plus a 6% commission on the employee's total sales. How much must the employee sell in order to obtain a weekly salary of $600?

33. *Number of Coins* You have 30 coins in dimes and quarters with a combined value of $5.55. Determine the number of coins of each type.

34. *Poll Results* Thirteen hundred people were surveyed in an opinion poll. Candidates A and B received the same number of votes. Candidate C received five-fourths as many votes as each of the other two candidates. How many votes did each candidate receive?

35. *Time* The average speed of a train is 60 miles per hour. How long will it take the train to travel 562 miles?

36. *Speed* A person can walk 20 kilometers in 3 hours and 47 minutes. What is the person's average speed?

37. *Distance* A jet plane has an average speed of 1,500 miles per hour. How far will it travel in $2\frac{1}{3}$ hours?

38. *Speed* For the first hour of a 350-mile trip your average speed is only 40 miles per hour. Determine the average speed that must be maintained for the remainder of the trip if you want the average speed for the entire trip to be 50 miles per hour.

39. *Work Rate* Find the time for two people working together to complete a task if working individually it takes them 5 hours and 6 hours, respectively.

40. *Work Rate* Suppose the person in Exercise 39 who can complete the task in 5 hours has already worked one hour on the project when the second individual starts. How long will they work together to complete the task?

41. *Number Problem* Find three consecutive multiples of 3 whose sum is 153.

42. *Number Problem* The sum of a number and twice the sum of the number and five is 106. Find the number.

43. *Dimensions of a Swimming Pool* The width of a rectangular swimming pool is 4 feet less than its length. Find the dimensions of the pool if its perimeter is 112 feet.

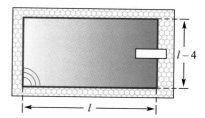

44. *Dimensions of a Triangle* The perimeter of an isosceles triangle is 65 centimeters. Find the length of the two equal sides if each is ten centimeters longer than the third side. (An isosceles triangle has two equal sides and the perimeter of a triangle is the sum of the lengths of its three sides.)

Simple Interest In Exercises 45–50, assume that the interest is determined using the formula for simple interest.

45. Find the total interest you will earn on a $1,000 corporate bond that matures in five years and has a 9.5% interest rate.

46. Find the annual interest rate on a certificate of deposit that pays $60 per year in interest on a principal of $750.

47. Find the principal required to have an annual interest income of $25,000 if the annual interest rate on the principal is 8.75%.

48. A corporation borrows two and one-quarter million dollars for two years to modernize one of its manufacturing facilities. The corporation pays an annual interest rate of 11%. What will be the total principal and interest that must be repaid?

49. An inheritance of $20,000 is divided between two investments earning 8% and 10.5% simple interest. How much is in each investment if the total interest for one year is $1,725?

50. You invest $2,500 in a certificate of deposit that has an annual interest rate of 7%. After six months the interest is computed and added to the principal. During the second six months the interest is computed using the original investment plus the interest earned during the first six months. What is the total interest earned during the first year of the investment?

51. Solve for t.
Simple Interest Balance: $A = P + Prt$

52. Solve for h.

Area of a Trapezoid: $A = \frac{1}{2}(a + b)h$

CHAPTER 4 TEST

Take this test as you would take a test in class. After you are done, check your work with the answers given in the back of the book.

1. Express the ratio of 40 inches to 2 yards as a fraction reduced to lowest terms. (Use the same units of measure for both quantities.)

2. Solve the proportion: $\dfrac{5}{8} = \dfrac{x}{12}$

3. Solve the proportion: $\dfrac{2x}{3} = \dfrac{x + 14}{5}$

4. One-half inch represents 30 miles on a map. Approximate the distance between two cities that are $2\frac{1}{2}$ inches apart on the map.

5. Express the fraction $\frac{3}{8}$ as (a) a percent and (b) a decimal.

6. What is 150% of 28?

7. 324 is 27% of what number?

8. 90 is what percent of 250?

9. The list price of a calculator is $112. The cost for the retailer is $80. Find the markup rate.

10. A person has 20 coins in dimes and quarters with a combined value of $3.80. Determine the number of coins of each type.

11. A floral shop creates a mixed arrangement of roses, which cost $18 per dozen, and carnations, which cost $9 per dozen. An arrangement of one dozen flowers costs $12. Determine the number of roses per dozen flowers in the arrangement.

12. The time required for a family to travel 264 miles to their vacation destination was $5\frac{1}{2}$ hours. Find their average speed.

13. Suppose you can paint a building in 9 hours. A novice painter would require 12 hours for the same task. Working together, how long will it take the two of you to paint the building?

14. Find three consecutive integers whose sum is 93.

15. Solve for R in the formula: $S = C + RC$.

16. Find the total interest for a six-year $1,000 bond paying an annual percentage rate of 8%. (Use the formula for simple interest.)

17. Find the principal required to earn $500 per year in simple interest, if the annual interest rate is 8%.

CUMULATIVE TEST: CHAPTERS 1–4

Take this test as you would take a test in class. After you are done, check your work with the answers given in the back of the book.

1. Place the correct symbol ($<$ or $>$) between the two real numbers.

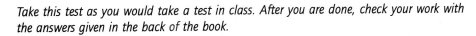

$$-\frac{3}{4} \quad \rule{1cm}{0.3cm} \quad \left|-\frac{7}{8}\right|$$

2. Evaluate: $(-200)(2)(-3)$ **3.** Evaluate: $\dfrac{3}{8} - \dfrac{5}{6}$ **4.** Evaluate: $\dfrac{-2}{9} \div \dfrac{8}{75}$ **5.** Evaluate: $-(-2)^3$

6. Use exponential form to write the product $3 \cdot (x + y) \cdot (x + y) \cdot 3 \cdot 3$.

7. Simplify: $(3x^3)(5x^4)$

8. Simplify: $(a^3b^2)(ab)^5$

9. Use the Distributive Property to expand $-2x(x - 3)$.

10. Simplify: $2x^2 - 3x + 5x^2 - (2 + 3x)$

11. Identify the rule of algebra illustrated by the equation

$$2 + (3 + x) = (2 + 3) + x.$$

12. Solve: $12x - 3 = 7x + 27$

13. Solve: $2x - \dfrac{5x}{4} = 13$

14. Solve: $2(x - 3) + 3 = 12 - x$

15. Solve the inequality and sketch the graph of its solution.

$$-1 \le \frac{x + 3}{2} < 2$$

16. Express the ratio "24 ounces to 2 pounds" as a fraction in reduced form.

17. The tax on a property with an assessed value of $95,000 is $2,350. Find the tax in the same area on a property with an assessed value of $120,000.

18. The suggested retail price of a camcorder is $1,150. The camcorder is to be on sale for "20% off" of the list price. Find the sale price.

19. The sum of two consecutive even integers is 494. Find the two numbers.

20. An inheritance of $12,000 is divided between two investments earning 7.5% and 9% simple interest. How much is in each investment if the total interest for one year is $960?

Polynomials

Chapter Overview

In this chapter we begin the study of algebraic expressions called polynomials. Much of what we do in algebra from this point on involves operations with polynomials and their use in real-world applications. Since polynomials are algebraic expressions, our work with polynomials is governed primarily by the rules of algebra given in Section 2.2.

Adding and Subtracting Polynomials
Basic Definitions   •   Adding Polynomials   •   Subtracting Polynomials

Basic Definitions

Remember that the *terms* of an algebraic expression are those parts separated by addition. An algebraic expression whose terms are all of the form ax^k, where a is any real number and k is a nonnegative integer, is called a **polynomial in one variable**, or simply a **polynomial**. Here are some examples of polynomials in one variable.

$$2x + 5, \qquad x^2 - 3x + 7, \quad \text{and} \quad x^3 + 8$$

In the term ax^k, a is called the **coefficient** of the term and k is called the **degree** of the term. Since a polynomial is an algebraic sum, the coefficients take on the signs between the terms. For instance,

$$x^4 + 2x^3 - 5x^2 + 7 = (1)x^4 + 2x^3 + (-5)x^2 + (0)x + 7$$

has coefficients 1, 2, -5, 0, and 7. For this polynomial, the last term, 7, is called the **constant term**. We usually write polynomials in the order of descending powers of the variable. This is called **standard form**. For instance, the standard form of $3x^2 - 5 - x^3 + 2x$ is

$$-x^3 + 3x^2 + 2x - 5. \qquad \text{Standard form}$$

The **degree of a polynomial** is the degree of the term with the highest power, and the coefficient of this term is called the **leading coefficient** of the polynomial. For instance, the polynomial

$$-3x^4 + 4x^2 + x + 7$$

is of fourth degree, and its leading coefficient is -3.

Definition of a Polynomial in x	Let $a_n, a_{n-1} \ldots , a_2, a_1, a_0$ be real numbers and let n be a *nonnegative integer*. A **polynomial in x** is an expression of the form $$a_n x^n + a_{n-1}x^{n-1} + \cdots + a_2 x^2 + a_1 x + a_0$$ where $a_n \neq 0$. The polynomial is of **degree** n, and the number a_n is called the **leading coefficient**. The number a_0 is called the **constant term**.

NOTE: The following are *not* polynomials for the reasons stated.

$$2x^{-1} + 5 \qquad \text{Exponent in } 2x^{-1} \text{ is } not \text{ nonnegative}$$
$$x^3 + 3x^{1/2} \qquad \text{Exponent in } 3x^{1/2} \text{ is } not \text{ an integer}$$

EXAMPLE 1   ■   Identifying Leading Coefficients and Degrees of Polynomials

Find the degree and leading coefficient for each of the following polynomials.

(a) $4x^2 - 5x^7 - 2 + 3x$ (b) $4 - 9x^2$

(c) 8 (d) $2 + x^3 - 5x^2$

Solution

Polynomial	Standard Form	Degree	Leading Coefficient
(a) $4x^2 - 5x^7 - 2 + 3x$	$-5x^7 + 4x^2 + 3x - 2$	7	-5
(b) $4 - 9x^2$	$-9x^2 + 4$	2	-9
(c) 8	8	0	8
(d) $2 + x^3 - 5x^2$	$x^3 - 5x^2 + 2$	3	1

In part (c) note that a polynomial that has *only* a constant term is considered to have a degree of zero.   ■

A polynomial with only one term is called a **monomial**. Polynomials with two *unlike* terms are called **binomials**, and those with three *unlike* terms are called **trinomials**. Here are some examples.

Monomial: $3x^2$

Binomial: $-3x + 1$

Trinomial: $4x^2 - 5x + 6$

NOTE: The prefix *poly* means many. For instance, a polygon is a many-sided figure. Similarly, the prefix *mono* means one, the prefix *bi* means two, and the prefix *tri* means three.

Adding Polynomials

As with algebraic expressions, the key to adding two polynomials is to recognize *like* terms. (Like terms are those having the same degree.) By the Distributive Property, we can then combine the like terms using either a horizontal or a vertical arrangement of terms. For instance, the polynomials $2x^2 + 3x + 1$ and $x^2 - 2x + 2$ can be added horizontally to obtain

$$(2x^2 + 3x + 1) + (x^2 - 2x + 2) = (2x^2 + x^2) + (3x - 2x) + (1 + 2)$$
$$= 3x^2 + x + 3$$

or they can be added vertically to obtain the same result.

$$\begin{array}{r} 2x^2 + 3x + 1 \\ \underline{x^2 - 2x + 2} \\ 3x^2 + x + 3 \end{array}$$

When you use this vertical arrangement to add polynomials, be sure that you line up the like terms.

EXAMPLE 2 ■ Adding Polynomials Horizontally

Use a horizontal arrangement to find the following sums.

(a) $(x^3 + 2x^2 + 4) + (3x^2 - x + 5)$

(b) $(2x^2 - x + 3) + (4x^2 - 7x + 2) + (-x^2 + x - 2)$

Solution

(a) $(x^3 + 2x^2 + 4) + (3x^2 - x + 5)$ 　　　　　Given polynomials

$= (x^3) + (2x^2 + 3x^2) + (-x) + (4 + 5)$ 　　Group like terms

$= x^3 + 5x^2 - x + 9$ 　　　　　　　　Standard form

(b) $(2x^2 - x + 3) + (4x^2 - 7x + 2) + (-x^2 + x - 2)$

$= (2x^2 + 4x^2 - x^2) + (-x - 7x + x) + (3 + 2 - 2)$

$= 5x^2 - 7x + 3$ 　　■

EXAMPLE 3 ■ Adding Polynomials Vertically

Use a vertical format to find the following sums.

(a) $(-4x^3 - 2x^2 + x - 5) + (2x^3 + 3x + 4)$

(b) $(5x^3 + 2x^2 - x + 7) + (3x^2 - 4x + 7) + (-x^3 + 4x^2 - 2x - 8)$

Solution

(a) $\begin{aligned}-4x^3 - 2x^2 + \ x - 5\\ \underline{2x^3 \qquad\ + 3x + 4}\\ -2x^3 - 2x^2 + 4x - 1\end{aligned}$ 　　(b) $\begin{aligned}5x^3 + 2x^2 - \ x + 7\\ 3x^2 - 4x + 7\\ \underline{-x^3 + 4x^2 - 2x - 8}\\ 4x^3 + 9x^2 - 7x + 6\end{aligned}$ 　■

Subtracting Polynomials

To subtract one polynomial from another, we must change the sign of each of the terms of the polynomial that is being subtracted and then add the resulting like terms. For instance, we can subtract the polynomial $(x^2 + x - 1)$ from the polynomial $(2x^2 + 3x - 4)$ as follows.

$$(2x^2 + 3x - 4) - (x^2 + x - 1) = 2x^2 + 3x - 4 - x^2 - x + 1$$
$$= (2x^2 - x^2) + (3x - x) + (-4 + 1)$$
$$= x^2 + 2x - 3$$

EXAMPLE 4 ■ Subtracting Polynomials Horizontally

Perform the following operations.

(a) $(3x^3 - 4x^2 + 3) - (x^3 + 3x^2 - x - 4)$

(b) $(x^2 - 2x + 1) - [(x^2 + x - 3) + (-2x^2 - 4x)]$

Solution

(a) $(3x^3 - 4x^2 + 3) - (x^3 + 3x^2 - x - 4)$ Given polynomials

$= 3x^3 - 4x^2 + 3 - x^3 - 3x^2 + x + 4$ Change signs

$= (3x^3 - x^3) + (-4x^2 - 3x^2) + (x) + (3 + 4)$ Group like terms

$= 2x^3 - 7x^2 + x + 7$ Standard form

(b) $(x^2 - 2x + 1) - [(x^2 + x - 3) + (-2x^2 - 4x)]$

$= (x^2 - 2x + 1) - [(x^2 - 2x^2) + (x - 4x) + (-3)]$

$= (x^2 - 2x + 1) - [-x^2 - 3x - 3]$

$= x^2 - 2x + 1 + x^2 + 3x + 3$

$= (x^2 + x^2) + (-2x + 3x) + (1 + 3)$

$= 2x^2 + x + 4$ ■

Be especially careful to use the correct signs when you are subtracting one polynomial from another. One of the most common mistakes in algebra is to forget to change signs correctly when subtracting one expression from another. Here is an example.

Wrong sign
↓

$(x^2 - 2x + 3) - (x^2 + 2x - 2) \neq x^2 - 2x + 3 - x^2 + 2x - 2$ Common error

↑
Wrong sign

Note that the error was forgetting to change two of the signs in the polynomial that was being subtracted. Here is the correct way to perform the subtraction.

Correct sign
↓

$(x^2 - 2x + 3) - (x^2 + 2x - 2) = x^2 - 2x + 3 - x^2 - 2x + 2$ Correct

↑
Correct sign

Just as we did for addition, we can use a vertical arrangement to subtract one polynomial from another. (The vertical arrangement doesn't work well with subtractions involving three or more polynomials.) When using a vertical arrangement, write the polynomial that is being subtracted underneath the one it is being subtracted from. Be sure to align the like terms in vertical columns.

EXAMPLE 5   ■   Subtracting Polynomials Vertically

Use a vertical arrangement to perform the following operations.

(a) $(3x^2 + 7x - 6) - (3x^2 + 7x)$

(b) $(4x^4 - 2x^3 + 5x^2 - x + 8) - (3x^4 - 2x^3 + 3x - 4)$

Solution

(a) $(3x^2 + 7x - 6)$ ⟹ $3x^2 + 7x - 6$

$\underline{-(3x^2 + 7x)}$ ⟹ $\underline{-3x^2 - 7x}$ Change signs and add

-6 Combine like terms

(b) $(4x^4 - 2x^3 + 5x^2 - x + 8)$ ⟹ $4x^4 - 2x^3 + 5x^2 - x + 8$

$\underline{-(3x^4 - 2x^3 + 3x - 4)}$ ⟹ $\underline{-3x^4 + 2x^3 - 3x + 4}$

$x^4 + 5x^2 - 4x + 12$ ∎

EXAMPLE 6   ∎   **Combining Polynomials**

Perform the indicated operations.

(a) $(3x^2 - 7x + 2) - (4x^2 + 6x - 1) + (-x^2 + 4x + 5)$

(b) $(-2x^2 + 4x - 3) - [(4x^2 - 5x + 8) - (-x^2 + x + 3)]$

(c) $3(x^2 - 2x + 1) - 2(x^2 + x - 3)$

Solution

(a) $(3x^2 - 7x + 2) - (4x^2 + 6x - 1) + (-x^2 + 4x + 5)$

$= 3x^2 - 7x + 2 - 4x^2 - 6x + 1 - x^2 + 4x + 5$ Change signs

$= (3x^2 - 4x^2 - x^2) + (-7x - 6x + 4x) + (2 + 1 + 5)$ Group like terms

$= -2x^2 - 9x + 8$ Standard form

(b) $(-2x^2 + 4x - 3) - [(4x^2 - 5x + 8) - (-x^2 + x + 3)]$

$= (-2x^2 + 4x - 3) - [4x^2 - 5x + 8 + x^2 - x - 3]$

$= (-2x^2 + 4x - 3) - [(4x^2 + x^2) + (-5x - x) + (8 - 3)]$

$= (-2x^2 + 4x - 3) - [5x^2 - 6x + 5]$

$= -2x^2 + 4x - 3 - 5x^2 + 6x - 5$

$= (-2x^2 - 5x^2) + (4x + 6x) + (-3 - 5)$

$= -7x^2 + 10x - 8$

(c) $3(x^2 - 2x + 1) - 2(x^2 + x - 3)$

$= 3x^2 - 6x + 3 - 2x^2 - 2x + 6$

$= (3x^2 - 2x^2) + (-6x - 2x) + (3 + 6)$

$= x^2 - 8x + 9$ ∎

DISCUSSION PROBLEM   ∎   **Adding Polynomials**

Is it possible to find two third-degree polynomials whose sum is a second-degree polynomial? (If it is possible, give an example. If it is not possible, state the reason.)

Is it possible to find two third-degree polynomials whose sum is a fourth-degree polynomial? (If it is possible, give an example. If it is not possible, state the reason.)

∎

Warm-Up

The following warm-up exercises involve skills that were covered in earlier sections. You will use these skills in the exercise set for this section.

In Exercises 1–4, use the Distributive Property to expand the given expression.

1. $10(x - 1)$

2. $4(3 - 2z)$

3. $-\dfrac{1}{2}(4 - 6x)$

4. $-25(2x - 3)$

In Exercises 5 and 6, list the terms of the algebraic expression.

5. $10x - 3y + 4$

6. $-2r + 8s - 6$

In Exercises 7 and 8, give the coefficient of the term.

7. $\dfrac{3}{4}x$

8. $-8y$

In Exercises 9 and 10, simplify the expression by combining like terms.

9. $8y - 2x + 7x - 10y$

10. $\dfrac{5}{6}x - \dfrac{2}{3}x + 8$

5.1 EXERCISES

In Exercises 1–10, write the polynomial in standard form. Then find its degree and leading coefficient.

1. $2x - 3$

2. $4x^2 + 9$

3. $9 - 2y^4$

4. $5x^3 - 3x^2 + 10$

5. $8x + 2x^5 - x^2 - 1$

6. $5z - 10z^2$

7. 10

8. -32

9. $v_0 t - 16t^2$
(v_0 is a constant)

10. $64 - \dfrac{1}{2}at^2$
(a is a constant)

In Exercises 11–16, determine whether the polynomial is a monomial, a binomial, or a trinomial.

11. $x^2 - 2x + 3$

12. $-6y$

13. $x^3 - 4$

14. $u^2 - 3u + 5$

15. 5

16. $16 - z^2$

In Exercises 17–20, determine whether the algebraic expression is a polynomial.

17. $\dfrac{6}{x}$ **18.** $6x$ **19.** $9 - z$ **20.** $9 - |z|$

In Exercises 21–26, give an example of a polynomial in one variable that satisfies the given conditions. (*Note:* Each problem has many correct answers.)

21. A binomial of degree 3   **22.** A trinomial of degree 4

23. A monomial of degree 2   **24.** A binomial of degree 5

25. A trinomial of degree 6   **26.** A monomial of degree 0

In Exercises 27–36, perform the addition using a vertical arrangement.

27. $(x^2 - 4) + (2x^2 + 6)$ **28.** $(x^3 + 2x - 3) + (4x + 5)$

29. $(3 - x^3) + (5 + 3x^3)$ **30.** $(2z^3 + 3z - 2) + (z^2 - 2z)$

31. $(2 - 3y) + (y^4 + 3y + 2)$ **32.** $(a^2 + 3a - 2) + (5a - a^2 - 6) + (a^2 + 2)$

33. $(x^2 - 2x + 2) + (x^2 + 4x) + 2x^2$ **34.** $(5y + 10) + (y^2 - 3y - 2)$

35. Add $6x^2 + 5$ to $3 - 2x^2$.   **36.** Add $2z - 8z^2 - 3$ to $z^2 + 5z$.

In Exercises 37–40, perform the addition using a horizontal arrangement.

37. $(3z^2 - z + 2) + (z^2 - 4)$ **38.** $(6x^4 + 8x) + (4x - 6)$

39. $(2a - 3) + (a^2 - 2a) + (4 - a^2)$ **40.** $(u - 3) + (4u + 1)$

In Exercises 41–50, perform the subtraction using a vertical arrangement.

41. $(2x^2 - x + 1) - (3x^2 + x - 1)$ **42.** $(y^4 - 2) - (y^4 + 2)$

43. $(2 - x^3) - (2 + x^3)$ **44.** $(4z^3 - 6) - (-z^3 + z - 2)$

45. $(4t^3 - 3t + 5) - (3t^2 - 3t - 10)$ **46.** $(-s^2 - 3) - (2s^2 + 10s)$

47. $(6x^3 - 3x^2 + x) - (x^3 + 3x^2 + 3)$ **48.** $(y^2 - y) - (2y^2 + y)$

49. Subtract $7x^3 - 4x + 5$ from $10x^3 + 15$.   **50.** Subtract $y^5 - y^4$ from $y^2 + 3y^4$.

In Exercises 51–56, perform the subtraction using a horizontal arrangement.

51. $(x^2 - x) - (x - 2)$ **52.** $(x^2 - 4) - (x^2 - 4)$

53. $(4 - 2x - x^3) - (3 - 2x + 2x^3)$ **54.** $(t^4 - 2t^2) - (3t^2 - t^4 - 5)$

55. $10 - (u^2 + 5)$ **56.** $(z^3 + z^2 + 1) - z^2$

In Exercises 57–70, perform the indicated operations.

57. $(6x - 5) - (8x + 15)$

58. $(2x^2 + 1) + (x^2 - 2x + 1)$

59. $-(x^3 - 2) + (4x^3 - 2x)$

60. $-(5x^2 - 1) - (-3x^2 + 5)$

61. $2(x^4 + 2x) + (5x + 2)$

62. $(z^4 - 2z^2) + 3(z^4 + 4)$

63. $(15x^2 - 6) - (-8x^3 - 14x^2 - 17)$

64. $(15x^4 - 18x - 19) - (-13x^4 - 5x + 15)$

65. $5z - [3z - (10z + 8)]$

66. $(y^3 + 1) - [(y^2 + 1) + (3y - 7)]$

67. $2(t^2 + 5) - 3(t^2 + 5) + 5(t^2 + 5)$

68. $-10(u + 1) + 8(u - 1) - 3(u + 6)$

69. $8v - 6(3v - v^2) + 10(10v + 3)$

70. $3(x^2 - 2x + 3) - 4(4x + 1) - (3x^2 - 2x)$

71. *Area* Find the area of the shaded region in the accompanying figure.

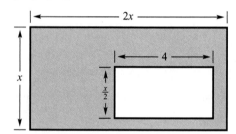

72. *Area* Find the area of the shaded region in the accompanying figure.

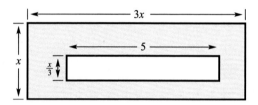

| SECTION 5.2 | **Multiplying Polynomials**
Monomial Multipliers • *Multiplying Binomials* • *Multiplying Polynomials* • *Special Products* |

Monomial Multipliers

To multiply polynomials we make use of many of the rules for simplifying algebraic expressions. You may want to review the rules of exponents (Section 2.1), the Distributive Property (Section 2.2), and the procedures for removing symbols of grouping and combining like terms (Section 2.3).

The simplest type of polynomial multiplication involves a monomial multiplier. The product is obtained by direct application of the Distributive Property. For instance, to multiply the monomial $2x$ by the polynomial $(3x^2 - 4x + 1)$, we multiply *each* of the terms of the polynomial by $2x$.

$$(2x)(3x^2 - 4x + 1) = (2x)(3x^2) - (2x)(4x) + (2x)(1)$$
$$= 6x^3 - 8x^2 + 2x$$

EXAMPLE 1 ■ Finding Products with Monomial Multipliers

Multiply the following.

(a) $(3x - 7)(-2x)$ (b) $3x^2(5x - x^3 + 2)$ (c) $(-x)(2x^2 - 3x)$

Solution

(a) $(3x - 7)(-2x) = 3x(-2x) - 7(-2x)$ Distributive Property

$\qquad = -6x^2 + 14x$ Standard form

(b) $3x^2(5x - x^3 + 2) = (3x^2)(5x) - (3x^2)(x^3) + (3x^2)(2)$ Distributive Property

$\qquad = 15x^3 - 3x^5 + 6x^2$

$\qquad = -3x^5 + 15x^3 + 6x^2$ Standard form

(c) $(-x)(2x^2 - 3x) = (-x)(2x^2) - (-x)(3x)$ Distributive Property

$\qquad = -2x^3 + 3x^2$ Standard form ■

Multiplying Binomials

To multiply two binomials, we make use of both (left and right) forms of the Distributive Property. For example, if we treat the binomial $(5x + 7)$ as a single quantity, we can multiply $(3x - 2)$ by $(5x + 7)$ as follows.

$(3x - 2)(5x + 7) = 3x(5x + 7) - 2(5x + 7)$

$\qquad = (3x)(5x) + (3x)(7) - (2)(5x) - 2(7)$

$\qquad = 15x^2 + 21x - 10x - 14$

Product of First terms	Product of Outer terms	Product of Inner terms	Product of Last terms

$\qquad = 15x^2 + 11x - 14$

With practice you should be able to multiply two binomials without writing out all of the above steps. In fact, the four products in the boxes above suggest that we can put the product of two binomials in the FOIL form in just one step. This is referred to as the **FOIL Method**. Note that the words *first*, *outer*, *inner*, and *last* refer to the position of the terms in the original product.

$(3x - 2)(5x + 7)$

First, Outer, Inner, Last

EXAMPLE 2 ■ Multiplying Binomials Using the Distributive Property
Multiply the following.

$$(x - 1)(x + 5)$$

Solution

$$(x - 1)(x + 5) = x(x + 5) - (1)(x + 5)$$ Right Distributive Property

$$= x^2 + 5x - x - 5$$ Left Distributive Property

$$= x^2 + (5x - x) - 5$$ Collect like terms

$$= x^2 + 4x - 5$$ Combine like terms ■

EXAMPLE 3 ■ Multiplying Binomials Using the FOIL Method
Use the FOIL Method to multiply the following.
(a) $(3x + 5)(2x + 1)$ (b) $(x - 4)(x + 4)$

Solution

F   O   I   L

(a) $(3x + 5)(2x + 1) = 6x^2 + 3x + 10x + 5$

$$= 6x^2 + 13x + 5$$ Combine like terms

(b) Note that the outer and inner products add up to zero in this case.

F   O   I   L

$$(x - 4)(x + 4) = x^2 + 4x - 4x - 16$$

$$= x^2 - 16$$ Combine like terms ■

EXAMPLE 4 ■ Simplifying Polynomial Expressions
Simplify the following expression and write the result in standard form.

$$5x(x^2 + 2x - 5) - (x^3 + 10x)$$

Solution

$$5x(x^2 + 2x - 5) - (x^3 + 10x)$$

$$= 5x^3 + 10x^2 - 25x - x^3 - 10x$$ Distributive Property

$$= 4x^3 + 10x^2 - 35x$$ Combine like terms ■

EXAMPLE 5 ■ Simplifying Polynomial Expressions
Simplify the following expressions and write the result in standard form.
(a) $(3x^2 - 2)(4x + 7) - (4x)^2$ (b) $(4x + 5)^2$

Solution

(a) $(3x^2 - 2)(4x + 7) - (4x)^2$

$$= 12x^3 + 21x^2 - 8x - 14 - (4x)^2 \qquad \text{Multiply binomials}$$

$$= 12x^3 + 21x^2 - 8x - 14 - 16x^2 \qquad \text{Square monomial}$$

$$= 12x^3 + 5x^2 - 8x - 14 \qquad \text{Combine like terms}$$

(b) Here we can consider raising a binomial to a power as repeated multiplication.

$$(4x + 5)^2 = (4x + 5)(4x + 5) \qquad \text{Repeated multiplication}$$

$$= 16x^2 + 20x + 20x + 25 \qquad \text{Multiply binomials}$$

$$= 16x^2 + 40x + 5 \qquad \text{Combine like terms} \blacksquare$$

Multiplying Polynomials

The FOIL Method for multiplying two binomials is simply a device for guaranteeing the following rule: *each term of one binomial must be multiplied by each term of the other binomial.*

$$(ax + b)(cx + d) = ax(cx) + ax(d) + b(cx) + b(d) \qquad \text{F} \quad \text{O} \quad \text{I} \quad \text{L}$$

The same rule applies to the product of two polynomials: *each term of one polynomial must be multiplied by each term of the other polynomial.* This can be done using either a horizontal or a vertical arrangement, as shown in the next two examples.

EXAMPLE 6   ■   Multiplying Polynomials (Horizontal Arrangement)

Use a horizontal arrangement of factors to find the following product.

$$(2x^2 - 7x + 1)(4x + 3)$$

Solution

To help assure that each term of one polynomial is multiplied by each term of the other, we use the Distributive Property in a horizontal arrangement.

$$(2x^2 - 7x + 1)(4x + 3)$$

$$= (2x^2 - 7x + 1)(4x) + (2x^2 - 7x + 1)(3) \qquad \text{Distributive Property}$$

$$= 8x^3 - 28x^2 + 4x + 6x^2 - 21x + 3 \qquad \text{Distributive Property}$$

$$= 8x^3 - 22x^2 - 17x + 3 \qquad \text{Combine like terms} \blacksquare$$

EXAMPLE 7 ■ Multiplying Polynomials (Vertical Arrangement)

Find the following product using a vertical arrangement of factors.

$$(3x^2 + x - 5)(2x - 1)$$

Solution

In the vertical arrangement, we line up like terms in the same vertical columns much like we align digits in whole-number multiplication.

$$
\begin{array}{r}
3x^2 + x - 5 \\
\times 2x - 1 \\
\hline
-3x^2 - x + 5 \\
6x^3 + 2x^2 - 10x \\
\hline
6x^3 - x^2 - 11x + 5
\end{array}
$$

Place polynomial with most terms on top

$-1(3x^2 + x - 5)$

$2x(3x^2 + x - 5)$

Combine like terms in columns ■

When multiplying two polynomials, it is best to write each in standard form before using either the horizontal or vertical formats. This is illustrated in the next example.

EXAMPLE 8 ■ Multiplying Polynomials

Multiply the following.

$$(x + 3x^2 - 4)(5 + 3x - x^2)$$

Solution

$$
\begin{array}{r}
3x^2 + x - 4 \\
\times -x^2 + 3x + 5 \\
\hline
15x^2 + 5x - 20 \\
9x^3 + 3x^2 - 12x \\
-3x^4 - x^3 + 4x^2 \\
\hline
-3x^4 + 8x^3 + 22x^2 - 7x - 20
\end{array}
$$

Standard form

Standard form

$5(3x^2 + x - 4)$

$3x(3x^2 + x - 4)$

$-x^2(3x^2 + x - 4)$

■

EXAMPLE 9 ■ Multiplying Polynomials

Multiply the following.

$$(x - 3)^3$$

Solution

To raise $(x - 3)$ to the third power, we use two steps. First, since $(x - 3)^3 = (x - 3)^2(x - 3)$, we find the square of $(x - 3)$, as follows.

$$(x - 3)(x - 3) = x^2 - 3x - 3x + 9 \qquad \text{Find } (x - 3)^2$$
$$= x^2 - 6x + 9 \qquad \text{Combine like terms}$$

Now, using a vertical arrangement, we find $(x - 3)^3$ as follows.

$$
\begin{array}{r}
x^2 - 6x + 9 \\
\times \quad\quad x - 3 \\
\hline
-3x^2 + 18x - 27 \\
x^3 - 6x^2 + 9x \quad\quad \\
\hline
x^3 - 9x^2 + 27x - 27
\end{array}
$$

Thus, $(x - 3)^3 = x^3 - 9x^2 + 27x - 27$.　■

MATH MATTERS

The Spectrum

The spectrum is the visible result produced when light is resolved into its various wavelengths or frequencies. For instance, when light is passed through a prism, as shown in the accompanying figure, the light is separated into a rainbow of colors ranging from red to violet.

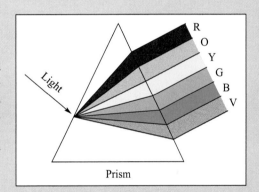

The frequencies and wavelengths of the colors of the rainbow are as follows.

Color of Light	Frequency (10^{14} hertz)	Wavelength (10^{-7} meters)
Red	4.284–4.634	6.470–7.000
Orange	4.634–5.125	5.850–6.470
Yellow	5.125–5.215	5.750–5.850
Green	5.215–6.104	4.912–5.750
Blue	6.104–7.115	4.240–4.912
Violet	7.115–7.495	4.000–4.240

Special Products

Some binomial products, like those in Example 3(b) and Example 5(b), have special forms that occur frequently in algebra. Let's look at those products again. We call the product $(x + 4)(x - 4)$ the **product of the sum and difference of two terms**. With such products, the two middle terms cancel as follows.

$$(x + 4)(x - 4) = x^2 - 4x + 4x - 16 \qquad \text{Sum and difference of two terms}$$

$$= x^2 - 16 \qquad \text{Product has no middle term}$$

Another common type of product is the **square of a binomial**. With this type of product, the middle term is always twice the product of the terms in the binomial.

$$(4x + 5)^2 = (4x + 5)(4x + 5) \qquad \text{Square of a binomial}$$

$$= 16x^2 + 20x + 20x + 25 \qquad \text{Outer and inner terms are equal}$$

$$= 16x^2 + 40x + 25 \qquad \text{Middle term is twice the product of the terms of the binomial}$$

You should learn to recognize the patterns of these two special products. We give the general form of these special products in the following statements. The FOIL Method can be used to verify each of the rules.

Special Products

Let a and b be real numbers, variables, or algebraic expressions.

Special Product	*Example*
Sum and Difference of Two Terms:	
$(a + b)(a - b) = a^2 - b^2$	$(2x - 5)(2x + 5) = 4x^2 - 25$
Square of a Binomial:	
$(a + b)^2 = a^2 + 2ab + b^2$	$(3x + 4)^2 = 9x^2 + 2(3x)(4) + 16$
	$= 9x^2 + 24x + 16$
$(a - b)^2 = a^2 - 2ab + b^2$	$(x - 7)^2 = x^2 - 2(x)(7) + 49$
	$= x^2 - 14x + 49$

NOTE: When squaring a binomial, the resulting middle term is always *twice* the product of the two terms.

$$(a + b)^2 = a^2 + \underbrace{2(ab)}_{} + b^2$$

$$\underbrace{}_{\text{Terms}} \qquad \underset{\substack{\text{Twice the}\\\text{product of}\\\text{the terms}}}{}$$

Be sure you don't forget the middle term. For instance, $(a + b)^2$ is *not* equal to $a^2 + b^2$.

EXAMPLE 10 ■ Finding the Product of the Sum and Difference of Two Terms

Find the following products.

(a) $(5x - 6)(5x + 6)$ (b) $(2 + 3x)(2 - 3x)$

Solution

(a) This special product represents the sum and difference of two terms. Therefore, the product has the form $(a - b)(a + b) = a^2 - b^2$.

$$(5x - 6)(5x + 6) = (5x)^2 - (6)^2$$
$$= 25x^2 - 36$$

(b) This special product also represents the sum and difference of two terms. Therefore, the product has the form $(a + b)(a - b) = a^2 - b^2$.

$$(2 + 3x)(2 - 3x) = (2)^2 - (3x)^2$$
$$= 4 - 9x^2 \qquad ■$$

EXAMPLE 11 ■ Squaring a Binomial

Find the following product.

$$(4x - 9)^2$$

Solution

Note that the *middle* term of the product is twice the product of the two terms of the binomial.

$$(4x - 9)^2 = (4x)^2 - 2(4x)(9) + (9)^2$$
$$= 16x^2 - 72x + 81 \qquad ■$$

DISCUSSION PROBLEM ■ Pascal's Triangle

The following triangular pattern of numbers shows the first seven rows of **Pascal's Triangle**, named after the French mathematician Blaise Pascal (1623–1662). Describe the pattern formed by the numbers in the triangle.

$$
\begin{array}{ccccccccccccc}
 & & & & & & 1 & & & & & & \\
 & & & & & 1 & & 1 & & & & & \\
 & & & & 1 & & 2 & & 1 & & & & \\
 & & & 1 & & 3 & & 3 & & 1 & & & \\
 & & 1 & & 4 & & 6 & & 4 & & 1 & & \\
 & 1 & & 5 & & 10 & & 10 & & 5 & & 1 & \\
1 & & 6 & & 15 & & 20 & & 15 & & 6 & & 1 \\
\end{array}
$$

$$(x + 1)^0 = 1$$
$$(x + 1)^1 = x + 1$$
$$(x + 1)^2 = x^2 + 2x + 1$$
$$(x + 1)^3 = x^3 + 3x^2 + 3x + 1$$
$$(x + 1)^4 = x^4 + 4x^3 + 6x^2 + 4x + 1$$
$$(x + 1)^5 = x^5 + 5x^4 + 10x^3 + 10x^2 + 5x + 1$$
$$(x + 1)^6 = x^6 + 6x^5 + 15x^4 + 20x^3 + 15x^2 + 6x + 1$$

How does Pascal's Triangle relate to the coefficients of the polynomial $(x + 1)^n$? Use this pattern to write out the expansion of $(x + 1)^7$. ■

Warm-Up

The following warm-up exercises involve skills that were covered in earlier sections. You will use these skills in the exercise set for this section.

In Exercises 1 and 2, rewrite the product in exponential form.

1. $(-3)(-3)(-3)(-3)$ **2.** $x \cdot x \cdot x$

In Exercises 3–6, rewrite the expression as a repeated multiplication.

3. $\left(\dfrac{4}{5}\right)^4$ **4.** $(4.5)^5$ **5.** $(x^2)^3$ **6.** $(y^4)^2$

In Exercises 7–10, simplify the algebraic expression.

7. $2(x - 4) + 5x$ **8.** $4(3 - y) + 2(y + 1)$

9. $-3(z - 2) - (z - 6)$ **10.** $(u - 2) - 3(2u + 1)$

5.2 EXERCISES

In Exercises 1–40, perform the indicated multiplication and simplify.

1. $x(-2x)$

2. $(-5y)(-3y)$

3. $\left(\dfrac{x}{4}\right)(10x)$

4. $9x\left(\dfrac{x}{12}\right)$

5. $t^2(4t)$

6. $3u(u^4)$

7. $(-2b^2)(-3b)$

8. $(-4m)(3m^2)$

9. $(x^4)^2$

10. $(2y)^3$

11. $(-3t)^3$

12. $(-4z)^2$

13. $2x(3x)^2$

14. $(4z)^3\left(\dfrac{z}{2}\right)^2$

15. $2x(6x^4) - 3x^2(2x^2)$

16. $-8y(-5y^4) - 2y^2(5y^3)$

17. $y(3 - y)$

18. $z(z - 3)$

19. $-x(x^2 - 4)$

20. $-t(10 - 3t)$

21. $3t(2t - 5)$

22. $-5u(u^2 + 4)$

23. $3x(x^2 - 2x + 1)$

24. $y^2(4y^2 + 2y - 3)$

25. $-4x(3 + 3x^2 - 6x^3)$

26. $5v(5 - 4v + 5v^2)$

27. $2x(x^2 - 2x + 8)$

28. $-3x^2(x - 3)$

29. $-2x(-3x)(5x + 2)$

30. $4x(-2x)(x^2 - 1)$

31. $(x + 3)(x + 4)$

32. $(x - 5)(x + 10)$

33. $(3x - 5)(2x + 1)$

34. $(7x - 2)(4x - 3)$

35. $(x + y)(x + 2y)$

36. $(2x - y)(x - 2y)$

37. $5x(x + 1) - 3x(2x - 4)$

38. $(2x - 3)(x + 3) + 3(2x - 1)$

39. $(s - 2t)(s + t) - (s - 2t)(s - t)$

40. $2u(u - v) + 3v(u + v)$

In Exercises 41–46, perform the indicated multiplication and simplify. (Use a horizontal arrangement.)

41. $(x - 2)(x^2 + 2x + 4)$

42. $(x + 1)(x^2 - x + 1)$

43. $(x^3 - 2x + 1)(x - 5)$

44. $(x^2 + 9)(x^2 - x - 4)$

45. $(x + 3)(x^2 - 6x + 2)$

46. $(2x^2 + 3)(x^2 - 2x + 3)$

In Exercises 47–52, perform the indicated multiplication and simplify. (Use a vertical arrangement.)

47. $(x + 3)(x^2 - 3x + 9)$

48. $(2x^2 + 3)(4x^4 - 6x^2 + 9)$

49. $(x^2 + x - 2)(x^2 - x + 2)$

50. $(x^2 + 2x + 5)(2x^2 - x - 1)$

51. $(x^3 + x + 3)(x^2 + 5x - 4)$

52. $(x - 1)(x^2 + x + 1)$

In Exercises 53–56, perform the indicated operations and simplify.

53. $(x + 1)(x^2 + 2x - 1)$

54. $(3s + 1)(3s + 4) - 3s(3s - 4)$

55. $(u - 1)(2u + 3)(2u + 1)$

56. $(2x + 5)(x - 2)(5x - 3)$

In Exercises 57–76, perform the indicated multiplication. (Use one of the special products.)

57. $(x + 2)(x - 2)$ **58.** $(y + 9)(y - 9)$ **59.** $(2u + 3)(2u - 3)$ **60.** $(3z + 4)(3z - 4)$

61. $(2x + 3y)(2x - 3y)$ **62.** $(5u + 12v)(5u - 12v)$ **63.** $(x + 6)^2$ **64.** $(x + 10)^2$

65. $(a - 2)^2$ **66.** $(t - 3)^2$ **67.** $(3x - 2)^2$ **68.** $(2x - 8)^2$

69. $(2x - 5y)^2$ **70.** $(4s + 3t)^2$ **71.** $(8 - 3z)^2$ **72.** $(1 - 5t)^2$

73. $[(x + 1) + y]^2$ **74.** $[(x - 3) - y]^2$ **75.** $[u - (v - 3)]^2$ **76.** $[2u + (v + 1)]^2$

In Exercises 77 and 78, perform the indicated multiplication and simplify.

77. $(x + 2)^2 - (x - 2)^2$

78. $(u + 5)^2 + (u - 5)^2$

In Exercises 79 and 80, verify the given equation by multiplying the left side of the equation.

79. $(a + b)^3 = a^3 + 3a^2b + 3ab^2 + b^3$

80. $(a - b)^3 = a^3 - 3a^2b + 3ab^2 - b^3$

In Exercises 81 and 82, perform the indicated multiplication by using the results of Exercises 79 and 80.

81. $(x + 2)^3$ **82.** $(x - 1)^3$

83. *Dimensions of a Sign* The height of a rectangular sign is twice its width w (see figure). Find (a) the perimeter and (b) the area of the sign.

84. *Area of a Clock Face* The radius of a circular clock face is $x + 2$ inches (see figure). Find the area A of the clock face.

Figure for 83   Figure for 84

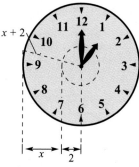

85. *Area of a Sail* The base of a triangular sail is $2x$ feet and its height is $x + 10$ feet (see figure). Find the area A of the sail.

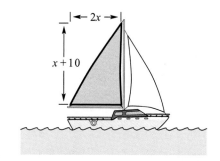

86. *Balance in an Account* After two years an investment of $500, compounded annually at an interest rate r, will yield an amount

$$500(1 + r)^2.$$

Find this product.

87. *FOIL Method* Add the areas of each rectangle in the accompanying figure to find the area of the entire figure. Notice that this demonstrates the FOIL Method for finding the product $(x + a)(x + b)$.

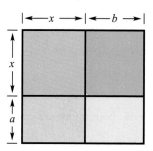

88. *Special Product* Add the areas of the four rectangular regions shown in the accompanying figure. What special product does this represent?

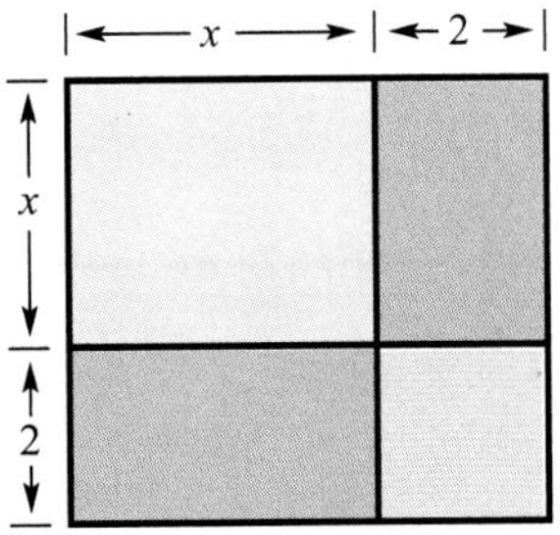

89. Perform each of the following multiplications.
(a) $(x - 1)(x + 1)$
(b) $(x - 1)(x^2 + x + 1)$
(c) $(x - 1)(x^3 + x^2 + x + 1)$
(d) Use the pattern formed in the first three products to guess the pattern for

$$(x - 1)(x^4 + x^3 + x^2 + x + 1).$$

Verify your guess by multiplying.

SECTION **5.3**

Dividing Polynomials

Dividing a Monomial by a Monomial • *Dividing a Polynomial by a Monomial* • *Dividing a Polynomial by a Binomial*

Dividing a Monomial by a Monomial

How we divide two polynomials depends on the number of terms in the divisor. In this section we limit our discussion to division of polynomials by monomials or binomials.

We begin by looking at division problems in which both the numerator *and* the denominator are monomials. To divide a monomial by a monomial, we make use of the *subtraction* property of exponents illustrated in the following examples. (In each of the following, we assume that the variable is *not zero*.)

By Reducing	*By Subtracting Exponents*
$\frac{x^4}{x^2} = \frac{x \cdot x \cdot \cancel{x} \cdot \cancel{x}}{\cancel{x} \cdot \cancel{x}} = x^2$	$\frac{x^4}{x^2} = x^{4-2} = x^2$
$\frac{y^3}{y^3} = \frac{\cancel{y} \cdot \cancel{y} \cdot \cancel{y}}{\cancel{y} \cdot \cancel{y} \cdot \cancel{y}} = 1$	$\frac{y^3}{y^3} = y^{3-3} = y^0 = 1$
$\frac{5y^7}{2y^5} = \frac{5 \cdot y \cdot y \cdot \cancel{y} \cdot \cancel{y} \cdot \cancel{y} \cdot \cancel{y} \cdot \cancel{y}}{2 \cdot \cancel{y} \cdot \cancel{y} \cdot \cancel{y} \cdot \cancel{y} \cdot \cancel{y}} = \frac{5y^2}{2}$	$\frac{5y^7}{2y^5} = \frac{5y^{7-5}}{2} = \frac{5y^2}{2}$
$\frac{2x^2}{x^5} = \frac{2 \cdot \cancel{x} \cdot \cancel{x}}{x \cdot x \cdot x \cdot \cancel{x} \cdot \cancel{x}} = \frac{2}{x^3}$	$\frac{2x^2}{x^5} = \frac{2}{x^{5-2}} = \frac{2}{x^3}$

The above examples show that we can divide one monomial by another (provided each has the same variable) by simply subtracting the smaller exponent from the larger exponent. The following list summarizes this technique.

Rules of Exponents	Let m and n be positive integers, and let a represent a real number, a variable, or an algebraic expression.

1. $\dfrac{a^m}{a^n} = a^{m-n}$, if $m > n$

2. $\dfrac{a^m}{a^n} = \dfrac{1}{a^{n-m}}$, if $n > m$

3. $\dfrac{a^n}{a^n} = 1 = a^0$

Note that we make a special definition for raising a *nonzero* quantity to the zero power. That is, if $a \neq 0$, then $a^0 = 1$.

NOTE: The subtraction property of exponents works only for monomials with the *same variable* for a base. For instance, the subtraction rule does not apply to x^5/y^3 because no cancellations can occur.

$$\frac{x^5}{y^3} = \frac{x \cdot x \cdot x \cdot x \cdot x}{y \cdot y \cdot y} \qquad \text{No simplifying is possible}$$

EXAMPLE 1 ■ Dividing a Monomial by a Monomial

Perform the following divisions. (In each case assume that $x \neq 0$.)

(a) $16x^4 \div 4x^2$ (b) $16x^4 \div 3x$

(c) $8x^3 \div \dfrac{1}{2}x^3$ (d) $12x^3 \div 4x^4$

Solution

(a) To divide the monomial $16x^4$ by the monomial $4x^2$, we use the subtraction property of exponents.

$$\frac{16x^4}{4x^2} = \frac{16}{4}(x^{4-2}) = 4x^2$$

(b) $\dfrac{16x^4}{3x} = \dfrac{16}{3}(x^{4-1}) = \dfrac{16}{3}x^3$

(c) $\dfrac{8x^3}{\frac{1}{2}x^3} = \dfrac{8}{\left(\frac{1}{2}\right)}\left(\dfrac{x^3}{x^3}\right) = 8\left(\dfrac{2}{1}\right)(1) = 16$

(d) $\dfrac{12x^3}{4x^4} = \dfrac{12}{4}\left(\dfrac{x^3}{x^4}\right) = 3\left(\dfrac{1}{x}\right) = \dfrac{3}{x}$ ■

Although the division problems in Example 1 are straightforward, you should study this example carefully. Be sure you can justify each step in the example. Also remember that there are often several ways to solve a given problem in algebra. As you gain practice and confidence, you will discover that you like some techniques better than others. For instance, which one of the following techniques seems best to you?

1. $\dfrac{6x^3}{2x} = \dfrac{3 \cdot \cancel{2} \cdot x \cdot x \cdot \cancel{x}}{\cancel{2} \cdot \cancel{x}} = 3x^2, \quad x \neq 0$

2. $\dfrac{6x^3}{2x} = \left(\dfrac{6}{2}\right)\left(\dfrac{x^3}{x}\right) = (3)(x^{3-1}) = 3x^2, \quad x \neq 0$

3. $\dfrac{6x^3}{2x} = \dfrac{3x^3}{x} = 3x^{3-1} = 3x^2, \quad x \neq 0$

When two different people (even math teachers) are writing out the steps of a solution, rarely will the steps be the same, so don't worry if your steps don't look exactly like someone else's. If you feel comfortable with writing more steps, then you should write more steps. Just be sure that each step can be justified by the rules of algebra.

Dividing a Polynomial by a Monomial

The preceding illustrations show how to divide a *monomial* by a monomial. To divide a *polynomial* by a monomial we use the reverse form of the rule for adding two fractions with a common denominator. In Section 1.3 we added two fractions with like denominators, using the rule

$$\frac{a}{c} + \frac{b}{c} = \frac{a+b}{c}.$$

Here we view the rule in the *reverse* order and divide a polynomial by a monomial by dividing each term of the polynomial by the monomial. That is,

$$\frac{a+b}{c} = \frac{a}{c} + \frac{b}{c}.$$

Here is an example.

$$\frac{x^3 - 5x^2}{x^2} = \frac{x^3}{x^2} - \frac{5x^2}{x^2} = x - 5, \quad x \neq 0$$

Note that the essence of this problem is to separate the original division problem into *two* division problems, each involving the division of a monomial by a monomial.

EXAMPLE 2   ■   Dividing a Polynomial by a Monomial

(a) $\dfrac{6x - 5}{3} = \dfrac{6x}{3} - \dfrac{5}{3} = 2x - \dfrac{5}{3}$

(b) $\dfrac{4x^2 - 3x}{3x} = \dfrac{4x^2}{3x} - \dfrac{3x}{3x} = \dfrac{4x}{3} - 1, \quad x \neq 0$

(c) $\dfrac{8x^3 - 6x^2 + 10x}{2x} = \dfrac{8x^3}{2x} - \dfrac{6x^2}{2x} + \dfrac{10x}{2x} = 4x^2 - 3x + 5, \quad x \neq 0$ ■

EXAMPLE 3 ■ Dividing a Polynomial by a Monomial

Perform the following division. (Assume $x \neq 0$.)

$(5x^3 - 4x^2 - x + 6) \div 2x$

Solution

$$\frac{5x^3 - 4x^2 - x + 6}{2x} = \frac{5x^3}{2x} - \frac{4x^2}{2x} - \frac{x}{2x} + \frac{6}{2x} \qquad \text{Divide each term separately}$$

$$= \frac{5x^2}{2} - 2x - \frac{1}{2} + \frac{3}{x} \qquad \text{Use rules for dividing monomials} \qquad ■$$

Dividing a Polynomial by a Binomial

To divide a polynomial by a *binomial*, we follow the *long division* pattern used for dividing whole numbers. Recall that we divide 6,982 by 27 in the following manner.

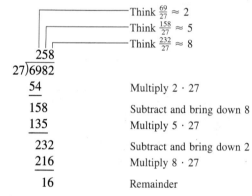

Think $\frac{69}{27} \approx 2$
Think $\frac{158}{27} \approx 5$
Think $\frac{232}{27} \approx 8$

$$
\begin{array}{r}
258 \\
27\overline{)6982} \\
\underline{54} \\
158 \\
\underline{135} \\
232 \\
\underline{216} \\
16
\end{array}
$$

Multiply $2 \cdot 27$
Subtract and bring down 8
Multiply $5 \cdot 27$
Subtract and bring down 2
Multiply $8 \cdot 27$
Remainder

Sometimes we express the remainder as a fractional part of the divisor and then the quotient is written as $258\frac{16}{27}$ or $258 + \frac{16}{27}$. Note in the next four examples how this long division algorithm is used to divide a polynomial by a binomial.

EXAMPLE 4 ■ Dividing a Polynomial by a Binomial

Divide $(x^2 + 3x + 5)$ by $(x + 1)$.

Solution

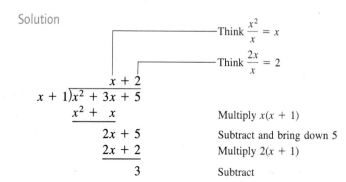

Think $\frac{x^2}{x} = x$
Think $\frac{2x}{x} = 2$

$$
\begin{array}{r}
x + 2 \\
x + 1\overline{)x^2 + 3x + 5} \\
\underline{x^2 + x} \\
2x + 5 \\
\underline{2x + 2} \\
3
\end{array}
$$

Multiply $x(x + 1)$
Subtract and bring down 5
Multiply $2(x + 1)$
Subtract

Considering the remainder as a fractional part of the divisor, we can write the quotient as

$$x + 2 + \frac{3}{x + 1}.$$

Thus, we have

$$\underbrace{\overbrace{\frac{x^2 + 3x + 5}{x + 1}}^{\text{Dividend}}}_{\text{Divisor}} = \overbrace{x + 2}^{\text{Quotient}} + \underbrace{\overbrace{\frac{3}{x + 1}}^{\text{Remainder}}}_{\text{Divisor}}.$$

EXAMPLE 5   ■   A Binomial Divisor

Divide $(6x^3 - 19x^2 + 16x - 4)$ by $(x - 2)$.

Solution

Think $\dfrac{6x^3}{x} = 6x^2$

Think $-\dfrac{7x^2}{x} = -7x$

Think $\dfrac{2x}{x} = 2$

$$
\begin{array}{r}
6x^2 - 7x + 2 \\
x - 2 \overline{\smash{)}6x^3 - 19x^2 + 16x - 4} \\
\underline{6x^3 - 12x^2} \\
-7x^2 + 16x \\
\underline{-7x^2 + 14x} \\
2x - 4 \\
\underline{2x - 4} \\
0
\end{array}
$$

Multiply $6x^2(x - 2)$

Subtract and bring down $16x$

Multiply $-7x(x - 2)$

Subtract and bring down -4

Multiply $2(x - 2)$

Remainder

Thus, we have

$$\frac{6x^3 - 19x^2 + 16x - 4}{x - 2} = 6x^2 - 7x + 2, \quad x \neq 2.$$

Note that the remainder in this problem is zero. In such cases we say that the denominator (or divisor) **divides evenly** into the numerator (or dividend).   ■

Ask her to do this

EXAMPLE 6   ■   A Binomial Divisor

Divide $(-13x^3 + 10x^4 + 8x - 7x^2 + 4)$ by $(3 - 2x)$.

Solution

We first write the divisor and dividend in standard polynomial form (decreasing powers of x).

$$
\begin{array}{r}
-5x^3 - x^2 + 2x - 1 \\
-2x + 3 \overline{)10x^4 - 13x^3 - 7x^2 + 8x + 4} \\
\underline{10x^4 - 15x^3} \\
2x^3 - 7x^2 \\
\underline{2x^3 - 3x^2} \\
- 4x^2 + 8x \\
\underline{- 4x^2 + 6x} \\
2x + 4 \\
\underline{2x + 3} \\
7
\end{array}
$$

Using the fractional form of the remainder, we can write

$$
\frac{10x^4 - 13x^3 - 7x^2 + 8x + 4}{-2x + 3} = -5x^3 - x^2 + 2x - 1 + \frac{7}{-2x + 3}. \qquad \blacksquare
$$

EXAMPLE 7 ■ A Binomial Divisor

Divide $x^3 - 1$ by $x - 1$.

Solution

Because there are no x^2- or x-terms in the dividend, we line up the subtractions by using *zero* coefficients (or by leaving a space) for these missing terms.

$$
\begin{array}{r}
x^2 + x + 1 \\
x - 1 \overline{)x^3 + 0x^2 + 0x - 1} \\
\underline{x^3 - x^2} \\
x^2 \\
\underline{x^2 - x} \\
x - 1 \\
\underline{x - 1} \\
0
\end{array}
$$

Thus, $x - 1$ divides evenly into $x^3 - 1$ and we can write

$$
\frac{x^3 - 1}{x - 1} = x^2 + x + 1, \quad x \neq 1. \qquad \blacksquare
$$

DISCUSSION PROBLEM ■ You Be the Instructor

Suppose you are tutoring a friend in algebra and you want to create some division problems for practice. Describe a method for finding a third-degree polynomial that is evenly divisible by a given first-degree polynomial. Demonstrate your method by finding a third-degree polynomial that is evenly divisible by $x + 2$. ■

Warm-Up

The following warm-up exercises involve skills that were covered in earlier sections. You will use these skills in the exercise set for this section.

In Exercises 1–4, write the fraction in reduced form.

1. $\dfrac{8}{12}$ **2.** $\dfrac{18}{144}$ **3.** $\dfrac{60}{150}$ **4.** $\dfrac{175}{42}$

In Exercises 5–10, find the product and simplify.

5. $-2x^2(5x^3)$ **6.** $4y(3y^2 - 1)$

7. $(2z + 1)(2z - 1)$ **8.** $(x + 4)(2x - 5)$

9. $(x + 7)^2$ **10.** $(x + 1)(x^2 - x + 1)$

5.3 EXERCISES

In Exercises 1–10, perform the indicated division in two ways: by cancellation and by subtracting exponents. (Assume that each denominator is not zero.)

1. $\dfrac{x^5}{x^2}$ **2.** $\dfrac{z^6}{z^4}$ **3.** $\dfrac{2^3 y^4}{2^2 y^2}$ **4.** $\dfrac{3^5 x^7}{3^3 x^4}$

5. $\dfrac{z^4}{z^7}$ **6.** $\dfrac{y^8}{y^3}$ **7.** $\dfrac{x^5}{x^5}$ **8.** $\dfrac{5^3 x}{5^3 x^2}$

9. $\dfrac{4^5 x^3}{4x^5}$ **10.** $\dfrac{6z^5}{6z^5}$

In Exercises 11–22, simplify the given expression. (Assume that each denominator is not zero.)

11. $\dfrac{-3x^2}{6x}$ **12.** $\dfrac{-4a^6}{-2a^2}$ **13.** $\dfrac{-12z^3}{-3z}$ **14.** $\dfrac{16y^5}{8y^3}$

15. $\dfrac{32b^4}{12b^3}$ **16.** $\dfrac{-7c^2}{8c^5}$ **17.** $\dfrac{-22y^2}{4y}$ **18.** $\dfrac{54x^2}{-24x^4}$

19. $\dfrac{-18s^4}{-12r^2 s}$ **20.** $-\dfrac{16v}{4v^2}$ **21.** $\dfrac{(-3z)^2}{18z^3}$ **22.** $\dfrac{(4ab)^3}{(-8a)^2 b^4}$

In Exercises 23–70, perform the indicated division and simplify. (Assume that each denominator is not zero.)

23. $\dfrac{3z + 3}{3}$ **24.** $\dfrac{7x + 7}{7}$ **25.** $\dfrac{4z - 12}{4}$ **26.** $\dfrac{8u - 24}{8}$

27. $(5x^2 - 2x) \div x$

28. $(16a^2 + 5a) \div a$

29. $\dfrac{25z^3 + 10z^2}{-5z}$

30. $\dfrac{12c^4 - 36c}{-6c}$

31. $\dfrac{8z^3 + 3z^2 - 2z}{z}$

32. $\dfrac{3x^4 + 5x^3 - 4x^2}{x^2}$

33. $-\dfrac{4x^2 - 3x}{x}$

34. $\dfrac{14y^4 + 21y^3}{-7y^3}$

35. $\dfrac{m^3 + 3m - 4}{m}$

36. $\dfrac{l^2 - 4l + 8}{l}$

37. $\dfrac{5x^2 + 3x}{x}$

38. $\dfrac{16r^3 - 12r^2}{4r^2}$

39. $\dfrac{x^2 - x - 2}{x + 1}$

40. $\dfrac{x^2 - 5x + 6}{x - 2}$

41. $(x^2 + 9x + 20) \div (x + 4)$

42. $(x^2 - 7x - 30) \div (x - 10)$

43. Divide $2z^2 + 5z - 3$ by $z + 3$.

44. Divide $4u^2 - 18u - 10$ by $u - 5$.

45. Divide $3y^2 + 4y - 4$ by $3y - 2$.

46. Divide $7t^2 - 10t - 8$ by $7t + 4$.

47. $(18t^2 - 21t - 4) \div (3t - 4)$

48. $(20t^2 + 32t - 16) \div (2t + 4)$

49. $\dfrac{9x^2 - 1}{3x + 1}$

50. $\dfrac{25y^2 - 4}{5y - 2}$

51. $\dfrac{x^3 - 8}{x - 2}$

52. $\dfrac{x^3 + 27}{x + 3}$

53. $\dfrac{x^3 - 7x + 6}{x - 2}$

54. $\dfrac{x^3 - 28x - 48}{x + 4}$

55. $(7x + 3) \div (x + 2)$

56. $(8x - 5) \div (2x + 1)$

57. $\dfrac{x^2 + 9}{x + 3}$

58. $\dfrac{y^2 + 3}{y + 3}$

59. $\dfrac{4x^2 + 3x - 3}{x + 1}$

60. $\dfrac{7x^2 + 4x + 3}{x + 2}$

61. $\dfrac{2x^2 + 7x + 8}{2x + 3}$

62. $\dfrac{6x^2 + x - 15}{3x - 4}$

63. $\dfrac{4z^2 + 6z}{2z - 1}$

64. $\dfrac{10y^2 - 3y}{y + 3}$

65. $\dfrac{x^3 - 4x^2 + 9x - 7}{x - 2}$

66. $\dfrac{2x^3 - 2x^2 + 3x + 9}{x + 1}$

67. $\dfrac{3t^3 + 7t^2 + 3t - 2}{t + 2}$

68. $\dfrac{8t^3 - 44t^2 - 19t - 36}{t - 6}$

69. Divide $x^4 - 1$ by $x - 1$.

70. Divide x^4 by $x - 1$.

In Exercises 71–74, simplify the given expression. (Assume that each denominator is not zero.)

71. $\dfrac{4x^3}{x^2} - 2x$

72. $\dfrac{25x^2y}{10x} + \dfrac{3xy^2}{2y}$

73. $\dfrac{8u^2v}{2u} + \dfrac{(uv)^2}{uv}$

74. $\dfrac{x^2 + 2x + 1}{x + 1} - (3x - 4)$

In Exercises 75–78, determine whether the cancellation shown is valid.

75. $\dfrac{3+4}{3} = \dfrac{\overset{1}{\cancel{3}}+4}{\underset{1}{\cancel{3}}} = 5$

76. $\dfrac{4+7}{4+11} = \dfrac{\cancel{4}+7}{\cancel{4}+11} = \dfrac{7}{11}$

77. $\dfrac{7 \cdot 12}{19 \cdot 7} = \dfrac{\cancel{7} \cdot 12}{19 \cdot \cancel{7}} = \dfrac{12}{19}$

78. $\dfrac{24}{43} = \dfrac{2\cancel{4}}{\cancel{4}3} = \dfrac{2}{3}$

SECTION
5.4

Negative Exponents and Scientific Notation
Negative Exponents • Rules of Exponents • Scientific Notation

---

Negative Exponents

In this section we extend the rules for exponential expressions to include **negative exponents**. Consider the rule

$$a^m \cdot a^n = a^{m+n}, \quad a \neq 0.$$

If this rule is to hold for negative exponents, then the statement

$$a^2 \cdot a^{-2} = a^{2+(-2)} = a^0 = 1$$

implies that a^{-2} is the *reciprocal* of a^2. In other words, it must be true that

$$a^{-2} = \frac{1}{a^2}.$$

This suggests the following general rule for negative exponents.

Negative Exponent Rule

Let n be an integer and let a be a real number, variable, or algebraic expression such that $a \neq 0$.

$$a^{-n} = \frac{1}{a^n}$$

NOTE: This rule allows us to move *factors* from numerator to denominator as in

$$3a^{-4} = \frac{3}{a^4}$$

or from denominator to numerator as in

$$\frac{1}{5a^{-3}} = \frac{a^3}{5}.$$

EXAMPLE 1   ■   Monomials Involving Negative Exponents

Rewrite each of the following so that the expression has no negative exponents.

(a) x^{-7} (b) $5x^{-4}$ (c) $\dfrac{1}{2x^{-3}}$ (d) $x^{-2}y^3$

Solution

(a) For this expression, the negative exponent is in the numerator. To eliminate the negative exponent, we move the x^{-7} from the numerator to the denominator, as follows.

$$x^{-7} = \frac{1}{x^7}$$

(b) In this expression the negative exponent applies *only* to the x, and not to the 5. Thus, to eliminate the negative exponent, we move the x^{-4} to the denominator (by changing the sign of the exponent). Note that we leave the 5 in the numerator, as follows.

$$5x^{-4} = \frac{5}{x^4}$$

(c) Here again, the negative exponent applies only to the x, and not to the 2. Hence, we eliminate the negative exponent, as follows.

$$\frac{1}{2x^{-3}} = \frac{x^3}{2}$$

(d) For this problem, the negative exponent applies to the x, so we can rewrite the problem, as follows.

$$x^{-2}y^3 = \frac{y^3}{x^2}$$ ■

The negative exponent rule allows us to move only *factors* in a numerator (or denominator), *not terms*. For example,

$$\frac{x^{-2} \cdot y}{4} = \frac{y}{4x^2}$$ x^{-2} is a *factor* of the numerator

whereas

$$\frac{x^{-2} + y}{4} \text{ DOES NOT EQUAL } \frac{y}{4x^2}.$$ x^{-2} is a *term* of the numerator

Rules of Exponents

All of the rules of exponents that we have studied so far can also apply to negative exponents. For convenience, we summarize all of the rules of exponents in the following list.

Rules of Exponents	Let m and n be integers, and let a and b be real numbers, variables, or algebraic expressions such that $a \neq 0$ and $b \neq 0$.

Property   *Example*

1. $a^m a^n = a^{m+n}$ $y^2 \cdot y^4 = y^{2+4}$

2. $\dfrac{a^m}{a^n} = a^{m-n}$ $\dfrac{x^7}{x^4} = x^{7-4} = x^3$

3. $a^{-n} = \dfrac{1}{a^n}$ $y^{-4} = \dfrac{1}{y^4}$

4. $a^0 = 1$ $(x^2 + 1)^0 = 1$

5. $(ab)^m = a^m b^m$ $(5x)^4 = 5^4 x^4$

6. $\left(\dfrac{a}{b}\right)^m = \dfrac{a^m}{b^m}$ $\left(\dfrac{2}{x}\right)^3 = \dfrac{2^3}{x^3}$

7. $(a^m)^n = a^{mn}$ $(y^3)^{-4} = y^{3(-4)} = y^{-12}$

EXAMPLE 2 ■ **Using Rules of Exponents**

Use rules of exponents to write each of the following without negative exponents. (Assume that each variable is not zero.)

(a) $(-3ab^4)(4ab^{-3})$ (b) $\dfrac{y^{-2}}{3y^{-5}}$

Solution

(a) $(-3ab^4)(4ab^{-3}) = (-3)(4)(a)(a)(b^4)(b^{-3})$ Regroup factors

$\qquad\qquad\qquad\quad = (-12)(a^{1+1})(b^{4-3})$ Apply rules of exponents

$\qquad\qquad\qquad\quad = -12a^2 b$ Simplify

(b) $\dfrac{y^{-2}}{3y^{-5}} = \dfrac{y^{-2}y^5}{3}$ Apply rules of exponents

$\qquad\quad = \dfrac{y^{-2+5}}{3}$

$\qquad\quad = \dfrac{y^3}{3}$ Simplify   ■

NOTE: There is more than one way to solve problems like those in Example 2. For instance, you might prefer to write Example 2(b) as

$$\frac{y^{-2}}{3y^{-5}} = \frac{y^5}{3y^2} = \frac{y^{5-2}}{3} = \frac{y^3}{3}.$$

EXAMPLE 3 ■ Using Rules of Exponents

Use rules of exponents to write each of the following without negative exponents. (Assume that each variable is not zero.)

(a) $3x^{-1}(-4x^2y)^0$ (b) $\left(\dfrac{5x^3}{y^{-1}}\right)^2$ (c) $\left(\dfrac{a^2}{3}\right)^{-2}$

Solution

(a) This problem is a little tricky. Note that the factor $(-4x^2y)$ is raised to the zero power. Since any number (other than zero) raised to the zero power is 1, we can write the following.

$$3x^{-1}(-4x^2y)^0 = 3x^{-1}(1) = \frac{3}{x}$$

(b) This problem can also be tricky. The important thing to realize is that the *entire fraction* $(5x^3/y^{-1})$ is raised to the second power. This means that we must apply the exponent 2 to each factor of the numerator and denominator, as follows.

$$\left(\frac{5x^3}{y^{-1}}\right)^2 = \frac{(5x^3)^2}{(y^{-1})^2} = \frac{5^2(x^3)^2}{y^{-2}} = \frac{25x^6}{y^{-2}} = 25x^6y^2$$

(c) $\left(\dfrac{a^2}{3}\right)^{-2} = \dfrac{a^{-4}}{3^{-2}} = \dfrac{3^2}{a^4} = \dfrac{9}{a^4}$ ■

Scientific Notation

Exponents provide an efficient way of writing and computing with the very large (or very small) numbers used in science. For instance, a drop of water contains more than 33 billion billion molecules. That is 33 followed by 18 zeros.

33,000,000,000,000,000,000

It is convenient to write such numbers in **scientific notation**. This notation has the form $c \times 10^n$, where $1 \le c < 10$ and n is an integer. Thus, the number of molecules in a drop of water can be written in scientific notation as follows.

$3.3 \times 10^{19} = 33,000,000,000,000,000,000$

19 places

The *positive* exponent 19 indicates that the number is large (10 or more) and that the decimal point has been moved 19 places. A *negative* exponent in scientific notation indicates that the number is *small* (less than 1). For instance, the mass (in grams) of one electron is approximately as follows.

$9.0 \times 10^{-28} = 0.00000000000000000000000000009$

28 places

EXAMPLE 4   ■   Converting from Decimal Notation to Scientific Notation

Write each of the following real numbers in scientific notation.

(a) 0.0000782   (b) 836,100,000

Solution

(a) To write the number in scientific notation, we move the decimal point five places, as follows.

$$0.0000782 = 7.82 \times 10^{-5}$$

Five places

(b) To write the number in scientific notation, we move the decimal point eight places, as follows.

$$836,100,000.0 = 8.361 \times 10^{8}$$

Eight places   ■

EXAMPLE 5   ■   Converting from Scientific Notation to Decimal Notation

Convert each of the following numbers from scientific notation to decimal notation.

(a) 1.345×10^{2} (b) 9.36×10^{-6}

Solution

(a) For this number, the exponent of the number 10 is a positive 2. Thus, to write the number in decimal notation, we move the decimal point two places, as follows.

$$1.345 \times 10^{2} = 134.5$$

Two places

(b) For this number, the exponent of the number 10 is a negative 6. Thus, to write the number in decimal notation, we move the decimal point six places, as follows.

$$9.36 \times 10^{-6} = 0.00000936$$

Six places   ■

NOTE: To convert decimal numbers to scientific notation, or conversely, make the following associations.

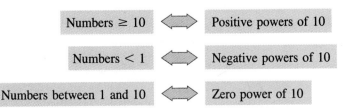

Numbers ≥ 10	⟺	Positive powers of 10
Numbers < 1	⟺	Negative powers of 10
Numbers between 1 and 10	⟺	Zero power of 10

Most scientific calculators automatically switch to scientific notation when they are displaying large (or small) numbers that exceed the display range. Try multiplying

$98,900,000 \times 5,000.$

If your calculator follows standard conventions, its display should be

| 4.945   11 | or | 4.945   E   11 |.

This means that $c = 4.945$ and the exponent of 10 is $n = 11$, which implies that the number is 4.945×10^{11}.

To *enter* numbers in scientific notation, your calculator should have an exponential entry key labeled [EE] or [EXP]. If you were to perform the preceding multiplication using scientific notation, you could begin by writing

$98,900,000 \times 5,000 = (9.89 \times 10^7)(5.0 \times 10^3)$

and then entering the following.*

9.89 [EXP] 7 [×] 5 [EXP] 3 [=]   Scientific

9.89 [EE] 7 [×] 5 [EE] 3 [ENTER]   Graphing

EXAMPLE 6 ■   **Using Scientific Notation**

Find the following using a scientific calculator.

(a) $78,000 \times 2,400,000,000$ (b) $0.000000748 \div 500$

Solution

(a) Since $78,000 = 7.8 \times 10^4$ and $2,400,000,000 = 2.4 \times 10^9$, we can multiply the two numbers using the following calculator steps.

7.8 [EXP] 4 [×] 2.4 [EXP] 9 [=]   Scientific

7.8 [EE] 4 [×] 2.4 [EE] 9 [ENTER]   Graphing

After entering these keystrokes, the calculator display should read | 1.872   14 |. Therefore, the product of the two numbers is

$(7.8 \times 10^4)(2.4 \times 10^9) = 1.872 \times 10^{14} = 187,200,000,000,000.$

(b) Since $0.000000748 = 7.48 \times 10^{-7}$ and $500 = 5.0 \times 10^2$, we can divide the two numbers as follows.

7.48 [EXP] 7 [+/−]   [÷] 5 [EXP] 2 [=]   Scientific

7.48 [EE] [(−)] 7 [÷] 5 [EE] 2 [ENTER]   Graphing

*The graphing calculator keystrokes given in this text correspond to the *TI-81* graphing calculator from *Texas Instruments*. For other graphing calculators, the keystrokes may differ.

After entering these keystrokes, the calculator display should read $\boxed{1.496 \quad -9}$. Therefore, the quotient of the two numbers is

$$\frac{7.48 \times 10^{-7}}{5.0 \times 10^{2}} = 1.496 \times 10^{-9} = 0.000000001496. \qquad \blacksquare$$

DISCUSSION PROBLEM   ■   **Exponential Expressions**

We know that 10^6 is a large number and 10^{-6} is a small number. However, the sizes of the exponential expressions

$$x^6 \quad \text{and} \quad x^{-6}$$

depend on the value of x. Can you find a value of x such that

$$x^6 = 0.000001?$$

Can you find a value of x such that

$$x^{-6} = 1,000,000? \qquad \blacksquare$$

Warm-Up

The following warm-up exercises involve skills that were covered in earlier sections. You will use these skills in the exercise set for this section.

In Exercises 1–6, evaluate the given quantity.

1. 4^3 **2.** $(-2)^5$ **3.** $\left(-\dfrac{2}{3}\right)^2$ **4.** -8^2

5. $25 - 3^2 \cdot 2$ **6.** $(12 - 9)^3 \cdot 4 + 36$

In Exercises 7–10, simplify the given expression. (Assume that each denominator is not zero.)

7. $x^2 \cdot x^3$ **8.** $(ab)^4 \div (ab)^3$ **9.** $\dfrac{u^4 v^2}{uv}$ **10.** $(y^2 z^3)(z^2)$

5.4   EXERCISES

In Exercises 1–10, rewrite the quantity so that it has no negative exponents and then evaluate.

1. 3^{-2} **2.** 5^{-3} **3.** $(-4)^{-3}$ **4.** $(-6)^{-2}$

5. $\dfrac{1}{16^{-1}}$ **6.** $\dfrac{4}{3^{-2}}$ **7.** $\dfrac{2^{-4}}{3^{-2}}$ **8.** $\dfrac{4^{-3}}{2}$

9. $\left(\dfrac{2}{3}\right)^{-2}$ **10.** $\left(\dfrac{3}{4}\right)^{-3}$

In Exercises 11–40, use the rules of exponents to write the expression without negative exponents and simplify. (Assume that each variable is not zero.)

11. $\dfrac{x^2}{x^{-3}}$

12. $\dfrac{z^4}{z^{-2}}$

13. $\dfrac{y^{-5}}{y}$

14. $\dfrac{x^{-3}}{x^2}$

15. $\dfrac{x^{-4}}{x^{-2}}$

16. $\dfrac{t^{-5}}{t^{-1}}$

17. $(y^{-3})^2$

18. $(z^{-2})^3$

19. $(s^2)^{-1}$

20. $(a^3)^{-3}$

21. $(2x^2)^{-2}$

22. $(2x^5)^0$

23. $\dfrac{b^2 \cdot b^{-3}}{b^4}$

24. $\dfrac{c^{-3} \cdot c^4}{c^{-1}}$

25. $(3x^2y)^{-2}$

26. $(4x^{-3}y^2)^{-3}$

27. $(4a^{-2}b^3)^{-3}$

28. $(-2s^{-1}t^{-2})^{-1}$

29. $(-2x^2)^3(4x^3)^{-1}$

30. $(4y^{-2})(3y^4)$

31. $(5x^2y^4z^6)^3(5x^2y^4z^6)^{-3}$

32. $(2x^2 + y^2)^4(2x^2 + y^2)^{-4}$

33. $\left(\dfrac{x}{10}\right)^{-1}$

34. $\left(\dfrac{4}{z}\right)^{-2}$

35. $\left(\dfrac{3z^2}{x}\right)^{-2}$

36. $\left(\dfrac{x^{-3}y^4}{5}\right)^{-3}$

37. $(2x^2y)^0$

38. $\dfrac{(3z)^{-2}}{(3z)^{-2}}$

39. $\dfrac{3}{2} \cdot \left(\dfrac{-2}{3}\right)^{-3}$

40. $\dfrac{(-x)^3}{-x^{-2}}$

In Exercises 41–50, write the given number in decimal form.

41. 1.09×10^6

42. 2.345×10^8

43. 6.21×10^0

44. 9.4675×10^4

45. 8.52×10^{-3}

46. 7.021×10^{-5}

47. 8.67×10^{-2}

48. 4.73×10^0

49. $(8 \times 10^3) + (3 \times 10^0) + (5 \times 10^{-2})$

50. $(6 \times 10^4) + (9 \times 10^3) + (4 \times 10^{-1})$

In Exercises 51–60, write the given number in scientific notation.

51. 93,000,000

52. 900,000,000

53. 1,637,000,000

54. 67.8

55. 0.000435

56. 0.008367

57. 0.004392

58. 0.0875

59. 16,000,000

60. 0.00000045

In Exercises 61–70, use a calculator to evaluate the given quantity.

61. $8,000,000 \times 623,000$

62. $93,200,000 \times 1,657,000$

63. $0.000345 \times 8,980,000,000$

64. $345,000 \times 0.000086$

65. $3,200,000^5$

66. $75,000,000^6$

67. $(3.28 \times 10^{-6})^4$

68. 0.000045^3

69. $\dfrac{848,000,000}{1,620,000}$

70. $\dfrac{67,000,000}{0.0052}$

71. *Distance to a Star* The star Beta Andromedae is approximately 76 light-years from Earth (see figure). (A light-year is the distance light can travel in one year.) Estimate the distance to this star if a light-year is approximately 5.88×10^{12} miles.

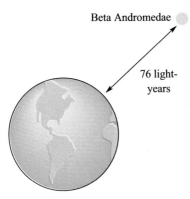

Beta Andromedae

76 light-years

72. *Time for Light to Travel* Light travels from the sun to the Earth in approximately

$$\frac{9.3 \times 10^7}{1.1 \times 10^7} \text{ minutes.}$$

Write this time in decimal form.

73. *Hydraulic Compression* The hydraulic cylinder in a large press contains 2 gallons of oil. When the cylinder is under full pressure the actual volume of oil will decrease by the following amount.

$$2(150)(20 \times 10^{-6}) \text{ gallons}$$

Write this volume in decimal form.

74. *Powers of 2* Complete the following table by evaluating the indicated powers of 2.

x	-1	-2	-3	-4	-5
2^x					

75. *An Infinite Sum* Successively remove from the unit square in the accompanying figure the fractional parts given in the table of Exercise 74. If this process were continued, what do you think would happen? Use this to estimate the sum.

$$\frac{1}{2} + \frac{1}{4} + \frac{1}{8} + \frac{1}{16} + \cdots.$$

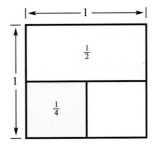

CHAPTER 5   SUMMARY

As you review and prepare for a test on this chapter, first try to obtain a global view of what was discussed. Then review the specific skills needed in each category.

Operations with Polynomials

- *Identify* the degree and leading coefficient of a polynomial.

  Section 5.1, Exercises 1–10

- *Add* two or more polynomials by combining like terms.

  Section 5.1, Exercises 27–36,
  Exercises 57–70

- *Subtract* polynomials by changing signs and combining like terms.

  Section 5.1, Exercises 51–70

- *Combine* polynomials by removing symbols of grouping.

  Section 5.1, Exercises 57–70

- *Multiply* two binomials, using a special product.

  Section 5.2, Exercises 57–60

- *Multiply* a polynomial by a monomial, a binomial, or another polynomial.

  Section 5.2, Exercises 1–56

- *Divide* a monomial by a monomial.

  Section 5.3, Exercises 1–22

- *Divide* a polynomial by a monomial.

  Section 5.3, Exercises 23–36

- *Divide* a polynomial by a binomial, using long division.

  Section 5.3, Exercises 37–70

Negative Exponents and Scientific Notation

- *Simplify* an algebraic expression by using rules of exponents to rewrite the expression so that it contains no negative exponents.

  Section 5.4, Exercises 11–40

- *Convert* a decimal number to scientific notation.

  Section 5.4, Exercises 41–50

- *Convert* a number in scientific notation to decimal form.

  Section 5.4, Exercises 51–60

- *Evaluate* a numerical expression that requires the use of scientific notation.

  Section 5.4, Exercises 61–74

Chapter 5 Review Exercises

In Exercises 1–10, change the right side so that it is equal to the left side of the equation or expression.

1. $\dfrac{3}{8} + \dfrac{1}{8} \neq \dfrac{4}{16}$

2. $\dfrac{1}{3}(2x) \neq \dfrac{2}{3} \cdot \dfrac{x}{3}$

3. $-3(x - 2) \neq -3x - 6$

4. $3x - (x + 5) \neq 2x + 5$

5. $(2x)^4 \neq 2x^4$

6. $(-x)^4 \neq -x^4$

7. $(x + 2)^2 \neq x^2 + 4$

8. $3^{-2} \neq -9$

9. $\dfrac{7 - x}{7} \neq 1 - x$

10. $(x + 3)(x - 3) \neq x^2 + 6x - 9$

In Exercises 11–40, perform the indicated operations and simplify. (Assume that each denominator is not zero.)

11. $(2x + 3) + (x - 4)$

12. $\left(\dfrac{1}{2}x + \dfrac{2}{3}\right) + \left(4x + \dfrac{1}{3}\right)$

13. $(4 - x^2) + 2(x - 2)$

14. $(3u + 4u^2) + 5(u + 1) + 3u^2$

15. $(t - 5) - (t^2 - t - 5)$

16. $(y^2 + 3) - (y^2 - 9)$

17. $(-x^3 - 3x) - 2(2x^3 + x + 1)$

18. $(z^2 + 6z) - 3(z^2 + 2z)$

19. $4y^2 - [y - 3(y^2 + 2)]$

20. $(6a^3 + 3a) - 2[a + (a^3 - 2)]$

21. $2x(x + 4)$

22. $3y(-4y)(y + 1)$

23. $(x + 3)(2x - 4)$

24. $(3y + 2)(4y - 3)$

25. $(x^2 + 5x + 2)(2x + 3)$

26. $(s^3 + 4s - 3)(s - 3)$

27. $2u(u - 5) - (u + 1)(u - 5)$

28. $(3v - 2)(-2v) + 2v(3v - 2)$

29. $(5x^2 + 15x) \div 5x$

30. $(8u^3 + 4u^2) \div 2u$

31. $7x^3 \div 3x^2$

32. $(x^2 + x) \div x$

33. $\dfrac{x^2 - x - 6}{x - 3}$

34. $\dfrac{x^2 + x - 20}{x + 5}$

35. $\dfrac{24x^2 - x - 8}{3x - 2}$

36. $\dfrac{4x + 7}{2x - 1}$

37. $\dfrac{2x^3 + 2x^2 - x + 2}{x - 1}$

38. $\dfrac{6x^4 - 4x^3 - 27x^2 + 18x}{3x - 2}$

39. $\dfrac{x^4 - 3x^2 + 2}{x^2 - 1}$

40. $\dfrac{3x^4}{x^2 - 1}$

In Exercises 41–50, find the given product.

41. $(x + 3)^2$

42. $(x - 5)^2$

43. $(4x - 7)^2$

44. $(9 - 2x)^2$

45. $(u - 6)(u + 6)$

46. $(3a + 8)(3a - 8)$

47. $(2t - 1)(3t + 1)$

48. $(4 - 3b)(1 - b)$

49. $[(a - 1) + b][(a - 1) - b]$

50. $[(a - 1) + b]^2$

In Exercises 51–60, evaluate the given quantity.

51. 4^{-2}

52. 3^{-4}

53. $6^{-4}6^2$

54. $(2^2 \cdot 3^2)^{-1}$

55. $\left(\dfrac{3}{5}\right)^{-3}$

56. $\left(\dfrac{2^{-2}}{3}\right)^2$

57. $(3 \times 10^3)^2$

58. $(4 \times 10^{-3})(5 \times 10^7)$

59. $\dfrac{1.85 \times 10^9}{5 \times 10^4}$

60. $\dfrac{1}{(4 \times 10^{-2})^3}$

In Exercises 61–66, rewrite the expression without negative exponents and simplify. (Assume that each variable is not zero.)

61. $(x^2 y^{-3})^2$

62. $5(x + 3)^0$

63. $\dfrac{t^{-4}}{t^{-1}}$

64. $\dfrac{a^3 \cdot a^{-2}}{a^{-1}}$

65. $\left(\dfrac{y}{5}\right)^{-2}$

66. $(3x^2 y^4)^3 (3x^2 y^4)^{-3}$

67. *Dimensions of a Wall* The length of a rectangular wall is x units, and its height is $x - 3$ units (see figure). Find (a) the perimeter of the wall and (b) the area of the wall.

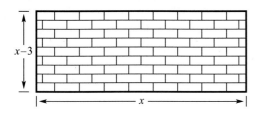

68. *Probability* The probability of three successes in five trials of an experiment is given by

$$10p^3(1 - p)^2. \quad \text{Find this product.}$$

69. *Metal Expansion* When the temperature of an iron steam pipe 150 feet long is increased by 100° C, the length of the pipe will increase by $100(150)(10 \times 10^{-6})$ feet, as shown in the accompanying figure. Write this amount in decimal form.

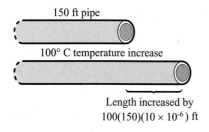

150 ft pipe

100° C temperature increase

Length increased by
$100(150)(10 \times 10^{-6})$ ft

70. *Special Product* The accompanying figure shows two squares with edges of lengths x and y.
  (a) Remove the small square with edge of length y from the larger square. What is the area of the remaining figure?
  (b) Slide and rotate the top rectangle so that it fits against the right side of the figure. What are the dimensions of the resulting rectangle? What is its area and what special product formula have you demonstrated geometrically?

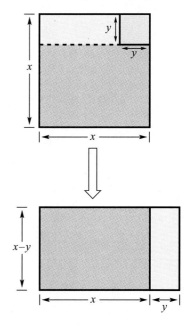

CHAPTER 5 TEST

Take this test as you would take a test in class. After you are done, check your work with the answers given in the back of the book.

1. Determine the degree and the leading coefficient of the polynomial
   $-3x^4 - 5x^2 + 2x - 10$.

2. Add: $(3z^2 - 3z + 7) + (8 - z^2)$

3. Subtract: $(8u^3 + 3u^2 - 2u - 1) - (u^3 + 3u^2 - 2u)$ 4. Simplify: $-5(x^2 - 1) + 3(4x + 7) - (x^2 + 26)$

5. Simplify: $6y - [2y - (3 + 4y - y^2)]$ 6. Multiply and simplify: $4x\left(\dfrac{3x}{2}\right)^2$

7. Multiply and simplify: $3a(2a^2 - 5)$

8. Multiply and simplify: $(5b + 3)(2b - 1)$

9. Multiply and simplify: $(z + 2)(2z^2 - 3z + 5)$

10. Multiply and simplify: $(x - 5)^2$

11. Multiply and simplify: $(2x - 3)(2x + 3)$

12. The length of the base of a triangle is $6x$ units and its height is $2x + 3$ units. Find the area A of the triangle if $A = \frac{1}{2}bh$.

13. Simplify: $\dfrac{-6a^2b}{-9ab}$

14. Divide: $\dfrac{15x + 25}{5}$

15. Divide: $\dfrac{x^3 - x - 6}{x - 2}$

16. Divide: $\dfrac{4x^3 + 10x^2 - 2x - 5}{2x + 1}$

17. Evaluate: $\dfrac{2^{-3}}{3^{-1}}$

18. Simplify: $(3x^{-2}y^3)^{-2}$

19. The mean distance from the Earth to the moon is 3.84×10^8 meters. Write the distance in decimal form.

20. The standard atmospheric pressure is 101,300 newtons per square meter. Write the pressure in scientific notation.

21. Evaluate without using a calculator: $(1.5 \times 10^5)^2$

CHAPTER SIX
Factoring

Chapter Overview

Learning algebra has much in common with learning to swim, ice skate, or roller-skate. It is better when we learn the skills both forwards and backwards. Much of what you will study in this chapter is the reverse of what you studied in Section 5.2. Instead of multiplying two polynomials to get a new one, you will learn to find factors whose product is a given polynomial. This factoring process is an important tool for solving equations and for simplifying algebraic expressions.

SECTION
6.1

Factoring Polynomials with Common Factors

Introduction • Greatest Common Factor • Common Monomial Factors • Factoring by Grouping

Introduction

In this chapter we switch from multiplying polynomials to the *reverse* process of **factoring polynomials**. For example, in Section 5.2 we used the Distributive Property to *multiply* and *remove* parentheses as follows.

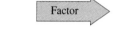

$$2x\,(7 - 3x) = 14x - 6x^2 \qquad \text{Distributive Property}$$

In this chapter we use the Distributive Property in the reverse direction to *factor* and *create* parentheses.

$$14x - 6x^2 = 2x(7 - 3x) \qquad \text{Distributive Property}$$

Factoring an expression changes a *sum of terms* into a *product of factors*. Later we will see that this is an important strategy for solving equations and for simplifying algebraic expressions.

Greatest Common Factor

To factor an expression efficiently, you need to understand the concept of the **greatest common factor** of two (or more) integers or terms. Recall that every positive integer greater than 1 can be factored as a product of **prime** numbers. (A prime number is a positive integer whose only factors are 1 and itself. The first several primes are 2, 3, 5, 7, 11, and 13.) Here are some examples of factoring a number as a product of prime numbers.

$$12 = 4 \cdot 3 = 2 \cdot 2 \cdot 3$$
$$30 = 6 \cdot 5 = 2 \cdot 3 \cdot 5$$

The **greatest common factor** of two or more integers is the greatest integer that is a factor of each number. For example, the greatest common factor of 12 and 30 and 6.

$$\left.\begin{array}{l} 12 = 2 \cdot 2 \cdot 3 \\ 30 = 2 \cdot 3 \cdot 5 \end{array}\right\} \quad 6 = 2 \cdot 3 \text{ is common to both integers}$$

In a similar way, an algebraic term can be factored into a product of prime factors. Some examples are as follows.

$$6x^2 = 2 \cdot 3 \cdot x \cdot x$$
$$15x^3y = 3 \cdot 5 \cdot x \cdot x \cdot x \cdot y$$

The greatest common factor of the expressions $6x^2$ and $15x^3y$ is $3x^2$.

$$6x^2 = 2 \cdot 3 \cdot x \cdot x = (3x^2)(2)$$
$$15x^3y = 3 \cdot 5 \cdot x \cdot x \cdot x \cdot y = (3x^2)(5xy)$$

$\left.\right\}$ $3x^2$ is common to both expressions

EXAMPLE 1 ■ Finding the Greatest Common Factor

Find the greatest common factor of each of the following pairs.

(a) 18 and 42   (b) $5x^2y^2$ and $30x^3y$

Solution

(a) Since $18 = 2 \cdot 3 \cdot 3$ and $42 = 2 \cdot 3 \cdot 7$, it follows that the greatest common factor of 18 and 42 is $2 \cdot 3 = 6$.

(b) From the factorizations

$$5x^2y^2 = 5 \cdot x \cdot x \cdot y \cdot y = (5x^2y)(y)$$
$$30x^3y = 2 \cdot 3 \cdot 5 \cdot x \cdot x \cdot x \cdot y = (5x^2y)(6x)$$

we conclude that the greatest common factor is $5x^2y$.   ■

EXAMPLE 2 ■ Finding the Greatest Common Factor

Find the greatest common factor of the following.

$$8x^5, \quad 20x^3, \quad \text{and} \quad 16x^4$$

Solution

From the factorizations

$$8x^5 = 2 \cdot 2 \cdot 2 \cdot x \cdot x \cdot x \cdot x \cdot x = (4x^3)(2x^2)$$
$$20x^3 = 2 \cdot 2 \cdot 5 \cdot x \cdot x \cdot x = (4x^3)(5)$$
$$16x^4 = 2 \cdot 2 \cdot 2 \cdot 2 \cdot x \cdot x \cdot x \cdot x = (4x^3)(4x)$$

we conclude that the greatest common factor is $4x^3$.   ■

Common Monomial Factors

Suppose we consider the three terms given in Example 2 as terms of the polynomial

$$8x^5 + 16x^4 + 20x^3.$$

The greatest common factor, $4x^3$, of these terms is called the **greatest common monomial factor** of the polynomial. When we use the Distributive Property to remove this factor from each term of the polynomial, we say that we are **factoring out** the greatest common monomial factor.

$$8x^5 + 16x^4 + 20x^3 = 4x^3(2x^2) + 4x^3(4x) + 4x^3(5) \qquad \text{Factor each term}$$
$$= 4x^3(2x^2 + 4x + 5) \qquad\qquad \text{Factor out common monomial factor}$$

EXAMPLE 3 ■ Factoring Out the Greatest Common Monomial Factor

Factor out the greatest common monomial factor from the following polynomial.

$$6x - 18$$

Solution

The greatest common factor of $6x$ and 18 is 6. Thus, we can factor the given polynomial as follows.

$$6x - 18 = 6(x) - 6(3) = 6(x - 3) \qquad\qquad ■$$

EXAMPLE 4 ■ Factoring Out the Greatest Common Monomial Factor

Factor out the greatest common monomial factor from the following polynomial.

$$10y^3 - 25y^2$$

Solution

The greatest common factor of $10y^3$ and $25y^2$ is $5y^2$. Thus, we can factor the given polynomial as follows.

$$10y^3 - 25y^2 = (5y^2)(2y) - (5y^2)(5) = 5y^2(2y - 5) \qquad\qquad ■$$

EXAMPLE 5 ■ Factoring Out the Greatest Common Monomial Factor

Factor out the greatest common monomial factor from the following polynomial.

$$45x^3 - 15x^2 - 15$$

Solution

The greatest common factor of $45x^3$, $15x^2$, and 15 is 15. Thus, we can factor the given polynomial as follows.

$$45x^3 - 15x^2 - 15 = 15(3x^3) - 15(x^2) - 15(1) = 15(3x^3 - x^2 - 1) \qquad ■$$

NOTE: If a polynomial (with integer coefficients) in x has a greatest common monomial factor of the form ax^n, then the following statements must be true.

1. The coefficient, a, of the greatest common monomial factor must be the greatest integer that *divides* each coefficient of the polynomial.

2. The variable factor, x^n, of the greatest common monomial factor has the *least power* of x of all terms of the polynomial.

EXAMPLE 6 ■ Common Monomial Factors

Factor each of the following.

(a) $35y^3 - 7y^2 - 14y$ (b) $6y^5 + 3y^3 - 2y^2$

Solution

(a) $35y^3 - 7y^2 - 14y = 7y(5y^2) - 7y(y) - 7y(2)$ 7 divides all coefficients
                                                         y has least power of terms

$= 7y(5y^2 - y - 2)$ Factor out common monomial factor

(b) $6y^5 + 3y^3 - 2y^2 = y^2(6y^3) + y^2(3y) - y^2(2)$ y^2 has least power of terms

$= y^2(6y^3 + 3y - 2)$ Factor out common monomial factor ■

We usually consider the greatest common monomial factor of a polynomial to have a positive coefficient. However, sometimes it is convenient to factor a negative number out of a polynomial. You can see how this is done in the next example.

EXAMPLE 7 ■ A Negative Common Monomial Factor

Factor the following polynomial in two ways: first, by factoring out a 2, and then by factoring out a -2.

$-2x^2 + 8x - 12$

Solution

To factor out the common monomial factor of 2, we write the following.

$-2x^2 + 8x - 12 = 2(-x^2) + 2(4x) - 2(6)$ Factor 2 out of each term

$= 2(-x^2 + 4x - 6)$ Factor 2 out of the polynomial

To factor -2 out of the polynomial, we write the following.

$-2x^2 + 8x - 12 = -2(x^2) + (-2)(-4x) + (-2)(6)$ Factor -2 out of each term

$= -2(x^2 - 4x + 6)$ Factor -2 out of the polynomial ■

NOTE: With experience, you should be able to omit writing the second step shown in Examples 6 and 7. For instance, to factor -2 out of $-2x^2 + 8x - 12$, you could simply write

$-2x^2 + 8x - 12 = -2(x^2 - 4x + 6).$

Factoring by Grouping

There are occasions when a common factor of a polynomial is not simply a monomial. For instance, the polynomial

$x^2(x - 2) + 3(x - 2)$

has the common *binomial* factor $(x - 2)$. Factoring out this common factor produces

$$x^2(x - 2) + 3(x - 2) = (x - 2)(x^2 + 3).$$

This type of factoring procedure often occurs with polynomial expressions involving four (or more) terms, and it involves a procedure called **factoring by grouping**.

EXAMPLE 8 ■ Common Binomial Factors

Factor each of the following.

(a) $5x^2(7x - 1) - 3(7x - 1)$ (b) $2x(3x - 4) + (3x - 4)$

Solution

(a) Each of the terms of this polynomial has a binomial factor of $(7x - 1)$. Factoring this binomial out of each term produces the following.

$$5x^2(7x - 1) - 3(7x - 1) = (7x - 1)(5x^2 - 3)$$

(b) Each of the terms of this polynomial has a binomial factor of $(3x - 4)$. Factoring this binomial out of each term produces the following.

$$2x(3x - 4) + (3x - 4) = (3x - 4)(2x + 1)$$

Be sure you see that when $(3x - 4)$ is factored out of itself, we are left with 1. This follows from the fact that $(3x - 4)(1) = (3x - 4)$. ■

In Example 8 the given polynomial was already grouped so that it was easy to determine the common binomial factor. In practice you will have to do the grouping as well as the factoring. To see how this works, let's consider the expression

$$x^3 - 2x^2 - 3x + 6$$

and try to *factor* it. We note first that there is no common monomial factor to take out of all four terms. But suppose we *group* the first two terms together and the last two terms together. Then we have the following.

$x^3 - 2x^2 - 3x + 6 = (x^3 - 2x^2) - (3x - 6)$	Group terms
$= x^2(x - 2) - 3(x - 2)$	Take out common factor in each group
$= (x - 2)(x^2 - 3)$	Take out common binomial factor

Note that a different grouping would yield the same result.

$x^3 - 2x^2 - 3x + 6 = (x^3 - 3x) - (2x^2 - 6)$	Group terms
$= x(x^2 - 3) - 2(x^2 - 3)$	Take out common factor in each group
$= (x^2 - 3)(x - 2)$	Take out common binomial factor

EXAMPLE 9 ■ Factoring by Grouping

Factor the following polynomial.

$$x^3 - 2x^2 - x + 2$$

Solution

By grouping the first two terms together and the third and fourth terms together, we obtain the following.

$$
\begin{aligned}
x^3 - 2x^2 - x + 2 &= (x^3 - 2x^2) - (x - 2) \\
&= x^2(x - 2) - (x - 2) \\
&= (x - 2)(x^2 - 1)
\end{aligned}
$$

■

EXAMPLE 10 ■ Factoring by Grouping

Factor the following polynomial.

$$3x^2 - 12x + 5x - 20$$

Solution

By grouping the first two terms together and the third and fourth terms together, we obtain the following.

$$
\begin{aligned}
3x^2 - 12x + 5x - 20 &= (3x^2 - 12x) + (5x - 20) \\
&= 3x(x - 4) + 5(x - 4) \\
&= (x - 4)(3x + 5)
\end{aligned}
$$

■

EXAMPLE 11 ■ Factoring by Grouping

Factor the following expression.

$$x^2y - 3y - x^3 + 3x$$

Solution

By grouping the first two terms together and the third and fourth terms together, we obtain the following.

$$
\begin{aligned}
x^2y - 3y - x^3 + 3x &= (x^2y - 3y) - (x^3 - 3x) \\
&= y(x^2 - 3) - x(x^2 - 3) \\
&= (x^2 - 3)(y - x)
\end{aligned}
$$

■

NOTE: You can always check to see that you have factored an expression correctly by multiplying out the factors and comparing the result to the original expression.

DISCUSSION PROBLEM ■ Factoring by Grouping

Some third-degree polynomials can be factored by grouping and some cannot. For instance, the polynomial

$$x^3 - 3x^2 - 2x + 6$$

can be factored by grouping, whereas the polynomial

$$x^3 - 3x^2 - 2x - 6$$

cannot be factored by grouping. Find several other third-degree polynomials, some of which can be factored by grouping and some of which cannot. ■

Warm-Up

The following warm-up exercises involve skills that were covered in earlier sections. You will use these skills in the exercise set for this section.

In Exercises 1–10, find the product.

1. $2(5 - 15)$ **2.** $-3(8 + 6)$

3. $12(2x - 3)$ **4.** $7(4 - 3x)$

5. $-6(10 - 7x)$ **6.** $-2y(y + 1)$

7. $-3t(t + 2)$ **8.** $8xy(xy - 3)$

9. $(2 - x)(2 + x)$ **10.** $(x + 4)^2$

6.1   EXERCISES

In Exercises 1–16, find the greatest common factor of the given expressions.

1. 24, 90   **2.** 20, 45   **3.** 18, 150, 100

4. 60, 80, 90   **5.** $z^2, -z^6$ **6.** t^4, t^7

7. $2x^2, 12x$ **8.** $36x^4, 18x^3$ **9.** u^2v, u^3v^2

10. $r^6s^4, -rs$ **11.** $9yz^2, -12y^2z^3$ **12.** $-15x^6y^3, 45xy^3$

13. $28a^4b^2, 14a^3b^3, 42a^2b^5$ **14.** $16x^2y, 12xy^2, 36x^2y^2$

15. $14x^2, 1, 7x^4$ **16.** $5y^4, 10x^2y^2, 15xy$

In Exercises 17–60, factor the expression by removing any common factors. (*Note:* Some of the expressions have no common factor.)

17. $3x + 3$

18. $5y + 5$

19. $6z - 6$

20. $3x - 3$

21. $8t - 16$

22. $3u + 12$

23. $-25x - 10$

24. $-14y - 7$

25. $24y^2 - 18$

26. $7z^3 + 21$

27. $x^2 + x$

28. $-s^3 - s$

29. $25u^2 - 14u$

30. $36t^4 + 24t^2$

31. $2x^4 + 6x^3$

32. $9z^6 + 27z^4$

33. $7s^2 + 9t^2$

34. $12x^2 - 5y^3$

35. $3x^2y^2 - xy$

36. $4uv + u^2v^2$

37. $-10r^3s^2 - 7rs^2$

38. $-9a^2b^4 + 12a^2b$

39. $16a^3b^3 + 24a^4b^3$

40. $6x^4y + 12x^2y$

41. $10abc + 10a^2bc$

42. $21x^2y^2z - 35xz$

43. $12x^2 + 16x - 8$

44. $9 - 3y - 15y^2$

45. $100 + 75z - 50z^2$

46. $42t^3 - 21t^2 + 7$

47. $9x^4 + 6x^3 + 18x^2$

48. $32a^5 - 2a^3 + 6a$

49. $x(x - 3) + 5(x - 3)$

50. $x(x + 6) + 3(x + 6)$

51. $t(s + 10) - 8(s + 10)$

52. $y(q - 5) - 10(q - 5)$

53. $a^2(b + 2) - a(b + 2)$

54. $x^3(x + 4) + x^2(x + 4)$

55. $z^3(z + 5)^2 + z^2(z + 5)$

56. $(a + b)(c + 7) - (a + b)(c + 7)^2$

57. $y^3(y - 8)^2 + y^2(y - 8)^3$

58. $(x + y)(x - y) - x(x - y)$

59. $5u^2v^4 + 5u^2v^2 + 5uv^2$

60. $11x^5y^3 - 22xy^2 + 11y^2$

In Exercises 61–70, factor the given expression by grouping.

61. $x^2 + 10x + x + 10$

62. $x^2 - 5x + x - 5$

63. $y^2 - 4y + 2y - 8$

64. $y^2 + 3y + 3y + 9$

65. $x^3 + 2x^2 + x + 2$

66. $x^3 - 5x^2 + x - 5$

67. $t^3 - 3t^2 + 2t - 6$

68. $3s^3 + 6s^2 + 2s + 4$

69. $z^3 + 3z^2 - 2z - 6$

70. $4u^3 - 2u^2 - 6u + 3$

In Exercises 71–74, write the polynomial in standard form. Then factor out the greatest common (negative) factor.

71. $5 - 10x$

72. $3 - x$

73. $4 + 2x - x^2$

74. $18 - 12x - 6x^2$

In Exercises 75–80, rewrite the given polynomial by factoring out the indicated fraction.

75. $\frac{1}{2}x + \frac{3}{4} = \frac{1}{4}(\underline{\quad\quad})$

76. $\frac{2}{3}x - \frac{1}{6} = \frac{1}{6}(\underline{\quad\quad})$

77. $\frac{7}{8}x + \frac{5}{16} = \frac{1}{16}(\underline{\quad\quad})$

78. $\frac{5}{12}u - \frac{5}{8} = \frac{1}{24}(\underline{\quad\quad})$

79. $2y - \frac{1}{5} = \frac{1}{5}(\underline{\quad\quad})$

80. $3z + \frac{3}{4} = \frac{1}{4}(\underline{\quad\quad})$

81. *Dimensions of a Microwave Oven* The area of the front of a microwave oven of height h is given by $44h - h^2$. Factor this expression to determine the length of the microwave in terms of h.

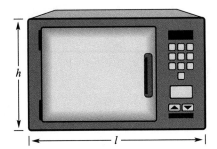

82. *Unit Price* The revenue R for selling x units of a product at a price of p dollars per unit is given by $R = xp$. For a particular product the revenue is given by

$$R = 900x - 0.1x^2.$$

Factor this expression for the revenue and determine an expression that represents the price p in terms of x.

83. *Simple Interest* The amount after t years when a principal of P dollars is invested at $r\%$ simple interest for t years is given by $P + Prt$. Factor this expression for simple interest.

84. *Chemical Reaction* The rate of change in a chemical reaction is given by

$$kQx - kx^2$$

where Q is the amount of the original substance, x is the amount of substance formed, and k is a constant of proportionality. Factor this expression.

SECTION

6.2

Factoring Polynomials of the Form $x^2 + bx + c$

Factoring Trinomials of the Form $x^2 + bx + c$ • *Factoring Trinomials in Two Variables* • *Factoring Completely*

Factoring Trinomials of the Form $x^2 + bx + c$

We know from Section 5.2 that the product of two binomials is often a trinomial. Here are some examples.

Factored Form	F	O	I	L	*Trinomial Form*

$$(x - 1)(x + 5) = x^2 + 5x - x - 5 = x^2 + 4x - 5$$
$$(x - 3)(x - 3) = x^2 - 3x - 3x + 9 = x^2 - 6x + 9$$
$$(3x + 5)(2x + 1) = 6x^2 + 3x + 10x + 5 = 6x^2 + 13x + 5$$

Try covering the factored forms in the left-hand column above. Can you determine the factored forms from the trinomial forms? Our goal in this section is to show how this can be done for factorable trinomials* of the form $x^2 + bx + c$.

*Some trinomials are not factorable. We will discuss this more in Section 6.3.

To begin, let's consider the following factorization.

$$x^2 + bx + c = (x + m)(x + n)$$

By multiplying the right-hand side, we obtain the following.

$$(x + m)(x + n) = x^2 + nx + mx + mn$$
$$= x^2 + \underbrace{(n + m)}_{}x + \underbrace{mn}_{}$$

Sum of   Product
terms   of terms
↓   ↓

$$= x^2 + \boxed{b}\, x + \boxed{c}$$

Thus, to *factor* a trinomial $x^2 + bx + c$ into a product of two binomials we use the following pattern.

$$x^2 + bx + c = (x + m)(x + n)$$

↑   ↑
m and *n* are factors
of *c* whose sum is *b*

Note how this strategy is used in the following examples.

EXAMPLE 1 ■ Factoring Trinomials

Factor each of the following into a product of two binomials.

(a) $x^2 + 5x + 6$ (b) $x^2 + 5x - 6$

Solution

(a) For this trinomial, we have $x^2 + bx + c = x^2 + 5x + 6$. Thus, $b = 5$ and $c = 6$. From our general strategy for factoring a trinomial, we need to find two numbers whose product is 6 and whose sum is 5. After testing some possibilities, we see that the numbers 2 and 3 work. Therefore, the factorization is as follows.

$$x^2 + 5x + 6 = (x + \underline{\quad})(x + \underline{\quad})$$

Think: We need factors of 6 whose sum is 5

$$= (x + 2)(x + 3)$$

$2(3) = 6, \quad 2 + 3 = 5$

(b) For this trinomial, we have $x^2 + bx + c = x^2 + 5x - 6$. Thus, $b = 5$ and $c = -6$, and we need to find two numbers whose product is -6 and whose sum is 5. After trying some possibilities, we see that the numbers 6 and -1 work. Therefore, the factorization is as follows.

$$x^2 + 5x - 6 = (x + \underline{\quad})(x + \underline{\quad})$$

Think: We need factors of -6 whose sum is 5

$$= (x + 6)(x - 1)$$

$6(-1) = -6, \quad 6 - 1 = 5$ ■

EXAMPLE 2   ■   Factoring Trinomials

Factor each of the following into a product of two binomials.

(a) $x^2 - x - 6$ (b) $x^2 - 5x + 6$

Solution

(a) For this trinomial, we have $x^2 + bx + c = x^2 - x - 6$. Thus, $b = -1$ and $c = -6$, and we need to find two numbers whose product is -6 and whose sum is -1.

$$x^2 - x - 6 = (x +)(x +)$$

Think: We need factors of -6 whose sum is -1

$$= (x + 2)(x - 3)$$ $(2)(-3) = -6,$ $2 - 3 = -1$

(b) For this trinomial, we have $x^2 + bx + c = x^2 - 5x + 6$. Thus, $b = -5$ and $c = 6$, and we need to find two numbers whose product is 6 and whose sum is -5.

$$x^2 - 5x + 6 = (x +)(x +)$$

Think: We need factors of 6 whose sum is -5

$$= (x - 2)(x - 3)$$ $(-2)(-3) = 6,$ $-2 - 3 = -5$ ■

NOTE: When the constant term of the trinomial is positive, its factors must have *like* signs; otherwise, its factors have *unlike* signs.

When factoring a trinomial of the form $x^2 + bx + c$, if you have trouble finding two factors of c whose sum is b, it may be helpful to make a list of all the distinct pairs of factors and then choose the appropriate pair from the list. For instance, consider the trinomial

$$x^2 - 5x - 24.$$

For this trinomial, we have $c = -24$ and $b = -5$. Thus, we need to find two factors of -24 whose sum is -5, as follows.

Factors of -24	*Sum of Factors*	
$(1)(-24)$	$1 - 24 = -23$	
$(-1)(24)$	$-1 + 24 = 23$	
$(2)(-12)$	$2 - 12 = -10$	
$(-2)(12)$	$-2 + 12 = 10$	
$(3)(-8)$	$3 - 8 = -5$	Correct choice
$(-3)(8)$	$-3 + 8 = 5$	
$(4)(-6)$	$4 - 6 = -2$	
$(-4)(6)$	$-4 + 6 = 2$	

With experience, you will be able to *mentally* narrow this list down to only two or three possibilities whose sums could then be tested to determine the correct factorization.

EXAMPLE 3   ■   Factoring a Trinomial

Factor the following trinomial.

$$x^2 - 2x - 15$$

Solution

To factor this trinomial, we need to find two factors of -15 whose sum is -2. After testing some possibilities, we see that 3 and -5 work. Thus, the factorization is as follows.

$$x^2 - 2x - 15 = (x +\ \ \ \ \ )(x +\ \ \ \ \ )$$

$$= (x + 3)(x - 5)$$

Think: We need factors of -15 whose sum is -2

$(3)(-5) = -15, \quad 3 - 5 = -2$

EXAMPLE 4   ■   Factoring a Trinomial

Factor the following trinomial.

$$x^2 + 15x + 36$$

Solution

To factor this trinomial, we need to find two factors of 36 whose sum is 15. After testing some possibilities, we see that 3 and 12 work. Thus, the factorization is as follows.

$$x^2 + 15x + 36 = (x +\ \ \ \ \ )(x +\ \ \ \ \ )$$

$$= (x + 3)(x + 12)$$

Think: We need factors of 36 whose sum is 15

$(3)(12) = 36, \quad 3 + 12 = 15$

EXAMPLE 5   ■   Factoring a Trinomial

Factor the following trinomial.

$$x^2 - 16x + 64$$

Solution

To factor this trinomial, we need to find two factors of 64 whose sum is -16. After testing some possibilities, we see that -8 and -8 work. Thus, the factorization is as follows.

$$x^2 - 16x + 64 = (x +\ \ \ \ \ )(x +\ \ \ \ \ )$$

$$= (x - 8)(x - 8)$$

Think: We need factors of 64 whose sum is -16

$(-8)(-8) = 64, \quad -8 - 8 = -16$

Factoring Trinomials in Two Variables

In the first five examples, we looked at factoring trinomials of the form $x^2 + bx + c$. We now look at trinomials of the form $x^2 + bxy + cy^2$.

Note that this trinomial has two variables, x and y. However, from the factorization

$$x^2 + bxy + cy^2 = (x + my)(x + ny) = x^2 + (m + n)xy + mny^2$$

we still need to find two factors of c whose sum is b. We demonstrate this type of factoring in Examples 6 and 7.

EXAMPLE 6   ■   Factoring a Trinomial in Two Variables

Factor the following trinomial.

$$x^2 - xy - 12y^2$$

Solution

For this trinomial, we have $x^2 + bxy + cy^2 = x^2 - xy - 12y^2$. Thus, $b = -1$ and $c = -12$, and we need to find two numbers whose product is -12 and whose sum is -1.

$x^2 - xy - 12y^2 = (x + y)(x + y)$ *Think:* We need factors
                                                              of -12 whose sum is -1

$ = (x - 4y)(x + 3y)$ $(-4)(3) = -12, \quad -4 + 3 = -1$ ■

EXAMPLE 7   ■   Factoring a Trinomial in Two Variables

Factor the following trinomial.

$$y^2 - 6xy + 8x^2$$

Solution

For this trinomial, we have $y^2 + byx + cx^2 = y^2 - 6yx + 8x^2$. Thus, $b = -6$ and $c = 8$, and we need to find two numbers whose product is 8 and whose sum is -6.

$y^2 - 6yx + 8x^2 = (y + x)(y + x)$ *Think:* We need factors
                                                              of 8 whose sum is -6

$ = (y - 2x)(y - 4x)$ $(-2)(-4) = 8, \quad -2 - 4 = -6$ ■

Factoring Completely

Some trinomials have a common monomial factor. In such cases we first factor out the common monomial factor. Then we try to factor the resulting trinomial by the methods in this section. We refer to such a multiple-stage factoring process as **factoring completely**. For instance, the trinomial

$$2x^2 - 4x - 6 = 2(x^2 - 2x - 3) = 2(x - 3)(x + 1)$$

is factored completely.

EXAMPLE 8   ■   Factoring Completely

Factor the following trinomial completely.

$$2x^2 - 12x + 10$$

Solution

To begin, we notice that the trinomial has a common monomial factor of 2. Thus, we start the factoring process by factoring 2 out of each term.

$$2x^2 - 12x + 10 = 2(x^2 - 6x + 5) \qquad \text{Take out common factor}$$

$$= 2(x + \underline{})(x + \underline{}) \qquad \textit{Think: } \text{We need factors of 5 whose sum is } -6$$

$$= 2(x - 5)(x - 1) \qquad (-5)(-1) = 5, \quad -5 - 1 = -6$$

■

EXAMPLE 9   ■   Factoring Completely

Factor the following trinomial completely.

$$3x^3 - 27x^2 + 54x$$

Solution

This trinomial has a common monomial factor of $3x$. Thus, we start the factoring process by factoring $3x$ out of each term.

$$3x^3 - 27x^2 + 54x = 3x(x^2 - 9x + 18) \qquad \text{Take out common factor}$$

$$= 3x(x + \underline{})(x + \underline{}) \qquad \textit{Think: } \text{We need factors of 18 whose sum is } -9$$

$$= 3x(x - 3)(x - 6) \qquad (-3)(-6) = 18, \; -3 - 6 = -9$$

■

EXAMPLE 10   ■   Factoring Completely

Factor the following trinomial completely.

$$4y^4 + 32y^3 + 28y^2$$

Solution

This trinomial has a common monomial factor of $4y^2$. Thus, we start the factoring process by factoring $4y^2$ out of each term.

$$4y^4 + 32y^3 + 28y^2 = 4y^2(y^2 + 8y + 7) \qquad \text{Take out common factor}$$

$$= 4y^2(y + \underline{})(y + \underline{}) \qquad \textit{Think: } \text{We need factors of 7 whose sum is 8}$$

$$= 4y^2(y + 1)(y + 7) \qquad (1)(7) = 7, \quad 1 + 7 = 8$$

■

DISCUSSION PROBLEM   ■   You Be the Instructor

Suppose you are tutoring someone in algebra and you want to create several trinomials for your student to factor. You probably want to start with several simple ones and then put in a few that are tougher. Write a short paragraph describing how to create factorable trinomials.   ■

Warm-Up

The following warm-up exercises involve skills that were covered in earlier sections. You will use these skills in the exercise set for this section.

In Exercises 1–10, find the product.

1. $-4(x - 6)$ **2.** $6(3x + 8)$

3. $y(y + 2)$ **4.** $-a^2(a - 1)$

5. $(x - 2)(x - 5)$ **6.** $(t + 3)(t + 6)$

7. $(u - 8)(u + 3)$ **8.** $(v - 1)(v - 6)$

9. $(z - 3)(z + 1)$ **10.** $(x - 4)(x + 7)$

6.2   EXERCISES

In Exercises 1–8, find the missing factor. Then check your answer by multiplying the two factors.

1. $x^2 + 4x + 3 = (x + 3)()$ **2.** $x^2 + 5x + 6 = (x + 3)()$

3. $a^2 + a - 6 = (a + 3)()$ **4.** $c^2 + 2c - 3 = (c + 3)()$

5. $y^2 - 2y - 15 = (y + 3)()$ **6.** $y^2 - 4y - 21 = (y + 3)()$

7. $z^2 - 5z + 6 = (z - 3)()$ **8.** $z^2 - 4z + 3 = (z - 3)()$

In Exercises 9 and 10, find all possible products of the form $(x + m)(x + n)$ where $m \cdot n$ is the specified product. (Assume m and n are integers.)

9. $m \cdot n = 12$

10. $m \cdot n = 18$

In Exercises 11–34, factor the given trinomial.

11. $x^2 + 6x + 8$

12. $x^2 + 13x + 12$

13. $x^2 - 13x + 40$

14. $x^2 - 9x + 14$

15. $x^2 - 7x + 12$

16. $x^2 + 10x + 24$

17. $x^2 - x - 6$

18. $x^2 + x - 6$

19. $x^2 + 2x - 15$

20. $x^2 - 2x - 15$

21. $x^2 + 3x - 70$

22. $x^2 + 21x + 108$

23. $x^2 - 17x + 72$

24. $x^2 - 8x - 240$

25. $x^2 + 19x + 60$

26. $x^2 - 30x + 216$

27. $x^2 + xy - 2y^2$

28. $x^2 - 5xy + 6y^2$

29. $x^2 + 8xy + 15y^2$

30. $u^2 - 4uv - 5v^2$

31. $a^2 + 2ab - 15b^2$

32. $y^2 + 4yz - 60z^2$

33. $x^2 - 7xz - 18z^2$

34. $x^2 + 15xy + 50y^2$

In Exercises 35–44, factor the given trinomial completely.

35. $3x^2 + 21x + 30$

36. $4x^2 - 32x + 60$

37. $x^3 - 13x^2 + 30x$

38. $3x^2 + 3x - 6$

39. $8x^2 - 16x - 24$

40. $-10x^2 + 30x + 100$

41. $x^3 + 5x^2y + 6xy^2$

42. $x^2y - 6xy^2 + 8y^3$

43. $2x^3y + 4x^2y^2 - 6xy^3$

44. $x^4y^2 + 3x^3y^3 + 2x^2y^4$

In Exercises 45–50, find all integers b such that the trinomial can be factored.

45. $x^2 + bx + 15$

46. $x^2 + bx + 10$

47. $x^2 + bx - 12$

48. $x^2 + bx - 18$

49. $x^2 + bx + 36$

50. $x^2 + bx - 48$

In Exercises 51–56, find *two* integers c such that the trinomial can be factored.

51. $x^2 + 3x + c$

52. $x^2 + 5x + c$

53. $x^2 - 6x + c$

54. $x^2 - 15x + c$

55. $x^2 - 9x + c$

56. $x^2 + 12x + c$

In Exercises 57–60, factor the trinomial $x^2 + 3x + 2 = (x + 1)(x + 2)$ and show the result using the geometric concept of area, as illustrated in the accompanying figure.

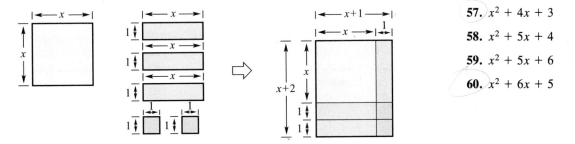

57. $x^2 + 4x + 3$

58. $x^2 + 5x + 4$

59. $x^2 + 5x + 6$

60. $x^2 + 6x + 5$

6.3

Factoring Polynomials of the Form $ax^2 + bx + c$

Factoring Trinomials of the Form $ax^2 + bx + c$ •   Factoring Completely   •   Factoring by Grouping

Factoring Trinomials of the Form $ax^2 + bx + c$

In this section we show how to factor a trinomial whose leading coefficient is *not* 1. The process is similar to that used in the preceding section, but it involves more testing (trial and error) of possible factors. Some examples of trinomials with $a \neq 1$ are

$$2x^2 + 5x - 3, \qquad 4x^2 + 13x + 10, \quad \text{and} \quad 9x^2 - 6x + 1.$$

To factor this type of trinomial we use the following pattern.

Factors of a

$$ax^2 + bx + c = (x +)(x +)$$

Factors of c

The goal is to find a combination of factors of a and c so that the outer and inner products add up to the middle term bx. For instance, in the trinomial $6x^2 + 11x + 3$, we have $a = 6$, $c = 3$, and $b = 11$. After some experimentation, we find that the factorization is

F   O   I   L
↓   ↓   ↓   ↓

$$(2x + 3)(3x + 1) = 6x^2 + 2x + 9x + 3 = 6x^2 + 11x + 3.$$

Note that the outer (O) and inner (I) products add up to $11x$.

EXAMPLE 1 ■ Factoring a Trinomial of the Form $ax^2 + bx + c$

Factor the trinomial $6x^2 + 5x - 4$.

Solution

For this trinomial, we have $ax^2 + bx + c = 6x^2 + 5x - 4$, which implies that $a = 6$, $c = -4$, and $b = 5$. The possible factors of 6 are (1)(6) and (2)(3), and the possible factors of -4 are $(-1)(4)$, $(1)(-4)$, and $(2)(-2)$. By trying the *many* different combinations of these factors, we obtain the following list.

$$(x + 1)(6x - 4) = 6x^2 + 2x - 4$$
$$(x - 1)(6x + 4) = 6x^2 - 2x - 4$$
$$(x + 4)(6x - 1) = 6x^2 + 23x - 4$$
$$(x - 4)(6x + 1) = 6x^2 - 23x - 4$$
$$(x + 2)(6x - 2) = 6x^2 + 10x - 4$$
$$(x - 2)(6x + 2) = 6x^2 - 10x - 4$$
$$(2x + 1)(3x - 4) = 6x^2 - 5x - 4$$
$$(2x - 1)(3x + 4) = 6x^2 + 5x - 4 \qquad \text{Correct factorization}$$
$$(2x + 4)(3x - 1) = 6x^2 + 10x - 4$$
$$(2x - 4)(3x + 1) = 6x^2 - 10x - 4$$
$$(2x + 2)(3x - 2) = 6x^2 + 2x - 4$$
$$(2x - 2)(3x + 2) = 6x^2 - 2x - 4$$

Thus, we conclude that the correct factorization is

$$6x^2 + 5x - 4 = (2x - 1)(3x + 4). \qquad ■$$

To help shorten the list of possible factorizations of a trinomial, we suggest the following guidelines.

Guidelines for Factoring a Trinomial

1. If the given trinomial has a *common monomial factor*, remove it before trying to find binomial factors. For instance, the trinomial $12x^2 + 10x - 8$ has a common factor of 2. By removing this common factor, we obtain $12x^2 + 10x - 8 = 2(6x^2 + 5x - 4)$.

2. Do not switch the signs of the factors of c unless the middle term is correct except in sign. For instance, in Example 1, after determining that the factorization $(x + 4)(6x - 1)$ is not correct, it is unnecessary to test $(x - 4)(6x + 1)$.

3. Do not use binomial factors that have a common monomial factor. Such a factor cannot be correct because the trinomial has no common monomial factor. (Any common monomial factor was already removed in Step 1.) For instance, in Example 1, it is unnecessary to test $(x + 1)(6x - 4) = 6x^2 + 2x - 4$, because the left side has a common factor of 2.

Using these suggestions, we shorten the list given in Example 1 to the following.

$$(x + 4)(6x - 1) = 6x^2 + 23x - 4$$
$$(2x + 1)(3x - 4) = 6x^2 - 5x - 4$$
$$(2x - 1)(3x + 4) = 6x^2 + 5x - 4 \qquad \text{Correct factorization}$$

Do you see why we could cut the list from 12 possible factorizations to only 3?

EXAMPLE 2   ■   Factoring a Trinomial of the Form $ax^2 + bx + c$

Factor the following trinomial.

$$2x^2 + x - 15$$

Solution

For this trinomial we have $a = 2$, which factors as $(1)(2)$, and $c = -15$, which factors as $(1)(-15)$, $(-1)(15)$, $(3)(-5)$, or $(-3)(5)$. We test the possible factors as follows.

$$(2x + 1)(x - 15) = 2x^2 - 29x - 15$$
$$(2x + 15)(x - 1) = 2x^2 + 13x - 15$$
$$(2x + 3)(x - 5) = 2x^2 - 7x - 15$$
$$(2x + 5)(x - 3) = 2x^2 - x - 15 \qquad \text{Middle term has incorrect sign}$$
$$(2x - 5)(x + 3) = 2x^2 + x - 15 \qquad \text{Correct factorization}$$

Therefore, the correct factorization is

$$2x^2 + x - 15 = (2x - 5)(x + 3). \qquad ■$$

NOTE: Notice in Example 2 that if the middle term is correct except in sign, we need only change the signs of the factors of c.

EXAMPLE 3   ■   Factoring a Trinomial of the Form $ax^2 + bx + c$

Factor the following trinomial.

$$4x^2 + 13x + 10$$

Solution

For this trinomial we have $a = 4$, which factors as $(1)(4)$ or $(2)(2)$, and $c = 10$, which factors as $(1)(10)$ or $(2)(5)$. We test the possible factors as follows.

$$(x + 10)(4x + 1) = 4x^2 + 41x + 10$$
$$(x + 2)(4x + 5) = 4x^2 + 13x + 10 \qquad \text{Correct factorization}$$

Therefore, the correct factorization is

$$4x^2 + 13x + 10 = (x + 2)(4x + 5). \qquad ■$$

EXAMPLE 4   ■   Factoring a Trinomial of the Form $ax^2 + bx + c$

Factor the following trinomial.

$$9x^2 - 6x + 1$$

Solution

For this trinomial we have $a = 9$, which factors as $(1)(9)$ or $(3)(3)$, and $c = 1$, which factors as $(1)(1)$. We test the possible factors as follows.

$$(x + 1)(9x + 1) = 9x^2 + 10x + 1$$
$$(3x + 1)(3x + 1) = 9x^2 + 6x + 1 \qquad \text{Middle term has incorrect sign}$$
$$(3x - 1)(3x - 1) = 9x^2 - 6x + 1 \qquad \text{Correct factorization}$$

Therefore, the correct factorization is

$$9x^2 - 6x + 1 = (3x - 1)(3x - 1).$$ ■

Factoring Completely

We already mentioned that if a trinomial has a common monomial factor, then the common monomial factor should be removed first. The next two examples illustrate this technique.

EXAMPLE 5   ■   Factoring Completely

Factor the following trinomial completely.

$$4x^2 - 30x + 14$$

Solution

This trinomial has a common monomial factor of 2, so we remove this factor first.

$$4x^2 - 30x + 14 = 2(2x^2 - 15x + 7)$$

Now, for the new trinomial $2x^2 - 15x + 7$, we have $a = 2$ and $c = 7$. The possible factorizations of this trinomial are as follows.

$$(2x - 7)(x - 1) = 2x^2 - 9x + 7$$
$$(2x - 1)(x - 7) = 2x^2 - 15x + 7 \qquad \text{Correct factorization}$$

Therefore, the complete factorization of the original trinomial is

$$4x^2 - 30x + 14 = 2(2x^2 - 15x + 7) = 2(2x - 1)(x - 7).$$ ■

EXAMPLE 6   ■   Factoring Completely

Factor the following trinomial completely.

$$18x^3 - 33x^2 - 30x$$

Solution

This trinomial has a common monomial factor of $3x$, so we remove this factor first.

$$18x^3 - 33x^2 - 30x = 3x(6x^2 - 11x - 10)$$

Now, for the new trinomial $6x^2 - 11x - 10$, we have $a = 6$ and $c = -10$. The possible factorizations of this trinomial are as follows.

$$
\begin{aligned}
(x + 10)(6x - 1) &= 6x^2 + 59x - 10 \\
(x + 2)(6x - 5) &= 6x^2 + 7x - 10 \\
(2x + 1)(3x - 10) &= 6x^2 - 17x - 10 \\
(2x + 5)(3x - 2) &= 6x^2 + 11x - 10 \qquad \text{Middle term has incorrect sign} \\
(2x - 5)(3x + 2) &= 6x^2 - 11x - 10 \qquad \text{Correct factorization}
\end{aligned}
$$

Therefore, the complete factorization of the original trinomial is

$$18x^3 - 33x^2 - 30x = 3x(6x^2 - 11x - 10) = 3x(2x - 5)(3x + 2). \qquad \blacksquare$$

When factoring a trinomial with a negative leading coefficient, we suggest that you first factor -1 out of the trinomial. This technique is demonstrated in the next example.

EXAMPLE 7   ■   Factoring a Trinomial with a Negative Leading Coefficient

Factor the following trinomial.

$$-5x^2 + 7x + 6$$

Solution

This trinomial has a negative leading coefficient, so we begin by factoring out -1.

$$-5x^2 + 7x + 6 = (-1)(5x^2 - 7x - 6)$$

Now, for the new trinomial $5x^2 - 7x - 6$, we have $a = 5$ and $c = -6$. The possible factorizations of this trinomial are as follows.

$$
\begin{aligned}
(x + 6)(5x - 1) &= 5x^2 + 29x - 6 \\
(x + 1)(5x - 6) &= 5x^2 - x - 6 \\
(x + 3)(5x - 2) &= 5x^2 + 13x - 6 \\
(x + 2)(5x - 3) &= 5x^2 + 7x - 6 \qquad \text{Middle term has incorrect sign} \\
(x - 2)(5x + 3) &= 5x^2 - 7x - 6 \qquad \text{Correct factorization}
\end{aligned}
$$

Thus, a correct factorization is

$$-5x^2 + 7x + 6 = (-1)(x - 2)(5x + 3) = (-x + 2)(5x + 3).$$

Another correct factorization would be

$$(x - 2)(-5x - 3). \qquad \blacksquare$$

Not all trinomials are factorable using only integers. For instance, to factor $x^2 + 3x + 5$ we need factors of 5 that add up to 3. This is not possible since the only integer factors of 5 are 1 and 5, and their sum is not 3. We say that such a trinomial is **not factorable**. Try factoring $2x^2 - 3x + 2$ to see that it is also not factorable. Watch for other trinomials that are not factorable in the exercises for this chapter.

Factoring by Grouping

In this section you have seen that factoring a trinomial can involve quite a bit of trial and error. An alternative technique that some people like to use is factoring by grouping. For instance, suppose we rewrite the trinomial $2x^2 + x - 15$ as

$$2x^2 + x - 15 = 2x^2 + 6x - 5x - 15.$$

Then, by grouping the first two terms and the third and fourth terms, we could factor the polynomial as follows.

$$\begin{aligned} 2x^2 + x - 15 &= 2x^2 + (6x - 5x) - 15 & \text{Rewrite middle term} \\ &= (2x^2 + 6x) - (5x + 15) & \text{Group terms} \\ &= 2x(x + 3) - 5(x + 3) & \text{Factor groups} \\ &= (2x - 5)(x + 3) & \text{Factor out common binomial factor} \end{aligned}$$

The key to this method of factoring is knowing how to rewrite the middle term. In general, *to factor a trinomial $ax^2 + bx + c$ by grouping, we choose factors of the product ac that add up to b and use these factors to rewrite the middle term.* This technique is illustrated in Example 8.

EXAMPLE 8   ■   Factoring a Trinomial by Grouping

Use factoring by grouping to factor the following trinomial.

$$2x^2 + 5x - 3$$

Solution

In the trinomial $2x^2 + 5x - 3$, we have $a = 2$ and $c = -3$, which implies that the product ac is -6. Because -6 factors as $(6)(-1)$ and $6 - 1 = 5 = b$, we rewrite the middle term as $5x = 6x - x$. This produces the following.

$$\begin{aligned} 2x^2 + 5x - 3 &= 2x^2 + 6x - x - 3 & \text{Rewrite middle term} \\ &= (2x^2 + 6x) - (x + 3) & \text{Group terms} \\ &= 2x(x + 3) - (x + 3) & \text{Factor groups} \\ &= (x + 3)(2x - 1) & \text{Factor out common binomial factor} \end{aligned}$$

Therefore, the given trinomial factors as

$$2x^2 + 5x - 3 = (x + 3)(2x - 1).$$ ■

What do you think of this grouping technique? Many people think that it is more efficient than the trial and error process, especially when the coefficients a and c have many factors.

DISCUSSION PROBLEM   ■   A Factoring Program

Try entering the following factoring program on a computer that uses the BASIC language. Use the program to factor the following trinomials.

$6x^2 + 5x - 4$	Example 1
$2x^2 + x - 15$	Example 2
$4x^2 + 13x + 10$	Example 3
$9x^2 - 6x + 1$	Example 4

Note that the coefficients of the trinomial should be substituted for those on line 210.

BASIC Program:

```
10 READ A,B,C
20 IF C=D THEN GOTO 160
30 FOR I= -ABS(A) TO ABS(A)
40 IF I=0 THEN GOTO 130
50 A1=I: A2=INT(A/I)
60 IF A1*A2<>A THEN GOTO 130
70 FOR J = -ABS(C) TO ABS(C)
80 FOR J=O THEN GOTO 120
90 C1=J: C2=INT(C/J)
100 IF C1*C2<>C THEN GOTO 120
110 IF A1*C2+A2*C1=B THEN GOTO 170
120 NEXT J
130 NEXT I
140 PRINT "TRINOMIAL IS NOT FACTORABLE"
150 GOTO 200
160 A1=-A: C1=-B: A2=-1: C2=0
170 PRINT "TRINOMIAL FACTORS AS (A1X + C1)(A2X + C2) WITH"
180 PRINT "A1 =";-A1;"C1=";-C1
190 PRINT "A2 =";-A2;"C2 =";-C2
200 END
210 DATA 2,1,0
```

■

MATH MATTERS

Prime Numbers

A prime number is a positive integer that is greater than 1 and cannot be factored as the product of smaller integers. For instance, the numbers 2, 3, and 5 are prime numbers, whereas the number 4 is not prime (because $4 = 2 \cdot 2$). Prime numbers have fascinated people for hundreds of years, and part of this fascination stems from the fact that no one has ever discovered a simple pattern for prime numbers. There are 168 prime numbers that are less than 1,000. Can you determine the next three prime numbers in the list? (The answer is given in the back of the text.)

2	3	5	7	11	13	17	19	23	29	31	37
41	43	47	53	59	61	67	71	73	79	83	89
97	101	103	107	109	113	127	131	137	139	149	151
157	163	167	173	179	181	191	193	197	199	211	223
227	229	233	239	241	251	257	263	269	271	277	281
283	293	307	311	313	317	331	337	347	349	353	359
367	373	379	383	389	397	401	409	419	421	431	433
439	443	449	457	461	463	467	479	487	491	499	503
509	521	523	541	547	557	563	569	571	577	587	593
599	601	607	613	617	619	631	641	643	647	653	659
661	673	677	683	691	701	709	719	727	733	739	743
751	757	761	769	773	787	797	809	811	821	823	827
829	839	853	857	859	863	877	881	883	887	907	911
919	929	937	941	947	953	967	971	977	983	991	997

The sieve of Eratosthenes. The first number held back on each level is prime.

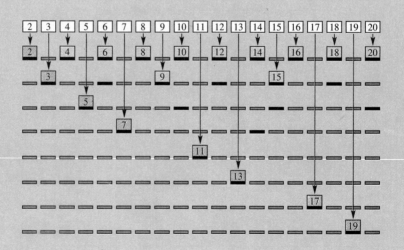

The following warm-up exercises involve skills that were covered in earlier sections. You will use these skills in the exercise set for this section.

In Exercises 1 and 2, rewrite the given polynomial by factoring out the indicated fraction.

1. $\dfrac{1}{3}x + \dfrac{5}{9} = \dfrac{1}{9}(\rule{1.5cm}{0.1mm})$

2. $\dfrac{5}{8}x - \dfrac{3}{2} = \dfrac{1}{8}(\rule{1.5cm}{0.1mm})$

In Exercises 3–10, factor the polynomial completely.

3. $6x + 12$

4. $4x - 12$

5. $x^2 y - xy^2$

6. $6x^3 - 3x^2 + 9x$

7. $x^2 + x - 42$

8. $x^2 + 13x + 42$

9. $x^3 - x^2 - 42x$

10. $2x^2 y + 8xy - 64y$

6.3 EXERCISES

In Exercises 1–8, find the missing factor.

1. $5x^2 + 18x + 9 = (x + 3)(\rule{1.2cm}{0.1mm})$

2. $5x^2 + 19x + 12 = (x + 3)(\rule{1.2cm}{0.1mm})$

3. $5a^2 + 12a - 9 = (a + 3)(\rule{1.2cm}{0.1mm})$

4. $5c^2 + 11c - 12 = (c + 3)(\rule{1.2cm}{0.1mm})$

5. $2y^2 - 3y - 27 = (y + 3)(\rule{1.2cm}{0.1mm})$

6. $3y^2 - y - 30 = (y + 3)(\rule{1.2cm}{0.1mm})$

7. $4z^2 - 13z + 3 = (z - 3)(\rule{1.2cm}{0.1mm})$

8. $6z^2 - 23z + 15 = (z - 3)(\rule{1.2cm}{0.1mm})$

In Exercises 9 and 10, find all possible products of the form $(5x + m)(x + n)$, where $m \cdot n$ is the specified product. (Assume m and n are integers.)

9. $m \cdot n = 12$

10. $m \cdot n = 36$

In Exercises 11–30, factor the given trinomial. [*Note:* Some of the trinomials cannot be factored (using integer coefficients).]

11. $2x^2 + 5x + 3$

12. $3x^2 + 7x + 2$

13. $2x^2 - x - 3$

14. $3x^2 + 5x - 2$

15. $2y^2 - 3y + 1$

16. $3a^2 - 5a + 2$

17. $15a^2 + 14a - 8$

18. $3z^2 - z - 2$

19. $18u^2 - 9u - 2$

20. $6v^2 + v - 2$

21. $10t^2 - 3t - 18$

22. $24s^2 + 37s - 5$

23. $2x^2 + x + 3$

24. $6x^2 - 10x + 5$

25. $15m^2 + 16m - 15$

26. $21b^2 - 40b - 21$

27. $16z^2 - 34z + 15$

28. $12x^2 - 41x + 24$

29. $5s^2 - 10s + 6$

30. $10t^2 + 43t - 9$

In Exercises 31–40, factor the trinomial by grouping.

31. $3x^2 + 7x + 2$ **32.** $2x^2 + 17x + 21$ **33.** $2x^2 + x - 3$ **34.** $5x^2 - 14x - 3$

35. $6x^2 + 5x - 4$ **36.** $12x^2 + 11x + 2$ **37.** $15x^2 - 11x + 2$ **38.** $12x^2 - 13x + 3$

39. $16x^2 + 2x - 3$ **40.** $20x^2 + 19x - 1$

In Exercises 41–46, factor the given trinomial.

41. $-2x^2 + x + 3$ **42.** $-5x^2 + x + 4$ **43.** $1 - 4x - 60x^2$ **44.** $2 + 5x - 12x^2$

45. $-6x^2 + 7x + 10$ **46.** $2 + x - 6x^2$

In Exercises 47–60, factor the polynomial completely. [*Note:* Some of the polynomials cannot be factored (using integer coefficients).]

47. $x^2 - 3x$ **48.** $3a^4 - 9a^3$ **49.** $u(u - 3) + 9(u - 3)$

50. $y^2(y + 1) - y(y + 1)$ **51.** $v^2 + v - 42$ **52.** $z^2 - 3z - 40$

53. $6x^2 + 8x - 8$ **54.** $6x^2 - 6x - 36$ **55.** $2x^2 + 2x + 1$

56. $-15x^2 - 2x + 8$ **57.** $15y^2 - 7y^3 - 2y^4$ **58.** $3x^3 + 4x^2 + 2x$

59. $u^2v^2 + 2uv^2 - 3v^2$ **60.** $3x^2 - 4x + 2$

In Exercises 61–64, find all integers b such that the trinomial can be factored.

61. $3x^2 + bx + 10$ **62.** $4x^2 + bx + 3$ **63.** $2x^2 + bx - 6$ **64.** $5x^2 + bx - 6$

In Exercises 65–68, find two integers c such that the trinomial can be factored.

65. $4x^2 + 3x + c$ **66.** $2x^2 + 5x + c$ **67.** $3x^2 - 10x + c$ **68.** $8x^2 - 3x + c$

In Exercises 69 and 70, factor the trinomial $2x^2 + 3x + 1 = (2x + 1)(x + 1)$ and show the result using the geometric concept of area, as illustrated in the accompanying figure.

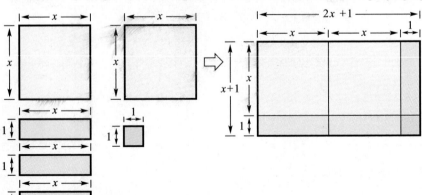

69. $2x^2 + 5x + 2$

70. $3x^2 + 4x + 1$

71. *Volume of a Sandbox* The sandbox shown in the figure has a height of x and a width of $x + 2$. The volume of the box is

$$2x^3 + 7x^2 + 6x.$$

Find the length of the box.

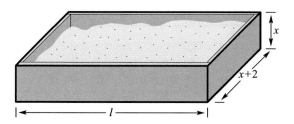

72. *Volume of a Cake Box* The cake box shown in the figure has a height of x and a width of $x + 1$. The volume of the box is

$$3x^3 + 4x^2 + x.$$

Find the length of the box.

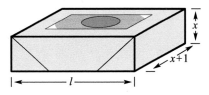

<table>
<tr><td>SECTION
6.4</td><td>**Factoring Polynomials with Special Forms**
Introduction • *Difference of Two Squares* • *Perfect Square Trinomials* •
Sum and Difference of Two Cubes</td></tr>
</table>

Introduction

Some polynomials have special forms that you should learn to recognize. Here are some examples of the special forms that we will study in this section.

Difference of two squares:	$x^2 - 4 = (x + 2)(x - 2)$
Perfect square trinomial:	$x^2 + 2x + 1 = (x + 1)^2$
Perfect square trinomial:	$x^2 - 6x + 9 = (x - 3)^2$
Sum of two cubes:	$x^3 + 125 = (x + 5)(x^2 - 5x + 25)$
Difference of two cubes:	$x^3 - 1 = (x - 1)(x^2 + x + 1)$

Difference of Two Squares

One of the easiest special polynomial forms to recognize and to factor is $a^2 - b^2$, called a **difference of two squares**. It factors according to the following pattern (see the special products in Section 5.2).

■ **Difference of Two Squares**	Let a and b be real numbers, variables, or algebraic expressions. $a^2 - b^2 = (a + b)(a - b)$ Difference Opposite signs

This pattern can be illustrated geometrically, as shown in Figure 6.1. Note that the area of the shaded region on the left is represented by $a^2 - b^2$ (the area of the larger square minus the area of the smaller square). On the right, the *same* area is represented by a rectangle whose width is $a + b$ and whose length is $a - b$.

FIGURE 6.1

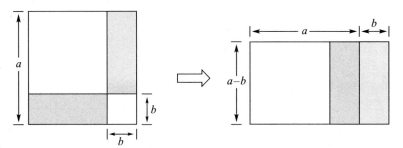

To recognize perfect squares, look for coefficients that are squares of integers and for variables raised to *even* powers. Here are some examples.

Given Polynomial		*Think: Difference of Squares*		*Factored Form*
$x^2 - 1$	➡	$(x)^2 - (1)^2$	➡	$(x + 1)(x - 1)$
$4x^2 - 9$	➡	$(2x)^2 - (3)^2$	➡	$(2x + 3)(2x - 3)$
$25 - 64x^4$	➡	$(5)^2 - (8x^2)^2$	➡	$(5 + 8x^2)(5 - 8x^2)$

EXAMPLE 1 ■ Factoring the Difference of Two Squares

Factor the following polynomial.

$$x^2 - 36$$

Solution

Because x^2 and 36 are both perfect squares, we recognize this polynomial to be the difference of two squares. Therefore, the polynomial factors as follows.

$$x^2 - 36 = x^2 - 6^2 \qquad \text{Write as difference of two squares}$$
$$= (x + 6)(x - 6) \qquad \text{Factored form}$$
■

When you factor a polynomial, remember that you can check your result by multiplying the factors. For instance, in Example 1 we can check that we have the correct factors as follows.

$$
\begin{array}{r}
x + 6 \\
x - 6 \\
\hline
-6x - 36 \\
x^2 + 6x \\
\hline
x^2 \quad\quad - 36
\end{array}
$$

EXAMPLE 2 ■   Factoring the Difference of Two Squares

Factor the following polynomial.

$$81x^2 - 49$$

Solution

Because $81x^2$ and 49 are both perfect squares, we recognize this polynomial to be the difference of two squares. Therefore, the polynomial factors as follows.

$$81x^2 - 49 = (9x)^2 - 7^2 \qquad \text{Write as difference of two squares}$$
$$= (9x + 7)(9x - 7) \qquad \text{Factored form} \quad ■$$

Remember that the rule $a^2 - b^2 = (a + b)(a - b)$ applies to polynomials or expressions in which a and b are themselves expressions. The next example illustrates this possibility.

EXAMPLE 3 ■   Factoring the Difference of Two Squares

Factor the following expression.

$$(x + 1)^2 - 4$$

Solution

Because $(x + 1)^2$ and 4 are both perfect squares, we recognize this expression to be the difference of two squares. Therefore, the expression factors as follows.

$$(x + 1)^2 - 4 = (x + 1)^2 - 2^2 \qquad \text{Write as difference of two squares}$$
$$= [(x + 1) + 2][(x + 1) - 2] \qquad \text{Factored form}$$
$$= (x + 3)(x - 1) \qquad \text{Simplify} \quad ■$$

Sometimes the difference of two squares can be hidden by the presence of a common monomial factor. Remember that with all factoring techniques, we should first remove any common monomial factors. This is demonstrated in Example 4.

EXAMPLE 4 ■ Removing a Common Monomial Factor First

Factor the following polynomials.

(a) $48 - 75x^2$ (b) $20x^3 - 5x$

Solution

(a) The polynomial $48 - 75x^2$ has a common monomial factor of 3. After removing this factor, we see that the remaining polynomial is the difference of two squares.

$$48 - 75x^2 = 3(16 - 25x^2)$$ Remove common monomial factor
$$= 3(4^2 - (5x)^2)$$ Write as difference of two squares
$$= 3(4 + 5x)(4 - 5x)$$ Factored form

(b) The polynomial $20x^3 - 5x$ has a common monomial factor of $5x$. After removing this factor, we see that the remaining polynomial is the difference of two squares.

$$20x^3 - 5x = 5x(4x^2 - 1)$$ Remove common monomial factor
$$= 5x((2x)^2 - 1^2)$$ Write as difference of two squares
$$= 5x(2x + 1)(2x - 1)$$ Factored form ■

To completely factor a polynomial, you should always check to see whether the factors obtained might themselves be factorable. That is, can any of the factors be factored? For instance, after factoring the polynomial $x^4 - 1$ once as the difference of two squares

$$x^4 - 1 = (x^2)^2 - 1^2 = (x^2 + 1)(x^2 - 1),$$

we see that the second factor is itself the difference of two squares. Thus, to factor the polynomial *completely*, we must continue the factoring process, as follows.

$$x^4 - 1 = (x^2 + 1)(x^2 - 1) = (x^2 + 1)(x + 1)(x - 1)$$

Another example of repeated factoring is given in the next example.

EXAMPLE 5 ■ Factoring Completely

Factor $x^4 - 16$ completely.

Solution

Recognizing $x^4 - 16$ as a difference of two squares, we write

$$x^4 - 16 = (x^2)^2 - 4^2 = (x^2 + 4)(x^2 - 4).$$

Note that the second factor $(x^2 - 4)$ is itself a difference of two squares and we therefore obtain

$$x^4 - 16 = (x^2 + 4)(x^2 - 4) = (x^2 + 4)(x + 2)(x - 2).$$ ■

Note in Example 5 that we did not attempt to further factor the *sum of two squares*. The reason for this is that a second-degree polynomial that is the sum of two squares cannot be factored as the product of binomials (using integers as coefficients). For instance, the second-degree polynomials $x^2 + 4$ and $4x^2 + 9$ cannot be factored (using integers as coefficients).

Perfect Square Trinomials

A **perfect square trinomial** is the square of a binomial. For instance,

$$x^2 + 4x + 4 = (x + 2)^2$$

is the square of the binomial $(x + 2)$, and

$$4x^2 - 12x + 9 = (2x - 3)^2$$

is the square of the binomial $(2x - 3)$. Perfect square trinomials come in two patterns: one in which the middle term is positive and the other in which the middle term is negative.

Perfect Square Trinomials	Let a and b be numbers, variables, or algebraic expressions.
	1. $a^2 + 2ab + b^2 = (a + b)^2$
	Same sign
	2. $a^2 - 2ab + b^2 = (a - b)^2$
	Same sign

To recognize a perfect square trinomial, remember that the first and last terms must be perfect squares and positive, and the middle term must be twice the product of a and b. (Note that the middle term can be positive or negative.)

EXAMPLE 6 ■ Identifying Perfect Square Trinomials

Identify which of the following polynomials are perfect square trinomials. If the polynomial is not a perfect square trinomial, give the reason.

(a) $m^2 - 4m + 4$ (b) $16x^2 + 40x + 25$

(c) $4x^2 - 2x + 1$ (d) $y^2 + 6y - 9$

Solution

(a) This polynomial is a perfect square trinomial because its first and last terms are perfect squares and positive, and the middle term is (the negative of) twice the product of m and 2.

$$m^2 - 4m + 4 = m^2 - 2(2m) + 2^2$$

(b) This polynomial is a perfect square trinomial because its first and last terms are perfect squares and positive, and the middle term is twice the product of $4x$ and 5.

$$16x^2 + 40x + 25 = (4x)^2 + 2(4x)(5) + 5^2$$

(c) This polynomial is *not* a perfect square trinomial because the middle term is not twice the product of $2x$ and 1.

$$4x^2 - 2x + 1 \neq (2x)^2 - 2(2x)(1) + 1^2$$

(d) This polynomial is *not* a perfect square trinomial because the last term, -9, is not positive. ∎

EXAMPLE 7 ■ Factoring Perfect Square Trinomials

(a) $y^2 - 6y + 9 = y^2 - 2(3y) + 3^2 = (y - 3)^2$

(b) $16x^2 + 40x + 25 = (4x)^2 + 2(4x)(5) + 5^2 = (4x + 5)^2$

(c) $9x^2 - 24xy + 16y^2 = (3x)^2 - 2(3x)(4y) + (4y)^2 = (3x - 4y)^2$ ∎

Sum and Difference of Two Cubes

The last type of special factoring that we will study in this section is the sum or difference of two cubes. The patterns for these two special forms are summarized as follows. In these patterns, pay particular attention to the signs of the terms.

Sum and Difference of Cubes

Let a and b be real numbers, variables, or algebraic expressions.

Like signs

1. $a^3 + b^3 = (a + b)(a^2 - ab + b^2)$

Unlike signs

Like signs

2. $a^3 - b^3 = (a - b)(a^2 + ab + b^2)$

Unlike signs

EXAMPLE 8   ■   Factoring Sums and Differences of Cubes

Factor the following polynomials.

(a) $y^3 + 27$ (b) $64 - x^3$

Solution

(a) This polynomial is the sum of two cubes because y^3 is the cube of y, and 27 is the cube of 3. Therefore, we can factor the polynomial as follows.

$$
\begin{aligned}
y^3 + 27 &= y^3 + 3^3 && \text{Write as sum of two cubes} \\
&= (y + 3)(y^2 - (y)(3) + 3^2) && \text{Factored form} \\
&= (y + 3)(y^2 - 3y + 9) && \text{Simplify}
\end{aligned}
$$

(b) This polynomial is the difference of two cubes because 64 is the cube of 4, and x^3 is the cube of x. Therefore, we can factor the polynomial as follows.

$$
\begin{aligned}
64 - x^3 &= 4^3 - x^3 && \text{Write as difference of two cubes} \\
&= (4 - x)(4^2 + 4x + x^2) && \text{Factored form} \\
&= (4 - x)(16 + 4x + x^2) && \text{Simplify} \qquad ■
\end{aligned}
$$

We find that it is easy to make arithmetic errors when applying the patterns for factoring the sum or difference of two cubes. When you use these patterns, we suggest that you check your work by multiplying the two factors. For instance, we can check the factors given in Example 8 as follows.

(a)

$$
\begin{array}{r}
y^2 - 3y + 9 \\
y + 3 \\
\hline
3y^2 - 9y + 27 \\
y^3 - 3y^2 + 9y \\
\hline
y^3 \qquad\qquad + 27
\end{array}
$$

(b)

$$
\begin{array}{r}
x^2 + 4x + 16 \\
-x + 4 \\
\hline
4x^2 + 16x + 64 \\
-x^3 - 4x^2 - 16x \\
\hline
-x^3 \qquad\qquad + 64
\end{array}
$$

EXAMPLE 9   ■   Removing a Common Monomial Factor First

Factor the following polynomials.

(a) $2x^2 - 20x + 50$ (b) $2x^3 - 16$

Solution

(a) This polynomial has a common monomial factor of 2. After removing this factor we find that the remaining polynomial is a perfect square trinomial.

$$
\begin{aligned}
2x^2 - 20x + 50 &= 2(x^2 - 10x + 25) && \text{Remove common monomial factor} \\
&= 2(x - 5)^2 && \text{Factor as perfect square trinomial}
\end{aligned}
$$

$$
\begin{aligned}
\text{(b)} \quad 2x^3 - 16 &= 2(x^3 - 8) && \text{Remove common monomial factor} \\
&= 2(x^3 - 2^3) && \text{Write as difference of cubes} \\
&= 2(x - 2)(x^2 + 2x + 4) && \text{Factored form} \qquad ■
\end{aligned}
$$

We have discussed the basic factoring techniques one at a time. From this point on, you must decide which technique to apply to any given problem situation. To assist you in this selection process, we provide the following guidelines.

Guidelines for Factoring Polynomials

1. Factor out any common factors.

2. Factor according to one of the special polynomial forms: difference of squares, sum or difference of cubes, or perfect square trinomials.

3. Factor trinomials, $ax^2 + bx + c$, using the methods for $a = 1$ or $a \neq 1$.

4. Factor by grouping—for polynomials with four terms.

5. Check to see whether the factors themselves can be factored further.

6. Check the results by multiplying the factors.

DISCUSSION PROBLEM ■   A Three-Dimensional View of a Special Product

Figure 6.2 shows two cubes: a large cube whose volume is a^3 and a smaller cube whose volume is b^3. If the smaller cube is removed from the larger, the remaining solid has a volume of $a^3 - b^3$ and is composed of three rectangular boxes, labeled Box 1, Box 2, and Box 3. Find the volume of each box and describe how these results are related to the following special product pattern.

$$a^3 - b^3 = (a - b)(a^2 + ab + b^2)$$
$$= (a - b)a^2 + (a - b)ab + (a - b)b^2$$

FIGURE 6.2

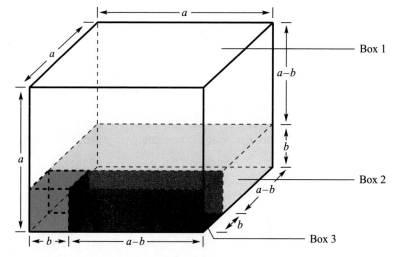

Warm-Up

The following warm-up exercises involve skills that were covered in earlier sections. You will use these skills in the exercise set for this section.

In Exercises 1–6, find the product.

1. $-12x(x - 3)$

2. $3z(-2z + 5)$

3. $(x + 10)(x - 10)$

4. $(y + a)(y - a)$

5. $(x + 5)^2$

6. $(3x - 2)^2$

In Exercises 7–10, factor the polynomial completely.

7. $10x^2 + 70$

8. $16x - 4x^2$

9. $3x^2 - 5x + 2$

10. $2x^2 - x - 1$

6.4   EXERCISES

In Exercises 1–10, factor the given difference of two squares.

1. $x^2 - 36$

2. $y^2 - 49$

3. $x^2 - \dfrac{1}{4}$

4. $x^2 - \dfrac{4}{9}$

5. $16y^2 - 9$

6. $9z^2 - 25$

7. $49 - 9y^2$

8. $100 - 49x^2$

9. $25 - (z + 5)^2$

10. $(x - 1)^2 - 4$

In Exercises 11–18, factor the polynomial completely.

11. $2x^2 - 72$

12. $3x^2 - 27$

13. $8 - 50x^2$

14. $a^3 - 16a$

15. $x^4 - 1$

16. $z^4 - 16$

17. $y^4 - 81$

18. $u^4 - 256$

In Exercises 19–32, factor the perfect square trinomial.

19. $x^2 - 4x + 4$

20. $x^2 + 10x + 25$

21. $z^2 + 6z + 9$

22. $a^2 - 12a + 36$

23. $4t^2 + 4t + 1$

24. $9x^2 - 12x + 4$

25. $25y^2 - 10y + 1$

26. $16z^2 + 24z + 9$

27. $b^2 + b + \dfrac{1}{4}$

28. $4t^2 - \dfrac{4}{3}t + \dfrac{1}{9}$

29. $x^2 - 6xy + 9y^2$

30. $u^2 + 8uv + 16v^2$

31. $4y^2 + 20yz + 25z^2$

32. $9x^2 - 18xy + 9y^2$

In Exercises 33–38, find two real numbers b so that the algebraic expression is a perfect square trinomial.

33. $x^2 + bx + 100$ **34.** $x^2 + bx + \dfrac{16}{25}$ **35.** $4x^2 + bx + \dfrac{1}{4}$ **36.** $9x^2 + bx + 64$

37. $x^2 + bxy + 4y^2$ **38.** $25x^2 + bxy + 4y^2$

In Exercises 39–42, find a real number c so that the algebraic expression is a perfect square trinomial.

39. $x^2 + 6x + c$ **40.** $x^2 + 10x + c$ **41.** $y^2 - 4y + c$ **42.** $z^2 - 14z + c$

In Exercises 43–50, factor the given sum or difference of cubes.

43. $x^3 - 8$ **44.** $x^3 - 27$ **45.** $y^3 + 64$ **46.** $z^3 + 125$

47. $1 - 8t^3$ **48.** $27s^3 + 1$ **49.** $8t^3 - 1$ **50.** $64x^3 - 125$

In Exercises 51–76, factor the polynomial completely.

51. $x^3 - 4x^2$

52. $6x^2 - 54$

53. $x^2 - 2x + 1$

54. $16 + 6x - x^2$

55. $1 - 4x + 4x^2$

56. $9x^2 - 6x + 1$

57. $2x^2 + 4x - 2x^3$

58. $2y^3 - 7y^2 - 15y$

59. $9x^2 + 10x + 1$

60. $13x + 6 + 5x^2$

61. $4x(2x - 1) + (2x - 1)$

62. $5(3 - 4x) - x(3 - 4x)$

63. $5 - x + 5x^2 - x^3$

64. $3x^3 + x^2 + 15x + 5$

65. $x^4 - 4x^3 + x^2 - 4x$

66. $4x^4 - 2x^3 - 6x^2 + 3x$

67. $25 - (z + 2)^2$

68. $(t - 1)^2 - 49$

69. $2t^3 - 16$

70. $24x^3 - 3$

71. $u^3 + 2u^2 + 3u$

72. $u^3 + 2u^2 - 3u$

73. $x^4 - 81$

74. $2x^4 - 32$

75. $1 - x^4$

76. $81 - y^4$

In Exercises 77–80, evaluate the given product in your head using the following two examples as models.

$$29^2 = (30 - 1)^2 = 30^2 - 2(30)(1) + 1^2 = 900 - 60 + 1 = 841$$
$$48 \cdot 52 = (50 - 2)(50 + 2) = 50^2 - 2^2 = 2{,}500 - 4 = 2{,}496$$

77. 21^2 **78.** 49^2 **79.** $59 \cdot 61$ **80.** $28 \cdot 32$

In Exercises 81 and 82, write the given polynomial as the sum of two squares.

81. $x^2 + 6x + 10 = (x^2 + 6x + 9) + 1 = \boxed{}^2 + \boxed{}^2$

82. $x^2 + 8x + 25 = (x^2 + 8x + 16) + 9 = \boxed{}^2 + \boxed{}^2$

83. *Dimensions of a Box* The box shown in the figure has a square base and a height of x. The volume of the box is

$$x^3 + 20x^2 + 100x \quad \text{cubic inches.}$$

Find the dimensions of the base of the box.

84. *Dimensions of a Building* The building shown in the figure has a square base and a height of x. The volume of the building is

$$x^3 - 80x^2 + 1,600x \quad \text{cubic feet.}$$

Find the dimensions of the base of the building.

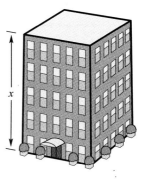

SECTION
6.5

Factoring and Problem Solving

The Zero-Factor Property • Solving Equations by Factoring • Applications

The Zero-Factor Property

We have spent nearly two chapters developing skills for *rewriting* (simplifying and factoring) polynomials. We now are ready to use these skills together with the following **Zero-Factor Property** to *solve* equations. The Zero-Factor Property indicates that if the product of two factors is zero, then one (or both) of the factors must also be zero.

Zero-Factor Property

Let a and b be real numbers, variables, or algebraic expressions. If a and b are factors such that

$$ab = 0$$

then $a = 0$ or $b = 0$. This property applies to three or more factors as well.

NOTE: The Zero-Factor Property is just another way of saying that the only way the product of two or more real numbers can be zero is if one or more of the real numbers is zero.

The Zero-Factor Property is the primary property for solving equations in algebra. For instance, to solve the equation

$$(x - 1)(x + 2) = 0$$

we use the Zero-Factor Property to conclude that either $(x - 1)$ or $(x + 2)$ must be zero. Setting the first factor equal to zero implies that $x = 1$ is a solution. That is,

$$x - 1 = 0 \implies x = 1.$$

(Note that the arrow $\implies$ is read as "implies that.") Similarly, setting the second factor equal to zero implies that $x = -2$ is a solution. That is,

$$x + 2 = 0 \implies x = -2.$$

Thus, the equation $(x - 1)(x + 2) = 0$ has exactly two solutions: 1 and -2.

Solving Equations by Factoring

In each of the examples in this section, note how we combine our factoring skills with the Zero-Factor Property to solve equations.

EXAMPLE 1 ■ Using Factoring to Solve an Equation

Solve the following equation.

$$x^2 - x - 6 = 0$$

Solution

The strategy here is to first check to see that the right side of the equation is zero. Next, we factor the left side of the equation. Finally, we apply the Zero-Factor Property to find the solutions.

$x^2 - x - 6 = 0$	Given equation (set equal to 0)
$(x + 2)(x - 3) = 0$	Factor left side of equation
$x + 2 = 0 \implies x = -2$	Set first factor equal to 0
$x - 3 = 0 \implies x = 3$	Set second factor equal to 0

Thus, the given equation has two solutions: -2 and 3.

Check First Solution

$x^2 - x - 6 = 0$	Given equation
$(-2)^2 - (-2) - 6 \stackrel{?}{=} 0$	Replace x by -2
$4 + 2 - 6 \stackrel{?}{=} 0$	
$0 = 0$	Solution checks

Check Second Solution

$$x^2 - x - 6 = 0 \qquad\qquad \text{Given equation}$$
$$3^2 - 3 - 6 \overset{?}{=} 0 \qquad\qquad \text{Replace } x \text{ by } 3$$
$$9 - 3 - 6 \overset{?}{=} 0$$
$$0 = 0 \qquad\qquad \text{Solution checks} \qquad\qquad ■$$

NOTE: Factoring and the Zero-Factor Property allow us to solve a polynomial equation of degree 2 (a **quadratic equation**) by converting it into two *linear* equations, which we already know how to solve. This is a common strategy of algebra—to break down a given problem into simpler parts, each solved by previously learned methods.

To use the Zero-Factor Property, a polynomial equation *must* be written in **standard form**. That is, the polynomial must be on one side of the equation and zero must be the only term on the other side of the equation. For instance, to write the equation $x^2 - 2x = 3$ in standard form, we must subtract 3 from both sides of the equation, as follows.

$$x^2 - 2x = 3 \qquad\qquad \text{Given equation (nonstandard form)}$$
$$x^2 - 2x - 3 = 3 - 3 \qquad\qquad \text{Subtract 3 from both sides}$$
$$x^2 - 2x - 3 = 0 \qquad\qquad \text{Standard form}$$

To solve this equation, we factor the left side as $(x - 3)(x + 1)$ and conclude that the solutions are 3 and -1.

EXAMPLE 2   ■   Solving a Polynomial Equation

Solve the following equation.

$$2x^2 + 5x = 12$$

Solution

$$2x^2 + 5x = 12 \qquad\qquad \text{Given equation}$$
$$2x^2 + 5x - 12 = 0 \qquad\qquad \text{Write in standard form}$$
$$(2x - 3)(x + 4) = 0 \qquad\qquad \text{Factor left side of equation}$$
$$2x - 3 = 0 \implies x = \frac{3}{2} \qquad\qquad \text{Set first factor equal to 0}$$
$$x + 4 = 0 \implies x = -4 \qquad\qquad \text{Set second factor equal to 0}$$

Therefore, the solutions are $\frac{3}{2}$ and -4. Check these solutions in the original equation.
■

In Examples 1 and 2, the given equations each involved a second-degree polynomial and each had *two different* solutions. Sometimes we encounter second-degree polynomial equations that have only one (repeated) solution. This occurs when the left side of the equation is a perfect square trinomial, as shown in Example 3.

EXAMPLE 3 ■ An Equation with a Repeated Solution

$$x^2 - 8x + 20 = 4 \qquad \text{Given equation}$$
$$x^2 - 8x + 16 = 0 \qquad \text{Write in standard form}$$
$$(x - 4)^2 = 0 \qquad \text{Factor}$$
$$x - 4 = 0 \implies x = 4 \qquad \text{Set factor equal to 0}$$

Note that even though the left side of this equation has two factors, the two factors are the same. Thus, we can conclude that the only solution of the equation is 4. Check this solution in the original equation. ■

In the next example, we look at an equation that has three different solutions. It is not a coincidence that the equation in the example is of degree three. In general, a polynomial equation can have *at most* as many solutions as its degree. For instance, a second-degree equation can have zero, one, or two real solutions, but it cannot have three or more solutions.

EXAMPLE 4 ■ Solving a Polynomial Equation with Three Factors

$$3x^3 = 12x^2 + 15x \qquad \text{Given equation}$$
$$3x^3 - 12x^2 - 15x = 0 \qquad \text{Standard form}$$
$$3x(x^2 - 4x - 5) = 0 \qquad \text{Remove common monomial factor}$$
$$3x(x - 5)(x + 1) = 0 \qquad \text{Factor completely}$$
$$3x = 0 \implies x = 0 \qquad \text{Set first factor equal to 0}$$
$$x - 5 = 0 \implies x = 5 \qquad \text{Set second factor equal to 0}$$
$$x + 1 = 0 \implies x = -1 \qquad \text{Set third factor equal to 0}$$

Therefore, the given equation has three solutions: 0, 5, and -1. Check these solutions in the original equation. ■

Be sure you see that the Zero-Factor Property can be applied only to a product that is equal to *zero*. For instance, we cannot conclude from the equation

$$x(x - 3) = 10$$

that $x = 10$ and $x - 3 = 10$ yield solutions. Instead, we must first write the equation in standard form and then factor the left side, as follows.

$$x^2 - 3x - 10 = 0 \implies (x - 5)(x + 2) = 0$$

Now, from the factored form we can see that the solutions are 5 and -2.

EXAMPLE 5 ■ Solving a Polynomial Equation

Solve the following equation.

$$(x + 3)(x + 6) = 4$$

Solution

To write this equation in standard form, we must first multiply the factors on the left side, as follows.

$(x + 3)(x + 6) = 4$	Given equation
$x^2 + 9x + 18 = 4$	Multiply factors
$x^2 + 9x + 14 = 0$	Standard form
$(x + 2)(x + 7) = 0$	Factor left side of equation
$x + 2 = 0 \implies x = -2$	Set first factor equal to 0
$x + 7 = 0 \implies x = -7$	Set second factor equal to 0

Therefore, the equation has two solutions: -2 and -7. Check these solutions in the original equation.   ■

Applications

In the last part of this section, we look at some applications that involve solving polynomial equations. The general approach to such problems is similar to that studied in Chapter 4.

EXAMPLE 6   ■   **An Application of a Polynomial Equation**

A rectangular family room has an area of 160 square feet. The length of the room is 6 feet more than its width. Find the dimensions of the room.

Solution

FIGURE 6.3

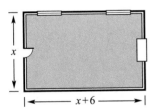

To begin, we make a sketch of the room, as shown in Figure 6.3. Note that we have labeled the width of the room as x, and the length of the room as $x + 6$, because the length is 6 feet more than the width.

Verbal model: Length · Width = Area

Labels: Width $= x$ (in feet)
Length $= x + 6$ (in feet)
Area $= 160$ (in square feet)

Equation:
$$x(x + 6) = 160$$
$$x^2 + 6x - 160 = 0$$
$$(x + 16)(x - 10) = 0$$
$$x = -16 \text{ or } 10$$

Note that the mathematical equation has two solutions: -16 and 10. However, for this application, the negative solution makes no sense, so we discard the negative solution (and use the positive solution). Thus, the width of the room is 10 feet, and the length of the room is 16 feet.

Check

To check this solution, we go back to the original statement of the problem. Note that a length of 16 feet is 6 feet more than a width of 10 feet. Moreover, a rectangular room with dimensions 16 feet by 10 feet has an area of 160 square feet. Thus, the solution checks. ∎

EXAMPLE 7 ■ **An Application of a Polynomial Equation**

The product of two consecutive positive integers is 56. What are the integers?

Solution

Verbal model: | First integer | · | Second integer | = | 56 |

Labels: First integer $= n$
Second integer $= n + 1$

Equation:
$$n(n + 1) = 56$$
$$n^2 + n - 56 = 0$$
$$(n + 8)(n - 7) = 0$$
$$n = -8 \text{ or } 7$$

Since the original statement of the problem required that the integers be positive, we discard -8 as a solution and choose $n = 7$. Thus, the two integers are $n = 7$ and $n + 1 = 8$. ∎

EXAMPLE 8 ■ **An Application of a Polynomial Equation**

A rock is dropped from the top of a 256-foot river gorge, as shown in Figure 6.4. The height (in feet) of the rock is given by the equation

$$\text{Height} = -16t^2 + 256$$

where t is the time measured in seconds. How long will it take the rock to hit the bottom of the gorge?

FIGURE 6.4

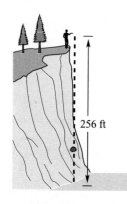

256 ft

Solution

From Figure 6.4, note that the bottom of the gorge corresponds to a height of 0 feet. Thus, we substitute a height of 0 into the given equation and solve for t.

$$0 = -16t^2 + 256 \qquad \text{Set height equal to 0}$$
$$16t^2 - 256 = 0 \qquad \text{Standard form}$$
$$16(t^2 - 16) = 0 \qquad \text{Remove common monomial factor}$$
$$16(t + 4)(t - 4) = 0 \qquad \text{Factor left side of equation}$$
$$t = -4 \text{ or } 4 \qquad \text{Solutions}$$

Since a time of -4 seconds doesn't make sense in this problem, we choose the positive solution and conclude that the rock hits the bottom of the gorge 4 seconds after it is dropped.   ∎

DISCUSSION PROBLEM   ∎   You Be the Instructor

Suppose you are tutoring a person in algebra and you want to create several equations for your student to practice solving. You probably want to start with several that have easy solutions, such as 1 and -2. Then you might want to create some equations with tougher solutions, such as $-\frac{3}{2}$ and $\frac{1}{6}$. Write a short paragraph describing how to do this and give some examples.   ∎

Warm-Up

The following warm-up exercises involve skills that were covered in earlier sections. You will use these skills in the exercise set for this section.

In Exercises 1–4, solve the given equation.

1. $4x - 48 = 0$

2. $3x + 9 = 0$

3. $\dfrac{x}{4} + \dfrac{x}{3} = \dfrac{1}{3}$

4. $\dfrac{3}{x} + \dfrac{3}{2} = 2$

In Exercises 5–10, factor the given polynomial.

5. $3x^2 + 7x$

6. $4x^2 - 25$

7. $16 - (x - 11)^2$

8. $x^2 + 7x - 18$

9. $10x^2 + 13x - 3$

10. $6x^2 - 73x + 12$

6.5   EXERCISES

In Exercises 1–10, solve the given equation.

1. $x(x - 5) = 0$

2. $z(z - 3) = 0$

3. $(y - 2)(y - 3) = 0$

4. $(s - 4)(s - 10) = 0$

5. $(a + 1)(a - 2) = 0$

6. $(t - 3)(t + 8) = 0$

7. $(2t - 5)(3t + 1) = 0$

8. $(5x + 3)(x - 8) = 0$

9. $x(x - 3)(x + 25) = 0$

10. $(y - 1)(2y + 3)(y + 12) = 0$

In Exercises 11–50, solve the given equation.

11. $x^2 - 16 = 0$ **12.** $x^2 - 144 = 0$ **13.** $3y^2 - 27 = 0$ **14.** $25z^2 - 100 = 0$

15. $1 - x^2 = 0$ **16.** $4 - x^2 = 0$ **17.** $(t - 3)^2 - 25 = 0$

18. $(s + 5)^2 - 49 = 0$ **19.** $6x^2 + 3x = 0$ **20.** $4x^2 - x = 0$

21. $x(x - 8) + 2(x - 8) = 0$ **22.** $u(u + 2) - 3(u + 2) = 0$ **23.** $x^2 - 2x - 8 = 0$

24. $x^2 - 8x - 9 = 0$ **25.** $m^2 - 2m + 1 = 0$ **26.** $a^2 + 6a + 9 = 0$

27. $x^2 + 14x + 49 = 0$ **28.** $x^2 - 10x + 25 = 0$ **29.** $4t^2 - 12t + 9 = 0$

30. $16x^2 + 56x + 49 = 0$ **31.** $3 + 5x - 2x^2 = 0$ **32.** $33 + 5y - 2y^2 = 0$

33. $6x^2 + 4x - 10 = 0$ **34.** $12x^2 + 7x + 1 = 0$ **35.** $x(x - 5) = 14$

36. $x(x - 1) = 6$ **37.** $y(2y + 1) = 3$ **38.** $x(5x - 14) = 3$

39. $(x + 1)(x + 4) = 4$ **40.** $(x + 1)(x - 2) = 4$ **41.** $(x - 7)(x - 3) = 5$

42. $(x - 9)(x + 2) = 12$ **43.** $x^3 + 5x^2 + 6x = 0$ **44.** $x^3 - 3x^2 - 10x = 0$

45. $2t^3 + 5t^2 - 12t = 0$ **46.** $3u^3 - 5u^2 - 2u = 0$ **47.** $x^2(x - 2) - 9(x - 2) = 0$

48. $y^2(y + 3) - (y + 3) = 0$ **49.** $a^3 + 2a^2 - 4a - 8 = 0$ **50.** $x^3 - x^2 - 16x + 16 = 0$

51. *Number Problem* Find two consecutive positive integers whose product is 72.

52. *Number Problem* Find two consecutive positive even integers whose product is 440.

53. *Dimensions of a Rectangle* The length of a rectangle is 3 inches greater than the width. The area of the rectangle is 108 square inches. Find the dimensions of the rectangle.

54. *Dimensions of a Rectangle* The length of a rectangle is one and one-half times the width. The area of the rectangle is 600 square inches. Find the dimensions of the rectangle.

55. *Height of an Object* An object is dropped from a weather balloon 1,600 feet above the ground (see figure). Find the time it takes for the object to reach the ground. The height (above ground) of the object is given by

$$\text{Height} = -16t^2 + 1,600$$

where the height is measured in feet and the time t is measured in seconds.

Figure for 55

1,600 ft

56. *Height of a Diver* A diver jumps from a diving board that is 32 feet above the water (see figure). The height of the diver is given by

$$\text{Height} = -16t^2 + 16t + 32$$

where the height is measured in feet and the time t is measured in seconds. How many seconds will it take for the diver to reach the water?

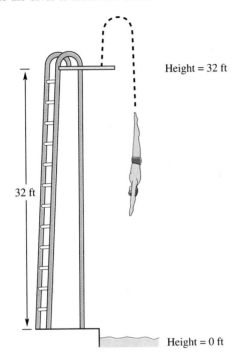

Height = 32 ft

32 ft

Height = 0 ft

57. *Constructing a Box* An open box is to be made from a square piece of cardboard by cutting 2-inch squares from each corner and turning up the sides (see figure).

(a) Show that the volume of the box is given by $V = 2x^2$.

(b) Complete the following table.

x	2	4	6	8
V				

(c) Find the dimensions of the original piece of cardboard if $V = 200$ cubic inches.

Figure for 57

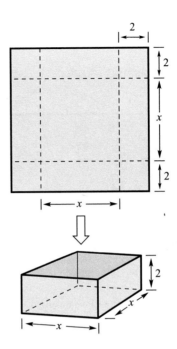

58. *Dimensions of a Box* An open box with a square base is constructed from 108 square inches of material. The height of the box is 3 inches. Find the dimensions of the base of the box. (*Hint:* The surface area is given by $S = x^2 + 4xh$.)

59. If a and b are nonzero real numbers, show that the equation $ax^2 + bx = 0$ must have two different solutions.

60. If a is a nonzero real number, find the two solutions of the equation $ax^2 - ax = 0$.

CHAPTER 6 SUMMARY

As you review and prepare for a test on this chapter, first try to obtain a global view of what was discussed. Then review the specific skills needed in each category.

Factoring Polynomials

■ *Calculate* the greatest common factor of an arithmetic expression.　　Section 6.1, Exercises 1–16

■ *Factor* a polynomial by taking out a common factor.　　Section 6.1, Exercises 17–60; Section 6.2, Exercises 35–44

■ *Factor* a polynomial by grouping.　　Section 6.1, Exercises 61–70

■ *Factor* a trinomial of the form $x^2 + bx + c$.　　Section 6.2, Exercises 11–34

■ *Factor* a trinomial of the form $ax^2 + bx + c$.　　Section 6.3, Exercises 11–30, Exercises 41–46

■ *Factor* a trinomial completely.　　Section 6.2, Exercises 35–44; Section 6.3, Exercises 47–60; Section 6.4, Exercises 51–71

■ *Factor* a trinomial by grouping.　　Section 6.3, Exercises 31–40

■ *Factor* the difference of two squares.　　Section 6.4, Exercises 1–18

■ *Factor* a perfect square trinomial.　　Section 6.4, Exercises 19–32

■ *Factor* the sum and difference of two cubes.　　Section 6.4, Exercises 43–50

Solving Equations by Factoring

■ *Solve* an equation by factoring and applying the Zero-Factor Property.　　Section 6.5, Exercises 1–50

■ *Create and solve* polynomial equations from verbal statements.　　Section 6.5, Exercises 51–60

Chapter 6 Review Exercises

In Exercises 1–14, factor the given polynomial.

1. $5x^2 + 10x^3$ **2.** $7y - 21y^4$ **3.** $8a - 12a^3$ **4.** $6u - 9u^2 + 15u^3$

5. $24(x + 1) - 18(x + 1)^2$ **6.** $(u + 1)(u - 2) + 3(u + 1)$

7. $a^2 - 100$ **8.** $16b^2 - 1$ **9.** $(u + v)^2 - 4$ **10.** $(y - 2)^2 - 9$

11. $x^2 - 8x + 16$ **12.** $y^2 + 24y + 144$ **13.** $9s^2 + 12s + 4$ **14.** $u^2 - 2uv + v^2$

In Exercises 15–20, find the missing factor(s).

15. $\frac{1}{3}x + \frac{5}{6} = \frac{1}{6}()$ **16.** $x^3 - x = x()()$

17. $3x^2 + 14x + 8 = (x + 4)()$ **18.** $2x^2 + 21x + 10 = (x + 10)()$

19. $x^4 - 2x^2 + 1 = (x + 1)^2()^2$ **20.** $u^4 - v^4 = (u^2 + v^2)()()$

In Exercises 21–40, factor the polynomial completely. (Some of the polynomials are not factorable.)

21. $x^2 - 3x - 28$ **22.** $x^2 - 3x - 40$ **23.** $6x^2 + 7x + 2$ **24.** $16x^2 + 13x - 3$

25. $6u^3 + 3u^2 - 30u$ **26.** $5t - 125t^3$ **27.** $-16a^3 - 16a^2 - 4a$ **28.** $8x^3 - 40x^2 + 32x$

29. $10x^2 + 9xy + 2y^2$ **30.** $4u^2 + uv - 5v^2$ **31.** $s^3t - st^3$ **32.** $y^3z + 4y^2z^2 + 4yz^3$

33. $2x^2 - 3x + 1$ **34.** $3x^2 + 8x + 4$ **35.** $27 - 8t^3$ **36.** $z^3 - 125$

37. $x^3 + 2x^2 + x + 2$ **38.** $x^3 - 5x^2 + 5x - 25$ **39.** $x^3 - 4x^2 - 4x + 16$ **40.** $x^3 + 6x^2 - x - 6$

In Exercises 41–46, find all integers b such that the trinomial is factorable.

41. $x^2 + bx + 9$ **42.** $x^2 + bx + 14$

43. $x^2 + bx - 24$ **44.** $2x^2 + bx - 16$

45. $3x^2 + bx - 20$ **46.** $3x^2 + bx + 1$

In Exercises 47–50, find *two* integers c such that the trinomial is factorable.

47. $x^2 + 6x + c$ **48.** $x^2 - 7x + c$

49. $2x^2 - 4x + c$ **50.** $5x^2 + 6x + c$

In Exercises 51–60, solve the given equation.

51. $x(2x - 3) = 0$

52. $3x(5x + 1) = 0$

53. $x^2 - 81 = 0$

54. $(x + 1)^2 - 16 = 0$

55. $x^2 - 12x + 36 = 0$

56. $2t^2 - 3t - 2 = 0$

57. $4s^2 + s - 3 = 0$

58. $y^3 - y^2 - 6y = 0$

59. $x(7 - x) = 12$

60. $a^3 - 3a^2 - a + 3 = 0$

61. *Height of an Object* A rock is thrown upward from a height of 48 feet with an initial velocity of 32 feet per second. The height (above water) of the rock is given by

$$\text{Height} = -16t^2 + 32t + 48$$

where the height is measured in feet and the time t is measured in seconds. Find the time for the rock to reach the water.

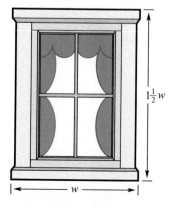

62. *Dimensions of a Window* The height of a rectangular window is one and one-half times its width. The area of the window is 2,400 square inches. Find the dimensions of the rectangle.

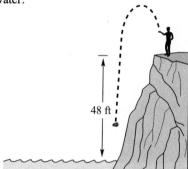

63. *Number Problem* The product of two consecutive positive even integers is 168. Find the two integers.

64. *Dimensions of a Box* A box with a square base is constructed from 300 square inches of material (see figure). The height of the box is 5 inches. Find the dimensions of the box. (*Hint:* The surface area is given by $S = x^2 + 4xh$.)

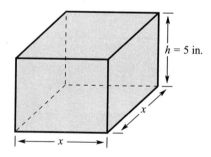

CHAPTER 6 TEST

Take this test as you would take a test in class. After you are done, check your work with the answers given in the back of the book.

1. Factor: $7x^2 - 14x^3$

2. Factor: $z(z + 7) - 3(z + 7)$

3. Factor: $t^2 - 4t - 5$

4. Factor: $6x^2 - 11x + 4$

5. Factor completely: $6y^3 + 45y^2 + 75y$

6. Factor: $4 - 25v^2$

7. Factor: $4x^2 - 20x + 25$

8. Factor and simplify: $16 - (z + 9)^2$

9. Factor by grouping: $x^3 + 2x^2 - 9x - 18$

10. Find the missing factor: $\dfrac{2}{3}x - \dfrac{3}{4} = \dfrac{1}{12}(\rule{2em}{0.6em})$

11. Solve: $(x + 4)(2x - 3) = 0$

12. Solve: $7x^2 - 14x = 0$

13. Solve: $3x^2 + 7x - 6 = 0$

14. Find all integers b such that $x^2 + bx + 5$ can be factored.

15. Find a real number c such that $x^2 + 12x + c$ is a perfect square trinomial.

16. The width of a rectangle is 5 inches less than the length. The area of the rectangle is 84 square inches. Find the dimensions of the rectangle.

17. The height of an object dropped from a height of 40 feet is given by

    $$\text{Height} = -16t^2 + 40$$

    where the height is measured in feet and the time t is measured in seconds. How long will it take the object to fall to a height of 15 feet?

CHAPTER SEVEN

Algebraic Fractions and Equations

Chapter Overview

In Chapters 5 and 6 we studied the algebra of polynomials. That is, we studied techniques for simplifying polynomials, adding and subtracting polynomials, multiplying and dividing polynomials, factoring polynomials, and solving polynomial equations. In this chapter we extend these skills to include algebraic fractions.

Simplifying Algebraic Fractions
Algebraic Fractions • *Simplifying Algebraic Fractions*

Algebraic Fractions

A fraction whose numerator and denominator are polynomials is called an **algebraic fraction** or a **rational expression**. Some examples are

$$\frac{5}{x-2}, \quad \frac{3x}{x^2-2x+1}, \quad \text{and} \quad \frac{x^3-4x}{x^2+x-2}.$$

In our work with algebraic fractions, we will assume that all real number values of the variable that make the denominator zero are excluded. For the three fractions above, we exclude $x = 2$ from the first fraction, $x = 1$ from the second, and both $x = 1$ and $x = -2$ from the third fraction. We refer to the *usable* (nonexcluded) values of the variable as the **domain** of the algebraic fraction.

EXAMPLE 1 ■ Finding the Domain of an Algebraic Fraction

Find the domain of each of the following by identifying the excluded values.

(a) $\dfrac{7}{x+3}$ (b) $\dfrac{x-5}{9}$ (c) $\dfrac{x}{x^2-16}$ (d) $\dfrac{x^2+1}{x^2+4x-5}$

Solution

(a) The denominator is zero when $x + 3 = 0$ or $x = -3$. Thus, the domain is all real values of x such that $x \neq -3$.

(b) The denominator, 9, is never zero. Thus, the domain is the set of *all* real numbers.

(c) For this algebraic fraction, the denominator

$$x^2 - 16 = (x+4)(x-4)$$

is zero when $x = -4$ or 4. Thus, the domain is all real values of x such that $x \neq -4$ and $x \neq 4$.

(d) For this algebraic fraction, the denominator

$$x^2 + 4x - 5 = (x+5)(x-1)$$

is zero when $x = -5$ or 1. Thus, the domain is all real values of x such that $x \neq -5$ and $x \neq 1$. ■

In applications involving algebraic fractions, it is often necessary to further restrict the domain. To indicate such a restriction, we write the domain to the right of the fraction. For instance, the domain of the algebraic fraction

$$\frac{x^2 + 10}{x + 2}, \quad x > 0$$

is restricted to be the set of *positive* real numbers.

EXAMPLE 2   ■   An Application Involving a Restricted Domain

A publisher asks a printing company to print copies of a book. The printing company charges $5,000 as a start-up fee, plus $4 per book. If x copies of the book are printed, the total cost (in dollars) of the printing is given by

Total cost $= 5,000 + 4x$.

The average cost per book depends on the number of books printed. If 100 books are printed, the average cost is $(5,000 + 4(100))/100 = \$54$ per book. On the other hand, if 10,000 books are printed, the average cost is $(5,000 + 4(10,000))/10,000 = \$4.50$ per book. In general, if x copies of the book are printed, then the average cost per book is given by

$$\text{Average cost per book} = \frac{5,000 + 4x}{x}.$$

What is the domain of this algebraic fraction?

Solution

If we were simply considering the fraction $(5,000 + 4x)/x$ as a mathematical quantity, we would say that the domain is all real values of x such that $x \neq 0$. However, because this fraction is a mathematical model representing a real-life situation, we must decide which values of x make sense in real life. For this model, the variable x represents the number of books printed. Since the number of books printed must be a positive integer, we conclude that the domain is the set of positive integers. That is,

Domain $= \{1, 2, 3, 4, \ldots\}$.   ■

Simplifying Algebraic Fractions

The rules for operating with algebraic fractions are like those for numerical fractions (see the properties of fractions in Section 1.5). As with numerical fractions, we say that an algebraic fraction is **simplified** or is in **reduced form** if its numerator and denominator have no common factors (other than 1). To write a fraction in reduced form, we apply the following **reduction law**.

Reduction Law for Fractions	Let a, b, and c represent real numbers, variables, or algebraic expressions such that $b \neq 0$ and $c \neq 0$. Then the following reduction law is valid.
	$$\frac{ac}{bc} = \frac{a}{b}$$

Be sure you see that this reduction law allows us only to divide out factors, *not* terms. For instance, consider the following.

$$\frac{\cancel{3} \cdot x}{\cancel{3}(x + 4)}$$
We CAN divide out common *factor* 3

$$\frac{\cancel{3} + x}{\cancel{3} + (x + 4)}$$
We CANNOT divide out common *term* 3

$$\frac{\cancel{x}(x^2 + 4)}{\cancel{x}}$$
We CAN divide out common *factor* x

$$\frac{x + (x^2 + 4)}{x}$$
We CANNOT divide out common *term* x

Using the reduction law to simplify an algebraic fraction requires two steps. First, we must completely factor the numerator and denominator. Then, we apply the reduction law to divide out any *factors* that are common to both the numerator and denominator. Thus, your success in simplifying algebraic fractions actually lies in your ability to completely factor the polynomials in both the numerator and denominator.

EXAMPLE 3 ■ Simplifying an Algebraic Fraction

Simplify the following algebraic fraction.

$$\frac{4x^3 - 8x^2 + 12x}{4x^2}$$

Solution

To begin, we completely factor both the numerator and denominator. If the numerator and denominator have any common factors, then we divide out the common factors to obtain the simplified form. If the numerator and denominator do not have any common factors, then the fraction is already in simplified form.

$$\frac{4x^3 - 8x^2 + 12x}{4x^2} = \frac{4x(x^2 - 2x + 3)}{4x^2}$$
Factor numerator and denominator

$$= \frac{4\cancel{x}(x^2 - 2x + 3)}{4\cancel{x}(x)}$$
Divide out common factor $4x$

$$= \frac{x^2 - 2x + 3}{x}$$
Simplified form ■

EXAMPLE 4 ■ Restricting the Domain After Simplifying

Simplify the following algebraic fraction.

$$\frac{x^2 + 4x - 12}{3x - 6}$$

Solution

$$\frac{x^2 + 4x - 12}{3x - 6} = \frac{(x + 6)(x - 2)}{3(x - 2)} \qquad \text{Factor numerator and denominator}$$

$$= \frac{(x + 6)(x - 2)}{3(x - 2)} \qquad \text{Cancel common factor } (x - 2)$$

$$= \frac{x + 6}{3}, \quad x \neq 2 \qquad \text{Simplified form}$$

The simplification process used in this example has a tricky technicality. This technicality arises from the fact that canceling common factors from the numerator and denominator of a fraction can change the domain of the fraction. For instance, in this example the domain of the original fraction is all real values of x such that $x \neq 2$. Thus, in order for us to equate the original fraction with the simplified fraction, we must restrict the domain of the simplified fraction to exclude 2. ■

EXAMPLE 5 ■ Simplifying an Algebraic Fraction

Simplify the following algebraic fraction.

$$\frac{x^3 - 4x}{x^2 + x - 2}$$

Solution

$$\frac{x^3 - 4x}{x^2 + x - 2} = \frac{x(x^2 - 4)}{(x + 2)(x - 1)} \qquad \text{Partially factor}$$

$$= \frac{x(x + 2)(x - 2)}{(x + 2)(x - 1)} \qquad \text{Factor completely}$$

$$= \frac{x(x + 2)(x - 2)}{(x + 2)(x - 1)} \qquad \text{Cancel common factor } (x + 2)$$

$$= \frac{x(x - 2)}{x - 1}, \quad x \neq -2 \qquad \text{Simplified form} \qquad ■$$

Be sure to *completely* factor the numerator and denominator of an algebraic fraction before concluding that there is no common factor. This may involve a change in signs to see if further reduction is possible. Watch for this in the next example.

EXAMPLE 6 ■ **Simplifying an Algebraic Fraction**

Simplify the following algebraic fraction.

$$\frac{12 + x - x^2}{2x^2 - 9x + 4}$$

Solution

$$\frac{12 + x - x^2}{2x^2 - 9x + 4} = \frac{(4 - x)(3 + x)}{(2x - 1)(x - 4)} \qquad \text{Factor completely}$$

$$= \frac{-(x - 4)(3 + x)}{(2x - 1)(x - 4)} \qquad (4 - x) = -1(x - 4)$$

$$= \frac{-(x - 4)(3 + x)}{(2x - 1)(x - 4)} \qquad \text{Divide out common factor } (x - 4)$$

$$= -\frac{3 + x}{2x - 1}, \quad x \neq 4 \qquad \text{Simplified form}$$

Be sure you see that when we divided out the common factor of $(x - 4)$, we kept the minus sign. In the simplified form of the fraction, we usually like to move the minus sign out in from of the fraction. However, this is merely a personal preference. Any of the following forms would be legitimate.

$$-\frac{3 + x}{2x - 1} = \frac{-(3 + x)}{2x - 1} = \frac{-3 - x}{2x - 1} = \frac{x + 3}{1 - 2x} \qquad ■$$

In the next example, we apply the reduction law to simplify an algebraic fraction that involves more than one variable.

EXAMPLE 7 ■ **Simplifying an Algebraic Fraction Involving Two Variables**

Simplify the following algebraic fraction.

$$\frac{x^2 - 2xy + y^2}{5x - 5y}$$

Solution

$$\frac{x^2 - 2xy + y^2}{5x - 5y} = \frac{(x - y)^2}{5(x - y)} \qquad \text{Factor numerator and denominator}$$

$$= \frac{(x - y)(x - y)}{5(x - y)} \qquad \text{Divide out common factor } (x - y)$$

$$= \frac{x - y}{5}, \quad x \neq y \qquad \text{Simplified form} \qquad ■$$

As you study the examples and work the exercises in this and the following two sections, keep in mind that you are *rewriting expressions in simpler forms*. You are not solving equations. We introduce equal signs in the steps of the solution only to indicate that the new form of the expression (fraction) is equivalent to the previous one.

DISCUSSION PROBLEM ■ Comparing Two Algebraic Fractions

Suppose you are asked to determine whether the algebraic fractions

$$\frac{x^2 + x + 2}{x^3 + 4x^2 + 5x + 6} \quad \text{and} \quad \frac{1}{x + 3}$$

are equivalent. Write a short paragraph describing how you could determine whether or not two fractions are equivalent. Then illustrate your answer by using the two given fractions. ■

Warm-Up

The following warm-up exercises involve skills that were covered in earlier sections. You will use these skills in the exercise set for this section.

In Exercises 1–4, fill in the missing numerator or denominator.

1. $-\dfrac{3}{4} = \dfrac{\rule{1cm}{0.4cm}}{12}$ **2.** $\dfrac{5}{8} = \dfrac{30}{\rule{1cm}{0.4cm}}$

3. $\dfrac{5}{3} = \dfrac{-35}{\rule{1cm}{0.4cm}}$ **4.** $-\dfrac{3}{7} = \dfrac{\rule{1cm}{0.4cm}}{42}$

In Exercises 5–8, write the fraction in reduced form.

5. $\dfrac{30}{45}$ **6.** $\dfrac{36}{84}$

7. $\dfrac{84}{98}$ **8.** $\dfrac{63}{99}$

In Exercises 9 and 10, factor the given polynomial.

9. $x^3 - 3x^2 + 4x$ **10.** $10x^4 + 3x^3 - 4x^2$

7.1 EXERCISES

In Exercises 1–10, find the domain of the given algebraic fraction.

1. $\dfrac{5}{x-4}$

2. $\dfrac{10}{x-6}$

3. $\dfrac{x}{x+2}$

4. $\dfrac{2z}{z+8}$

5. $\dfrac{3}{x^2+4}$

6. $\dfrac{y^2-1}{5}$

7. $\dfrac{4t}{t^2-25}$

8. $\dfrac{z+2}{z^2-4}$

9. $\dfrac{x^2}{x^2-x-2}$

10. $\dfrac{y+5}{4y^2+y-3}$

In Exercises 11–20, find the missing factor.

11. $\dfrac{5}{2x}=\dfrac{5\;\rule{1cm}{0.3mm}}{6x^2}$

12. $\dfrac{3x}{2}=\dfrac{3x\;\rule{1cm}{0.3mm}}{4x^2}$

13. $\dfrac{3}{4}=\dfrac{3\;\rule{1cm}{0.3mm}}{4(x+1)}$

14. $\dfrac{11}{16}=\dfrac{11\;\rule{1cm}{0.3mm}}{32(x-4)^2}$

15. $\dfrac{x}{2}=\dfrac{x(x+2)}{2\;\rule{1cm}{0.3mm}}$

16. $\dfrac{5x}{8}=\dfrac{25x^2}{8\;\rule{1cm}{0.3mm}}$

17. $\dfrac{x+1}{x}=\dfrac{(x+1)\;\rule{1cm}{0.3mm}}{x(x-2)}$

18. $\dfrac{3y-4}{y+1}=\dfrac{(3y-4)\;\rule{0.6cm}{0.3mm}\,y-1}{y^2-1}$

19. $\dfrac{3x}{x-3}=\dfrac{3x\;\rule{1cm}{0.3mm}}{x^2-x-6}$

20. $\dfrac{1-z}{z^2}=\dfrac{(1-z)\;\rule{1cm}{0.3mm}}{z^3+z^2}$

In Exercises 21–50, simplify the given algebraic fraction.

21. $\dfrac{4x}{12}$

22. $\dfrac{18y}{36}$

23. $\dfrac{2y^2}{y}$

24. $\dfrac{5z^3}{z}$

25. $\dfrac{15x^2}{10x}$

26. $\dfrac{18y^2}{60y^5}$

27. $\dfrac{x^2(x+1)}{x(x+1)}$

28. $\dfrac{ab(b-3)}{b(b-3)^2}$

29. $\dfrac{x-5}{2x-10}$

30. $\dfrac{x^2-25}{x-5}$

31. $\dfrac{5-x}{2x-10}$

32. $\dfrac{x^2-25}{5-x}$

33. $\dfrac{3xy}{xy+x}$

34. $\dfrac{x+x^2y}{xy+1}$

35. $\dfrac{y^2-16}{3y+12}$

36. $\dfrac{x^2-25z^2}{x-5z}$

37. $\dfrac{a+2}{a^2+4a+4}$

38. $\dfrac{u^2-6u+9}{u-3}$

39. $\dfrac{x^2-5x}{x^2-10x+25}$

40. $\dfrac{z^2+12z+36}{5z+30}$

41. $\dfrac{y^2-4}{y^2+3y-10}$

42. $\dfrac{x^2-7x}{x^2-8x+7}$

43. $\dfrac{3-x}{x^2-5x+6}$

44. $\dfrac{y^2-7y+12}{y^2+3y-18}$

45. $\dfrac{x^2+8x-20}{x^2+11x+10}$

46. $\dfrac{z^2-3z-18}{z^2+2z-48}$

47. $\dfrac{x^3+5x^2+6x}{x^2-4}$

48. $\dfrac{t^3-t}{t^3+5t^2-6t}$

49. $\dfrac{x^3-2x^2+x-2}{x-2}$

50. $\dfrac{x^2-9}{x^3+x^2-9x-9}$

51. *Average Cost* A machine shop has a setup cost of $3,000 for the production of a new product. The cost of labor and materials for producing each unit is $7.50.
   (a) Write an algebraic fraction that gives the average cost per unit when x units are produced.
   (b) Determine the domain of the fraction in part (a).
   (c) Find the average cost per unit when 100 units are produced.

52. *Average Cost* A machine shop has a setup cost of $5,000 for the production of a new product. The cost of labor and materials for producing each unit is $12.50.
   (a) Write an algebraic fraction that gives the average cost per unit when x units are produced.
   (b) Determine the domain of the fraction in part (a).
   (c) Find the average cost per unit when 200 units are produced.

53. *Air Pollution* A utility company burns coal to produce electricity. The cost (in dollars) of removing p percent of the air pollutants in the stack emission of the utility company is given by the algebraic fraction

$$\frac{80{,}000p}{1 - p}.$$

Determine the domain of the algebraic fraction. (Assume that the percentage p is written in decimal form.)

54. *Comparing Distances* Suppose you start a trip and drive at an average speed of 50 miles per hour. Two hours later a friend starts a trip on the same road and drives at an average speed of 60 miles per hour.
   (a) Find polynomial expressions that represent the distance each of you has driven when your friend has been driving for t hours.
   (b) Use the result of part (a) to determine the ratio of the distance your friend has driven to the distance you have driven.
   (c) Find the ratio described in part (b) when $t = 5$ and when $t = 10$.

SECTION

7.2

Multiplying and Dividing Algebraic Fractions

Multiplying Algebraic Fractions • *Dividing Algebraic Fractions* • *Compound Fractions*

Multiplying Algebraic Fractions

The rule for multiplying algebraic fractions is the same as the rule for multiplying numerical fractions. That is, we *multiply numerators, multiply denominators, and simplify.* Here is an example.

$$\frac{x - 2}{x + 1} \cdot \frac{x + 1}{x + 2} = \frac{(x - 2)(x + 1)}{(x + 1)(x + 2)} \qquad \text{Multiply numerators and denominators}$$

$$= \frac{(x - 2)(x + 1)}{(x + 1)(x + 2)} \qquad \text{Cancel common factor}$$

$$= \frac{x - 2}{x + 2}, \quad x \neq -1 \qquad \text{Simplified form}$$

In order to recognize common factors in the product of fractions, you should write the numerators and denominators in completely factored form.

EXAMPLE 1   ■   Multiplying Algebraic Fractions

Multiply the following algebraic fractions.

$$\frac{3x^2y}{2xy^2} \cdot \frac{-10xy^3}{6x^3}$$

Solution

$$\frac{3x^2y}{2xy^2} \cdot \frac{-10xy^3}{6x^3} = \frac{-3(10)x^3y^4}{2(6)x^4y^2}$$ Multiply numerators and denominators

$$= \frac{-3(2)(5)(x^3)(y^2)(y^2)}{2(3)(2)(x^3)(x)(y^2)}$$ Cancel common factors

$$= -\frac{5y^2}{2x}, \quad y \neq 0$$ Simplified form   ■

EXAMPLE 2   ■   Multiplying Algebraic Fractions

Multiply the following algebraic fractions.

$$\frac{4x}{x^2 - 9} \cdot \frac{x - 3}{8x^2 + 12x}$$

Solution

$$\frac{4x}{x^2 - 9} \cdot \frac{x - 3}{8x^2 + 12x} = \frac{4x(x - 3)}{(x + 3)(x - 3)(4x)(2x + 3)}$$ Multiply and factor

$$= \frac{4x(x - 3)}{(x + 3)(x - 3)(4x)(2x + 3)}$$ Cancel common factors

$$= \frac{1}{(x + 3)(2x + 3)}, \quad x \neq 0, x \neq 3$$ Simplified form   ■

EXAMPLE 3   ■   Multiplying Algebraic Fractions

Multiply the following algebraic fractions.

$$\frac{x^2 - 3x}{x^2 - 5x + 6} \cdot \frac{(x - 2)^2}{2x}$$

Solution

$$\frac{x^2 - 3x}{x^2 - 5x + 6} \cdot \frac{(x - 2)^2}{2x}$$

$$= \frac{x(x - 3)(x - 2)^2}{(x - 2)(x - 3)(2x)}$$ Multiply and factor

$$= \frac{x(x - 3)(x - 2)(x - 2)}{(x - 2)(x - 3)(2x)}$$ Cancel common factors

$$= \frac{x - 2}{2}, \quad x \neq 0, x \neq 2, x \neq 3$$ Simplified form   ■

The rule for multiplying fractions can be extended to cover products involving expressions that are not in fractional form. To do this, we rewrite the (nonfractional) expression as a fraction whose denominator is 1. Here is a simple example.

$$\frac{x+2}{x+4} \cdot (3x) = \frac{x+2}{x+4} \cdot \frac{3x}{1} = \frac{(x+2)(3x)}{x+4} = \frac{3x(x+2)}{x+4}$$

EXAMPLE 4   ■   Multiplying Algebraic Fractions

Multiply the following expressions.

$$\frac{x}{2x^2-x-3} \cdot (2x-3)$$

Solution

$$\frac{x}{2x^2-x-3} \cdot (2x-3) = \frac{x}{2x^2-x-3} \cdot \frac{2x-3}{1} \qquad \text{Write in fractional form}$$

$$= \frac{x(2x-3)}{(2x-3)(x+1)} \qquad \text{Multiply and factor}$$

$$= \frac{x\cancel{(2x-3)}}{\cancel{(2x-3)}(x+1)} \qquad \text{Cancel common factor}$$

$$= \frac{x}{x+1}, \quad x \neq \frac{3}{2} \qquad \text{Simplified form} \qquad ■$$

In the next example note how to cancel factors that differ only in sign.

EXAMPLE 5   ■   Multiplying Algebraic Fractions

Multiply the following algebraic fractions.

$$\frac{x-y}{6x+4y} \cdot \frac{3x+2y}{y^2-x^2}$$

Solution

$$\frac{x-y}{6x+4y} \cdot \frac{3x+2y}{y^2-x^2} = \frac{(x-y)(3x+2y)}{2(3x+2y)(y+x)(y-x)} \qquad \text{Multiply and factor}$$

$$= \frac{(x-y)(3x+2y)}{2(3x+2y)(y+x)(-1)(x-y)} \qquad (y-x) = (-1)(x-y)$$

$$= \frac{\cancel{(x-y)}\cancel{(3x+2y)}}{2\cancel{(3x+2y)}(y+x)(-1)\cancel{(x-y)}} \qquad \text{Cancel common factors}$$

$$= -\frac{1}{2(y+x)}, \quad x \neq y, x \neq -\frac{2}{3}y \qquad \text{Simplified form} \qquad ■$$

The rule for multiplying algebraic fractions can be extended to cover the product of three or more fractions. This procedure is illustrated in Example 6.

EXAMPLE 6 ■ **Multiplying Three Algebraic Fractions**

Multiply the following algebraic fractions.

$$\frac{x}{x-2} \cdot \frac{2x+4}{x^2-5x} \cdot \frac{x^2-3x+2}{x+2}$$

Solution

$$\frac{x}{x-2} \cdot \frac{2x+4}{x^2-5x} \cdot \frac{x^2-3x+2}{x+2}$$

$$= \frac{(x)(2)(x+2)(x-1)(x-2)}{(x-2)(x)(x-5)(x+2)} \qquad \text{Multiply and factor}$$

$$= \frac{\cancel{(x)}(2)\cancel{(x+2)}(x-1)\cancel{(x-2)}}{\cancel{(x-2)}\cancel{(x)}(x-5)\cancel{(x+2)}} \qquad \text{Cancel common factors}$$

$$= \frac{2(x-1)}{x-5}, \quad x \neq 0, x \neq 2, x \neq -2 \qquad \text{Simplified form}$$ ■

Dividing Algebraic Fractions

To divide two algebraic fractions we simply *invert the divisor and multiply*. That is, we multiply the first fraction by the *reciprocal* of the second. For instance, to perform the following division

$$\frac{x}{x+1} \div \frac{x}{x+2}$$

we invert the fraction $x/(x+2)$ and multiply as follows.

$$\frac{x}{x+1} \div \frac{x}{x+2} = \frac{x}{x+1} \cdot \frac{x+2}{x} = \frac{\cancel{x}(x+2)}{(x+1)\cancel{(x)}} = \frac{x+2}{x+1}, \quad x \neq 0, x \neq -2$$

EXAMPLE 7 ■ **Dividing Algebraic Fractions**

Perform the following division.

$$\frac{x}{x+4} \div \frac{x+3}{x+4}$$

Solution

$$\frac{x}{x+4} \div \frac{x+3}{x+4} = \frac{x}{x+4} \cdot \frac{x+4}{x+3} \qquad \text{Invert and multiply}$$

$$= \frac{(x)(x+4)}{(x+4)(x+3)} \qquad \text{Multiply numerators and denominators}$$

$$= \frac{(x)\cancel{(x+4)}}{\cancel{(x+4)}(x+3)} \qquad \text{Cancel common factor}$$

$$= \frac{x}{x+3}, \quad x \neq -4 \qquad \text{Simplified form}$$ ■

EXAMPLE 8 ■ **Dividing Algebraic Fractions**

Perform the following division.

$$\frac{x^2 - 2x}{x^2 - 6x + 8} \div \frac{2x}{3x - 12}$$

Solution

$$\frac{x^2 - 2x}{x^2 - 6x + 8} \div \frac{2x}{3x - 12} = \frac{x^2 - 2x}{x^2 - 6x + 8} \cdot \frac{3x - 12}{2x}$$ Invert and multiply

$$= \frac{(x)(x - 2)(3)(x - 4)}{(x - 2)(x - 4)(2x)}$$ Multiply and factor

$$= \frac{(x)\cancel{(x-2)}(3)\cancel{(x-4)}}{\cancel{(x-2)}\cancel{(x-4)}(2x)}$$ Cancel common factors

$$= \frac{3}{2}, \quad x \neq 0, \ x \neq 2, \ x \neq 4$$ Simplified form ■

Compound Fractions

Problems involving the division of two algebraic fractions are sometimes written as **compound fractions**. The rules for dividing fractions still apply in such cases. For instance, consider the following compound fraction.

$$\frac{\left(\dfrac{x + 2}{3}\right)}{\left(\dfrac{x - 2}{x}\right)} \quad \begin{array}{l} \text{Numerator fraction} \\ \rightarrow \text{Main fraction line} \\ \text{Denominator fraction} \end{array}$$

(Note that for compound fractions we make the main fraction line slightly longer than the fraction lines in the numerator and denominator.) To perform the division implied by this compound fraction, we invert the denominator and multiply, as follows.

$$\frac{\left(\dfrac{x + 2}{3}\right)}{\left(\dfrac{x - 2}{x}\right)} = \frac{x + 2}{3} \cdot \frac{x}{x - 2} = \frac{x(x + 2)}{3(x - 2)}, \quad x \neq 0$$

EXAMPLE 9 ■ **Simplifying a Compound Fraction**

Simplify the following compound fraction.

$$\frac{\left(\dfrac{x^2 + 4x + 3}{x - 2}\right)}{2x + 6}$$

Solution

In this case the denominator is not written in fractional form, so we begin by doing that.

$$\frac{\left(\dfrac{x^2 + 4x + 3}{x - 2}\right)}{2x + 6} = \frac{\left(\dfrac{x^2 + 4x + 3}{x - 2}\right)}{\left(\dfrac{2x + 6}{1}\right)} \qquad \text{Rewrite denominator}$$

$$= \frac{x^2 + 4x + 3}{x - 2} \cdot \frac{1}{2x + 6} \qquad \text{Invert and multiply}$$

$$= \frac{(x + 1)(x + 3)}{(x - 2)(2)(x + 3)} \qquad \text{Multiply and factor}$$

$$= \frac{(x + 1)\cancel{(x + 3)}}{(x - 2)(2)\cancel{(x + 3)}} \qquad \text{Cancel common factor}$$

$$= \frac{x + 1}{2(x - 2)}, \quad x \neq -3 \qquad \text{Simplified form} \qquad ■$$

DISCUSSION PROBLEM   ■   **Changing Units of Measure**

Suppose a car is traveling at 60 miles per hour. Write a short paragraph describing how you would find its speed in *feet per second*. Suppose you were driving at this rate and saw an animal in the road 100 feet ahead of you. Do you think you could stop in time to avoid hitting the animal?   ■

Warm-Up

The following warm-up exercises involve skills that were covered in earlier sections. You will use these skills in the exercise set for this section.

In Exercises 1–6, multiply or divide, as indicated, and simplify your answer.

1. $\dfrac{5}{16} \cdot \dfrac{4}{5}$

2. $\dfrac{7}{12} \cdot \dfrac{9}{14}$

3. $-\dfrac{12}{35} \cdot \dfrac{-25}{54}$

4. $\dfrac{-225}{-448} \cdot \dfrac{28}{-105}$

5. $\dfrac{16}{3} \div \dfrac{32}{45}$

6. $\dfrac{-7}{45} \div \dfrac{2}{3}$

In Exercises 7–10, factor the given polynomial.

7. $4x^3 - x$

8. $9x^2 - 4$

9. $15x^2 - 11x - 14$

10. $4x^2 - 28x + 49$

7.2 EXERCISES

In Exercises 1–6, find the missing factor.

1. $\dfrac{5}{2x} = \dfrac{5x^2}{2x\rule{1cm}{0.4pt}}$

2. $\dfrac{x}{x+1} = \dfrac{2x(x+1)}{(x+1)\rule{1cm}{0.4pt}}$

3. $\dfrac{3ab}{7} = \dfrac{3ab\rule{1cm}{0.4pt}}{7ab}$

4. $\dfrac{3t+6}{t} = \dfrac{(3t+6)\rule{1cm}{0.4pt}}{5t^2}$

5. $\dfrac{2x}{2+x} = \dfrac{2x\rule{1cm}{0.4pt}}{4-x^2}$

6. $\dfrac{x^2}{5-x} = \dfrac{x^2\rule{1cm}{0.4pt}}{5x-x^2}$

In Exercises 7–30, multiply the given algebraic fractions and simplify your answer.

7. $\dfrac{8x^2}{3} \cdot \dfrac{9}{16x}$

8. $\dfrac{6x}{5} \cdot \dfrac{1}{x}$

9. $\dfrac{12x^2}{6y} \cdot \dfrac{12y}{8x^2}$

10. $\dfrac{25y^2}{8x} \cdot \dfrac{8x}{5y}$

11. $\dfrac{8}{2+3x} \cdot (8+12x)$

12. $\dfrac{5-4x}{4} \cdot \dfrac{48}{10-8x}$

13. $3(a+2) \cdot \dfrac{1}{3a+6}$

14. $\dfrac{x+1}{2} \cdot \dfrac{2}{x+1}$

15. $\dfrac{1-r}{3} \cdot \dfrac{3}{r-1}$

16. $\dfrac{1-z}{1+z} \cdot \dfrac{z+1}{z-1}$

17. $\dfrac{5}{x-1} \cdot \dfrac{x-1}{25(x-2)}$

18. $\dfrac{(x+5)(x-3)}{x+2} \cdot \dfrac{1}{(x+5)(x+2)}$

19. $\dfrac{(x-9)(x+7)}{x+1} \cdot \dfrac{x}{9-x}$

20. $\dfrac{x+1}{x^3(3-x)} \cdot \dfrac{x(x-3)}{5}$

21. $\dfrac{r}{r-1} \cdot \dfrac{r^2-1}{r^2}$

22. $\dfrac{4y-16}{5y+15} \cdot \dfrac{2y+6}{y-4}$

23. $\dfrac{t^2-t-6}{t^2+6t+9} \cdot \dfrac{t+3}{t^2-4}$

24. $\dfrac{y^2-16}{2y^3} \cdot \dfrac{4y}{y^2-6y+8}$

25. $(x^2-4y^2) \cdot \dfrac{xy}{(x-2y)^2}$

26. $(u-2v)^2 \cdot \dfrac{u+2v}{u-2v}$

27. $\dfrac{x^2+xy-2y^2}{x^3+x^2y} \cdot \dfrac{x}{x^2+3xy+2y^2}$

28. $\dfrac{x-y}{x+y} \cdot \dfrac{x^2+y^2}{x^2-y^2}$

29. $\dfrac{a+1}{a-1} \cdot \dfrac{a^2-2a+1}{a} \cdot (3a^2+3a)$

30. $\dfrac{z^2-z-2}{z} \cdot \dfrac{2z^2+3z}{2z+3} \cdot \dfrac{z}{z-2}$

In Exercises 31–44, perform the indicated division and simplify your answer.

31. $\dfrac{7x^2}{10} \div \dfrac{14x^3}{15}$

32. $\dfrac{3(x+1)^2}{5x} \div \dfrac{9(x+1)}{10x^2}$

33. $\dfrac{3(x+4)}{4} \div \dfrac{x+4}{2}$

34. $\dfrac{x+2}{5(x-3)} \div \dfrac{x-2}{5(x-3)}$

35. $\dfrac{(xy)^2}{(x+y)^2} \div \dfrac{xy}{(x+y)^3}$

36. $\dfrac{x^2-y^2}{xy} \div \dfrac{(x-y)^2}{xy}$

37. $\dfrac{\left(\dfrac{x^3}{4}\right)}{\left(\dfrac{x}{8}\right)}$

38. $\dfrac{\left(\dfrac{x^2}{6}\right)}{\left(\dfrac{x^2}{3}\right)}$

39. $\dfrac{\left(\dfrac{a+5}{6}\right)}{(a+5)}$

40. $\dfrac{\left(\dfrac{x}{x-4}\right)}{\left(\dfrac{x}{4-x}\right)}$

41. $\dfrac{\left(\dfrac{2x^2-x-6}{2x^2+x-6}\right)}{\left(\dfrac{x-2}{x+2}\right)}$

42. $\dfrac{\left(\dfrac{6x^2-13x-5}{5x^2+5x}\right)}{\left(\dfrac{x(2x-5)}{5x+5}\right)}$

43. $\dfrac{(x+y)^2}{x^2+y^2} \div \dfrac{x+y}{x^3y+xy^3}$

44. $\dfrac{x^2-4y^2}{2x} \div \dfrac{x+2y}{4x}$

In Exercises 45–50, perform the indicated operations and simplify your answer.

45. $\left(\dfrac{x^2}{5} \cdot \dfrac{x+y}{2}\right) \div \dfrac{x}{30}$

46. $\left(\dfrac{2u^2v}{3} \cdot \dfrac{5}{u}\right) \div \dfrac{6u^2}{v^2}$

47. $\left(\dfrac{5x}{3}\right)^2 \div \left(\dfrac{5x}{2}\right)^3$

48. $\left(\dfrac{x}{3y}\right)^3 \div \left(\dfrac{2x}{y}\right)^2$

49. $\left[\left(\dfrac{x+2}{3}\right)^2 \cdot \left(\dfrac{x+1}{2}\right)^2\right] \div \dfrac{(x+1)(x+2)}{36}$

50. $\left[\left(\dfrac{4}{x-1}\right)^2 \cdot \left(\dfrac{x+1}{3}\right)^3\right] \div \dfrac{(x+1)^2}{27(x-1)}$

In Exercises 51 and 52, evaluate the expression at the given values of x. (If not possible, state the reason.)

Expression   *Values*

51. $\dfrac{x-5}{3x}$ (a) $x=5$ (b) $x=0$ (c) $x=-5$ (d) $x=6$

52. $\dfrac{4x}{x-3}$ (a) $x=0$ (b) $x=6$ (c) $x=2$ (d) $x=3$

53. *Photocopy Rate* A photocopier produces copies at a rate of 12 pages per minute.
 (a) Determine the time required to copy one page.
 (b) Determine the time required to copy x pages.
 (c) Determine the time required to copy 32 pages.

54. *Pump Rate* The pump in a well can pump water at the rate of 24 gallons per minute.
 (a) Determine the time required to pump one gallon.
 (b) Determine the time required to pump x gallons.
 (c) Determine the time required to pump 120 gallons.

55. *Area of a Region* Determine the area of the shaded portion of the rectangle in the accompanying figure.

56. *Area of a Circular Sector* The circle shown in the accompanying figure has a radius of r. The magnitude of the central angle of the shaded sector is x degrees. Use the fact that there are 360° in a circle to write the area of the sector as an algebraic fraction containing x.

Figure for 55

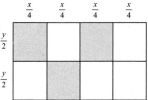

Figure for 56

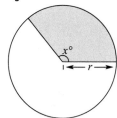

Adding and Subtracting Algebraic Fractions

Combining Algebraic Fractions with Like Denominators • *The Least Common Multiple of Polynomials* • *Combining Algebraic Fractions with Unlike Denominators* • *Compound Fractions*

Combining Algebraic Fractions with Like Denominators

As with numerical fractions, the procedure used to add (or subtract) two algebraic fractions depends upon whether the fractions have *like* or *unlike* denominators. To add (or subtract) two fractions with *like* denominators, we simply combine their numerators and place the result over the common denominator. Here are some examples.

$$\frac{3}{5} + \frac{8}{5} = \frac{3+8}{5} = \frac{11}{5}$$

$$\frac{3}{x} - \frac{8}{x} = \frac{3-8}{x} = -\frac{5}{x}$$

$$\frac{x-2}{7} - \frac{x}{7} = \frac{(x-2) - (x)}{7} = -\frac{2}{7}$$

$$\frac{x}{x+1} - \frac{4}{x+1} = \frac{x-4}{x+1}$$

In each of these cases, we used one of the following rules.

Adding or Subtracting Fractions with Like Denominators

Let a, b, and c be numbers, variables, or algebraic expressions such that $c \neq 0$. Then the following rules are valid.

1. $\dfrac{a}{c} + \dfrac{b}{c} = \dfrac{a+b}{c}$ Add fractions with like denominators

2. $\dfrac{a}{c} - \dfrac{b}{c} = \dfrac{a-b}{c}$ Subtract fractions with like denominators

---

EXAMPLE 1 ■ Adding Fractions with Like Denominators

Add the fractions $\dfrac{x}{3} + \dfrac{2-x}{3}$.

Solution

For these two fractions, the denominators are the same. Therefore, we can add the fractions by simply adding their numerators, as follows.

$$\frac{x}{3} + \frac{2-x}{3} = \frac{x + (2-x)}{3} = \frac{2}{3}$$

■

EXAMPLE 2 ■ Subtracting Fractions with Like Denominators

Subtract the fractions $\dfrac{5}{x+4} - \dfrac{2x}{x+4}$.

Solution

Since the denominators are equal, we use the rule for subtracting fractions with like denominators, as follows.

$$\frac{5}{x+4} - \frac{2x}{x+4} = \frac{5-2x}{x+4}$$ ■

After adding or subtracting two (or more) fractions, you should check the resulting fraction to see if it can be simplified. This procedure is illustrated in the next example.

EXAMPLE 3 ■ Adding Fractions and Simplifying

$$\frac{x}{x^2-4} + \frac{2}{x^2-4} = \frac{x+2}{x^2-4}$$ Add numerators

$$= \frac{x+2}{(x+2)(x-2)}$$ Factor completely

$$= \frac{\cancel{(x+2)} \cdot 1}{\cancel{(x+2)}(x-2)}$$ Cancel common factor

$$= \frac{1}{x-2}, \quad x \neq -2$$ Simplified form ■

The rules for adding and subtracting fractions with like denominators can be extended to cover sums and differences involving three or more fractions. Example 4 demonstrates this procedure.

EXAMPLE 4 ■ Combining Three Fractions with Like Denominators

$$\frac{x^2+6}{x+1} + \frac{2x-3}{x+1} - \frac{1-x}{x+1} = \frac{(x^2+6)+(2x-3)-(1-x)}{x+1}$$

$$= \frac{x^2+6+2x-3-1+x}{x+1}$$

$$= \frac{x^2+3x+2}{x+1}$$

$$= \frac{\cancel{(x+1)}(x+2)}{\cancel{x+1}}$$

$$= x+2, \quad x \neq -1$$ ■

The Least Common Multiple of Polynomials

To add or subtract fractions with *unlike* denominators, we must first rewrite each fraction using the **least common multiple** of the denominators of the individual fractions. The least common multiple of two (or more) polynomials is the simplest polynomial that is a multiple of each of the original polynomials. Here are some examples.

Polynomials	*Least Common Multiple*
9 and 15	$\overbrace{3 \cdot 3}^{9} \cdot 5 = 3^2 \cdot 5 = 45$ $\underbrace{3 \cdot 5}_{15}$
8 and 12	$\overbrace{2 \cdot 2 \cdot 2}^{8} \cdot 3 = 2^3 \cdot 3 = 24$ $\underbrace{ }_{12}$
x and $x + 2$	$x \cdot (x + 2) = x(x + 2)$
x^2 and $x(x + 2)$	$\overbrace{x \cdot x}^{x^2} \cdot (x + 2) = x^2(x + 2)$ $\underbrace{}_{x(x + 2)}$
$x^2 - 1$ and $x + 1$	$\overbrace{(x + 1) \cdot (x - 1)}^{x^2 - 1} = x^2 - 1$ $\underbrace{(x + 1)}_{x + 1}$

These examples illustrate the following guidelines for finding the least common multiple of two or more polynomials.

Guidelines for Finding the Least Common Multiple of Polynomials	1. Factor each polynomial completely. 2. List all the *different* factors of the polynomials. 3. For each factor, determine the highest power (of the factor) that occurs in the polynomials. 4. The least common multiple is the product of the different factors, each raised to the appropriate power.

EXAMPLE 5   ■   Finding Least Common Multiples

Find the least common multiple of each of the following sets of polynomials.

(a) $4, 6, 9$

(b) $5x, x^2$

(c) $x^2 - x, x - 1$

(d) $3x^2 + 6x, x^2 + 4x + 4$

Solution

(a) These three numbers factor as follows.

$$4 = 2 \cdot 2 = 2^2, \qquad 6 = 2 \cdot 3, \quad \text{and} \quad 9 = 3^2$$

The different factors are 2 and 3. Using the highest powers of these factors, we conclude that the least common multiple is $2^2 \cdot 3^2 = 36$.

(b) These two polynomials factor as follows.

$$5x = 5 \cdot x \quad \text{and} \quad x^2 = x \cdot x$$

The different factors are 5 and x. Using the highest powers of these factors, we conclude that the least common multiple is $5x^2$.

(c) These two polynomials factor as follows.

$$x^2 - x = x(x - 1) \quad \text{and} \quad x - 1$$

The different factors are x and $x - 1$. Using the highest powers of these factors, we conclude that the least common multiple is $x(x - 1)$.

(d) These two polynomials factor as follows.

$$3x^2 + 6x = 3x(x + 2) \quad \text{and} \quad x^2 + 4x + 4 = (x + 2)^2$$

The different factors are 3, x, and $x + 2$. Using the highest powers of these factors, we conclude that the least common multiple is $3x(x + 2)^2$.   ■

Combining Algebraic Fractions with Unlike Denominators

To add or subtract fractions with *unlike* denominators, we must first rewrite the fractions so that they have *like* denominators. The (like) denominator that we use is the least common multiple of the original denominators, and it is called the **least common denominator** of the original fractions. Once the fractions have been written with like denominators, we simply add the fractions using the rules given at the beginning of this section. This procedure is demonstrated in the next several examples.

EXAMPLE 6   ■   **Adding Fractions with Unlike Denominators**

Add the fractions $\dfrac{5}{3x} + \dfrac{7}{4x}$.

Solution

By factoring the denominators, $3x = 3 \cdot x$ and $4x = 2^2 \cdot x$, we conclude that the least common denominator is $12x$. Therefore, we can add the two fractions as follows.

$$\frac{5}{3x} + \frac{7}{4x} = \frac{5(4)}{3x(4)} + \frac{7(3)}{4x(3)} \qquad \text{Rewrite fractions using least common denominator}$$

$$= \frac{20}{12x} + \frac{21}{12x} \qquad \text{Like denominators}$$

$$= \frac{20 + 21}{12x} \qquad \text{Add fractions}$$

$$= \frac{41}{12x} \qquad \text{Simplified form} \qquad ■$$

EXAMPLE 7   ■   **Subtracting Fractions with Unlike Denominators**

Subtract the fractions $\dfrac{4}{x - 2} - \dfrac{2}{x + 1}$.

Solution

The only factors of the denominators are $(x - 2)$ and $(x + 1)$. Therefore, the least common denominator is $(x - 2)(x + 1)$.

$$\frac{4}{x - 2} - \frac{2}{x + 1} = \frac{4(x + 1)}{(x - 2)(x + 1)} - \frac{2(x - 2)}{(x - 2)(x + 1)} \qquad \text{Rewrite fractions using least common denominator}$$

$$= \frac{(4x + 4) - (2x - 4)}{(x - 2)(x + 1)} \qquad \text{Subtract numerators}$$

$$= \frac{4x + 4 - 2x + 4}{(x - 2)(x + 1)} \qquad \text{Remove parentheses}$$

$$= \frac{2x + 8}{(x - 2)(x + 1)} \qquad \text{Simplified form} \qquad ■$$

EXAMPLE 8   ■   **Adding Fractions with Unlike Denominators**

Add the fractions $\dfrac{2x}{x^2 - 9} + \dfrac{1}{3 - x}$.

Solution

The factors in the denominators are $x^2 - 9 = (x + 3)(x - 3)$ and $3 - x$. Since $(3 - x) = -1(x - 3)$, we can rewrite the original addition problem as a subtraction problem, as follows.

$$\frac{2x}{x^2 - 9} + \frac{1}{(-1)(x - 3)} = \frac{2x}{(x + 3)(x - 3)} - \frac{1}{x - 3}$$

Now, using the least common denominator of $(x + 3)(x - 3)$, we subtract the fractions as follows.

$$\frac{2x}{(x + 3)(x - 3)} - \frac{1}{x - 3}$$

$$= \frac{2x}{(x + 3)(x - 3)} - \frac{x + 3}{(x + 3)(x - 3)} \qquad \text{Rewrite fractions using least common denominator}$$

$$= \frac{2x - (x + 3)}{(x + 3)(x - 3)} \qquad \text{Subtract numerators}$$

$$= \frac{2x - x - 3}{(x + 3)(x - 3)} \qquad \text{Remove parentheses}$$

$$= \frac{(x - 3) \cdot 1}{(x + 3)(x - 3)} \qquad \text{Cancel common factor}$$

$$= \frac{1}{x + 3}, \quad x \neq 3 \qquad \text{Simplified form} \qquad \blacksquare$$

In the next example, we show how a least common denominator can be used to combine three fractions.

EXAMPLE 9   ■   Combining Fractions with Unlike Denominators

Combine the following fractions.

$$\frac{2x - 5}{6x + 9} - \frac{4}{2x^2 + 3x} + \frac{1}{x}$$

Solution

The denominators factor as $6x + 9 = 3(2x + 3)$, $2x^2 + 3x = x(2x + 3)$, and x. Therefore, we conclude that the least common denominator is $3x(2x + 3)$.

$$\frac{2x - 5}{6x + 9} - \frac{4}{2x^2 + 3x} + \frac{1}{x}$$

$$= \frac{2x - 5}{3(2x + 3)} - \frac{4}{x(2x + 3)} + \frac{1}{x} \qquad \text{Factor denominators}$$

$$= \frac{(2x - 5)(x)}{3(2x + 3)(x)} - \frac{(4)(3)}{x(2x + 3)(3)} + \frac{3(2x + 3)}{(x)(3)(2x + 3)} \qquad \text{Rewrite fractions using least common denominator}$$

$$= \frac{2x^2 - 5x - 12 + 6x + 9}{3x(2x + 3)} \qquad \text{Combine numerators}$$

$$= \frac{2x^2 + x - 3}{3x(2x + 3)} \qquad \text{Combine like terms}$$

$$= \frac{(x - 1)(2x + 3)}{3x(2x + 3)} \qquad \text{Factor and cancel}$$

$$= \frac{x - 1}{3x}, \quad x \neq -\frac{3}{2} \qquad \text{Simplified form} \qquad \blacksquare$$

MATH MATTERS

Numbers in Other Languages

The words used for numbers in most European languages have similarities. The following list shows the words used for the numbers 1 to 10 in English, French, German, Italian, Spanish, Swedish, and Latin.

Number	English	French	German	Italian	Spanish	Swedish	Latin
1	one	un	eins	uno	uno	en	unus
2	two	deux	zwei	due	dos	twa	duo
3	three	trois	drei	tre	tres	tre	tres
4	four	quatre	vier	quattro	cuatro	fyra	quattuar
5	five	cinq	fünf	cinque	cinco	fem	quinque
6	six	six	sechs	sei	seis	sex	sex
7	seven	sept	sieben	sette	siete	sju	septem
8	eight	huit	acht	otto	ocho	atta	octo
9	nine	neuf	neun	nove	nueve	nio	novem
10	ten	dix	zehn	dieci	diez	tio	decem

Note that the Latin words for 7, 8, 9, and 10 correspond to the ninth, tenth, eleventh, and twelfth months of the year. Do you know why these Latin words do not correspond to the seventh, eighth, ninth, and tenth months of the year? (The answer is given in the back of the text.)

Compound Fractions

Compound fractions can have numerators and/or denominators that are sums or differences of fractions. Here are a couple of examples.

$$\frac{\left(\dfrac{2}{x} - 3\right)}{\left(1 - \dfrac{1}{x}\right)} \quad \text{and} \quad \frac{\left(\dfrac{1}{x + 2} - \dfrac{1}{x}\right)}{\left(\dfrac{5}{x} + \dfrac{2}{x^2}\right)}$$

To simplify a compound fraction we first combine its numerator and its denominator into single fractions. Then we divide as demonstrated in the previous section (by inverting the denominator and multiplying).

EXAMPLE 10   ■   Simplifying a Compound Fraction

$$\frac{\left(\dfrac{x}{3} + \dfrac{2}{3}\right)}{\left(1 - \dfrac{2}{x}\right)} = \frac{\left(\dfrac{x}{3} + \dfrac{2}{3}\right)}{\left(\dfrac{x}{x} - \dfrac{2}{x}\right)} \qquad \text{Find least common denominator}$$

$$= \frac{\left(\dfrac{x + 2}{3}\right)}{\left(\dfrac{x - 2}{x}\right)} \qquad \text{Add fractions in numerator and denominator}$$

$$= \frac{x + 2}{3} \cdot \frac{x}{x - 2} \qquad \text{Invert and multiply}$$

$$= \frac{x(x + 2)}{3(x - 2)}, \quad x \neq 0 \qquad \text{Simplified form} \qquad ■$$

Another way of simplifying the compound fraction given in Example 10 is to multiply the numerator and denominator by the least common denominator of *every* fraction in the numerator and denominator. For this fraction, notice what happens when we multiply the numerator and denominator by $3x$.

$$\frac{\left(\dfrac{x}{3} + \dfrac{2}{3}\right)}{\left(1 - \dfrac{2}{x}\right)} = \frac{\left(\dfrac{x}{3} + \dfrac{2}{3}\right)}{\left(1 - \dfrac{2}{x}\right)} \cdot \frac{3x}{3x} = \frac{\dfrac{x}{3}(3x) + \dfrac{2}{3}(3x)}{(1)(3x) - \dfrac{2}{x}(3x)} = \frac{x^2 + 2x}{3x - 6}, \quad x \neq 0$$

EXAMPLE 11   ■   Compound Fractions

To simplify the compound fraction below, we multiply the numerator and denominator by the least common denominator of all fractions occurring in the numerator and denominator.

$$\frac{\left(\dfrac{2}{x + 2} - 1\right)}{\left(\dfrac{3}{x + 2} + \dfrac{2}{x}\right)} = \frac{\left(\dfrac{2}{x + 2} - 1\right)(x)(x + 2)}{\left(\dfrac{3}{x + 2} + \dfrac{2}{x}\right)(x)(x + 2)}$$

$$= \frac{2x - (x)(x + 2)}{3x + 2(x + 2)}$$

$$= \frac{2x - x^2 - 2x}{3x + 2x + 4}$$

$$= -\frac{x^2}{5x + 4}, \quad x \neq -2, x \neq 0 \qquad ■$$

DISCUSSION PROBLEM ■ Adding Fractions

One way to add two fractions is to use the rule

$$\frac{a}{b} + \frac{c}{d} = \frac{ad + bc}{bd}.$$

For instance,

$$\frac{2}{3} + \frac{4}{5} = \frac{2(5) + 3(4)}{3(5)} = \frac{22}{15}.$$

This rule can also be used to add algebraic fractions, as follows.

$$\frac{x + 2}{x} + \frac{x + 1}{2x} = \frac{2x(x + 2) + x(x + 1)}{x(2x)}$$

$$= \frac{2x^2 + 4x + x^2 + x}{2x^2}$$

$$= \frac{3x^2 + 5x}{2x^2}$$

$$= \frac{x(3x + 5)}{2x(x)}$$

$$= \frac{3x + 5}{2x}$$

Can you see any disadvantage to using this method rather than the least common denominator technique described in this section? ■

Warm-Up

The following warm-up exercises involve skills that were covered in earlier sections. You will use these skills in the exercise set for this section.

In Exercises 1–10, perform the indicated operations and simplify your answer.

1. $\dfrac{7}{9} + \dfrac{2}{9}$ 　　　　　 **2.** $\dfrac{3}{32} + \dfrac{5}{32}$ 　　　　　 **3.** $\dfrac{11}{15} - \dfrac{2}{15}$

4. $\dfrac{3}{10} - \dfrac{8}{10}$ 　　　　　 **5.** $\dfrac{3}{7} + \dfrac{3}{14}$ 　　　　　 **6.** $\dfrac{5}{24} + \dfrac{3}{16}$

7. $-\dfrac{2}{3} + \dfrac{46}{75}$ 　　　　　 **8.** $\dfrac{22}{5} - \dfrac{8}{35}$ 　　　　　 **9.** $\dfrac{3}{4} - \dfrac{7}{8} + \dfrac{7}{12}$

10. $\dfrac{5}{9} - \dfrac{2}{3} - \dfrac{5}{18}$

7.3 EXERCISES

In Exercises 1–14, perform the indicated operations and simplify your answer.

1. $\dfrac{9}{x} - \dfrac{4}{x}$

2. $\dfrac{7}{z^2} + \dfrac{10}{z^2}$

3. $\dfrac{y}{4} + \dfrac{3y}{4}$

4. $\dfrac{10x^2}{3} - \dfrac{4x^2}{3}$

5. $\dfrac{5}{3a} + \dfrac{9}{3a}$

6. $\dfrac{16}{5z} - \dfrac{11}{5z}$

7. $\dfrac{x}{3} + \dfrac{1-x}{3}$

8. $\dfrac{2x}{5} - \dfrac{7x}{5}$

9. $\dfrac{4z}{3} - \dfrac{4z-3}{3}$

10. $\dfrac{-16u}{7} - \dfrac{14-16u}{7}$

11. $\dfrac{5}{x-1} + \dfrac{x}{x-1}$

12. $\dfrac{2x-1}{x+3} + \dfrac{1-x}{x+3}$

13. $\dfrac{5y+2}{y-1} - \dfrac{4y+1}{y-1}$

14. $\dfrac{7s-5}{s+5} - \dfrac{2s-10}{s+5}$

In Exercises 15–20, find the least common multiple of the given polynomials.

15. $2x, \quad x^3$

16. $2t^2, \quad 24t$

17. $x, \quad 3(x+5)$

18. $18y^2, \quad 27y(y-3)$

19. $x^2-4, \quad x(x+2)$

20. $t^2+6t+9, \quad t^2-9$

In Exercises 21–26, rewrite the given fractions so that they have the same denominator.

21. $\dfrac{x+5}{3x-6}, \quad \dfrac{10}{x-2}$

22. $\dfrac{8x}{x+2}, \quad \dfrac{3}{x^2+x-2}$

23. $\dfrac{2}{x^2}, \quad \dfrac{5}{x(x+3)}$

24. $\dfrac{5t}{(t-3)^2}, \quad \dfrac{4}{t(t-3)}$

25. $\dfrac{x-8}{x^2-16}, \quad \dfrac{9x}{x^2-8x+16}$

26. $\dfrac{3y}{y^2-y-6}, \quad \dfrac{y+2}{y^2-3y}$

In Exercises 27–60, perform the specified operations and simplify your answer.

27. $\dfrac{1}{5x} - \dfrac{3}{5}$

28. $\dfrac{3}{2b} + \dfrac{5}{2b^2}$

29. $\dfrac{5}{z} + \dfrac{6}{z^2}$

30. $\dfrac{7}{6u^2} - \dfrac{2}{9u}$

31. $\dfrac{4}{x-3} + \dfrac{4}{3-x}$

32. $\dfrac{5}{6-t} - \dfrac{4}{t-6}$

33. $\dfrac{2x}{x-5} - \dfrac{5}{5-x}$

34. $\dfrac{3}{x-2} + \dfrac{5}{2-x}$

35. $6 - \dfrac{5}{x+3}$

36. $\dfrac{3}{x-1} - 5$

37. $\dfrac{1}{x-1} - \dfrac{1}{x+2}$

38. $\dfrac{3}{2x-1} - \dfrac{2}{x+1}$

39. $\dfrac{3}{2(x-4)} - \dfrac{1}{2x}$

40. $\dfrac{1}{2(x+1)} + \dfrac{1}{2(x-1)}$

41. $\dfrac{x}{x^2-9} + \dfrac{3}{x+3}$

42. $\dfrac{6}{z} - \dfrac{z-3}{z+2}$

43. $\dfrac{5v}{v(v+4)} + \dfrac{2v}{v^2}$

44. $\dfrac{t}{t+1} - \dfrac{t^2}{t(t+1)}$

45. $-\dfrac{1}{x} + \dfrac{2}{x^2+1}$

46. $\dfrac{2}{x+1} + \dfrac{1-x}{x^2-2x-3}$

47. $\dfrac{2}{x^2-4} - \dfrac{1}{x^2-3x+2}$

48. $\dfrac{x}{x^2+x-2} - \dfrac{1}{x+2}$

49. $\dfrac{3}{x-3} + \dfrac{9}{(x-3)^2}$

50. $\dfrac{2}{x-1} - \dfrac{1}{(x-1)^2}$

51. $\dfrac{2x}{x^2+1} - \dfrac{1}{x}$

52. $\dfrac{4}{t^2+2} - \dfrac{4}{t^2}$

53. $\dfrac{3}{x} - \dfrac{1}{x^2} + \dfrac{1}{x+1}$

54. $\dfrac{5}{2(x+1)} - \dfrac{1}{2x} - \dfrac{3}{2(x+1)^2}$

55. $x - \dfrac{1}{x+1} + \dfrac{5}{x-1}$

56. $x + 2 + \dfrac{4}{3(x-2)} + \dfrac{1}{2x}$

57. $\dfrac{y}{x^2+xy} - \dfrac{x}{xy+y^2}$

58. $\dfrac{5}{x+y} + \dfrac{5}{x-y}$

59. $\dfrac{3u}{u^2-2uv+v^2} + \dfrac{2}{u-v}$

60. $\dfrac{1}{x} - \dfrac{3}{y} + \dfrac{3x-y}{xy}$

In Exercises 61–70, simplify the given compound fraction.

61. $\dfrac{\left(\dfrac{3}{x}\right)}{\left(\dfrac{6}{x^2}\right)}$

62. $\dfrac{\left(\dfrac{2}{3}\right)}{\left(\dfrac{u}{v}\right)}$

63. $\dfrac{\left(1+\dfrac{3}{y}\right)}{y}$

64. $\dfrac{\left(\dfrac{1}{2}\right)}{\left(\dfrac{5}{x}+2\right)}$

65. $\dfrac{\left(\dfrac{x}{y}-1\right)}{x-y}$

66. $\dfrac{x-y}{\left(\dfrac{x}{y}-\dfrac{y}{x}\right)}$

67. $\dfrac{\left(z-\dfrac{4}{z}\right)}{\left(\dfrac{1}{z}-4\right)}$

68. $\dfrac{\left(\dfrac{u}{v}-2\right)}{\left(\dfrac{1}{v}+2u\right)}$

69. $\dfrac{\left(\dfrac{1}{x}-\dfrac{1}{x+1}\right)}{\left(\dfrac{1}{x+1}\right)}$

70. $\dfrac{\left(\dfrac{5}{y}-\dfrac{6}{2y+1}\right)}{\left(\dfrac{5}{y}+4\right)}$

71. *Work Rate* After two people work together for t hours on a common task, the fractional parts of the job done by each of the two workers are $t/8$ and $t/7$. What fractional part of the task has been completed?

72. *Work Rate* After two people work together for t hours on a common task, the fractional parts of the job done by each of the two workers are $t/6$ and $t/9$. What fractional part of the task has been completed?

73. *Average of Two Numbers* Determine the average of the two real numbers given by $x/5$ and $x/6$.

74. *Average of Two Numbers* Determine the average of the two real numbers given by $2x/3$ and $3x/5$.

75. *Equal Lengths* Find three real numbers that divide the real number line between $x/9$ and $x/6$ into four parts of equal lengths (see figure).

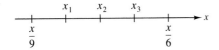

76. *Equal Lengths* Find two real numbers that divide the real number line between $x/3$ and $5x/4$ into three parts of equal lengths (see figure).

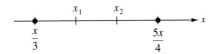

77. *Electrical Resistance* When two resistors are connected in parallel, the total resistance is given by

$$\frac{1}{\left(\dfrac{1}{R_1} + \dfrac{1}{R_2}\right)}.$$

Simplify this compound fraction.

Solving Equations Involving Fractions
Introduction • Equations Containing Constant Denominators •
Equations Containing Variable Denominators • Applications

Introduction

In this section we switch from *rewriting* fractional expressions to *solving* equations that involve fractions. The least common denominator continues to play a key role in the solution process.

Equations that involve fractions are common in real-life experiences. Consider the following example.

> You and your friend travel to separate colleges in the same amount of time. You drive 380 miles and your friend drives 400 miles. Your friend's average speed is three miles per hour faster than your average speed. What is your average speed and what is your friend's average speed?

One solution to this problem involves the fractional equation

$$\frac{380}{r} = \frac{400}{r + 3}.$$

We will discuss this application further in Example 7 in this section.

Equations Containing Constant Denominators

In Section 3.3 we showed how to solve equations containing fractions with *constant* denominators. We review that procedure here because it is the basis for solving more general equations involving fractions. Recall from Section 3.3 that we can clear an equation of fractions by multiplying both sides of the equation by the least common denominator of the fractions in the equation. Note how this is done in Example 1.

EXAMPLE 1 ■ An Equation Containing Constant Denominators

Solve the following equation.

$$\frac{x}{5} = 6 - \frac{x}{10}$$

Solution

For this equation, the least common denominator is 10. Therefore, we begin by multiplying both sides of the equation by 10.

$$\frac{x}{5} = 6 - \frac{x}{10}$$ Given equation

$$10\left(\frac{x}{5}\right) = 10\left(6 - \frac{x}{10}\right)$$ Multiply both sides by 10

$$2x = 60 - x$$ Distribute and simplify

$$3x = 60$$ Add x to both sides

$$x = 20$$ Solution

Therefore, the solution of the equation is 20. Check this solution in the original equation. ■

EXAMPLE 2 ■ An Equation Containing Constant Denominators

Solve the following equation.

$$\frac{x+6}{9} - \frac{x-2}{5} = \frac{4}{15}$$

Solution

For this equation, the least common denominator is 45. Therefore, we begin by multiplying both sides of the equation by 45.

$$\frac{x+6}{9} - \frac{x-2}{5} = \frac{4}{15}$$ Given equation

$$45\left(\frac{x+6}{9} - \frac{x-2}{5}\right) = 45\left(\frac{4}{15}\right)$$ Multiply both sides by 45

$$5(x+6) - 9(x-2) = 3(4)$$ Distribute and simplify

$$5x + 30 - 9x + 18 = 12 \qquad \text{Distributive Property}$$
$$-4x + 48 = 12 \qquad \text{Combine like terms}$$
$$-4x = -36 \qquad \text{Subtract 48 from both sides}$$
$$x = 9 \qquad \text{Solution}$$

Therefore, the solution of the equation is 9. Check this solution in the original equation.

■

Equations Containing Variable Denominators

As stated in Section 7.1, we *exclude* those values of a variable that make the denominator of an algebraic fraction zero. This is especially critical for solving equations that contain variable denominators. You will see why in the examples that follow.

EXAMPLE 3 ■ **An Equation Containing Variable Denominators**

Solve the equation $\dfrac{1}{x} - \dfrac{2}{3} = \dfrac{3}{x}$.

Solution

For this equation the least common denominator is $3x$. Therefore, we begin by multiplying both sides of the equation by $3x$.

$$\frac{1}{x} - \frac{2}{3} = \frac{3}{x} \qquad \text{Given equation}$$

$$3x\left(\frac{1}{x} - \frac{2}{3}\right) = 3x\left(\frac{3}{x}\right) \qquad \text{Multiply both sides by } 3x$$

$$\frac{3x}{x} - \frac{6x}{3} = \frac{9x}{x} \qquad \text{Distributive Property}$$

$$3 - 2x = 9, \quad x \neq 0 \qquad \text{Simplify fractions}$$

$$-2x = 6 \qquad \text{Subtract 3 from both sides}$$

$$x = -3 \qquad \text{Solution}$$

Therefore, the solution appears to be -3. We check this solution in the original equation as follows.

Check

$$\frac{1}{x} - \frac{2}{3} = \frac{3}{x} \qquad \text{Given equation}$$

$$\frac{1}{-3} - \frac{2}{3} \stackrel{?}{=} \frac{3}{-3} \qquad \text{Replace } x \text{ by } -3$$

$$-\frac{1}{3} - \frac{2}{3} \stackrel{?}{=} -1$$

$$-1 = -1 \qquad \text{Solution checks}$$

After checking, we conclude that the solution of the equation is -3.

■

Throughout the text we have emphasized the importance of checking solutions. Up to this point the main reason for checking has been to make sure that we didn't make arithmetic errors in the solution process. In the next example we will see that there is another reason for checking solutions in the original equation. That is, even with no mistakes in the solution process, it can happen that a trial solution does not satisfy the original equation. This type of solution is called **extraneous**. (An extraneous solution of an equation does not, by definition, satisfy its original equation, and must therefore not be listed as an actual solution.)

EXAMPLE 4 ■ An Equation with No Solution

Solve the following equation.

$$\frac{2x}{x + 3} = 1 - \frac{6}{x + 3}$$

Solution

The least common denominator of this equation is $x + 3$. Therefore, we begin by multiplying both sides of the equation by $x + 3$.

$\dfrac{2x}{x + 3} = 1 - \dfrac{6}{x + 3}$	Given equation
$(x + 3)\left(\dfrac{2x}{x + 3}\right) = (x + 3)\left(1 - \dfrac{6}{x + 3}\right)$	Multiply both sides by $x + 3$
$2x = (x + 3) - 6, \quad x \neq -3$	Distribute and simplify
$2x = x - 3$	Combine like terms
$x = -3$	Trial solution

At this point, the solution appears to be -3. However, by performing the following check, we see that this trial solution is extraneous.

Check

$\dfrac{2x}{x + 3} = 1 - \dfrac{6}{x + 3}$	Given equation
$\dfrac{2(-3)}{-3 + 3} \stackrel{?}{=} 1 - \dfrac{6}{-3 + 3}$	Replace x by -3
$\dfrac{-6}{0} \stackrel{?}{=} 1 - \dfrac{6}{0}$	Division by zero is undefined

Since the check resulted in *division by zero*, we conclude that -3 is extraneous. Therefore, the given equation has no solution. ■

NOTE: In Example 4 can you see why $x = -3$ is extraneous? By looking back at the original equation we can see that -3 is excluded from the domain of two of the fractions that occur in the equation.

EXAMPLE 5   ■   An Equation Containing Variable Denominators

Solve the following equation.

$$\frac{6}{x-1} + \frac{2x}{x-2} = 2$$

Solution

The least common denominator for this equation is $(x-1)(x-2)$. Therefore, we begin by multiplying both sides of the equation by $(x-1)(x-2)$.

$$\frac{6}{x-1} + \frac{2x}{x-2} = 2$$

$$(x-1)(x-2)\left(\frac{6}{x-1} + \frac{2x}{x-2}\right) = 2(x-1)(x-2)$$

$$6(x-2) + 2x(x-1) = 2(x^2 - 3x + 2), \quad x \neq 1, x \neq 2$$

$$6x - 12 + 2x^2 - 2x = 2x^2 - 6x + 4$$

$$10x = 16$$

$$x = \frac{8}{5}$$

After checking, we conclude therefore, the solution is $\frac{8}{5}$. Check this solution in the original equation.   ■

So far in this section each of the equations we have solved had one (or no) solution. In the next example, we look at an equation that has two solutions.

EXAMPLE 6   ■   An Equation that Has Two Solutions

Solve the following equation.

$$\frac{2x}{x+2} = \frac{1}{x^2 - 4} + 1$$

Solution

The least common denominator for this equation is $(x+2)(x-2) = x^2 - 4$. Therefore, we begin by multiplying both sides of the equation by $x^2 - 4$.

$$\frac{2x}{x+2} = \frac{1}{x^2 - 4} + 1 \qquad \text{Given equation}$$

$$(x^2 - 4)\left(\frac{2x}{x+2}\right) = (x^2 - 4)\left(\frac{1}{x^2 - 4} + 1\right) \qquad \text{Multiply both sides by } x^2 - 4$$

$$(x-2)(2x) = 1 + (x^2 - 4), \quad x \neq \pm 2 \qquad \text{Distribute and simplify}$$

$$2x^2 - 4x = 1 + x^2 - 4 \qquad \text{Distributive Property}$$

$$x^2 - 4x + 3 = 0 \qquad \text{Standard form}$$

$$(x - 3)(x - 1) = 0 \qquad \text{Factor}$$

$$x - 3 = 0 \implies x = 3 \qquad \text{Set first factor equal to 0}$$

$$x - 1 = 0 \implies x = 1 \qquad \text{Set second factor equal to 0}$$

After checking each solution, we conclude therefore, the equation has two solutions: 1 and 3. Check these solutions in the original equation. ■

Applications

At the beginning of this section we looked at an application involving distance, rates, and time. Now let's see how to solve this problem using an equation that involves fractions.

EXAMPLE 7   ■   **An Application: Average Speeds**

You and your friend travel to separate colleges in the same amount of time. You drive 380 miles and your friend drives 400 miles. Your friend's average speed is 3 miles per hour faster than your average speed. What is your average speed and what is your friend's average speed?

Solution

We will follow the problem-solving strategy discussed in Chapter 4.

Verbal model: Your time $=$ Your friend's time

Formula: Distance $=$ (Rate)(Time) $\implies$ Time $= \dfrac{\text{Distance}}{\text{Rate}}$

Labels: You: Distance $= 380$ miles,   Rate $= r$ miles per hour
Your friend: Distance $= 400$ miles,   Rate $= r + 3$ miles per hour

Equation: $$\frac{380}{r} = \frac{400}{r + 3}$$

$$380(r + 3) = 400(r), \quad r \neq 0, r \neq -3$$

$$380r + 1{,}140 = 400r$$

$$1{,}140 = 20r$$

$$57 = r$$

Therefore, your average speed is 57 miles per hour and your friend's average speed is $57 + 3 = 60$ miles per hour. Check this solution in the original statement of the problem. ■

EXAMPLE 8 ■ **An Application: Work Rates**

With only the cold water valve open, it takes 7 minutes to fill the tub of an automatic washer. With both the hot and cold water valves fully open, it takes only 4 minutes to fill the tub (see Figure 7.1). How long will it take to fill the tub with only the hot water valve open?

FIGURE 7.1

Cold: 7 minutes   Hot and cold: 4 minutes

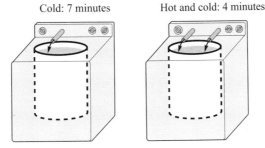

Solution

This problem is similar to that given in Example 8 in Section 4.4. You might want to review that problem before proceeding.

Verbal model: | Rate for cold water | + | Rate for hot water | = | Rate for warm water |

Labels: Warm water: Time = 4 minutes,   Rate = $\frac{1}{4}$ tub per minute

Cold water: Time = 7 minutes,   Rate = $\frac{1}{7}$ tub per minute

Hot water: Time = t minutes,   Rate = $1/t$ tub per minute

Equation:

$$\frac{1}{7} + \frac{1}{t} = \frac{1}{4}$$

$$28t\left(\frac{1}{7} + \frac{1}{t}\right) = 28t\left(\frac{1}{4}\right)$$

$$4t + 28 = 7t, \quad t \neq 0$$

$$28 = 3t$$

$$\frac{28}{3} = t$$

Thus, it will take about $9\frac{1}{3}$ minutes to fill the tub with hot water alone. Check this solution in the original statement of the problem. ■

EXAMPLE 9 ■ An Application: Batting Average

In this year's playing season, a baseball player has been up to bat 280 times and has hit the ball successfully 70 times. Thus, the batting average for the player is $70/280 = 0.250$. How many additional *consecutive* times must the player hit the ball to obtain a batting average of 0.300?

Solution

Verbal model:	Batting average	=	Total hits	÷	Total times at bat

Labels: Current times at bat $= 280$
Current hits $= 70$
Additional consecutive hits $= x$

Equation:

$$0.300 = \frac{x + 70}{x + 280}$$

$$0.300(x + 280) = x + 70, \quad x \neq -280$$

$$0.3x + 84 = x + 70$$

$$14 = 0.7x$$

$$20 = x$$

Thus, the player must successfully hit the ball for the next 20 times at bat. After that, the player's batting average will be $90/300 = 0.300$. ■

DISCUSSION PROBLEM ■ A Mathematical Fallacy

In mathematics, a **fallacy** is an argument or proof that is not logically sound. Here is an example. The following steps appear to prove that 1 is equal to 0. Of course this is not true, so there must be something wrong with the algebra. What is wrong?

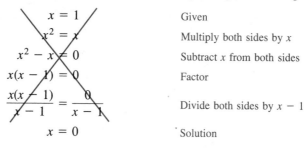

$x = 1$	Given
$x^2 = x$	Multiply both sides by x
$x^2 - x = 0$	Subtract x from both sides
$x(x - 1) = 0$	Factor
$\dfrac{x(x - 1)}{x - 1} = \dfrac{0}{x - 1}$	Divide both sides by $x - 1$
$x = 0$	Solution

■

Warm-Up *The following warm-up exercises involve skills that were covered in earlier sections. You will use these skills in the exercise set for this section.*

In Exercises 1–10, solve the given equation.

1. $5x = 8$ **2.** $-2x = 15$

3. $16x - 3 = 29$ **4.** $4 - 32x = 0$

5. $(x - 4)(x + 10) = 0$ **6.** $(2x + 3)(x - 9) = 0$

7. $x(x - 8) = 0$ **8.** $x(8 - x) = 16$

9. $x^2 + x - 56 = 0$ **10.** $t^2 - 5t = 0$

7.4 EXERCISES

In Exercises 1–4, determine whether the given value of x is a solution of the equation.

Equation	*Values*

1. $\dfrac{x}{5} - \dfrac{3}{x} = \dfrac{7}{10}$ (a) $x = 0$ (b) $x = \dfrac{-5}{2}$ (c) $x = \dfrac{1}{6}$ (d) $x = 6$

2. $\dfrac{3x}{5} + \dfrac{x^2}{2} = \dfrac{4}{5}$ (a) $x = 0$ (b) $x = -2$ (c) $x = \dfrac{4}{5}$ (d) $x = 2$

3. $\dfrac{5}{2x} - \dfrac{4}{x} = 3$ (a) $x = -\dfrac{1}{2}$ (b) $x = 4$ (c) $x = 0$ (d) $x = \dfrac{1}{4}$

4. $3 + \dfrac{1}{x + 2} = 4$ (a) $x = -1$ (b) $x = -2$ (c) $x = 0$ (d) $x = 5$

In Exercises 5–50, solve the given equation.

5. $\dfrac{x}{3} = \dfrac{2}{3}$ **6.** $\dfrac{x}{5} = \dfrac{12}{5}$ **7.** $\dfrac{x}{2} = \dfrac{1}{5}$ **8.** $\dfrac{y}{7} = \dfrac{3}{4}$

9. $\dfrac{z - 4}{3} - \dfrac{z}{8} = 0$ **10.** $3 + \dfrac{y}{5} = y - 1$ **11.** $\dfrac{t}{3} = 25 - \dfrac{t}{6}$ **12.** $\dfrac{x}{10} + \dfrac{x}{5} = 20$

13. $\dfrac{a + 3}{4} - \dfrac{a - 1}{6} = \dfrac{4}{3}$ **14.** $\dfrac{u - 5}{10} + \dfrac{u + 8}{15} = \dfrac{7}{10}$ **15.** $\dfrac{2}{x} = 8$ **16.** $\dfrac{5}{t} = -\dfrac{2}{3}$

17. $\dfrac{1}{3 - y} = -\dfrac{1}{10}$ **18.** $\dfrac{5}{u + 8} = \dfrac{1}{4}$ **19.** $3 - \dfrac{16}{a} = \dfrac{5}{3}$ **20.** $\dfrac{2}{b} + 1 = 9$

21. $\dfrac{10}{y + 3} + \dfrac{10}{3} = 6$ **22.** $35 - \dfrac{12}{x - 4} = 26$ **23.** $\dfrac{3}{x} = \dfrac{9}{2(x + 2)}$ **24.** $\dfrac{5}{x + 4} = \dfrac{5}{3(x + 1)}$

25. $\dfrac{3}{x+5} = \dfrac{2}{x+1}$

26. $\dfrac{7}{x+1} = \dfrac{5}{x-3}$

27. $\dfrac{4}{x+2} - \dfrac{1}{x} = \dfrac{1}{x}$

28. $\dfrac{3}{x+5} + \dfrac{5}{x} = \dfrac{10}{x}$

29. $10 - \dfrac{13}{x} = 4 + \dfrac{5}{x}$

30. $\dfrac{15}{x} - 4 = \dfrac{6}{x} + 3$

31. $\dfrac{3}{x(x-3)} + \dfrac{4}{x} = \dfrac{1}{x-3}$

32. $3 = 2 + \dfrac{2}{z+2}$

33. $\dfrac{x}{x+4} + \dfrac{4}{x+4} + 2 = 0$

34. $\dfrac{2}{(x-4)(x-2)} = \dfrac{1}{x-4} + \dfrac{2}{x-2}$

35. $\dfrac{1}{x-3} + \dfrac{1}{x+3} = \dfrac{10}{x^2-9}$

36. $\dfrac{1}{x-2} + \dfrac{3}{x+3} = \dfrac{4}{x^2+x-6}$

37. $2 = \dfrac{18}{x^2}$

38. $6 = \dfrac{150}{z^2}$

39. $\dfrac{50}{t} = 2t$

40. $\dfrac{18}{u} = \dfrac{u}{2}$

41. $x + 4 = \dfrac{-4}{x}$

42. $\dfrac{25}{t} = 10 - t$

43. $2y = \dfrac{y+6}{y+1}$

44. $\dfrac{3x}{x+1} = \dfrac{2}{x-1}$

45. $\dfrac{20-x}{x} = x$

46. $\dfrac{x+30}{x} = x$

47. $x + \dfrac{1}{x} = \dfrac{5}{2}$

48. $\dfrac{4}{x} - \dfrac{x}{6} = \dfrac{5}{3}$

49. $x = \dfrac{1 + \dfrac{3}{x}}{1 - \dfrac{1}{x}}$

50. $\dfrac{2x}{3} = \dfrac{1 - \dfrac{1}{x}}{\dfrac{1}{x}}$

51. *Number Problem* Find a number such that the sum of the number and its reciprocal is $\frac{10}{3}$.

52. *Number Problem* Find a number such that the sum of the number and three times its reciprocal is $\frac{28}{5}$.

53. *Wind Speed* An executive flew in the corporate jet to a meeting in a city 1,500 miles away (see figure). After traveling the same amount of time on the return flight, the pilot mentioned that they still had 300 miles to go. If the plane has a speed of 600 miles per hour in still air, how fast was the wind blowing? (Assume that the wind direction was parallel to the flight path and constant all day.)

54. *Wind Speed* A small plane has a speed of 170 miles per hour in still air. Find the speed of the wind if the plane traveled a distance of 400 miles with a tail wind in the same time it took to travel 280 miles into a head wind.

Figure for 53

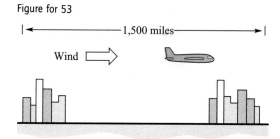

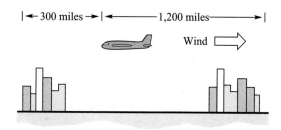

55. *Group Partnership* A group of outdoor enthusiasts plans to buy a piece of property for $48,000 by sharing the cost equally. To ease the financial burden, they look for two additional partners in order to reduce the per-person cost by $800. Determine the number of partners they would like to have in the group.

56. *Average Cost* The average cost for producing x units of a product is given by

$$\text{Average cost} = \frac{1}{2} + \frac{5{,}000}{x}.$$

Determine the number of units that must be produced to obtain an average cost of $2.50 per unit.

57. *Air Pollution* A utility company burns oil to generate electricity. The cost in dollars of removing p percent of the air pollution in the stack emission of the utility company is given by

$$\text{Cost} = \frac{80{,}000p}{1 - p}.$$

Determine the percentage of the stack emission that can be removed for $240,000. (Assume the percentage p is in decimal form.)

58. *Game Population* The game commission introduces 50 deer into newly acquired state game lands. The population N of the herd is given by the model

$$N = \frac{250(5 + 3t)}{25 + t}$$

where t is the time in years. Find the time required for the herd to increase to 125 deer.

Work Rate In Exercises 59 and 60, the first two columns of the table give the times required by each of two people working *alone* to complete a task. Complete the table by finding the time required for the two people to complete the task working *together*. (Assume that when they work together their individual rates do not change.)

59.

Person #1	Person #2	Together
4 days	4 days	
4 hours	6 hours	
4 hours	$2\frac{1}{2}$ hours	

60.

Person #1	Person #2	Together
30 minutes	30 minutes	
$6\frac{1}{2}$ hours	4 hours	
a days	b days	

61. *Average Speed* One car makes a trip of 400 miles. Another car takes the same amount of time to make a trip of 480 miles. The average speed of the second car is 10 miles per hour faster than the average speed of the first car. What is the average speed of each car?

62. *Average Speed* One car makes a trip of 440 miles. Another car takes the same amount of time to make a trip of 416 miles. The average speed of the second car is three miles per hour slower than the average speed of the first car. What is the average speed of each car?

63. *Batting Average* After 40 times at bat, a baseball player has a batting average of 0.300. How many additional consecutive times must the player hit the ball to obtain a batting average of 0.440?

64. *Batting Average* After 50 times at bat, a baseball player has a batting average of 0.160. How many additional consecutive times must the player hit the ball to obtain a batting average of 0.250?

65. *Speed* A car leaves a town 20 minutes after a fully loaded truck. The speed of the truck is approximately 10 miles per hour slower than the car. After traveling 100 miles, the car overtakes the truck. Find the speed of each vehicle.

CHAPTER 7 SUMMARY

As you review and prepare for a test on this chapter, first try to obtain a global view of what was discussed. Then review the specific skills needed in each category.

Operations with Algebraic Fractions

- *Find* the domain of an algebraic expression. Section 7.1, Exercises 1–10

- *Find* the missing factor in an equivalent algebraic fraction. Section 7.1, Exercises 11–20

- *Simplify and/or reduce* an algebraic fraction. Section 7.1, Exercises 21–50

- *Create* an algebraic fraction from a verbal application. Section 7.1, Exercises 51–54

- *Multiply and/or divide* two algebraic fractions. Section 7.2, Exercises 7–36

- *Simplify* a compound fraction by dividing. Section 7.2, Exercises 37–50

- *Add and/or subtract* two algebraic fractions with *like* denominators. Section 7.3, Exercises 1–14

- *Find* the least common multiple of two or more polynomials. Section 7.3, Exercises 15–26

- *Add and/or subtract* two algebraic fractions with *unlike* denominators. Section 7.3, Exercises 27–60

- *Simplify* compound fractions by multiplying the numerator and denominator by the LCD. Section 7.3, Exercises 61–70

Solving Fractional Equations

- *Solve* a fractional equation that contains constant denominators. Section 7.4, Exercises 5–14

- *Solve* a fractional equation that contains variable denominators. Section 7.4, Exercises 15–50

- *Check* for an extraneous solution to an equation. Section 7.4, Exercises 15–50

- *Create and solve* a fractional equation from a verbal statement. Section 7.4, Exercises 51–64

Chapter 7 Review Exercises

In Exercises 1–4, find the domain of the given algebraic fraction.

1. $\dfrac{8x}{x-5}$

2. $\dfrac{y+1}{y+3}$

3. $\dfrac{t}{t^2-3t+2}$

4. $\dfrac{x-10}{x(x^2-4)}$

In Exercises 5–10, simplify the given algebraic fraction.

5. $\dfrac{7x^2y}{21xy^2}$

6. $\dfrac{2(yz)^2}{6yz^4}$

7. $\dfrac{3b-6}{4b-8}$

8. $\dfrac{5a}{10a+25}$

9. $\dfrac{4x-4y}{y-x}$

10. $\dfrac{x-2}{x^2+x-6}$

In Exercises 11–50, perform the indicated operation and simplify your answer.

11. $y\cdot\dfrac{2}{y+3}$

12. $\dfrac{5}{8}\cdot\dfrac{x}{y}\cdot\dfrac{y^2}{x^2}$

13. $\dfrac{5x^2y}{4}\cdot\dfrac{6x}{10y^3}$

14. $(x^2y^3)\cdot\dfrac{3}{(xy)^3}$

15. $\dfrac{z}{z+1}\cdot\dfrac{z^2-1}{5}$

16. $\dfrac{1}{6}(x^2-36)\cdot\dfrac{3}{x^2-12x+36}$

17. $\dfrac{u}{u-1}\cdot\dfrac{u-u^2}{3u^2}$

18. $x\cdot\dfrac{x+1}{x^2-x}\cdot\dfrac{5x-5}{x^2+6x+5}$

19. $\dfrac{\left(\dfrac{4}{x}\right)}{\left(\dfrac{1}{x^2}\right)}$

20. $\dfrac{5x}{\left(\dfrac{x}{y}\right)}$

21. $\dfrac{\left(\dfrac{3x^2y}{4}\right)}{1}$

22. $\dfrac{0}{\left(\dfrac{5x}{y}\right)}$

23. $10y^2\div\dfrac{y}{5}$

24. $\dfrac{5}{z^2}\div 3z^2$

25. $\dfrac{u^2}{u^2-9}\div\dfrac{u}{u+3}$

26. $\dfrac{v+5}{v^2}\div\dfrac{v+5}{v^2}$

27. $\dfrac{x^2-8x}{x-1}\div\dfrac{x^2-16x+64}{x^2-1}$

28. $\left(\dfrac{2x}{y}\right)^2\div\left(\dfrac{x}{4y}\right)^3$

29. $\left(\dfrac{x}{y}\cdot\dfrac{x+1}{y+1}\right)\div\dfrac{x}{y^2-y}$

30. $xy\cdot\dfrac{8x}{y+4}\div\dfrac{y}{y^2+16}$

31. $\dfrac{5}{8}-\dfrac{3}{8}$

32. $\dfrac{4}{9}+\dfrac{11}{9}$

33. $\dfrac{4}{x+2}+\dfrac{x}{x+2}$

34. $\dfrac{3y+4}{2y+1}-\dfrac{y+3}{2y+1}$

35. $\dfrac{5}{16}-\dfrac{5}{24}$

36. $\dfrac{1}{8}+\dfrac{5}{6}-\dfrac{1}{12}$

37. $\dfrac{1}{x+2}-\dfrac{1}{x+1}$

38. $\dfrac{4}{x-3}+\dfrac{1}{x+4}$

39. $x-1+\dfrac{1}{x+2}+\dfrac{1}{x-1}$

40. $2x+\dfrac{3}{2(x-4)}-\dfrac{1}{2(x+2)}$

41. $\dfrac{1}{x}-\dfrac{x-1}{x^2+1}$

42. $\dfrac{1}{x-1}+\dfrac{1-x}{x^2+x+1}$

43. $\dfrac{1}{x-2}+\dfrac{1}{(x-2)^2}+\dfrac{1}{x+2}$

44. $\dfrac{2}{x} - \dfrac{3}{x-1} + \dfrac{4}{x+1}$

45. $\dfrac{\left(\dfrac{x^3}{x^2+x-2}\right)}{\left(\dfrac{x^2}{x^2-1}\right)}$

46. $\dfrac{\left(\dfrac{4x-6}{(x-1)^2}\right)}{\left(\dfrac{2x^2-3x}{x^2+2x-3}\right)}$

47. $\dfrac{x}{\left(1-\dfrac{1}{x}\right)}$

48. $\dfrac{\left(1+\dfrac{2}{x}-x\right)}{\left(\dfrac{2}{x}-1\right)}$

49. $\dfrac{\left(\dfrac{1}{x}-\dfrac{1}{y}\right)}{x^2-y^2}$

50. $\dfrac{\left(\dfrac{1}{x}-\dfrac{1}{y}\right)}{\left(\dfrac{1}{x}+\dfrac{1}{y}\right)}$

In Exercises 51–60, solve the given equation. (Be sure to check for extraneous solutions.)

51. $\dfrac{x}{4} = -2$

52. $\dfrac{t+3}{2} = \dfrac{1}{2}$

53. $3\left(1 - \dfrac{5}{t}\right) = 0$

54. $\dfrac{1}{y-2} = \dfrac{-2}{y+1}$

55. $\dfrac{7}{x} - 2 = \dfrac{3}{x} + 6$

56. $4 - \dfrac{3}{x+2} = \dfrac{5}{2}$

57. $\dfrac{t}{t-4} + \dfrac{3}{t-2} = 0$

58. $9 - x^{-2} = 0$

59. $\dfrac{2}{x} - \dfrac{x}{6} = \dfrac{2}{3}$

60. $x = \dfrac{x+20}{x}$

61. *Average Speed* Suppose you drive 72 miles one way on a service call for your company. The return trip takes 10 minutes less because you drive an average of 6 miles per hour faster. What is your average speed on the return trip?

62. *Average Speed* Each week a salesperson must make a 180-mile trip to pick up supplies. If this person's average speed could be increased by five miles per hour, the trip would take 24 minutes less. What is the salesperson's original speed?

63. *Forming a Partnership* A group of farmers agree to share equally in the cost of a $48,000 piece of machinery. If they could find two more farmers to join the group, each person's share of the cost would decrease by $4,000. How many farmers are currently in the group?

64. *Forming a Partnership* An individual is planning to start a small business that will require $24,000 before any income can be generated. Because it is difficult to borrow money for new ventures, the individual wants a group of friends to divide the cost equally for a future share of the profit. The person has found some investors, but three more are needed so that the price per person will be $1,800 less. How many investors are currently in the group?

65. *Work Rate* Your supervisor can do a task in 8 minutes that takes you 10 minutes to complete. Determine the time required to complete the task if you work together.

CHAPTER 7 TEST

Take this test as you would take a test in class. After you are done, check your work with the answers given in the back of the book.

1. Find the domain of the algebraic fraction $\dfrac{x}{x - 10}$.

2. Find the missing algebraic expression in the equation $\dfrac{2x^2}{x + 1} = \dfrac{2x^2(\rule{1.5cm}{0.3cm})}{x(x + 1)^2}$.

3. Simplify: $\dfrac{8x^2(x + 1)}{x(x + 1)^2}$

4. Simplify: $\dfrac{x^2 - 64}{x^2 - 3x - 40}$

5. Multiply and simplify: $\dfrac{18x}{5} \cdot \dfrac{15}{3x^3}$

6. Multiply and simplify: $(x + 2)^2 \cdot \dfrac{x - 2}{x^3 + 2x^2}$

7. Divide and simplify: $\dfrac{3x^2}{4} \div \dfrac{9x^3}{10}$

8. Divide and simplify: $\dfrac{\left(\dfrac{t}{t - 5}\right)}{\left(\dfrac{t^2}{5 - t}\right)}$

9. Simplify: $\left[\left(\dfrac{x}{x - 3}\right)^2 \cdot \dfrac{x^2}{x^2 - 3x}\right] \div (x - 3)^5$

10. Find the least common multiple of the polynomials $6x(x + 3)^2$, $9x^3$, and $12(x + 3)$.

11. Add and simplify: $\dfrac{8}{3u^2} + \dfrac{3}{u}$

12. Subtract and simplify: $\dfrac{3}{x + 2} - 6$

13. Subtract and simplify: $\dfrac{2}{x + 1} - \dfrac{2x}{x^2 + 2x + 1}$

14. Simplify: $\dfrac{4}{\left(\dfrac{2}{x} + 8\right)}$

15. Determine whether the given value of x is a solution of the equation $\dfrac{x}{4} + \dfrac{2}{x} = \dfrac{3}{2}$.

(a) $x = 1$ (b) $x = 2$ (c) $x = -\dfrac{1}{2}$ (d) $x = 4$

16. Solve: $5 + \dfrac{t}{3} = t + 2$

17. Solve: $\dfrac{5}{x + 1} - \dfrac{1}{x} = \dfrac{3}{x}$

18. Solve: $2\left(x + \dfrac{1}{x}\right) = 5$

19. The capacity of a pump is 80 gallons per minute. Determine the time required to pump (a) 1 gallon, (b) x gallons, and (c) 16 gallons.

20. Partners in a small business plan to buy new equipment for $18,000 by sharing the cost equally. To ease the financial burden, they look for three additional partners so that the cost per person will be reduced by $1,000. Determine the number of partners currently in the group.

CUMULATIVE TEST: CHAPTERS 5–7

Take this test as you would take a test in class. After you are done, check your work with the answers given in the back of the book.

1. Subtract: $(x^3 - 3x^2) - (x^3 + 2x^2 - 5)$

2. Multiply: $(6z)(-7z)(z^2)$

3. Multiply: $(3x + 5)(x - 4)$

4. Multiply: $(5x - 3)(5x + 3)$

5. Expand: $(5x + 6)^2$

6. Divide: $(6x^2 + 72x) \div 6x$

7. Divide: $\dfrac{x^2 - 3x - 2}{x - 4}$

8. Evaluate: $(3^2 \cdot 4^{-1})^2$

9. Rewrite the expression $\left(\dfrac{x}{2}\right)^{-2}$ using positive exponents.

10. Factor: $2u^2 - 6u$

11. Factor and simplify: $(x - 2)^2 - 16$

12. Factor completely: $x^3 + 8x^2 + 16x$

13. Factor completely: $x^3 + 2x^2 - 4x - 8$

14. Simplify: $\dfrac{5x - 25}{x^2 - 25}$

15. Multiply: $\dfrac{c}{c - 1} \cdot \dfrac{c^2 + 9c - 10}{c^3}$

16. Divide: $\dfrac{6}{(c - 1)^2} \div \dfrac{8}{c^3 - c^2}$

17. Add: $\dfrac{3}{x - 2} + \dfrac{x}{4 - x^2}$

18. Simplify: $\dfrac{\left(a^2 - \dfrac{1}{a}\right)}{\left(\dfrac{1}{2} + \dfrac{1}{a}\right)}$

$\dfrac{2a^2 - 2}{a + 2} = \dfrac{2(a^2 - 1)}{a + 2} = \dfrac{2(a+1)(a-1)}{a + 2}$

19. Solve: $5x^2 - 12x - 9 = 0$

20. Solve: $\dfrac{3}{x} - 4 = \dfrac{1}{x} + 1$

21. Suppose you drive 100 miles one way on a service call for your company. The return trip takes 30 minutes less because you drive an average of 10 miles per hour faster. What is your average speed on the return trip?

22. A new employee takes twice as long as an experienced employee to complete a task. Together they can complete the task in three hours. Determine the time it takes them to do the task individually.

CHAPTER EIGHT

Graphs and Linear Equations

Chapter Overview

*In this chapter we show how a coordinate system can be used to give a geometric picture of an algebraic equation in two variables. This kind of mathematics, called **analytic geometry**, ties together features of both geometry and algebra. So far, our overall picture of algebra has had three major features: learning the basic rules of algebra, simplifying algebraic expressions, and solving algebraic equations. Now, we add a fourth feature: sketching the graphs of algebraic equations.*

The Rectangular Coordinate System

The Rectangular Coordinate System • *Ordered Pairs as Solutions of Equations* • *Applications*

The Rectangular Coordinate System

Just as we can represent real numbers by points on the real number line, so can we represent ordered pairs of real numbers by points in a plane. This plane is called a **rectangular coordinate system** or the **Cartesian plane**, after the French mathematician René Descartes (1596–1650).

A rectangular coordinate system is formed by two real lines intersecting at right angles, as shown in Figure 8.1. The horizontal number line is usually called the **x-axis**, and the vertical number line is usually called the **y-axis**. (The plural of axis is *axes*.) The point of intersection of the two axes is called the **origin**, and the axes separate the plane into four regions called **quadrants**.

FIGURE 8.1

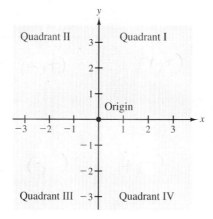

Each point in the plane corresponds to an **ordered pair** (x, y) of real numbers x and y, called the **coordinates** of the point. The first number (or **x-coordinate**) tells how far to the left or right the point is from the vertical axis, and the second number (or **y-coordinate**) tells how far up or down the point is from the horizontal axis, as shown in Figure 8.2.

NOTE: A positive x-coordinate implies that the point lies to the *right* of the vertical axis; a negative x-coordinate implies that the point lies to the *left* of the vertical axis; and an x-coordinate of zero implies that the point lies *on* the vertical axis. Similarly, a positive y-coordinate implies that the point lies *above* the horizontal axis; a negative y-coordinate implies that the point lies *below* the horizontal axis; and a y-coordinate of zero implies that the point lies *on* the horizontal axis.

FIGURE 8.2

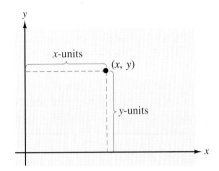

Locating a given point in a plane is called **plotting** the point. We show how this is done in Example 1.

EXAMPLE 1   ■   **Plotting Points in a Rectangular Coordinate System**

Plot the points $(-1, 2)$, $(3, 0)$, $(2, -1)$, $(3, 4)$, $(0, 0)$, and $(-2, -3)$ in a rectangular coordinate system.

Solution

The point $(-1, 2)$ is 1 unit to the *left* of the vertical axis and 2 units *above* the horizontal axis. Similarly, the point $(3, 0)$ is 3 units to the right of the vertical axis and *on* the horizontal axis. (It is on the horizontal axis because the y-coordinate is zero.) The other four points can be plotted in a similar way, as shown in Figure 8.3.

FIGURE 8.3

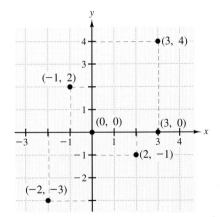

In Example 1 we were given the coordinates of several points and asked to plot the points in a rectangular coordinate system. In Example 2 we look at the reverse problem. That is, we are given points in a rectangular coordinate system and asked to determine their coordinates.

EXAMPLE 2   ■   Finding Coordinates of Points

Determine the coordinates for each of the points shown in Figure 8.4.

Solution

FIGURE 8.4

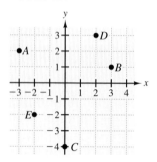

Point A lies 3 units to the *left* of the vertical axis, and 2 units *above* the horizontal axis. Therefore, point A must be given by the ordered pair $(-3, 2)$. The coordinates of the other four points can be determined in a similar way, and we summarize the results as follows.

Point	Position	Coordinates
A	3 units *left*, 2 units *up*	$(-3, 2)$
B	3 units *right*, 1 unit *up*	$(3, 1)$
C	0 units *left* (or *right*), 4 units *down*	$(0, -4)$
D	2 units *right*, 3 units *up*	$(2, 3)$
E	2 units *left*, 2 units *down*	$(-2, -2)$

The value of a rectangular coordinate system is that it allows us to visualize relationships between two variables. Today, Descartes's ideas are in common use in every scientific and business-related field. A rectangular coordinate system can give us a picture of how total sales are related to the price of an item, how temperature is related to time of day, how interest rates are related to months of a year, and so on.

EXAMPLE 3   ■   An Application of the Rectangular Coordinate System

The scores of the winning and losing football teams for the Super Bowl games from 1977 to 1990 are given in Table 8.1. (*Source:* National Football League.) Plot these points on a rectangular coordinate system.

TABLE 8.1

Year	1977	1978	1979	1980	1981	1982	1983
Winning Score	32	27	35	31	27	26	27
Losing Score	14	10	31	19	10	21	17

Year	1984	1985	1986	1987	1988	1989	1990
Winning Score	38	38	46	39	42	20	55
Losing Score	9	16	10	20	10	16	10

Solution

For these points, we choose to plot the years on the x-axis and the winning and losing scores on the y-axis. In Figure 8.5 the winning scores are shown as black dots, and the losing scores are shown as blue dots. Note that the break in the x-axis indicates that we have omitted the numbers between 0 and 1976.

FIGURE 8.5

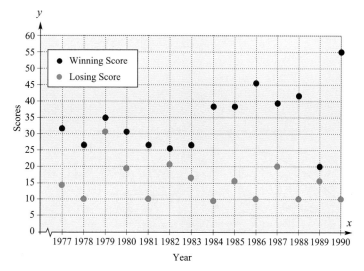

Ordered Pairs as Solutions of Equations

In Example 3 the relationship between the year and the Superbowl scores was given by a **table of values**. In mathematics the relationship between the variables x and y is often given by an equation. From the equation, we must then construct our own table of values. For instance, consider the equation

$$y = 2x + 1.$$

To construct a table of values for this equation, we choose several x-values and then calculate the corresponding y-values. For example, if we choose $x = 1$, then the y-value is

$$y = 2(1) + 1 = 3.$$

The corresponding ordered pair $(x, y) = (1, 3)$ is called a **solution point** (or simply a **solution**) of the equation. Table 8.2 shows a table of values (and the corresponding solution points) using x-values of -3, -2, -1, 0, 1, 2, and 3.

TABLE 8.2

Choose x	Calculate y from $y = 2x + 1$	Solution Points
$x = -3$	$y = 2(-3) + 1 = -5$	$(-3, -5)$
$x = -2$	$y = 2(-2) + 1 = -3$	$(-2, -3)$
$x = -1$	$y = 2(-1) + 1 = -1$	$(-1, -1)$
$x = 0$	$y = 2(0) + 1 = 1$	$(0, 1)$
$x = 1$	$y = 2(1) + 1 = 3$	$(1, 3)$
$x = 2$	$y = 2(2) + 1 = 5$	$(2, 5)$
$x = 3$	$y = 2(3) + 1 = 7$	$(3, 7)$

2 variables in the first degree are a line

Once we have constructed a table of values, we can get a visual idea of the relationship between the variables x and y by plotting the solution points in a rectangular coordinate system. For instance, the solution points shown in Table 8.2 are plotted in Figure 8.6.

FIGURE 8.6

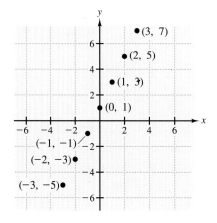

When making up a table of values for an equation, it is helpful to first solve the equation for y. For instance, the equation $4x + 2y = 7$ can be solved for y as follows.

$4x + 2y = 7$	Given equation
$2y = -4x + 7$	Subtract $4x$ from both sides
$y = -2x + \dfrac{7}{2}$	Solve for y

This procedure is further demonstrated in Example 4.

EXAMPLE 4   ■   **Making Up a Table of Values**

Make up a table of values showing five solution points for the equation

$$6x - 2y = 4.$$

Then plot the solution points in the rectangular coordinate system. (Choose x-values of -2, -1, 0, 1, and 2.)

Solution

To begin, we solve the given equation for y.

$$6x - 2y = 4$$
$$-2y = -6x + 4$$
$$y = 3x - 2$$

Now, using the equation $y = 3x - 2$, we construct the table of values, as shown in Table 8.3.

TABLE 8.3

x	-2	-1	0	1	2
$y = 3x - 2$	-8	-5	-2	1	4
Solution Points	$(-2, -8)$	$(-1, -5)$	$(0, -2)$	$(1, 1)$	$(2, 4)$

Finally, from Table 8.3, we plot the five solution points in the rectangular coordinate system, as shown in Figure 8.7.

FIGURE 8.7

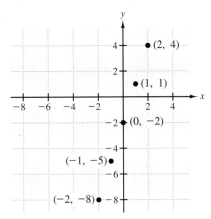

In the next example, you are given several ordered pairs and asked to determine whether they are solutions of the given equation. To do this you need to substitute the values of x and y into the equation. If the substitution produces a true equation, then the ordered pair (x, y) is a solution. Otherwise, the ordered pair is not a solution.

EXAMPLE 5   ■   Verifying Solutions of an Equation

Determine which of the following ordered pairs is a solution of $x + 3y = 6$.

(a) $(1, 2)$ (b) $\left(-2, \dfrac{8}{3}\right)$ (c) $(-6, 0)$ (d) $(0, 2)$ (e) $\left(\dfrac{8}{3}, \dfrac{10}{9}\right)$

Solution

(a) For the ordered pair $(x, y) = (1, 2)$, we substitute $x = 1$ and $y = 2$ into the given equation.

$$x + 3y = 6 \qquad \text{Given equation}$$
$$1 + 3(2) \stackrel{?}{=} 6 \qquad \text{Substitute } x = 1 \text{ and } y = 2$$
$$7 \neq 6 \qquad \text{Not a solution}$$

Since the substitution did not satisfy the given equation, we conclude that the ordered pair $(1, 2)$ is *not* a solution of the given equation.

(b) The ordered pair $\left(-2, \frac{8}{3}\right)$ *is* a solution of the given equation because

$$-2 + 3\left(\frac{8}{3}\right) = -2 + 8 = 6.$$

(c) The ordered pair $(-6, 0)$ is *not* a solution of the given equation because

$$-6 + 3(0) = -6 \neq 6.$$

(d) The ordered pair $(0, 2)$ *is* a solution of the given equation because

$$0 + 3(2) = 6.$$

(e) The ordered pair $\left(\frac{8}{3}, \frac{10}{9}\right)$ *is* a solution of the given equation because

$$\frac{8}{3} + 3\left(\frac{10}{9}\right) = \frac{8}{3} + \frac{10}{3} = \frac{18}{3} = 6. \qquad \blacksquare$$

Applications

As our last example in this section, we look at a real-life application in which you are asked to construct an equation.

EXAMPLE 6   ■   **A Business Application**

Suppose you are setting up a small manufacturing business in which you are assembling computer keyboards. Your initial investment in the business is $120,000, and your unit cost for assembling each keyboard is $40. Write an equation that relates your total cost to the number of keyboards produced. Then plot the total cost of producing 1,000, 2,000, 3,000, 4,000, and 5,000 keyboards.

Solution

For this problem we let x represent the number of keyboards and C represent the total cost of producing x keyboards. The total cost equation must represent both the initial cost (which is fixed) and the unit cost (which depends on the number of keyboards produced). For instance, to produce one keyboard, your total cost is

$$C = 40(1) + 120,000 = \$120,040.$$

Similarly, to produce two keyboards, your total cost is

$$C = 40(2) + 120,000 = \$120,080.$$

From this pattern, we can see that the total cost of producing x keyboards is given by the equation

$$C = 40x + 120,000.$$

Finally, using this equation, we construct the table of values shown in Table 8.4. The information given in the table is plotted in Figure 8.8.

TABLE 8.4

Number of Keyboards, x	Total Cost, C
1,000	$160,000
2,000	200,000
3,000	240,000
4,000	280,000
5,000	320,000

FIGURE 8.8

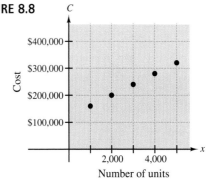

DISCUSSION PROBLEM ■ A Misleading Graph

While graphs can help us visualize relationships between two variables, they can also mislead people. The graphs shown in Figure 8.9 represent the same data points. Which of the two graphs is misleading, and why?

FIGURE 8.9

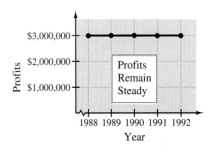

■ Warm-Up

The following warm-up exercises involve skills that were covered in earlier sections. You will use these skills in the exercise set for this section.

In Exercises 1–6, place the correct inequality symbol ($<$ or $>$) between the two real numbers, and plot each number on the real number line.

1. -4 ▭ 3 **2.** $\dfrac{8}{3}$ ▭ 2 **3.** -2 ▭ -6

4. -8 ▭ 0 **5.** $\dfrac{15}{4}$ ▭ $-\dfrac{1}{2}$ **6.** $\dfrac{2}{5}$ ▭ $\dfrac{15}{16}$

In Exercises 7–10, evaluate the given expression.

7. $4 - |-3|$ **8.** $-10 - (4 - 18)$

9. $\dfrac{3 - (5 - 20)}{4}$ **10.** $\dfrac{|3 - 18|}{3}$

8.1 EXERCISES

In Exercises 1–6, plot the points in a rectangular coordinate system.

1. (3, 2), (−4, 2), (2, −4)

2. (−1, 6), (−1, −6), (4, 6)

3. (−10, −4), (4, −4), (4, 3)

4. (−6, 4), (0, 0), (3, −2)

5. $\left(\frac{3}{2}, -1\right), \left(-3, \frac{3}{4}\right), \left(\frac{1}{2}, -\frac{1}{2}\right)$

6. $\left(-\frac{2}{3}, 4\right), \left(\frac{1}{2}, -\frac{5}{2}\right), \left(-4, -\frac{5}{4}\right)$

In Exercises 7 and 8, determine the quadrant in which each point is located.

7. (a) $\left(-3, \frac{1}{8}\right)$ (b) (4.2, −3.05)

8. (a) (−100, −365.6)   (b) $\left(\frac{3}{11}, \frac{7}{8}\right)$

In Exercises 9 and 10, label the approximate coordinates of the points shown on the coordinate system.

9.

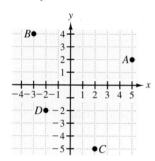

10.

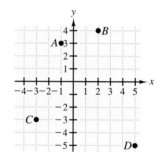

In Exercises 11–16, plot the given points and connect them with line segments to form the indicated figure. (*Note:* A rhombus is a four-sided figure whose sides have the same lengths.)

11. Triangle: (−1, 1), (2, −1), (3, 4)

12. Triangle: (0, 3), (−1, −2), (4, 8)

13. Square: (2, 4), (5, 1), (2, −2), (−1, 1)

14. Rectangle: (2, 1), (4, 2), (−1, 7), (1, 8)

15. Parallelogram: (5, 2), (7, 0), (1, −2), (−1, 0)

16. Rhombus: (0, 0), (1, 2), (2, 1), (3, 3)

In Exercises 17–20, determine the quadrant or quadrants in which the points must be located.

17. (−5, y),   y is a real number

18. (x, 3),   x is a real number

19. (x, y),   xy < 0

20. (x, y),   x < 0 and y < 0

In Exercises 21–24, plot the points whose coordinates are given in the table.

21. *Study Time* The accompanying table gives the hours x that a student studied for five different algebra exams and the resulting score y.

x	3.5	1	8	4.5	0.5
y	72	67	95	81	53

22. *Stopping Distance* The accompanying table gives the speed of a car x (in kilometers per hour) and the approximate stopping distance y in meters.

x	50	70	90	110	130
y	20	35	60	95	148

23. *Average Temperature* The accompanying table gives the normal temperature y (Fahrenheit) for Anchorage, Alaska, for each month of the year. The months are numbered 1 to 12, with 1 corresponding to January.

x	1	2	3	4	5	6
y	13	18	24	35	46	54
x	7	8	9	10	11	12
y	58	56	48	35	22	14

24. *Price per Share* The accompanying table gives the price y per common share of stock of a corporation (on December 31) for the years 1984 through 1989. The time in years is given by x.

x	y
1984	$25.50
1985	35.50
1986	44.125
1987	42.25
1988	47.875
1989	68.00

In Exercises 25–30, determine whether the ordered pairs are solution points of the given equation.

25. $y = 2x + 4$ (a) $(3, 10)$ (c) $(0, 0)$
           (b) $(-1, 3)$ (d) $(-2, 0)$

26. $y = 5x - 2$ (a) $(2, 0)$ (c) $(6, 28)$
           (b) $(-2, -12)$ (d) $(1, 1)$

27. $2y - 3x + 1 = 0$ (a) $(1, 1)$ (c) $(-3, -1)$
           (b) $(5, 7)$ (d) $\left(-2, -\dfrac{7}{2}\right)$

28. $x - 8y + 100 = 0$ (a) $(10, -10)$ (c) $(100, 0)$
           (b) $(-60, 5)$ (d) $(20, -10)$

29. $y = \dfrac{2}{3}x$ (a) $(6, 6)$ (c) $(0, 0)$
           (b) $(-9, -6)$ (d) $\left(-1, \dfrac{2}{3}\right)$

30. $y = \dfrac{5}{8}x - 2$ (a) $(8, 3)$ (c) $(-16, -12)$
           (b) $(0, 0)$ (d) $(32, 18)$

In Exercises 31–34, complete the given table by finding the y-coordinates of the solution points. (Plot your results in a rectangular coordinate system.)

31.

x	−2	0	2	4	6
$y = 3x - 4$					

32.

x	−2	0	2	4	6
$y = \frac{1}{4}x + 1$					

33.

x	−5	$\frac{3}{2}$	5	10	20
$y = -\frac{3}{2}x + 5$					

34.

x	−2	−1	0	$\frac{1}{2}$	$\frac{5}{4}$
$y = -\frac{1}{2}x + \frac{5}{4}$					

35. *Cost* The cost *y* of producing *x* units of a product is given by $y = 35x + 5,000$. Complete the following table to determine the cost of producing the specified number of units. (Plot your results in a rectangular coordinate system.)

x	$y = 35x + 5,000$
100	
150	
200	
250	
300	

36. *Hourly Wages* When an employee produces *x* units per hour, the hourly wage is given by $y = 0.50x + 10$. Complete the following table to determine the hourly wages for producing the specified number of units. (Plot your results in a rectangular coordinate system.)

x	2	5	8	10	20
$y = 0.50x + 10$					

37. Plot the points (3, 2), (−5, 4), and (6, −4) in a rectangular coordinate system. Then change the sign of the *x*-coordinate of each point and plot the three new points in the same rectangular coordinate system. What inference can you make about the location of a point when the sign of the *x*-coordinate is changed?

38. Plot the points (3, 2), (−5, 4), and (6, −4) in a rectangular coordinate system. Then change the sign of the *y*-coordinate of each point and plot the three new points in the same rectangular coordinate system. What inference can you make about the location of a point when the sign of the *y*-coordinate is changed?

Housing Starts In Exercises 39–42, use the accompanying bar chart showing new housing starts in the United States from 1976 to 1988. (*Source:* Department of Commerce.)

Figure for 39–42

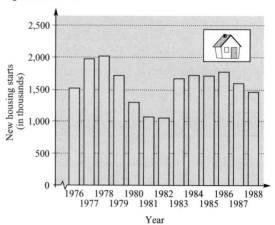

39. Estimate the number of new housing starts in 1981.

40. Estimate the number of new housing starts in 1986.

41. Estimate the increase in housing starts from 1982 to 1988.

42. Estimate the decrease in housing starts from 1976 to 1982.

Personal Income In Exercises 43–46, use the accompanying bar chart showing the personal income in the United States from 1980 to 1988. (*Source:* U.S. Bureau of the Census.)

Figure for 43–46

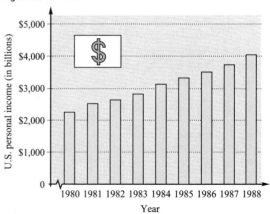

43. Estimate the personal income in the U.S. in 1982.

44. Estimate the personal income in the U.S. in 1986.

45. Estimate the increase in personal income in the U.S. from 1980 to 1984.

46. Estimate the increase in personal income in the U.S. from 1984 to 1988.

Money Spent on Education In Exercises 47 and 48, use the accompanying bar chart, which relates the amount of government spending per student to the per capita income of the country. (*Source: U.S. News & World Report.*)

47. Estimate the percentage of government spending per student to the per capita income in Sweden.

48. Estimate the percentage of government spending per student to the per capita income in the United States.

Figure for 47 and 48

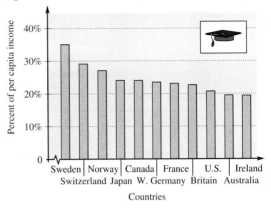

Graphs of Equations and Functions

*The Graph of an Equation   •   Intercepts: An Aid to Sketching Graphs   •   Applications   •
Functions and Their Graphs*

The Graph of an Equation

We have already seen that the solutions of an equation involving two variables can be represented by points in a rectangular coordinate system. The set of all solutions of an equation is called its **graph**. In Section 8.1 we took the first step in sketching the graph of an equation. That is, we constructed a table of values and plotted the corresponding solution points in a rectangular coordinate system. In this section we take the process one step further by connecting the solution points with a smooth curve or line. To see how this works, let's make a sketch of the graph of the equation

$$y = 2x - 1.$$

To begin, we construct a table of values, as shown in Table 8.5.

TABLE 8.5

x	-3	-2	-1	0	1	2	3
$y = 2x - 1$	-7	-5	-3	-1	1	3	5
Solution	$(-3, -7)$	$(-2, -5)$	$(-1, -3)$	$(0, -1)$	$(1, 1)$	$(2, 3)$	$(3, 5)$

Next, from the table of values we plot the solution points in a rectangular coordinate system, as shown in Figure 8.10(a). The seven points plotted in this figure represent only a few of the points on the graph of the equation. To complete our sketch, we try to determine whether the representative points form a pattern. In this case all seven points appear to lie on a line, so we complete the sketch by drawing a straight line through the points, as shown in Figure 8.10(b).

FIGURE 8.10   (a)

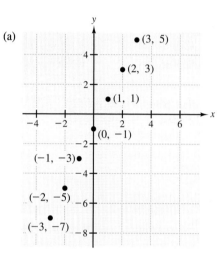

(b)

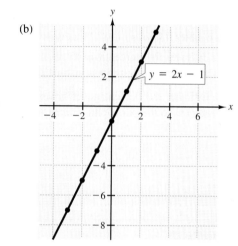

This method for sketching the graph of an equation is called the **point-plotting method**. We summarize the basic steps of the method as follows.

The Point-Plotting Method of Sketching a Graph

To sketch the graph of an equation by point plotting, use the following steps.

1. If possible, rewrite the equation by isolating one of the variables.

2. Make up a table of values showing several solution points.

3. Plot these points in a rectangular coordinate system.

4. Connect the points with a smooth curve or line.

EXAMPLE 1   ■   Sketching the Graph of an Equation

Sketch the graph of the following equation.

$$3x + y = 5$$

Solution

To begin, we rewrite the equation so that y is isolated on the left.

$$y = -3x + 5$$

Next, we make up a table of values, as shown in Table 8.6. (The choice of x-values to use in the table is somewhat arbitrary. However, the more x-values we choose, the easier it is to recognize a pattern.)

TABLE 8.6

x	-2	-1	0	1	2	3
$y = -3x + 5$	11	8	5	2	-1	-4
Solution	$(-2, 11)$	$(-1, 8)$	$(0, 5)$	$(1, 2)$	$(2, -1)$	$(3, -4)$

Now we plot the six solution points, as shown in Figure 8.11(a). It appears that all six points lie on a line, so we complete the sketch by drawing a line through the six points, as shown in Figure 8.11(b).

FIGURE 8.11

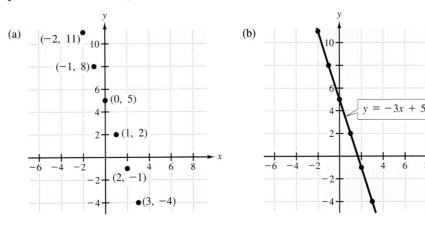

The equation in Example 1 is an example of a **linear equation** in two variables—it is of first degree in both variables and its graph is a straight line. (We will study this type of equation in Sections 8.3 and 8.4.) The graph of a nonlinear equation is not a straight line. The next two examples show how we can use the point-plotting method to sketch the graph of a nonlinear equation.

EXAMPLE 2 ■ Sketching the Graph of a Nonlinear Equation

Sketch the graph of the following equation.

$$x^2 + y = 4$$

Solution

To begin, we rewrite the equation so that y is isolated on the left.

$$y = -x^2 + 4$$

Next, we make up a table of values, as shown in Table 8.7. Watch out for the signs of the numbers when making up such a table. For instance, when $x = -3$, the value of y is

$$y = -(-3)^2 + 4 = -9 + 4 = -5.$$

TABLE 8.7

x	-3	-2	-1	0	1	2	3
$y = -x^2 + 4$	-5	0	3	4	3	0	-5
Solution	$(-3, -5)$	$(-2, 0)$	$(-1, 3)$	$(0, 4)$	$(1, 3)$	$(2, 0)$	$(3, -5)$

Now we plot the seven solution points, as shown in Figure 8.12(a). Finally, we connect the points with a smooth curve, as shown in Figure 8.12(b).

FIGURE 8.12 (a)

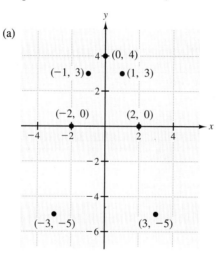

(b)

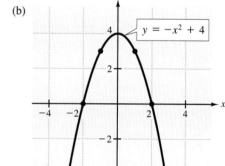

The graph of the equation given in Example 2 is called a **parabola**. We will study this type of graph in detail in Chapter 11.

In Example 3 we look at the graph of an equation that involves an absolute value. Remember from Chapter 1 that to find the absolute value of a number, we disregard the sign of the number. For instance, $|-5| = 5$, $|2| = 2$, and $|0| = 0$.

EXAMPLE 3 ■ The Graph of an Equation Involving an Absolute Value

{not going to stress}

Sketch the graph of the following equation.

$$y = |x - 1|$$

Solution

This equation is already written in a form with y isolated on the left, so we begin by making up a table of values, as shown in Table 8.8. Be sure to check the values in this table to see that you understand how the absolute value is working. For instance, when $x = -2$, the value of y is

$$y = |-2 - 1| = |-3| = 3.$$

Similarly, when $x = 2$, the value of y is

$$y = |2 - 1| = |1| = 1.$$

TABLE 8.8

x	-2	-1	0	1	2	3	4
$y = |x - 1|$	3	2	1	0	1	2	3
Solution	$(-2, 3)$	$(-1, 2)$	$(0, 1)$	$(1, 0)$	$(2, 1)$	$(3, 2)$	$(4, 3)$

Next, we plot the seven solution points, as shown in Figure 8.13(a). It appears that the points lie in a "V-shaped" pattern, with the point (1, 0) lying at the bottom of the "V." Following this pattern, we connect the points to form the graph shown in Figure 8.13(b).

FIGURE 8.13   (a)

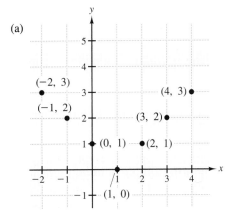

(b)

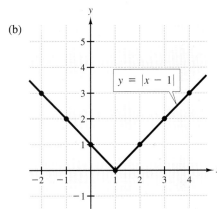

Intercepts: An Aid to Sketching Graphs

Two types of solution points that are especially useful are those having zero as either the x- or y-coordinate. Such points are called **intercepts** because they are the points at which the graph intersects the x- or y-axis.

Definition of Intercepts

The point $(a, 0)$ is called an **x-intercept** of the graph of an equation if it is a solution point of the equation. To find the x-intercepts, let y be zero and solve the equation for x.

The point $(0, b)$ is called a **y-intercept** of the graph of an equation if it is a solution point of the equation. To find the y-intercepts, let x be zero and solve the equation for y.

EXAMPLE 4 ■ Finding the Intercepts of a Graph

Find the intercepts and sketch the graph of the following equation.

$$y = 2x - 5$$

Solution

To determine whether the graph has any x-intercepts, we let y be zero and solve the resulting equation for x.

$y = 2x - 5$	Given equation
$0 = 2x - 5$	Let $y = 0$
$\dfrac{5}{2} = x$	Solve equation for x

Therefore, the graph has one x-intercept, which occurs at the point $\left(\frac{5}{2}, 0\right)$. To determine whether the graph has any y-intercepts, we let x be zero and solve the resulting equation for y.

$y = 2x - 5$	Given equation
$y = 2(0) - 5$	Let $x = 0$
$y = -5$	Solve equation for y

Therefore, the graph has one y-intercept, which occurs at the point $(0, -5)$. Now, to sketch the graph of the equation, we make up a table of values, as shown in Table 8.9. (Notice that the two intercepts are incorporated into the table.) Finally, using the solution points given in the table, we sketch the graph of the equation, as shown in Figure 8.14.

TABLE 8.9

x	-1	0	1	2	$\frac{5}{2}$	3	4
$y = 2x - 5$	-7	-5	-3	-1	0	1	3
Solution	$(-1, -7)$	$(0, -5)$	$(1, -3)$	$(2, -1)$	$\left(\frac{5}{2}, 0\right)$	$(3, 1)$	$(4, 3)$

FIGURE 8.14

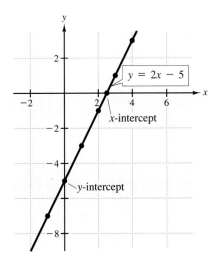

It is possible for a graph to have no intercepts or several intercepts. For instance, consider the three graphs in Figure 8.15.

FIGURE 8.15

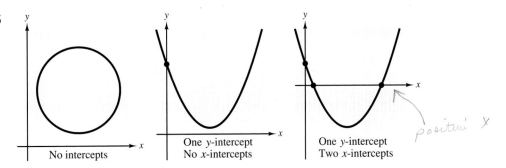

One y-intercept
No x-intercepts

One y-intercept
Two x-intercepts

No intercepts

EXAMPLE 5   ■   A Graph that Has Two *x*-Intercepts

Find the intercepts and sketch the graph of the following equation.

$$y = x^2 - 4x + 3$$

Solution

To determine whether the graph has any *x*-intercepts, we let *y* be zero and solve the resulting equation for *x*.

$y = x^2 - 4x + 3$	Given equation
$0 = x^2 - 4x + 3$	Let $y = 0$
$0 = (x - 1)(x - 3)$	Factor

By setting each of these factors equal to zero, we conclude that the equation $0 = x^2 - 4x + 3$ has two solutions: 1 and 3. Therefore, it follows that the graph has two x-intercepts: the points $(1, 0)$ and $(3, 0)$.

To determine whether the graph has any y-intercepts, we let x be zero and solve the resulting equation for y.

$y = x^2 - 4x + 3$	Given equation
$y = 0^2 - 4(0) + 3$	Let $x = 0$
$y = 3$	Solve equation for y

Therefore, the graph has one y-intercept, which occurs at the point $(0, 3)$.

Now, to sketch the graph of the equation, we make up a table of values, as shown in Table 8.10. (Notice that the three intercepts are incorporated into the table.) Finally, using the solution points given in the table, we sketch the graph of the equation, as shown in Figure 8.16.

TABLE 8.10

x	-1	0	1	2	3	4	5
$y = x^2 - 4x + 3$	8	3	0	-1	0	3	8
Solution Points	$(-1, 8)$	$(0, 3)$	$(1, 0)$	$(2, -1)$	$(3, 0)$	$(4, 3)$	$(5, 8)$

FIGURE 8.16

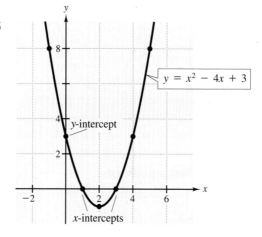

Applications

Newspapers and news magazines frequently use graphs to show time-of-year comparisons with the rate of inflation, housing costs, wholesale prices, or the unemployment rate. Industrial firms and businesses use graphs to report monthly or yearly production and sales statistics. Such graphs provide a simple geometric picture of the way one quantity changes with respect to another. In the following example, we show how a graph can help us visualize the concept of straight-line depreciation.

EXAMPLE 6 ■ An Application

A small business purchases a new delivery van for $25,500. For income tax purposes, the owner of the business decides to depreciate the van over a 10-year period. At the end of the 10 years, the salvage value of the van is expected to be $1,500. Find an equation that relates the value of the van to the number of years. Then sketch the graph of the equation. (Assume that the depreciation is the same each year. This type of depreciation is called **straight-line depreciation**.)

Solution

The total depreciation over the 10-year period is $25,500 - 1,500 = \$24,000$. Since the same amount is depreciated each year, it follows that the annual depreciation is $24,000/10 = \$2,400$. Thus, after one year, the value of the van is

Value after one year $= 25,500 - 2,400 = \$23,100.$

By similar reasoning, we can see that the value after two years is

Value after two years $= 25,500 - 2,400(2) = \$20,700.$

Now, by letting y represent the value of the van after t years, we follow the pattern given by the first two years and conclude that the equation is

$$y = 25,500 - 2,400t.$$

A sketch of the graph of this equation is shown in Figure 8.17.

FIGURE 8.17

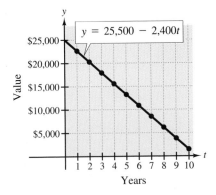

Functions and Their Graphs

In the first three examples discussed in this section, we were able to isolate the variable y on the left side of the equation. That is, we were able to write the equations in the form

$$y = -3x + 5, \qquad y = -x^2 + 4, \quad \text{or} \quad y = |x - 1|.$$

Equations like these are of special importance in mathematics, and we say that each of these equations represents y as a **function** of x. If you go on to higher-level courses in mathematics, you will study the concept of a function in detail. For purposes of this course, the important thing to realize about a function is this: If an equation represents y as a function of x, then no two different solution points can have the same x-value. Another way of describing this is to say that an equation represents y as a function of x if a given value of x determines only *one* value of y.

Definition of a Function of x	An equation in two variables x and y represents y **as a function of** x if a given value of x determines at most one value of y.

An easy way to determine whether an equation represents y as a function of x is by looking at its graph. If the graph has the property that no vertical line intersects the graph at more than one point, then the equation represents y as a function of x. On the other hand, if you can find a vertical line that intersects the graph at two (or more) points, then the equation does not represent y as a function of x. We call this graphical test the **vertical line test** for y as a function of x. Here are a couple of examples. The graphs of the equations

$$y = x + 1 \quad \text{and} \quad x = y^2 - 1$$

are shown in Figure 8.18. Notice that for the first equation, there is no vertical line that intersects the graph more than once. Therefore, the equation $y = x + 1$ *does* represent y as a function of x. However, for the second equation, there are some vertical lines that intersect the graph twice. Therefore, the equation $x = y^2 - 1$ *does not* represent y as a function of x.

FIGURE 8.18

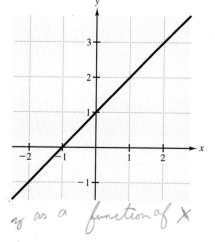

y as a function of x

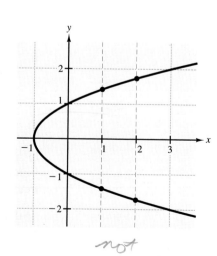

not

EXAMPLE 7 ■ Testing for *y* as a Function of *x*

Which of the following equations represent(s) *y* as a function of *x*?

(a) $x^2 = y + 5$ (b) $x = y^2 - 5$ (c) $y = 2$

Solution

(a) To begin, we rewrite the equation in the form $y = x^2 - 5$.

Since we were able to isolate *y* on the left side of the equation, it appears that the equation does represent *y* as a function of *x*. We can confirm this conclusion by the vertical line test. Notice in Figure 8.19 that no vertical line intersects the graph more than once. Therefore, this equation *does* represent *y* as a function of *x*.

(b) The graph of this equation is shown in Figure 8.20. Note that it is possible for a vertical line to intersect the graph at more than one point. Therefore, this equation *does not* represent *y* as a function of *x*.

Remember that if an equation represents *y* as a function of *x*, then no two different solution points of the equation can have the same *x*-value. For this equation, we see that the points $(-4, 1)$ and $(-4, -1)$ are both solution points of the equation. The fact that these two points have the same *x*-value is another way of seeing that the equation $x = y^2 - 5$ does not represent *y* as a function of *x*.

(c) At first glance, you might be tempted to say that this equation does not represent *y* as a function of *x* because *x* does not appear in the equation. However, in Figure 8.21, we see that the graph of this equation is simply a horizontal line (two units above the *x*-axis). Since this graph passes the vertical line test, we conclude that the equation

$$y = 2$$

does represent *y* as a function of *x*.

FIGURE 8.19

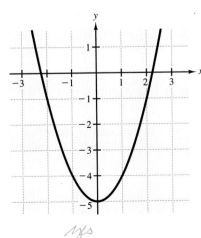

FIGURE 8.20

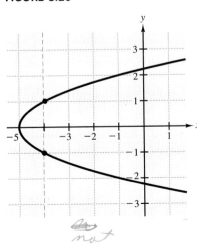

FIGURE 8.21

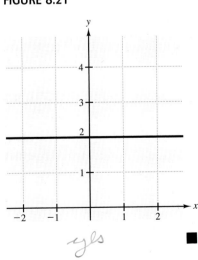

DISCUSSION PROBLEM ■ Interpreting a Graph

The graph in Figure 8.22 shows the speed of a delivery van on a 20-minute trip to deliver an order to a customer. Write a short paragraph that could describe the trip.

FIGURE 8.22

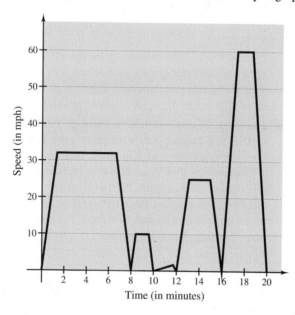

Time (in minutes)

Warm-Up

The following warm-up exercises involve skills that were covered in earlier sections. You will use these skills in the exercise set for this section.

In Exercises 1–4, plot the given points in a rectangular coordinate system.

1. $(-6, 4)$, $(-3, -4)$　　　　　　　**2.** $(4, 6)$, $(8, -2)$

3. $\left(\dfrac{7}{2}, \dfrac{9}{2}\right)$, $\left(\dfrac{4}{3}, -3\right)$　　　　　　**4.** $\left(-\dfrac{3}{4}, -\dfrac{7}{4}\right)$, $\left(-1, \dfrac{5}{2}\right)$

In Exercises 5 and 6, solve the given equation.

5. $\dfrac{5}{6}x - 7 = 0$　　　　　　　　**6.** $16 - \dfrac{2}{3}x = 0$

In Exercises 7–10, find the unknown coordinate of the solution point of the given equation.

7. $y = \dfrac{4}{5}x + 2$, $x = 15$　　　　　**8.** $y = 3 - \dfrac{5}{6}x$, $x = 12$

9. $y = 8 - 0.75x$, $y = -1$　　　　　**10.** $y = 2 + 0.6x$, $y = 4.4$

8.2   EXERCISES

In Exercises 1–4, complete the table and use the resulting solution points to sketch the graph of the equation.

1. $y = 9 - x$

x	-2	-1	0	1	2
y					
(x, y)					

2. $y = x - 1$

x	-2	-1	0	1	2
y					
(x, y)					

3. $x + 2y = 4$

x	-2	0	2	4	6
y					
(x, y)					

4. $3x - 2y = 6$

x	-2	0	2	4	6
y					
(x, y)					

In Exercises 5–18, find the x- and y-intercepts (if any) of the graph of the equation.

5. $x - y = 1$

6. $x + y = 10$

7. $2x + y - 4 = 0$

8. $3x - 2y + 6 = 0$

9. $y = \frac{3}{8}x + 15$

10. $y = 14 - \frac{2}{3}x$

11. $y = x^2 - 16$

12. $y = x^2 + 8x + 16$

13. $y = -4$

14. $x = 7$

15. $y = x^2 + 3$

16. $y = x^2 + 8$

17. $y = 2x^2 - x - 1$

18. $y = x(x - 3) - 5(x - 3)$

In Exercises 19–36, sketch the graph of the equation and show the coordinates of at least three solution points (including any intercepts) on the graph.

19. $y = 2 - x$

20. $y = x + 3$

21. $y = 2x - 1$

22. $y = -3x + 9$

23. $y = 3$

24. $x = 5$

25. $2x - 3y = 12$

26. $2x + 5y = 10$

27. $3x + 5y = 15$

28. $5x - 7y = 35$

29. $4x + y = 2$

30. $y - 2x = 3$

31. $x + 5 = 0$

32. $y - 8 = 0$

33. $y = x^2 - 4$

34. $y = x^2 - 2x$

35. $y = |x| - 4$

36. $y = |x - 4|$

In Exercises 37–42, use the vertical line test to determine whether y is a function of x.

37. $y = x - 2$

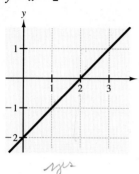

yes

38. $x + y = 4$

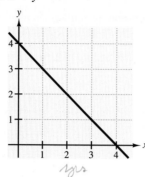

yes

39. $y = (x - 2)^2$

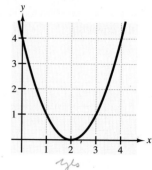

yes

40. $x - y^2 = 0$

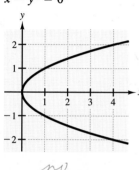

no

41. $x^2 + y^2 = 9$

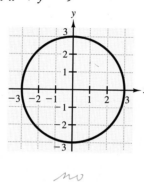

no

42. $y = x^3 - x$

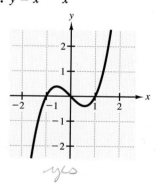

yes

In Exercises 43–48, sketch the graph of the equation. Use the vertical line test to determine whether y is a function of x.

43. $y = -3x + 2$

44. $8x - 6y = 24$

45. $y = x^2$

46. $y = x$

47. $y^2 = x - 1$

48. $|y| = x$

49. *Distance Traveled* Let y represent the distance traveled by a car that is moving at a constant speed of 35 miles per hour. Let t represent the number of hours that the car has been traveling. Write an equation that gives the distance y in terms of the time t and sketch the graph of the equation.

50. *Total Cost* The inventor of a new game estimates that the variable cost for producing the game is $0.95 per unit and the fixed costs are $6,000. Let C represent the total cost of producing x games. Write an equation that gives the total cost C in terms of the number of games produced x and sketch the graph of the equation.

SECTION

8.3

Graphing Calculators

Introduction • *Basic Graphing* • *Special Features*

SKIP

Introduction

In Section 8.2 we introduced the point-plotting method for sketching the graph of an equation. One of the disadvantages of the point-plotting method is that in order to get a good idea about the shape of a graph we need to plot *many* points. By plotting only a few points, we could badly misrepresent the graph. For instance, consider the equation

$$y = x^3.$$

Suppose we plotted only three points, $(-1, -1)$, $(0, 0)$, and $(1, 1)$, as shown in Figure 8.23(a). From these three points, a person might assume that the graph of the equation is a straight line. That, however, is not correct. By plotting several more points, we can see that the actual graph is not straight at all! [See Figure 8.23(b).]

FIGURE 8.23   (a)

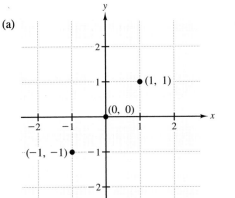

(b)

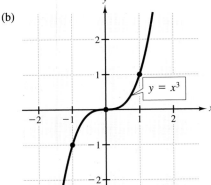

Thus, the point-plotting method leaves us with a dilemma. On the one hand, the method can be very inaccurate if only a few points are plotted. But, on the other hand, it is very time-consuming to plot a dozen (or more) points. Technology can help us solve this dilemma. Plotting several points (even hundreds of points) in a rectangular coordinate system is something that a computer or calculator can do easily.

The point-plotting method is used by *all* graphing packages for computers and *all* graphing calculators. Each computer or calculator screen is made up of a grid of hundreds or thousands of small areas called **pixels**. Screens that have many pixels per inch are said to have a higher **resolution** than screens that don't have as many. For instance, the screen shown in Figure 8.24(a) has a higher resolution than the screen shown in Figure 8.24(b). Note that the "graph" of the line on the first screen looks more like a line than the "graph" on the second screen.

FIGURE 8.24 (a)

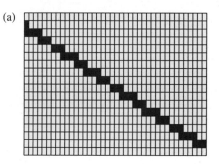

(b)

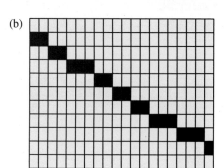

Screens on most graphing calculators have 48 pixels per inch, whereas screens on computer monitors typically have between 32 and 100 pixels per inch.

EXAMPLE 1 ■ Using Pixels to Sketch a Graph

Use the grid shown in Figure 8.25 to sketch a graph of the line $y = \frac{1}{2}x$. Each pixel on the grid must be either on (shaded black) or off (left unshaded).

Solution

To shade the grid we use the following rule. If a pixel contains a plotted point of the graph, then it will be "on;" otherwise the pixel will be "off." Using this rule, the graph of the line looks like that shown in Figure 8.26.

FIGURE 8.25

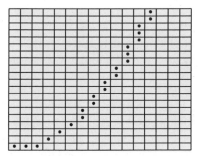

FIGURE 8.26

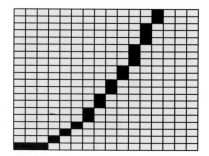

■

Basic Graphing

There are many different graphing packages for computers and graphing calculators you can buy. In this section we describe only one—the *TI-81* by *Texas Instruments*. The keys for the *TI-81* are shown in Figure 8.27. The keys on the top row are used for graphing. The steps used to sketch a simple graph are summarized in the following list.

Basic Graphing Steps for the *TI-81* Graphing Calculator	To sketch the graph of an equation involving x and y on the *TI-81*, use the following steps. (Before performing these steps, you should set your calculator so that all of the standard defaults are active. Press $\boxed{\text{ZOOM}}$, cursor to "Standard," and press $\boxed{\text{ENTER}}$.)

1. Rewrite the equation so that y is written as a function of x. In other words, rewrite the equation so that y is isolated on the left side of the equation.

2. Press the $\boxed{\text{Y=}}$ key. Then enter the right side of the equation on the first line of the display. (The first line is labeled as $Y_1=$.)

3. Press the $\boxed{\text{GRAPH}}$ key.

FIGURE 8.27

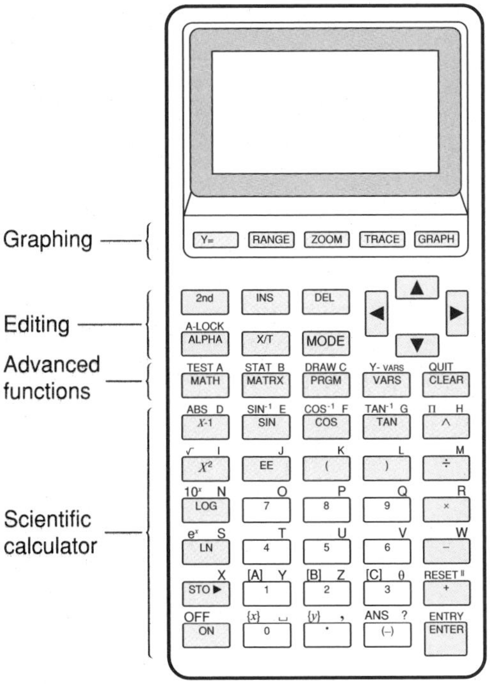

EXAMPLE 2   ■   **Sketching the Graph of a Linear Equation**

Sketch the graph of $2y + x = 4$.

Solution

To begin, we solve the given equation for y in terms of x.

$2y + x = 4$	Given equation
$2y = -x + 4$	Subtract x from both sides
$y = -\dfrac{1}{2}x + 2$	Divide both sides by 2

Next, after pressing the $\boxed{Y=}$ key, we enter the following keystrokes.

$\boxed{(-)}$ $\boxed{X|T}$ $\boxed{\div}$ 2 $\boxed{+}$ 2

The top row of the display should now be as follows.

$Y_1 = -X/2 + 2$

Now press the $\boxed{GRAPH}$ key, and the screen should look like the one shown in Figure 8.28.

FIGURE 8.28

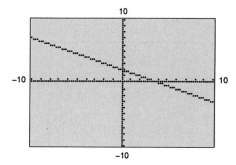

■

In Figure 8.28 notice that the calculator screen does not label the tic marks on the x-axis or the y-axis. To see what the tic marks represent, you can press $\boxed{RANGE}$. If you set your calculator to the standard graphing defaults before solving Example 2, the screen should be as follows.

RANGE	
Xmin=−10	The minimum x-value is −10.
Xmax=10	The maximum x-value is 10.
Xscl=1	The x-scale is one unit per tic mark.
Ymin=−10	The minimum y-value is −10.
Ymax=10	The maximum y-value is 10.
Yscl=1	The y-scale is one unit per tic mark.
Xres=1	The x-resolution is one plotted point per *one* pixel.

These settings are summarized visually in Figure 8.29. We suggest leaving the *x*-resolution as 1, although it can be set at any integer between 1 and 8. For instance, if you set the *x*-resolution as 2, then you will obtain one plotted point per *two* pixels. By doing this you will have increased the speed at which the graph is sketched, but you will have decreased the accuracy of the graph.

FIGURE 8.29

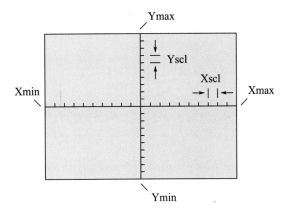

EXAMPLE 3 ■ Sketching the Graph of an Equation Involving Absolute Value

Sketch the graph of $y = |x - 3|$.

Solution

This equation is already written so that *y* is isolated on the left side of the equation. Thus, after pressing the [Y=] key, we enter the following keystrokes.

[ABS] [(] [X|T] [−] 3 [)]

The top row of the display should now be as follows.

$Y_1 = \text{abs}(X - 3)$

Now press the [GRAPH] key, and the screen should look like the one shown in Figure 8.30.

FIGURE 8.30 $y = |x - 3|$

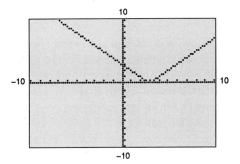

■

MATH MATTERS

The Speeds of Animals

The maximum speeds at which various animals can run, fly, or swim varies significantly. Here are several examples of approximate maximum speeds for several types of animals.

Mammal	Speed	Bird	Speed	Fish	Speed
Cheetah	60 mph	Falcon	225 mph	Sailfish	65 mph
Gazelle	50 mph	Eagle	96 mph	Tuna	46 mph
Racehorse	45 mph	Pigeon	60 mph	Dolphin	44 mph
Greyhound	40 mph	Swift	60 mph	Killer Whale	35 mph
Fox	40 mph	Swan	55 mph	Barracuda	30 mph
Kangaroo	40 mph	Martin	50 mph	Salmon	23 mph
Elephant	25 mph	Owl	40 mph	Trout	5 mph
Rat	6 mph	Robin	30 mph	Goldfish	4 mph

The fastest that humans can run a mile is about four minutes. How many miles an hour is this? (The answer is given in the back of the text.)

Special Features

In order to be able to use your graphing calculator to its best advantage, you must be able to use the RANGE key and the ZOOM key. The next two examples show how these keys can be used.

EXAMPLE 4   ■   Resetting the Scales on an Axis

Sketch the graph of $y = x^2 + 12$.

Solution

We begin as usual. Press Y= and enter $y = x^2 + 12$ on the first line.

X|T   x²   +   12

Press the GRAPH key, and you will notice that nothing appears on the screen (provided your calculator is set to standard defaults). The reason for this is that the lowest point on the graph of $y = x^2 + 12$ occurs at the point (0, 12). Using the standard range settings, we obtain a screen whose largest y-value is 10. In other words, none of the graph is visible on a screen that ranges between -10 and 10 for its x- and y-values, as shown in Figure 8.31(a).

To change these settings, press RANGE and change the Ymax=10 to Ymax=30. Then change the Yscl=1 to Yscl=5. Now press GRAPH, and you will obtain the graph shown in Figure 8.31(b). On this graph, note that each tic mark on the y-axis represents five units because we changed the y-scale to 5. Also note that the highest point on the y-axis is now 30 because we changed the maximum value of y to 30.

FIGURE 8.31 $y = x^2 + 12$

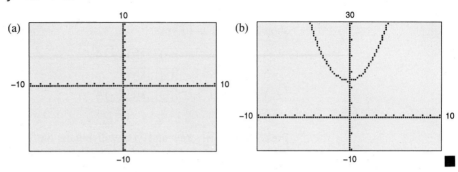

(a)

(b)

NOTE: If you changed the y-maximum and y-scale on your calculator as indicated in Example 4, you should return to the standard settings before working Example 5. To do this, press ZOOM , cursor to "Standard," and press ENTER .

EXAMPLE 5 ■ Using the Zoom Key

Sketch the graph of $y = x$. The graph of this equation is a straight line that makes a 45° angle with the x-axis and y-axis. From the graph on your calculator, does the angle appear to be 45°?

Solution

We begin as usual. Press Y= and enter $y = x$ on the first line. Press the GRAPH key and you will obtain the graph shown in Figure 8.32. Note that the angle the line makes with the x-axis doesn't appear to be 45°. The reason for this is that the screen is wider than it is tall. This has the effect of making the tic marks on the x-axis farther apart than the tic marks on the y-axis. To obtain the same distance between tic marks on both axes, we can change the graphing settings from "standard" to "square." To do this, press ZOOM , cursor to "Square," and press ENTER . The screen should look like that shown in Figure 8.33. Note in this figure that the square setting has changed the range settings so that the x-values vary between -15 and 15.

FIGURE 8.32 $y = x$ **FIGURE 8.33** $y = x$

EXAMPLE 6   ■   **Sketching More than One Graph on the Same Screen**

Sketch the graphs of $y = -x + 4$, $y = -x$, and $y = -x - 4$ on the same screen.

Solution

To begin, press $\boxed{Y=}$ and enter all three equations on the first three lines. The display should now be as follows.

$$Y_1 = -X + 4 \qquad \boxed{(-)} \ \boxed{X|T} \ \boxed{+} \ 4$$
$$Y_2 = -X \qquad\quad \boxed{(-)} \ \boxed{X|T}$$
$$Y_3 = -X - 4 \qquad \boxed{(-)} \ \boxed{X|T} \ \boxed{-} \ 4$$
$$Y_4 =$$

Press the $\boxed{\text{GRAPH}}$ key and you will obtain the graph shown in Figure 8.34. Note that the graph of each equation is a straight line, and that the lines are parallel to each other.

FIGURE 8.34 $y_1 = -x + 4$

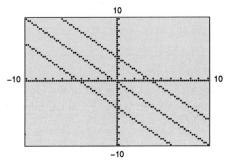

                                            ■

DISCUSSION PROBLEM   ■   **A Misleading Graph**

Set your calculator to its standard setting and sketch the graph of $y = x^2 - 12x$. The graph appears to be a straight line, as shown in Figure 8.35. However, this is misleading because the screen doesn't show some important portions of the graph. Can you find a range setting that reveals a better view of this graph?

FIGURE 8.35 $y = x^2 - 12x$

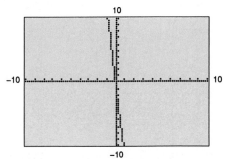

                                            ■

Warm-Up

The following warm-up exercises involve skills that were covered in earlier sections. You will use these skills in the exercise set for this section.

In Exercises 1–10, solve for y in terms of x.

1. $3x + y = 4$ **2.** $x - y = 0$

3. $5 - y + x = 0$ **4.** $4 - y + x = 0$

5. $2x + 2y = 10$ **6.** $3x - 3y = 15$

7. $2x + 3y = 2$ **8.** $4x - 5y = -2$

9. $3x + 4y - 5 = 0$ **10.** $-2x - 3y + 6 = 0$

8.3 EXERCISES

*In Exercises 1–20, use a graphing calculator to sketch the graph of the given equation. (Use a standard setting on each graph.)

```
RANGE
Xmin=-10
Xmax=10
Xscl=1
Ymin=-10
Ymax=10
Yscl=1
Xres=1
```

1. $y = x + 1$ **2.** $y = x - 4$ **3.** $y = -3x$ **4.** $y = -2x$

5. $y = -\dfrac{1}{3}x$ **6.** $y = \dfrac{1}{2}x$ **7.** $y = \dfrac{3}{4}x - 6$ **8.** $y = -3x + 2$

9. $y = \dfrac{1}{4}x^2$ **10.** $y = -\dfrac{2}{3}x^2$ **11.** $y = -2x^2 + 5$ **12.** $y = \dfrac{1}{8}x^2 - 2$

13. $y = x^2 - 4x + 2$ **14.** $y = -0.5x^2 - 2x + 2$ **15.** $y = -0.3x^2 + x + 8$ **16.** $y = x^2 + 0.5x - 7.5$

17. $y = |x - 3|$ **18.** $y = |x + 4|$ **19.** $y = |x + 1| - 2$ **20.** $y = 4 - |x - 2|$

---

*The symbol ▦ indicates exercises that are designed to be solved with a graphing calculator or some other graphing facility.

In Exercises 21–24, solve the equation for y in terms of x and use a graphing calculator to sketch the graph of the resulting equation. (Use a standard setting.)

21. $3x - 2y - 4 = 0$ **22.** $2x + 5y - 10 = 0$ **23.** $2y + x^2 - 4 = 0$ **24.** $x^2 - 3y - 6 = 0$

In Exercises 25–28, use a graphing calculator to sketch the graph of the given equation. (Use the indicated setting to sketch the graph.)

25. $y = 25 - 3x^2$ **26.** $y = 3x^2 + 2x - 3$ **27.** $y = 3x^2 - 9x$ **28.** $y = x^2 - 4x + 16$

RANGE
Xmin=−4
Xmax=4
Xscl=1
Ymin=−5
Ymax=25
Yscl=5
Xres=1

RANGE
Xmin=−4
Xmax=2
Xscl=1
Ymin=−5
Ymax=20
Yscl=2
Xres=1

RANGE
Xmin=−2
Xmax=4
Xscl=1
Ymin=−8
Ymax=8
Yscl=2
Xres=1

RANGE
Xmin=−3
Xmax=7
Xscl=1
Ymin=−5
Ymax=30
Yscl=5
Xres=1

In Exercises 29–32, find a setting on a graphing calculator so that the graph of the equation is approximately the same as the one in the accompanying figure.

29. $y = \frac{1}{2}x + 2$

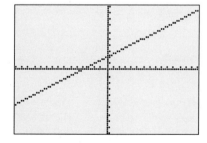

30. $y = 2x - 1$

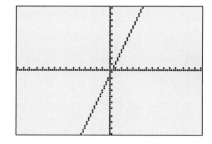

31. $y = \frac{1}{4}x^2 - 4x + 12$

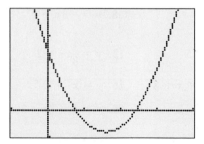

32. $y = 16 - 4x - x^2$

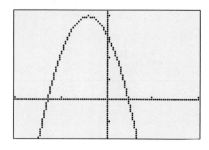

In Exercises 33 and 34, use a graphing calculator to determine the number of x-intercepts of the graph of the equation.

33. $y = -x^2 + 15$

34. $y = x^2 - 4x + 1$

In Exercises 35–38, use a graphing calculator to match the given equation with its graph. [The graphs are labeled (a), (b), (c), and (d).]

35. $y = x^2$

36. $y = 2x^2$

37. $y = \frac{1}{4}x^2$

38. $y = -\frac{1}{4}x^2$

(a)
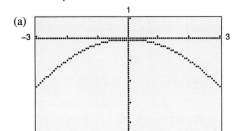

(b)
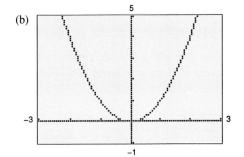

(c)
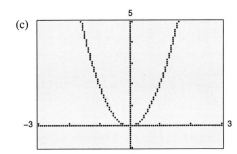

(d)
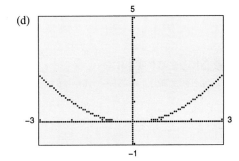

In Exercises 39–42, use a graphing calculator to sketch the graphs of the given equations on the same screen. Using the "square setting," determine the geometrical shape bounded by the graphs.

39. $y = -4$
$y = -|x|$

40. $y = |x|$
$y = 5$

41. $y = |x| - 8$
$y = -|x| + 8$

42. $y = -\frac{1}{2}x + 7$
$y = \frac{8}{3}(x + 5)$
$y = \frac{2}{7}(3x - 4)$

In Exercises 43 and 44, approximate the *x*-intercepts of the graph using the following method. After the graph is drawn, press the $\boxed{\text{TRACE}}$ key. Then press the left and right cursor keys until the cross hairs move near the point where the graph crosses the *x*-axis. Read the coordinates of the point from the bottom of the screen.

43. $y = x^2 - 3x - 1$

44. $y = 3x^2 - 8x + 2$

Number of Births and Deaths In Exercises 45 and 46, use the following models, which give the number of births and the number of deaths in the United States from 1960 to 1989.

$$y = 0.003x^2 - 0.112x + 4.302, \quad 0 \le x \le 29, \quad \text{Births}$$

$$y = 0.0002x^2 + 0.011x + 1.722, \quad 0 \le x \le 29, \quad \text{Deaths}$$

In these models, *x* is the year with $x = 0$ corresponding to 1960 and *y* is the number of births and deaths in millions. (*Source:* National Center for Health Statistics.)

45. Use a graphing calculator to sketch the graph of the equations using the following settings.

Calculator settings for 45

```
RANGE
Xmin=0
Xmax=30
Xscl=5
Ymin=1
Ymax=5
Yscl=1
Xres=1
```

46. In viewing the graphs of the equations, what conclusions can be made about the rate of growth of the population of the United States during the years 1960–1975? (This does not show any changes in population due to immigration.)

SECTION 8.4

The Slope of a Line

The Slope of a Line • Slope: An Aid to Sketching Lines • Parallel and Perpendicular Lines

The Slope of a Line

The **slope** of a nonvertical line is the number of units the line rises or falls vertically for each unit of horizontal change from left to right. For example, the line in Figure 8.36 rises two units for each unit of horizontal change from left to right, and we say that this line has a slope of $m = 2$.

As indicated in the following definition, we usually need two points on a line to determine the slope of the line.

Definition of the Slope of a Line	The **slope** m of the nonvertical line passing through the points (x_1, y_1) and (x_2, y_2) is

$$m = \frac{y_2 - y_1}{x_2 - x_1} = \frac{\text{Change in } y}{\text{Change in } x}$$

where $x_1 \ne x_2$ (see Figure 8.37).

FIGURE 8.36

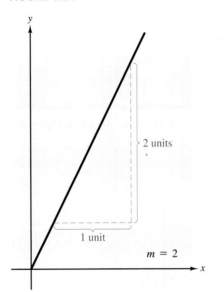

$m = 2$

FIGURE 8.37

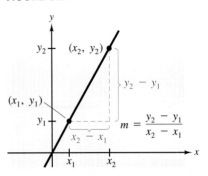

$$m = \frac{y_2 - y_1}{x_2 - x_1}$$

When the formula for slope is used, the *order of subtraction* is important. Given two points on a line, we are free to label either of them as (x_1, y_1), and the other as (x_2, y_2). However, once this is done, we must form the numerator and denominator using the same order of subtraction.

$$m = \frac{y_2 - y_1}{x_2 - x_1} \qquad m = \frac{y_1 - y_2}{x_1 - x_2} \qquad m = \frac{y_2 - y_1}{x_1 - x_2}$$

      Correct   Correct   Incorrect

EXAMPLE 1 ■ **Finding the Slope of a Line Passing Through Two Points**

Find the slopes of the lines passing through the following pairs of points.

(a) $(-2, 0)$ and $(3, 1)$ (b) $(-1, 2)$ and $(2, 2)$ (c) $(0, 0)$ and $(1, -1)$

Solution

(a) The slope of the line through $(x_1, y_1) = (-2, 0)$ and $(x_2, y_2) = (3, 1)$ is

$$m = \frac{y_2 - y_1}{x_2 - x_1} \qquad \Longleftarrow \quad \text{Difference in } y\text{-values}$$
$$\Longleftarrow \quad \text{Difference in } x\text{-values}$$

$$= \frac{1 - 0}{3 - (-2)}$$

$$= \frac{1}{3 + 2}$$

$$= \frac{1}{5}.$$

(b) The slope of the line through $(-1, 2)$ and $(2, 2)$ is

$$m = \frac{2 - 2}{2 - (-1)} = \frac{0}{3} = 0.$$

(c) The slope of the line through $(0, 0)$ and $(1, -1)$ is

$$m = \frac{-1 - 0}{1 - 0} = \frac{-1}{1} = -1.$$

The graphs of the three lines are shown in Figure 8.38.

FIGURE 8.38

(a)

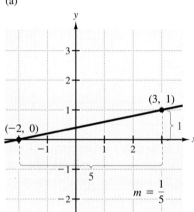

(b)

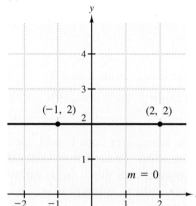

(c)

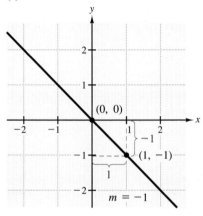

Note that the definition of slope does not apply to vertical lines. For instance, consider the points $(2, 4)$ and $(2, 1)$ on the vertical line shown in Figure 8.39. Applying the formula for slope, we have

$$\frac{4 - 1}{2 - 2} = \frac{3}{0}. \qquad \text{Undefined division by zero}$$

Since division by zero is not defined, we do not define the slope of a vertical line.

From the slopes of the lines shown in Figures 8.38 and 8.39, we make the following generalizations about the slope of a line.

1. A line with positive slope $(m > 0)$ *rises* from left to right.

2. A line with negative slope $(m < 0)$ *falls* from left to right.

3. A line with zero slope $(m = 0)$ is *horizontal*.

4. A line with undefined slope is *vertical*.

FIGURE 8.39

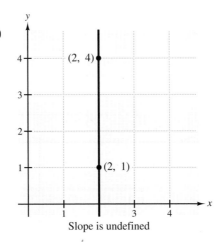

Slope is undefined

EXAMPLE 2   ■   Using Slope to Describe Lines

Use slope to determine whether the line through each of the following pairs *rises*, *falls*, is *horizontal*, or is *vertical*.

(a) $(3, -2)$, $(3, 3)$ (b) $(-2, 5)$, $(1, 4)$

(c) $(-4, -3)$, $(0, -3)$ (d) $(1, 0)$, $(4, 6)$

Solution

(a) Since

$$m = \frac{3 - (-2)}{3 - 3} = \frac{5}{0}$$ Undefined slope

is undefined, the line is vertical.

(b) Since

$$m = \frac{4 - 5}{1 - (-2)} = -\frac{1}{3} < 0$$ Negative slope

the line falls (from left to right).

(c) Since

$$m = \frac{-3 - (-3)}{0 - (-4)} = \frac{0}{4} = 0$$ Zero slope

the line is horizontal.

(d) Since

$$m = \frac{6 - 0}{4 - 1} = \frac{6}{3} = 2 > 0$$ Positive slope

the line rises (from left to right).

The graphs of these four lines are shown in Figure 8.40.

FIGURE 8.40

(a) Vertical line
   Undefined slope

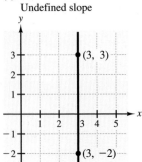

(b) Line falls
   Negative slope

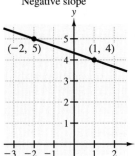

(c) Horizontal line
   Zero slope

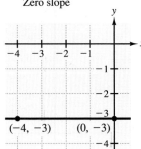

(d) Line rises
   Positive slope

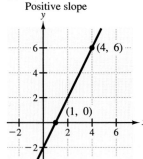

Any two points on a nonvertical line can be used to calculate its slope. This is demonstrated in the next example.

EXAMPLE 3 ■ Finding the Slope of a Line

Sketch the graph of the line given by the equation

$$3x - 2y = 4.$$

Then find the slope of the line. (Choose two different pairs of points on the line and show that the same slope is obtained from either pair.)

Solution

We begin by solving the given equation for y.

$$y = \frac{3}{2}x - 2$$

Then we construct the table of values shown in Table 8.11.

TABLE 8.11

x	-2	0	2	4
$y = \frac{3}{2}x - 2$	-5	-2	1	4
Solution Points	$(-2, -5)$	$(0, -2)$	$(2, 1)$	$(4, 4)$

From the solution points shown in the table, we sketch the graph of the line, as shown in Figure 8.41. To calculate the slope of the line using two different sets of points, we first use the points $(-2, -5)$ and $(0, -2)$, and obtain a slope of

$$m = \frac{-2 - (-5)}{0 - (-2)} = \frac{3}{2}.$$

Next, we use the points $(2, 1)$ and $(4, 4)$ to obtain a slope of

$$m = \frac{4 - 1}{4 - 2} = \frac{3}{2}.$$

Try some other pairs of points on the line to see that you obtain a slope of $m = \frac{3}{2}$ regardless of which two points you use.

FIGURE 8.41

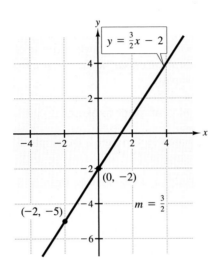

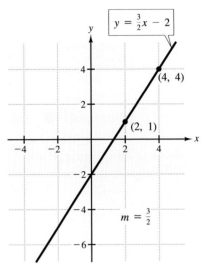

Slope: An Aid to Sketching Lines

We have seen that, before making up a table of values for an equation, we should first solve the equation for y. When we do this for a linear equation, we obtain some very useful information. Consider the results of Example 3.

$$3x - 2y = 4 \qquad \text{Given equation}$$
$$-2y = -3x + 4$$
$$y = \frac{3}{2}x - 2 \qquad \text{Solve for } y$$

Observe that the coefficient of x is the slope of the graph for this equation (see Example 3). Moreover, the constant term, -2, gives the y-intercept of the graph.

$$y = \boxed{\frac{3}{2}} \, x + \boxed{-2}$$

  ↑   ↑
 Slope y-intercept $(0, -2)$

We call this the **slope-intercept** form of the equation of the line.

Slope-Intercept Form of the Equation of a Line	The graph of the equation

$$y = mx + b$$

is a line whose slope is m and whose y-intercept is $(0, b)$.

So far, we have been plotting several points in order to sketch the equation of a line. However, now that we can recognize equations of lines, we don't have to plot as many points—two points are enough. (You might remember from geometry that *two points are all that are necessary to determine a line*.) We demonstrate this technique in the next two examples.

EXAMPLE 4   ■   Using the Slope and y-Intercept to Sketch a Line

Use the slope and y-intercept to sketch the graph of $y = \frac{1}{3}x + 2$.

Solution

The equation is already in slope-intercept form.

$$y = mx + b$$

$$y = \frac{1}{3}x + 2$$

Thus, we see that the slope of the line is $m = \frac{1}{3}$ and the y-intercept is $(0, b) = (0, 2)$. Knowing this information, we can sketch the graph of the line as follows. First, we plot the y-intercept, as shown in Figure 8.42(a). Then, using a slope of $\frac{1}{3}$,

$$m = \frac{1}{3} = \frac{\text{Change in } y}{\text{Change in } x},$$

we locate a second point on the line by moving three units to the right and one unit up (or one unit up and three units to the right), as shown in Figure 8.42(b). Finally, we obtain the graph by drawing the line that passes through the two points.

FIGURE 8.42   (a)

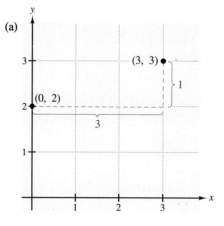

(b)

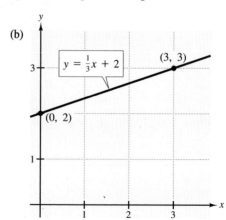

■

EXAMPLE 5 ■ Using the Slope and *y*-Intercept to Sketch a Line

Use the slope and *y*-intercept to sketch the graph of the following equation.

$$5x + 2y - 3 = 0$$

Solution

We begin by writing the equation in slope-intercept form.

$$5x + 2y - 3 = 0 \qquad \text{Given equation}$$

$$2y = -5x + 3$$

$$y = -\frac{5}{2}x + \frac{3}{2} \qquad \text{Slope-intercept form}$$

From this form, we see that the slope is $m = -\frac{5}{2}$ and the *y*-intercept is $(0, b) = \left(0, \frac{3}{2}\right)$. Thus, to sketch the line, we first plot the *y*-intercept, as shown in Figure 8.43(a). Then, using a slope of $-\frac{5}{2}$,

$$m = \frac{-5}{2} = \frac{\text{Change in } y}{\text{Change in } x},$$

we locate a second point on the line by moving two units to the right and five units down, as shown in Figure 8.43(b). (Note that we moved down because the slope is negative.)

FIGURE 8.43 (a)

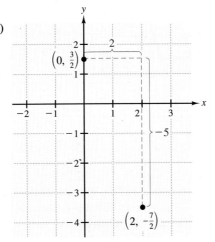

(b)

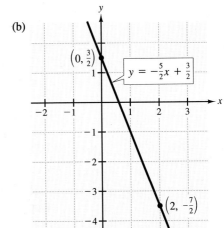

Parallel and Perpendicular Lines

We know from geometry that two lines in a plane are *parallel* if they do not intersect. What this means in terms of their slopes is suggested by the following example.

EXAMPLE 6 ■ Lines that Have the Same Slope

On the same set of axes, sketch the two lines given by the following equations.

$$y = 3x \quad \text{and} \quad y = 3x - 4$$

Solution

For the line given by

$$y = 3x$$

the slope is $m = 3$ and the y-intercept is $(0, 0)$. For the line

$$y = 3x - 4$$

the slope is also $m = 3$ and the y-intercept is $(0, -4)$. The graphs of these two lines are shown in Figure 8.44.

FIGURE 8.44

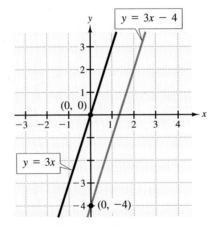

In Example 6 notice that the two lines have the same slope *and* that the two lines appear to be parallel. The following rule tells us that this is always the case. That is, two (nonvertical) lines are parallel *if and only if* they have the same slope.

Parallel Lines	Two distinct nonvertical lines are parallel if and only if they have the same slope.

NOTE: The phrase "if and only if" in this rule is used in mathematics as a way of writing two statements in one. The first statement says that *if two distinct nonvertical lines have the same slope, then they must be parallel*. The second statement says that *if two distinct nonvertical lines are parallel, then they must have the same slope*.

Another rule resulting from geometry is that two lines in a plane are *perpendicular* if they intersect at right (90°) angles. In terms of their slopes, this means that two nonvertical lines are perpendicular if their slopes are negative reciprocals of each other. For instance, the line given by

$$y = 2x$$

has a slope of $m_1 = 2$, and the line given by

$$y = -\frac{1}{2}x - 3$$

has a slope of $m_2 = -\frac{1}{2}$. Note that the slope of the second line is the negative reciprocal of the slope of the first line. That is,

$$m_2 = -\frac{1}{2} = -\frac{1}{m_1}.$$

From Figure 8.45, it appears that the two lines are perpendicular. We generalize this result in the following rule.

FIGURE 8.45

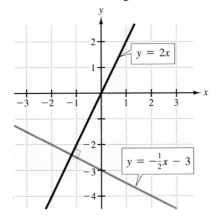

Perpendicular Lines

Consider two nonvertical lines whose slopes are m_1 and m_2. The two lines are perpendicular if and only if their slopes are *negative reciprocals* of each other. That is,

$$m_1 = -\frac{1}{m_2}.$$

EXAMPLE 7 ■ Determining Whether Lines are Parallel or Perpendicular

Determine whether the following pairs of lines are parallel, perpendicular, or neither.

(a) Line 1: $y = -3x - 2$ Line 2: $y = \frac{1}{3}x + 1$

(b) Line 1: $y = \frac{1}{2}x + 1$ Line 2: $y = \frac{1}{2}x - 1$

Solution

(a) The first line has a slope of $m_1 = -3$, and the second line has a slope of $m_2 = \frac{1}{3}$. Since these slopes are negative reciprocals of each other, the two lines must be perpendicular, as shown in Figure 8.46.

(b) Each of these two lines has a slope of $m = \frac{1}{2}$. Therefore, the two lines must be parallel, as shown in Figure 8.47.

FIGURE 8.46

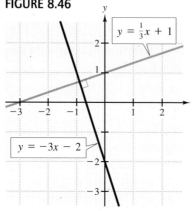

$y = \frac{1}{3}x + 1$

$y = -3x - 2$

FIGURE 8.47

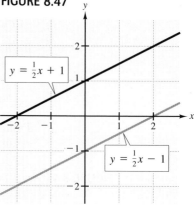

$y = \frac{1}{2}x + 1$

$y = \frac{1}{2}x - 1$

DISCUSSION PROBLEM ■ You Be the Instructor

Suppose you are teaching an algebra class and are writing a multiple choice test. One question asks students to find the slope of the line through the points $(2, -1)$ and $(-3, 4)$. Make up four "answers." Make one of the answers correct and make the other three represent common errors that you think a student might make. ■

Warm-Up

The following warm-up exercises involve skills that were covered in earlier sections. You will use these skills in the exercise set for this section.

In Exercises 1–4, evaluate the given expression.

1. $\dfrac{4 - 2}{7 - 3}$ **2.** $\dfrac{-2 - (-5)}{10 - 1}$ **3.** $\dfrac{4 - (-5)}{-3 - (-1)}$ **4.** $\dfrac{-5 - 8}{0 - (-3)}$

In Exercises 5 and 6, find $-1/m$ for the given value of m.

5. $m = -\dfrac{5}{4}$ **6.** $m = \dfrac{7}{8}$

In Exercises 7–10, solve for y in terms of x.

7. $2x - 3y = 5$ **8.** $4x + 2y = 0$

9. $y - (-4) = 3[x - (-1)]$ **10.** $y - 7 = \dfrac{2}{3}(x - 3)$

EXERCISES

In Exercises 1–6, estimate the slope of the given line from its graph.

1.

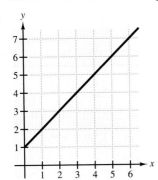

2.

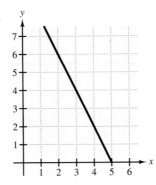

3.

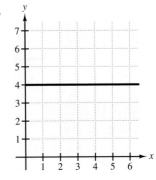

4.

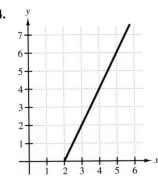

5.

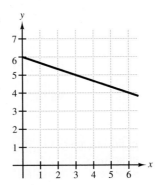

6.

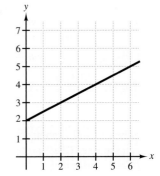

In Exercises 7 and 8, match the line in the figure with its slope.

7. (a) $m = \dfrac{3}{2}$

(b) $m = 0$

(c) $m = -2$

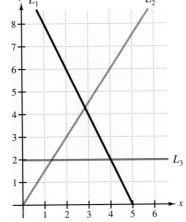

8. (a) $m = -\dfrac{3}{4}$

(b) m is undefined

(c) $m = 3$

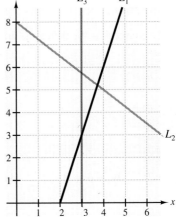

In Exercises 9–26, plot the points and find the slope (if possible) of the line passing through each pair of points. State whether the line through the pair of points rises, falls, is horizontal, or is vertical.

9. (0, 0), (4, 5)

10. (0, 0), (−1, −3)

11. (−3, −2), (1, 6)

12. (0, −6), (8, 0)

13. (0, 6), (8, 0)

14. (2, 4), (4, −4)

15. (−6, −1), (−6, 4)

16. (0, −10), (−4, 0)

17. (3, −4), (8, −4)

18. (1, 2), (−2, −2)

19. $\left(\frac{1}{4}, \frac{3}{2}\right)$, $\left(\frac{9}{2}, -3\right)$

20. $\left(-\frac{5}{4}, -\frac{1}{4}\right)$, $\left(\frac{7}{8}, \frac{3}{4}\right)$

21. (3.2, −1), (−3.2, 4)

22. (1.4, 0), (1.4, 3)

23. (3.5, −1), (5.75, 4.25)

24. (0, 6.4), (5, 6.4)

25. $(a, 3), (4, 3), \quad a \neq 4$

26. $(4, a), (4, 2), \quad a \neq 2$

In Exercises 27–30, find the unknown coordinate so that the line through the two points will have the given slope.

Points	Slope		Points	Slope

27. (3, 4), (x, 6) $m = -\frac{2}{5}$

28. (x, −2), (4, 0) $m = \frac{2}{3}$

29. (−3, y), (8, 2) $m = \frac{3}{4}$

30. (3, −2), (0, y) $m = -8$

In Exercises 31–40, a point on a line and the slope of the line are given. Find two additional points on the line. (Each problem has many correct answers.)

31. (2, 1), $m = 0$

32. (−3, 4), m is undefined

33. (1, −6), $m = 2$

34. (−2, −4), $m = 1$

35. (0, 1), $m = -2$

36. (−2, 4), $m = -3$

37. (−4, 0), $m = \frac{2}{3}$

38. (−1, −1), $m = -\frac{1}{4}$

39. (−8, 1), m is undefined

40. (−3, −1), $m = 0$

In Exercises 41–46, sketch the graph of the line through the point (0, 2) having the given slope.

41. $m = 3$

42. $m = 0$

43. m is undefined

44. $m = -1$

45. $m = -\frac{2}{3}$

46. $m = \frac{3}{4}$

In Exercises 47–50, plot the x- and y-intercepts and sketch the graph of the line.

47. $2x - 3y + 6 = 0$

48. $3x + 4y + 12 = 0$

49. $-5x + 2y - 10 = 0$

50. $3x - 7y - 14 = 0$

In Exercises 51–60, write the given equation in slope-intercept form. Use the slope and
y-intercept to sketch the graph of the line. (See Examples 4 and 5 earlier in this section.)

51. $2x - y - 3 = 0$ **52.** $x - y + 2 = 0$ **53.** $x + y = 0$ **54.** $x - y = 0$

55. $x + 2y - 2 = 0$ **56.** $3x - 2y - 2 = 0$ **57.** $3x - 4y + 2 = 0$ **58.** $10x + 6y - 3 = 0$

59. $y - 3 = 0$ **60.** $y + 5 = 0$

In Exercises 61–64, determine whether the lines L_1 and L_2 passing through the given pairs
of points are parallel, perpendicular, or neither.

61. L_1: $(0, -1)$, $(5, 9)$
L_2: $(0, 3)$, $(4, 1)$

62. L_1: $(-2, -1)$, $(1, 5)$
L_2: $(1, 3)$, $(5, -5)$

63. L_1: $(3, 6)$, $(-6, 0)$
L_2: $(0, -1)$, $\left(5, \frac{7}{3}\right)$

64. L_1: $(4, 8)$, $(-4, 2)$
L_2: $(3, -5)$, $\left(-1, \frac{1}{3}\right)$

In Exercises 65–68, sketch the graphs of the two lines on the same rectangular coordinate
system. Determine whether the lines are parallel, perpendicular, or neither.

65. L_1: $y = 2x - 3$
L_2: $y = 2x + 1$

66. L_1: $y = -\frac{1}{3}x - 3$
L_2: $y = -\frac{1}{3}x + 1$

67. L_1: $y = 2x - 3$
L_2: $y = -\frac{1}{2}x + 1$

68. L_1: $y = -\frac{1}{3}x - 3$
L_2: $y = 3x + 1$

69. *Stock Earnings* The accompanying graph gives the
earnings per share of common stock for TCBY Enterprises, Inc. (The Country's Best Yogurt) for the years
1983 through 1989. Find the slope between the points
for consecutive years. In what year(s) did earnings
increase (a) most slowly, and (b) most rapidly? (*Source:*
TCBY Enterprises, Inc.)

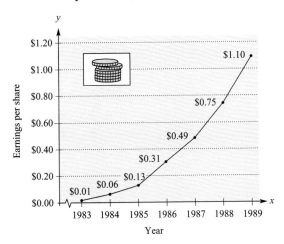

70. *Stock Dividends* The accompanying graph gives the
declared dividends per share of common stock for Wm.
Wrigley Jr. Company for 1983 through 1989. Find the
slope between the points for consecutive years. In what
year(s) was the increase in dividends (a) the greatest,
and (b) the least? (*Source:* Wm. Wrigley Jr. Co.)

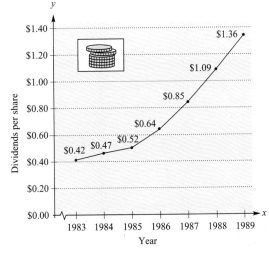

71. Whenever a quantity y is increasing or decreasing at a
constant rate with respect to time t, the graph of y versus
t is a line. What is the rate of change called?

Equations of Lines
The Point-Slope Equation of a Line   •   Equations of Horizontal and Vertical Lines   •   Applications

The Point-Slope Equation of a Line

There are two basic types of problems in analytic geometry.

1. Given an algebraic equation, sketch its geometric graph.

2. Given some geometric conditions about a graph, construct its algebraic equation.

So far in this chapter we have been working primarily with the first type of problem. In this section we look at the second problem. Specifically, if you know the slope of a line *and* you know the coordinates of one point on the line, then you can find an equation for the line. Before giving a general formula for doing this, let's look at an example.

EXAMPLE 1   ■   Finding an Equation of a Line Given Its Slope and a Point on the Line

A line has a slope of $\frac{5}{3}$ and passes through the point (2, 1). Find an equation for this line.

Solution

Using the methods of the previous section, we sketch the graph of the line, as shown in Figure 8.48. The slope of a line is the same through any two points on the line. Thus, to find an equation of the line, we let (x, y) represent *any* point on the line. Now, using the representative point (x, y) and the given point (2, 1), it follows that the slope of the line is given by

$$m = \frac{y - 1}{x - 2}.$$

◁ Difference in *y*-values
◁ Difference in *x*-values

FIGURE 8.48

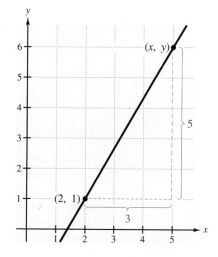

Since the slope of the line is given to be $m = \frac{5}{3}$, we can rewrite this equation as follows.

$$\frac{5}{3} = \frac{y - 1}{x - 2}$$ Slope formula

$$5(x - 2) = 3(y - 1)$$ Cross multiply

$$5x - 10 = 3y - 3$$

$$5x - 3y = 7$$ Equation of line

Therefore, we have found an equation for the line. It is $5x - 3y = 7$.   ■

The procedure in Example 1 can be used to derive a *formula* for the equation of a line, given its slope and a point on the line. In Figure 8.49, let (x_1, y_1) be a given point on the line whose slope is m. If (x, y) is any *other* point on the line, then it follows that

$$\frac{y - y_1}{x - x_1} = m.$$

This equation in variables x and y can be rewritten in the form

$$y - y_1 = m(x - x_1)$$

which is called the **point-slope form** of the equation of a line.

FIGURE 8.49

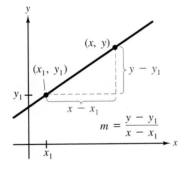

Point-Slope Form of the Equation of a Line

The **point-slope form** of the equation of the line with slope m and passing through the point (x_1, y_1) is

$$y - y_1 = m(x - x_1).$$

EXAMPLE 2   ■   The Point-Slope Form of the Equation of a Line

Find an equation of the line with slope 3 and passing through the point $(1, -2)$.

Solution

Using the point-slope form with $(x_1, y_1) = (1, -2)$ and $m = 3$, we have the following.

$$y - y_1 = m(x - x_1)$$ Point-slope form

$$y - (-2) = 3(x - 1)$$ Substitute $y_1 = -2$, $x_1 = 1$, and $m = 3$

$$y + 2 = 3x - 3$$

$$y = 3x - 5$$ Equation of line

The graph of this line is shown in Figure 8.50.

FIGURE 8.50

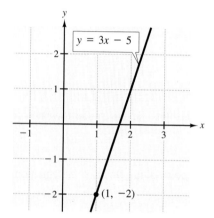

The point-slope form can be used to find the equation of a line passing through two points (x_1, y_1) and (x_2, y_2). First, we use the formula for the slope of the line passing through two points.

$$m = \frac{y_2 - y_1}{x_2 - x_1}$$

Then, once we know the slope, we use the point-slope form to obtain the equation

$$y - y_1 = \frac{y_2 - y_1}{x_2 - x_1}(x - x_1).$$

This is sometimes called the **two-point form** of the equation of a line.

EXAMPLE 3   ■   Finding an Equation of a Line Passing Through Two Points

Find an equation of the line that passes through the points $(3, 1)$ and $(-3, 4)$.

Solution

If we let $(x_1, y_1) = (3, 1)$ and $(x_2, y_2) = (-3, 4)$, then we can apply the formula for the slope of a line passing through two points as follows.

$$m = \frac{y_2 - y_1}{x_2 - x_1} = \frac{4 - 1}{-3 - 3} = \frac{3}{-6} = -\frac{1}{2}$$

Now, using the point-slope form, we find the equation of the line.

$$y - y_1 = m(x - x_1)$$

$$y - 1 = -\frac{1}{2}(x - 3)$$

$$y - 1 = -\frac{1}{2}x + \frac{3}{2}$$

$$y = -\frac{1}{2}x + \frac{5}{2}$$

The graph of this line is shown in Figure 8.51.

FIGURE 8.51

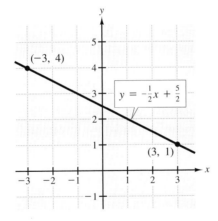

In Example 3 it did not matter which of the two points we labeled as (x_1, y_1) and (x_2, y_2). Try switching these labels to $(x_1, y_1) = (-3, 4)$ and $(x_2, y_2) = (3, 1)$ and reworking the problem to see that you obtain the same equation.

Equations of Horizontal and Vertical Lines

From the slope-intercept form of the equation of a line, we see that a horizontal line $(m = 0)$ has an equation of the form

$$y = (0)x + b \quad \text{or} \quad y = b. \qquad \text{Horizontal line}$$

This is consistent with the fact that each point on a horizontal line through $(0, b)$ has a y-coordinate of b, as shown in Figure 8.52.

In a similar way, each point on a vertical line through $(a, 0)$ has an x-coordinate of a, as shown in Figure 8.53. Hence, a vertical line has an equation of the form

$x = a$.   Vertical line

FIGURE 8.52

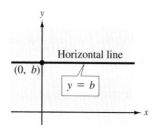

FIGURE 8.53

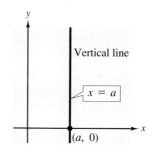

The equation of a vertical line cannot be written in the slope-intercept form because the slope of a vertical line is undefined. However, *every* line has an equation that can be written in the **general form**

$ax + by + c = 0$ General form

where a and b are not *both* zero.

For convenience, we summarize our discussion of lines in the following list.

Summary of Equations of Lines		
	1. Slope of line through $(x_1\ y_1)$ and (x_2, y_2):	$m = \dfrac{y_2 - y_1}{x_2 - x_1}$
	2. General form of equation of line:	$ax + by + c = 0$
	3. Equation of vertical line:	$x = a$
	4. Equation of horizontal line:	$y = b$
	5. Slope-intercept form of equation of line:	$y = mx + b$
	6. Point-slope form of equation of line:	$y - y_1 = m(x - x_1)$
	7. Parallel lines have *equal* slopes:	$m_1 = m_2$
	8. Perpendicular lines have *negative reciprocal* slopes:	$m_2 = -\dfrac{1}{m_1}$

EXAMPLE 4   ■   **Writing Equations of Horizontal and Vertical Lines**

Write an equation for each of the following lines.

(a) Vertical line through $(-3, 2)$

(b) Line passing through $(-1, 2)$ and $(4, 2)$

(c) Line passing through $(0, 2)$ and $(0, -2)$

Solution

(a) Since the line is vertical and passes through the point $(-3, 2)$, we know that every point on the line has an x-coordinate of -3. Therefore, the equation of the line is

$$x = -3.$$

(b) The line through $(-1, 2)$ and $(4, 2)$ is horizontal. Thus, its equation is

$$y = 2.$$

(c) The line through $(0, 2)$ and $(0, -2)$ is vertical. Thus, its equation is

$$x = 0. \qquad ■$$

EXAMPLE 5   ■   **Equations of Parallel Lines**

Find an equation of the line that passes through the point $(2, -1)$ and is parallel to the line $2x - 3y = 5$, as shown in Figure 8.54.

Solution

To begin, we write the given equation in slope-intercept form.

$$2x - 3y = 5 \qquad \text{Given equation}$$
$$-3y = -2x + 5$$
$$3y = 2x - 5$$
$$y = \frac{2}{3}x - \frac{5}{3} \qquad \text{Slope-intercept form}$$

Therefore, the given line has a slope of $m = \frac{2}{3}$. Since any line parallel to the given line must also have a slope of $\frac{2}{3}$, the required line through $(2, -1)$ has the following equation.

$$y - y_1 = m(x - x_1)$$

$$y - (-1) = \frac{2}{3}(x - 2)$$

$$y + 1 = \frac{2}{3}x - \frac{4}{3}$$

$$y = \frac{2}{3}x - \frac{7}{3}$$

FIGURE 8.54

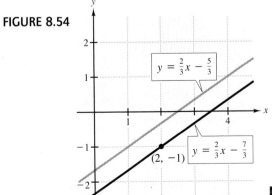

EXAMPLE 6   ■   Equations of Perpendicular Lines

Find an equation of the line that passes through the point $(2, -1)$ and is perpendicular to the line $2x - 3y = 5$, as shown in Figure 8.55.

FIGURE 8.55

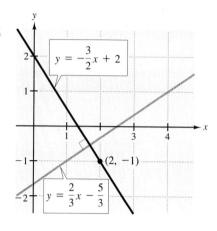

$$y = -\frac{3}{2}x + 2$$

$$y = \frac{2}{3}x - \frac{5}{3}$$

$(2, -1)$

Solution

From Example 5, the given line has slope $\frac{2}{3}$. Hence, any line perpendicular to this line must have a slope of $-\frac{3}{2}$. $\left(\text{Note that } -\frac{3}{2} \text{ is the negative reciprocal of } \frac{2}{3}.\right)$ Therefore, the equation of the required line through $(2, -1)$ has the following form.

$$y - y_1 = m(x - x_1)$$

$$y - (-1) = -\frac{3}{2}(x - 2)$$

$$y + 1 = -\frac{3}{2}x + 3$$

$$y = -\frac{3}{2}x + 2$$

■

EXAMPLE 7   ■   A Graphing Calculator Problem

Use a graphing calculator to sketch the graphs of the lines given by

$$y = x + 1 \quad \text{and} \quad y = -x + 3.$$

Sketch both graphs on the same display screen. The lines are supposed to be perpendicular (they have slopes of $m_1 = 1$ and $m_2 = -1$). Do they appear to be perpendicular on the calculator display?

Solution

If the zoom feature of your calculator is set on *standard*, then the tic marks on both the x-axis and y-axis will vary between -10 and 10, as shown in Figure 8.56. However, because the display screen of the calculator is not square, the lines will *not* appear to be perpendicular. In other words, the graphs of two perpendicular lines will only appear to be perpendicular if the tic marks on the *x*-axis have the same spacing as the tic marks on the *y*-axis. To correct this problem, set the zoom feature on your calculator to *square*, press ENTER, and draw the graph again. You should obtain the graph shown in Figure 8.57.

FIGURE 8.56 $y_1 = x + 1, y_2 = -x + 3$

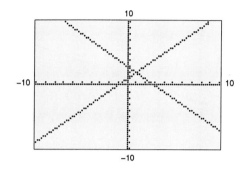

FIGURE 8.57 $y_1 = x + 1, y_2 = -x + 3$

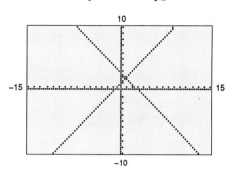

Applications

Linear equations are used frequently as mathematical models in business. The next example gives you one idea of how useful such a model can be.

EXAMPLE 8 ■ An Application: Total Sales

During the first year of operation, a company had total sales of $146 million. During the second year, the company had total sales of $154 million. Using this information only, what would you estimate the total sales to be during the third year?

Solution

FIGURE 8.58

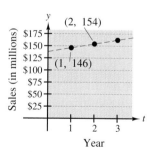

To solve this problem, we will use a *linear model*, with *y* representing the total sales and *t* representing the year. That is, in Figure 8.58 we let (1, 146) and (2, 154) be two points on the line representing the total sales for the company. The slope of the line passing through these points is

$$m = \frac{154 - 146}{2 - 1} = 8.$$

Now, using the point-slope form, we find the equation of the line as follows.

$$y - y_1 = m(t - t_1)$$
$$y - 146 = 8(t - 1)$$
$$y = 8t + 138$$

Finally, we estimate the total sales during the third year ($t = 3$) to be

$$y = 8(3) + 138 = \$162 \text{ million.}$$

The estimation method illustrated in Example 8 is called **linear extrapolation**. Note in Figure 8.59 that for linear extrapolation, the estimated point lies *to the right* of the given points. When the estimated point lies *between* two given points, we call the procedure **linear interpolation**, as also shown in Figure 8.59.

FIGURE 8.59   **Linear Extrapolation**

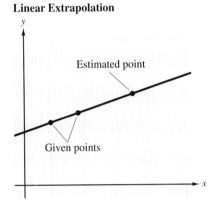

Linear Interpolation

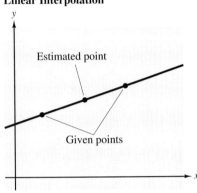

DISCUSSION PROBLEM ■   **Comparing Forms of Equations of Lines**

The two most commonly used forms of equations of lines are

$$y = mx + b \qquad \text{Slope-intercept form}$$

and

$$y - y_1 = m(x - x_1). \qquad \text{Point-slope form}$$

Write a short paragraph describing which of these two forms is better suited for use in the following two types of problems.

(a) Sketching the line represented by a *given* equation.

(b) Writing the equation of a line that passes through a *given point* with a *given slope*. ■

Warm-Up   *The following warm-up exercises involve skills that were covered in earlier sections. You will use these skills in the exercise set for this section.*

In Exercises 1–4, determine the slope of the line passing through the given points.

1. $(-3, 2)$, $(5, 0)$

2. $(-10, -5)$, $(4, 10)$

3. $(-4, -4)$, $(-4, 6)$

4. $\left(\dfrac{2}{3}, \dfrac{1}{2}\right)$, $\left(\dfrac{5}{6}, \dfrac{3}{4}\right)$

In Exercises 5 and 6, sketch the graph of the lines through the given point with the indicated slope. Make all four sketches in the same rectangular coordinate system.

Point	Slope			
5. (2, 3)	(a) 0	(b) 1	(c) 2	(d) $-\dfrac{1}{3}$
6. (−4, 1)	(a) 3	(b) −3	(c) $\dfrac{1}{2}$	(d) undefined

In Exercises 7–10, write the equation in the form $y = mx + b$.

7. $y - 8 = -\dfrac{2}{5}(x - 4)$

8. $y + 3 = \dfrac{4}{3}(x - 5)$

9. $y - (-1) = \dfrac{3 - (-1)}{2 - 4}(x - 4)$

10. $y - 5 = \dfrac{3 - 5}{0 - 2}(x - 2)$

8.5   EXERCISES

In Exercises 1–10, find the slope of the given line.

1. $y - 2 = 5(x + 3)$

2. $y + 3 = -2(x - 6)$

3. $y + \dfrac{5}{6} = \dfrac{2}{3}(x + 4)$

4. $y - \dfrac{1}{4} = \dfrac{5}{8}\left(x - \dfrac{13}{5}\right)$

5. $y = \dfrac{3}{8}x - 4$

6. $y = -3x + 10$

7. $2x - y = 0$

8. $x + 5 = 0$

9. $3x - 2y + 10 = 0$

10. $5x + 4y - 8 = 0$

In Exercises 11–16, find an equation of the line that passes through the given point and has the specified slope, and sketch the graph of the line. (Write your answer in slope-intercept form.)

Point	Slope		Point	Slope
11. (0, 0)	$m = -2$		**12.** (0, 0)	$m = \dfrac{1}{2}$
13. (0, −2)	$m = 3$		**14.** (0, 10)	$m = -\dfrac{1}{4}$
15. $\left(0, \dfrac{3}{2}\right)$	$m = -6$		**16.** $\left(0, -\dfrac{5}{2}\right)$	$m = \dfrac{3}{4}$

In Exercises 17–26, find an equation of the line through the given points and sketch the graph of the line. (Write your answer in general form.)

17. $(0, 0)$, $(4, 4)$ **18.** $(0, 0)$, $(2, -4)$ **19.** $(2, 3)$, $(6, 0)$ **20.** $(-7, -2)$, $(3, 3)$

21. $(-6, 2)$, $(3, 5)$ **22.** $(0, 3)$, $(5, 0)$ **23.** $(0, 3)$, $(5, 3)$ **24.** $(4, 1)$, $(4, 8)$

25. $\left(\frac{5}{2}, -1\right)$, $\left(\frac{9}{2}, 7\right)$ **26.** $\left(4, \frac{5}{3}\right)$, $\left(-1, \frac{1}{3}\right)$

In Exercises 27–36, find an equation of the line passing through the given point with the specified slope. (Write your answer in general form.)

27. $(0, -4)$, $m = \frac{1}{2}$ **28.** $(0, 7)$, $m = -1$ **29.** $(-3, 6)$, $m = -2$ **30.** $(0, 0)$, $m = 4$

31. $(4, 0)$, $m = -\frac{1}{3}$ **32.** $(-2, -5)$, $m = \frac{3}{4}$

33. $(6, -1)$, m is undefined   **34.** $(-10, -4)$, $m = 0$

35. $\left(4, \frac{5}{2}\right)$, $m = \frac{4}{3}$ **36.** $\left(-\frac{1}{2}, \frac{3}{2}\right)$, $m = -3$

In Exercises 37–48, find an equation of the line passing through the given points. (Write your answer in general form.)

37. $(5, -1)$, $(-5, 5)$ **38.** $(4, 3)$, $(-4, 4)$ **39.** $(2, 3)$, $(3, 2)$ **40.** $(0, 1)$, $(-2, 0)$

41. $(5, 4)$, $(-2, -4)$ **42.** $(3, 5)$, $(1, 6)$ **43.** $\left(2, \frac{1}{2}\right)$, $\left(\frac{1}{2}, \frac{5}{4}\right)$ **44.** $(-1, 4)$, $(6, 4)$

45. $(-8, 1)$, $(-8, 7)$ **46.** $(1, 1)$, $\left(6, -\frac{2}{3}\right)$ **47.** $(1, 0.6)$, $(2, -0.6)$ **48.** $(-8, 0.6)$, $(2, -2.4)$

In Exercises 49–54, write an equation of the line through the indicated point (a) parallel to the given line and (b) perpendicular to the given line.

	Point	Line		Point	Line
49.	$(2, 1)$	$4x - 2y = 3$	**50.**	$(-3, 2)$	$x + y = 7$
51.	$(-6, 4)$	$3x + 4y = 7$	**52.**	$\left(\frac{7}{8}, \frac{3}{4}\right)$	$5x + 3y = 0$
53.	$(-1, 0)$	$y + 3 = 0$	**54.**	$(2, 5)$	$x - 4 = 0$

55. Match each of the following situations with one of the accompanying graphs labeled (i), (ii), (iii), and (iv). Also determine the slope and how it is interpreted in the given situation.
(a) A person is paying $10 per week to a friend to repay a $100 loan.
(b) An employee is paid $12.50 per hour plus $1.50 for each unit produced per hour.
(c) A sales representative receives $20 per day for food, plus $0.25 for each mile traveled.
(d) A typewriter that was purchased for $600 depreciates $100 per year.

(i)

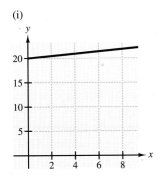

(ii)

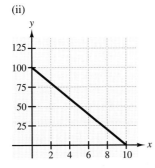

(iii)

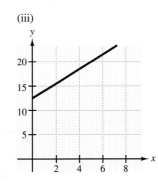

(iv)

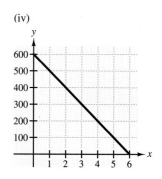

56. *Temperature* The linear relationship between the Fahrenheit and Celsius temperature scales is given by $F = \frac{9}{5}C + 32$. Use this equation to complete the following table.

C		$-10°$	10°			177°
F	0°			68°	90°	

57. *Salary Plus Commission* A sales representative receives a salary of $2,000 per month plus a commission of 2% of the total monthly sales. Write a linear equation giving the wages W in terms of sales S.

58. *Travel Reimbursement* A sales representative is reimbursed $110 per day for lodging and meals plus $0.25 per mile driven. Write a linear equation giving the daily cost C to the company in terms of x, the number of miles driven.

59. *Discount* A store is offering a 20% discount on all items in its inventory. Write a linear equation giving the sale price S for an item in terms of its list price L.

60. *Depreciation* A small business purchases a photocopier for $5,400. It is estimated that after four years its depreciated value will be $1,000. Assuming straight-line depreciation, write a linear equation giving the value V of the copier in terms of time t.

61. *Exam Score* Suppose you are able to obtain a score of 70 on an exam with no additional studying. For each additional hour of studying, you can increase your score by four points. Write a linear equation giving your estimated score y in terms of the number of hours t of additional studying.

62. *College Enrollment* A small college had an enrollment of 1,200 students in 1980. During the next 10 years the enrollment increased by approximately 50 students per year. Write a linear equation giving the enrollment N in terms of the year t. (Let $t = 0$ correspond to the year 1980.) If this constant rate of growth continues, predict the enrollment in the year 2000.

63. *Rental Occupancy* A real estate office handles an apartment complex with 50 units. When the rent per unit is $480 per month, all 50 units are rented. However, when the rent is $525 per month, the average number of rented units drops to 47. Assume the relationship between the monthly rent and the number of rented units is linear.
  (a) Write the equation giving the number of rented units x in terms of the rent p.
  (b) *(Linear Extrapolation)* Use this equation to predict the number of rented units if the rent is $555.
  (c) *(Linear Interpolation)* Use this equation to predict the number of rented units if the rent is $495.

64. *Test Scores* An instructor gives both 20-point quizzes and 100-point exams in a mathematics course. The average quiz and test scores for six students are given as ordered pairs (x, y), where x is the average quiz score and y is the average test score. The ordered pairs are (12, 78), (20, 89), (15, 86), (18, 95), (16, 80), and (10, 68).
  (a) Plot the points.
  (b) Use a ruler to sketch the "best-fitting" line through the points.
  (c) Find an equation for the line sketched in part (b).
  (d) Use the equation of part (c) to estimate the average test score for a student with an average quiz score of 14.

In Exercises 65 and 66, use a graphing calculator to sketch the graphs of the pair of lines. Set the zoom feature to the *square* setting (see Example 7). Determine whether the lines are parallel, perpendicular, or neither.

65. $y = -0.4x + 3$

$y = \dfrac{5}{2}x - 1$

66. $y = \dfrac{2x - 3}{3}$

$y = \dfrac{4x + 3}{6}$

67. Sketch the graphs of the following equations on the same screen. (Set the zoom feature to the *square* setting.)

(a) $y = \dfrac{1}{3}x + 2$

(b) $y = 4x + 2$

(c) $y = -3x + 2$

(d) $y = -\dfrac{1}{4}x + 2$

68. *Straight-Line Depreciation* A business purchases a new machine for $200,000. It is estimated that in five years its depreciated value will be $50,000. Assume the depreciation of the machine can be approximated by a straight line.

(a) Write a linear equation giving the value y of the machine in terms of time t.

(b) Use a graphing calculator to sketch the graph of the linear equation of part (a). (Adjust the RANGE settings.)

(c) Use the TRACE feature and the right and left cursor keys to approximate the value of the machine after three years.

SECTION 8.6	**Graphs of Linear Inequalities**
	Linear Inequalities in Two Variables • *The Graph of a Linear Inequality in Two Variables*

Linear Inequalities in Two Variables

A **linear inequality** in variables x and y is an inequality that can be written in one of the following forms.

$$ax + by < c \qquad ax + by \le c$$
$$ax + by > c \qquad ax + by \ge c$$

Here are some examples.

$$x - y > 2, \qquad 3x - 2y \le 6, \qquad x \ge 5, \quad \text{and} \quad y < -1$$

An ordered pair (x_1, y_1) is a **solution** of a linear inequality in x and y if the inequality is true when x_1 and y_1 are substituted for x and y, respectively.

EXAMPLE 1 ■ Verifying Solutions of Linear Inequalities

Determine whether or not the following points are solutions of the linear inequality $3x - y \ge -1$.

(a) $(0, 0)$ (b) $(1, 4)$ (c) $(-1, 2)$

Solution

(a) To determine whether the point $(0, 0)$ is a solution of the inequality, we substitute the coordinates of the point into the inequality as follows.

$$3x - y \geq -1 \qquad \text{Given inequality}$$
$$3(0) - 0 \overset{?}{\geq} -1 \qquad \text{Replace } x \text{ by } 0 \text{ and } y \text{ by } 0$$
$$0 \geq -1 \qquad \text{Inequality is satisfied}$$

Since the inequality is satisfied, we conclude that the point $(0, 0)$ is a solution.

(b) By substituting the coordinates of the point $(1, 4)$ into the given inequality, we obtain $3(1) - 4 = -1$. Since -1 is greater than or equal to -1, we conclude that the point $(1, 4)$ is also a solution of the inequality.

(c) By substituting the coordinates of the point $(-1, 2)$ into the given inequality, we obtain $3(-1) - 2 = -5$. Since -5 is less than -1, we conclude that the point $(-1, 2)$ is *not* a solution of the inequality.   ■

The Graph of a Linear Inequality in Two Variables

The **graph** of an inequality is the collection of all solution points of the inequality. To sketch the graph of a linear inequality such as $3x - 2y < 6$, we begin by sketching the graph of the *corresponding linear equation* $3x - 2y = 6$. Use a *dashed* line for the inequalities $<$ and $>$, and a *solid* line for the inequalities $\leq$ and $\geq$. The graph of the equation (corresponding to a given linear inequality) separates the plane into two regions called **half-planes**. In each half-plane, one of the following must be true.

1. All points in the half-plane are solutions of the inequality.

2. No point in the half-plane is a solution of the inequality.

Thus, we can determine whether the points in an entire half-plane satisfy the inequality by simply testing *one* point in the region. We summarize this graphing procedure as follows.

Sketching the Graph of a Linear Inequality in Two Variables	1. Replace the inequality sign by an equal sign, and sketch the graph of the resulting equation. (Use a dashed line for $<$ and $>$, and a solid line for $\leq$ and $\geq$.) 2. Test one point in each of the half-planes formed by the graph in Step 1. If the point satisfies the inequality, then shade the entire half-plane to denote that every point in the region satisfies the inequality.

EXAMPLE 2   ■   Sketching the Graph of a Linear Inequality

Sketch the graphs of the following linear inequalities.   (a) $x > -2$ (b) $y \leq 3$

Solution

(a) The graph of the corresponding equation $x = -2$ is a vertical line. The points that satisfy the inequality $x > -2$ are those lying to the right of this line, as shown in Figure 8.60.

(b) The graph of the corresponding equation $y = 3$ is a horizontal line. The points that satisfy the inequality $y \leq 3$ are those lying below (or on) this line, as shown in Figure 8.61.

FIGURE 8.60

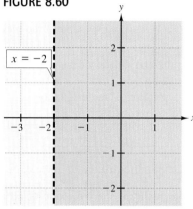

FIGURE 8.61

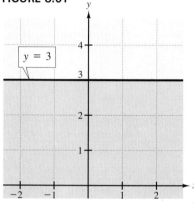

EXAMPLE 3   ■   Sketching the Graph of a Linear Inequality

Sketch the graph of the linear inequality $x - y < 2$.

Solution

The graph of the corresponding equation $x - y = 2$ is a line, as shown in Figure 8.62. Since the origin $(0, 0)$ satisfies the inequality, the graph consists of the half-plane lying above the line. (Try checking a point below the line. Regardless of which point you choose, you will see that it does not satisfy the inequality.)

FIGURE 8.62

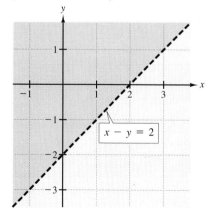

For a linear inequality in two variables, we can sometimes simplify the graphing procedure by writing the inequality in *slope-intercept* form. For instance, by writing $x - y < 2$ in the form

$$y > x - 2$$

we can see that the solution points lie *above* the line $y = x - 2$, as shown in Figure 8.63. Similarly, by writing the inequality $3x - 2y > 5$ in the form

$$y < \frac{3}{2}x - \frac{5}{2}$$

we see that the solutions lie *below* the line $y = \frac{3}{2}x - \frac{5}{2}$, as shown in Figure 8.63.

FIGURE 8.63

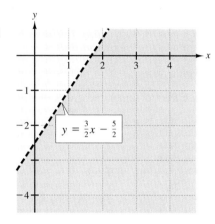

$$y = \frac{3}{2}x - \frac{5}{2}$$

EXAMPLE 4   ■   Sketching the Graph of a Linear Inequality

Use the slope-intercept form of a linear equation as an aid in sketching the graph of the inequality

$$5x + 4y \leq 12.$$

Solution

To begin, we rewrite the inequality in slope-intercept form.

$5x + 4y \leq 12$	Given inequality
$4y \leq -5x + 12$	Subtract $5x$ from both sides
$y \leq -\dfrac{5}{4}x + 3$	Slope-intercept form

From this form, we can conclude that the solution is the half-plane lying *on* or *below* the line

$$y = -\frac{5}{4}x + 3.$$

The graph is shown in Figure 8.64.

FIGURE 8.64

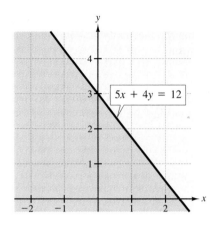

5x + 4y = 12

■

EXAMPLE 5   ■   A Graphing Calculator Problem

Use a graphing calculator to sketch the graph of the following inequality.

$$y \leq -x - 2$$

Solution

We can begin by sketching the graph of $y = -x - 2$. Press the $\boxed{Y=}$ key, enter $y_1 = -x - 2$, and press the $\boxed{GRAPH}$ key.* The display should look like that shown in Figure 8.65. From this graph, we can see that the solution of the inequality $y \leq -x - 2$ is the set of all points that lie below or on the line. To shade this portion of the display screen, press $\boxed{DRAW}$, cursor to "Shade(," press $\boxed{ENTER}$, and enter the following.

Shade(-10,-X-2)

After pressing $\boxed{ENTER}$, your display screen should look like the one shown in Figure 8.66. Note that the "Shade(" instruction on the *TI-81* has the form

Lower   Upper
bound   bound
↓   ↓

Shade(▨▨▨ , ▨▨▨).

For this particular graph, the lower bound is $y = -10$ (provided your zoom mode is set on *standard*) and the upper bound is $y = -x - 2$.

*The graphing calculator keystrokes given in this text correspond to the *TI-81* graphing calculator from *Texas Instruments*. However, you can also use other graphing calculators and computer graphing software for these examples and exercises.

FIGURE 8.65 $y = -x - 2$

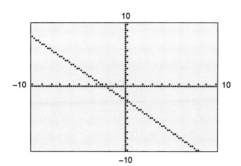

FIGURE 8.66 $y = -x - 2$

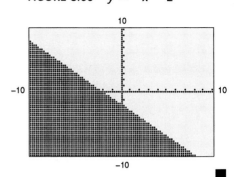

DISCUSSION PROBLEM ▩ A Nonlinear Inequality

Write a short paragraph describing how you would sketch the graph of the nonlinear inequality

$$y < x^2.$$

■

Warm-Up

The following warm-up exercises involve skills that were covered in earlier sections. You will use these skills in the exercise set for this section.

In Exercises 1–6, sketch the line represented by the given equation.

1. $x + 5 = 0$ **2.** $x + 2y = 0$

3. $y = -\dfrac{1}{3}x + 3$ **4.** $y = \dfrac{1}{5}x - 2$

5. $2x + y - 1 = 0$ **6.** $2y - 5 = 0$

In Exercises 7–10, determine whether the given points are solution points of the equation.

Equation	*Points*	
7. $8x + 3y - 15 = 0$	(a) $(1, 1)$	(b) $(0, 5)$
8. $4x - 5y + 20 = 0$	(a) $(5, 8)$	(b) $(0, 0)$
9. $\dfrac{1}{2}x - 2y + 6 = 0$	(a) $(8, 11)$	(b) $\left(2, \dfrac{7}{2}\right)$
10. $x + \dfrac{4}{5}y - 8 = 0$	(a) $\left(2, \dfrac{15}{2}\right)$	(b) $\left(1, \dfrac{5}{2}\right)$

8.6 EXERCISES

In Exercises 1–4, determine whether the points are solutions of the inequality.

Inequality   *Points*

1. $x + y > 5$ (a) $(0, 0)$ (b) $(3, 6)$ (c) $(-6, 20)$ (d) $(3, 2)$

2. $2x - y > 3$ (a) $(3, 0)$ (b) $(2, 6)$ (c) $(-6, -20)$ (d) $(3, 3)$

3. $-3x + 5y \leq 12$ (a) $(1, 2)$ (b) $(2, -3)$ (c) $(1, 3)$ (d) $(2, 8)$

4. $5x + 3y < 100$ (a) $(25, 10)$ (b) $(6, 10)$ (c) $(0, -12)$ (d) $(4, 5)$

In Exercises 5–8, state whether the boundary of the graph of the given inequality should be dashed or solid.

5. $2x + 3y < 6$ **6.** $2x + 3y \leq 6$ **7.** $2x + 3y \geq 6$ **8.** $2x + 3y > 6$

In Exercises 9–14, match the given inequality with its graph. [The graphs are labeled (a), (b), (c), (d), (e), and (f).]

9. $x + y < 4$ **10.** $x + y \leq 4$ **11.** $x + y > 4$ **12.** $x + y \geq 4$

13. $x > 1$ **14.** $y < 1$

(a)

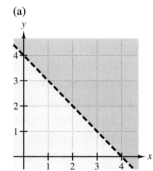

(b)

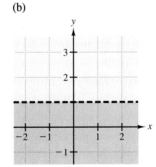

(c)

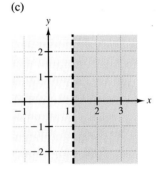

(d)

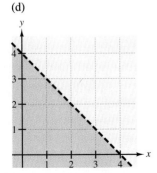

(e)

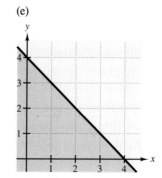

(f)

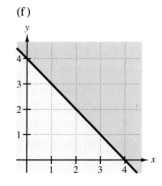

In Exercises 15–40, sketch the half-plane determined by the inequality.

15. $y \geq 3$

16. $x \leq 0$

17. $x > \dfrac{3}{2}$

18. $y < -2$

19. $x - y < 0$

20. $2x + y > 0$

21. $y \leq x - 2$

22. $y \geq -x + 3$

23. $y > x - 2$

24. $y < -x + 3$

25. $y \geq \dfrac{2}{3}x + \dfrac{1}{3}$

26. $y \leq -\dfrac{3}{4}x + 2$

27. $y > -2x + 10$

28. $y < 3x + 1$

29. $-3x + 2y - 6 < 0$

30. $x + 4y + 2 \geq 2$

31. $2x + y - 3 \geq 3$

32. $5x + 2y > 5$

33. $5x + 2y < 5$

34. $x - 2y + 6 \leq 0$

35. $x \geq 3y - 5$

36. $x > -2y + 10$

37. $y - 3 < \dfrac{1}{2}(x - 4)$

38. $y + 1 < -2(x - 3)$

39. $\dfrac{x}{3} + \dfrac{y}{4} < 1$

40. $\dfrac{x}{-2} + \dfrac{y}{2} > 1$

In Exercises 41–46, find an inequality that represents the graph.

41.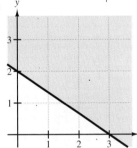

42.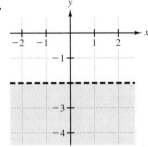

43.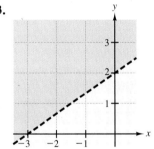

44.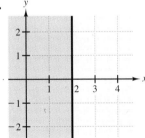

45.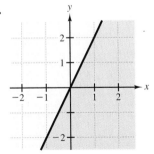

46.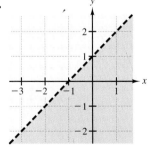

47. *Weekly Pay* Suppose you have two part-time jobs. One is at a grocery store which pays $6 per hour, and the other is mowing lawns, which pays $4 per hour. Between the two jobs you want to earn at least $120 a week. Write an inequality that shows the different numbers of hours you can work at each job, and sketch the graph of the inequality. From the graph, find several ordered pairs with positive integer coordinates that are solutions of the inequality.

48. *Number of Coins* A cash register must have at least $25 in change consisting of d dimes and q quarters. Write an inequality that shows the different numbers of coins that can be in the cash register, and sketch the graph of the inequality. From the graph, find several ordered pairs with positive integer coordinates that are solutions of the inequality.

49. *Furniture Production* Each table produced by a furniture company requires one hour in the assembly center, and a matching chair requires one and one-half hours in the assembly center. Twelve hours per day are available in the assembly center. Write an inequality that shows the different numbers of hours that can be spent assembling tables and chairs, and sketch the graph of the inequality. From the graph, find several ordered pairs with positive integer coordinates that are solutions of the inequality.

50. *Inventory Scheduling* A store sells two models of a certain type of computer. The cost to the store for these two models is $2,000 and $3,000, and the owner of the store does not want more than $30,000 invested in the inventory for these two models. Write an inequality that represents the different numbers of each model that can be held in inventory, and sketch the graph of the inequality. From the graph, find several ordered pairs with positive integer coordinates that are solutions of the inequality.

In Exercises 51–54, use a graphing calculator to sketch the graph of the inequality. (See Example 5.)

51. $y \le -2x + 4$

52. $y \ge x - 3$

53. $y \ge \frac{1}{2}x + 2$

54. $y \le -\frac{2}{3}x + 6$

CHAPTER 8   SUMMARY

As you review and prepare for a test on this chapter, first try to obtain a global view of what was discussed. Then review the specific skills needed in each category.

Aids to Sketching Graphs

- *Plot* points in a rectangular coordinate system. — Section 8.1, Exercises 1–16, 21–24
- *Test* if a specified ordered pair is a solution of a given equation. — Section 8.1, Exercises 25–30
- *Construct* a table of values for a given equation. — Section 8.1, Exercises 31–36
- *Sketch* the graph of an equation, using point-plotting. — Section 8.2, Exercises 1–4
- *Sketch* the graph of an equation and show the intercepts. — Section 8.2, Exercises 5–36
- *Determine* whether an equation represents y as a function of x. — Section 8.2, Exercises 37–48
- *Estimate* the slope of a line from its graph. — Section 8.4, Exercises 1–8
- *Sketch* a line, given its slope and one of its points. — Section 8.4, Exercises 41–46
- *Calculate* the slope of a line through two points. — Section 8.4, Exercises 9–26
- *Sketch* a line, using slope as an aid. — Section 8.4, Exercises 27–46
- *Sketch* a line, using the slope-intercept form of its equation. — Section 8.4, Exercises 51–60
- *Use* slope to determine whether pairs of lines are parallel, perpendicular, or neither. — Section 8.4, Exercises 61–68
- *Use* a graphing calculator to sketch the graph of a function. — Section 8.3, Exercises 1–24
- *Use* the RANGE and ZOOM keys on a graphing calculator. — Section 8.3, Exercises 25–32, 39–42

Writing Equations of Straight Lines

- *Write* the equation of a line using the point-slope form of the equation. — Section 8.5, Exercises 11–36
- *Write* the equation of a horizontal or vertical line. — Section 8.5, Exercises 37–48
- *Write* the equation of a line through a specified point and parallel or perpendicular to a given line. — Section 8.5, Exercises 49–54
- *Create* a linear equation from verbal statements. — Section 8.5, Exercises 56–64

Sketching Graphs of Inequalities

- *Test* whether a specified point is a solution point for a linear inequality. — Section 8.6, Exercises 1–4
- *Match* a linear inequality with its graph. — Section 8.6, Exercises 9–14
- *Sketch* the graph of a linear inequality. — Section 8.6, Exercises 15–40
- *Construct* a linear inequality from its graph. — Section 8.6, Exercises 41–46
- *Construct* a linear inequality from verbal statements. — Section 8.6, Exercises 47–50

Chapter 8 Review Exercises

In Exercises 1 and 2, plot the points in a rectangular coordinate system.

1. $(-2, 0)$, $\left(\frac{3}{2}, 4\right)$, $(-1, -3)$

2. $\left(3, -\frac{5}{2}\right)$, $\left(-5, 2\frac{3}{4}\right)$, $(4, 6)$

In Exercises 3–6, label the intercepts of the given graph and use the vertical line test to determine whether y is a function of x.

3. $x = y^2 - 4$

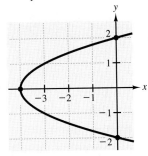

4. $y = x^3 - 3x + 2$

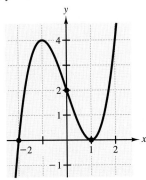

5. $y = x^2 - 4x$

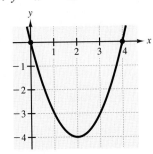

6. $x = y^3 - y$

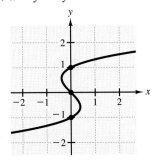

In Exercises 7–16, sketch the graph of the given equation and determine the intercepts of the graph.

7. $y = 4 - \frac{1}{2}x$

8. $y = \frac{3}{2}x - 1$

9. $y - 2x - 3 = 0$

10. $3x + 2y + 6 = 0$

11. $y = 3 - |x|$

12. $y = x(4 - x)$

13. $y - x^2 = 1$

14. $x = |y - 2|$

15. $y = (x - 4)^2$

16. $y = (x - 4)^2 - 1$

In Exercises 17–26, find the slope of the line through the given points.

17. $(2, 1)$, $(14, 6)$ **18.** $(-2, 2)$, $(3, -10)$ **19.** $(-1, 0)$, $(6, 2)$ **20.** $(1, 6)$, $(4, 2)$

21. $(4, 0)$, $(4, 6)$ **22.** $(1, 3)$, $(4, 3)$ **23.** $(-2, 5)$, $(1, 1)$ **24.** $(-6, 1)$, $(10, 5)$

25. $(1, -4)$, $(5, 10)$ **26.** $(-3, 3)$, $(8, 6)$

In Exercises 27–30, plot the given points and connect them with line segments to form the indicated figure.

27. Right triangle: $(-1, -1)$, $(10, 7)$, $(2, 18)$ **28.** Right triangle: $(-2, 6)$, $(1, 0)$, $(5, 2)$

29. Parallelogram: $(1, 1)$, $(8, 2)$, $(9, 5)$, $(2, 4)$ **30.** Square: $(-4, 0)$, $(1, -3)$, $(4, 2)$, $(-1, 5)$

In Exercises 31–34, a point on a line and the slope of the line are given. Find two additional points on the line. (Each problem has many correct answers.)

31. $(3, -1)$, $m = -2$ **32.** $\left(-2, \dfrac{3}{2}\right)$, $m = 2$ **33.** $(2, 3)$, $m = \dfrac{3}{4}$ **34.** $\left(-3, \dfrac{5}{2}\right)$, $m = -\dfrac{2}{3}$

In Exercises 35–44, find an equation of the line passing through the given point with the specified slope. (Write your answer in general form.)

35. $(4, -1)$, $m = 2$ **36.** $(-5, 2)$, $m = 3$ **37.** $(1, 2)$, $m = -4$ **38.** $(7, -3)$, $m = -1$

39. $(-1, -2)$, $m = \dfrac{4}{5}$ **40.** $(2, -4)$, $m = -\dfrac{1}{6}$ **41.** $\left(\dfrac{7}{2}, 3\right)$, $m = -\dfrac{8}{3}$ **42.** $\left(4, -\dfrac{4}{3}\right)$, $m = \dfrac{5}{2}$

43. $(3, 8)$, m is undefined   **44.** $(-2, 6)$, $m = 0$

In Exercises 45–50, find an equation of the line passing through the two given points. (Write your answer in general form.)

45. $(-4, 0)$, $(0, -2)$ **46.** $(-3, -2)$, $(4, 6)$ **47.** $(0, 8)$, $(6, 8)$ **48.** $(-1, 2)$, $(4, 7)$

49. $\left(\dfrac{2}{3}, \dfrac{1}{6}\right)$, $\left(3, \dfrac{5}{6}\right)$ **50.** $\left(\dfrac{1}{2}, 1\right)$, $\left(\dfrac{1}{2}, 5\right)$

In Exercises 51–54, find an equation of the line through the given point (a) parallel to the given line and (b) perpendicular to the given line. (Write your answer in general form.)

	Point	*Line*		*Point*	*Line*
51.	$(-1, 3)$	$2x + 3y = 1$	**52.**	$\left(\dfrac{1}{5}, -\dfrac{4}{5}\right)$	$5x + y = 2$
53.	$\left(\dfrac{5}{8}, 4\right)$	$4x + 3y = 16$	**54.**	$(-2, 1)$	$5x = 2$

In Exercises 55–60, sketch the half-plane determined by the inequality.

55. $x - 2 \geq 0$ **56.** $y + 3 < 0$ **57.** $2x + y < 1$ **58.** $3x - 4y > 2$

59. $x \leq 4y - 2$ **60.** $(y - 3) \geq \dfrac{2}{3}(x - 5)$

61. *Velocity* The velocity (in feet per second) of a ball thrown upward from ground level is given by $v = -32t + 48$, where t is the time in seconds (see figure).
(a) Find the velocity when $t = 1$.
(b) Find the time when the ball reaches its maximum height. (*Hint:* Find the time when $v = 0$.)
(c) Find the velocity when $t = 2$.

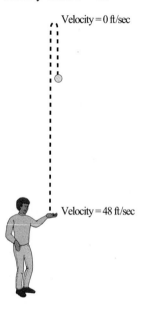

62. *Cost and Profit* A company produces a product for which the variable cost is $5.35 per unit and fixed costs are $16,000. The product is sold for $8.20 per unit. (Let x represent the number of units produced and sold.)
(a) Find the total cost C as a function of x.
(b) Find the profit P as a function of x.

63. *Area of a Rectangle* A wire 24 inches long is to be cut into four pieces to form a rectangle whose shortest side has a length of x (see figure). Express the area A of the rectangle as a function of x.

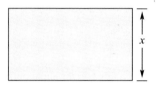

64. *Power Generation* The power generated by a wind turbine is given by the function

$$P = kw^3$$

where P is the number of kilowatts produced at a wind speed of w miles per hour.
(a) Find k when $P = 750$ and $w = 25$.
(b) Use the result of part (a) to find the power generated by a wind speed of 40 miles per hour.

Take this test as you would take a test in class. After you are done, check your work with the answers given in the back of the book.

1. Plot the points $(-1, 2)$, $(1, 4)$, and $(2, -1)$ in a rectangular coordinate system. Connect the points with line segments to form a right triangle.

2. Determine the quadrants in which the point (x, y) is located if $xy > 0$.

3. When an employee produces x units per hour, the hourly wage is given by $y = 0.75x + 4$. Complete the following table to determine the hourly wages for producing the specified number of units. Plot the results in a rectangular coordinate system.

x	2	4	6	8	10	12
$y = 0.75x + 4$						

4. Find the x- and y-intercepts (if any) of the graph of the equation $3x - 4y + 12 = 0$.

5. Find the x- and y-intercepts (if any) of the graph of the equation $y = (x - 2)(x + 1)$.

6. Sketch the graph of the equation $x - 2y = 6$.

7. Sketch the graph of the equation $x - y^2 = 1$.

8. A line with slope $m = \frac{3}{4}$ passes through the points $(1, -2)$ and $(9, y)$. Find the unknown coordinate y.

9. A line with slope $m = -2$ passes through the point $(-3, 4)$. Find two additional points on the line. (The problem has many correct answers.)

10. Sketch the graph of the line through the point $(2, 5)$ if the slope is undefined.

11. Determine the slope of the line passing through the points $(-5, 0)$ and $\left(2, \frac{3}{2}\right)$.

12. Find an equation of the line that passes through the point $(0, 6)$ with slope $m = -\frac{3}{8}$.

13. Find an equation of the line through the points $(2, 1)$ and $(6, 6)$. Write your answer in general form.

14. Determine the slope of the line *perpendicular* to the line given by the equation $3x - 5y + 2 = 0$.

15. Determine which points are solutions of the inequality $3x + 5y \le 16$.
    (a) $(2, 2)$ (b) $(6, -1)$ (c) $(-2, 4)$ (d) $(7, -1)$

16. Sketch the half-plane determined by the inequality $y \ge -2$.

17. Sketch the half-plane determined by the inequality $y < 5 - 2x$.

18. The value of a new plain paper copier is $4,200. It is estimated that during the first five years the copier will depreciate at the rate of $800 per year. Write the value V of the copier as a function of time t in years where $0 < t \le 5$. Find the value of the copier after three years.

CHAPTER NINE

Systems of Linear Equations

Chapter Overview

*In this chapter we will study **systems of linear equations** in two variables. The first three sections of the chapter introduce three different solution techniques: graphing, substitution, and elimination. The last section looks at several applications of systems of linear equations.*

Solving Systems of Linear Equations by Graphing

Introduction   •   Systems of Linear Equations   •   Solving a System of Linear Equations by Graphing

Introduction

Up to this point in the text, most problems have involved just one equation in either one or two variables. However, many problems in science, business, health services, and government involve two or more equations in two or more variables. For example, consider the following problem.

> A total of $12,000 is invested in two funds paying 9% and 11% simple interest. The combined annual interest for the two funds is $1,180. How much of the $12,000 is invested at each rate?

Letting x and y denote the amount (in dollars) in each fund, we can translate this problem into the following pair of linear equations in two variables.

$$x + y = 12{,}000$$
$$0.09x + 0.11y = 1{,}180$$

Such equations considered together form a **system of linear equations**. (We will solve this particular system of linear equations in the next section.)

Systems of Linear Equations

In this chapter we focus on problems involving two linear equations in two variables. A **solution** of such a system is an ordered pair (a, b) that satisfies each of the equations. For instance, to check that the point $(2, -1)$ is a solution of the system

$$3x + 2y = 4 \qquad \text{Equation 1}$$
$$-x + 3y = -5 \qquad \text{Equation 2}$$

we substitute the coordinates of the point into each equation. In the first equation, the substitution produces

$$3(2) + 2(-1) = 4 \qquad \text{Replace } x \text{ by 2 and } y \text{ by } -1$$

and in the second equation, the substitution produces

$$-(2) + 3(-1) = -5. \qquad \text{Replace } x \text{ by 2 and } y \text{ by } -1$$

Since the solution checks in *both* equations, we can conclude that it is a solution of the given system of linear equations.

EXAMPLE 1   ■   Checking Solutions of a System of Linear Equations

Consider the following system of linear equations.

$$x + y = 6 \qquad \text{Equation 1}$$
$$2x - 5y = -2 \qquad \text{Equation 2}$$

Which of the given points is a solution of this system of linear equations?

(a) (3, 3)   (b) (4, 2)

Solution

(a) To determine whether the point (3, 3) is a solution of the given system of linear equations, we must substitute the coordinates of the point into *each* of the given equations. Substituting into Equation 1 produces

$x + y = 6$	Equation 1
$3 + 3 \overset{?}{=} 6$	Replace x by 3 and y by 3
$6 = 6.$	Solution checks in Equation 1

Similarly, substituting into Equation 2 produces

$2x - 5y = -2$	Equation 2
$2(3) - 5(3) \overset{?}{=} -2$	Replace x by 3 and y by 3
$-9 \neq -2.$	Solution does not check in Equation 2

Since the point (3, 3) fails to check in *both* equations, we conclude that it is *not* a solution of the given system of linear equations.

(b) By substituting the coordinates of the point (4, 2) into the two given equations, we find that the point is a solution of the first equation

$4 + 2 = 6$	Replace x by 4 and y by 2

and is also a solution of the second equation

$2(4) - 5(2) = -2.$	Replace x by 4 and y by 2

Therefore, we conclude that the point *is* a solution of the given system of linear equations.   ∎

In Figure 9.1 we have sketched the two lines representing the two linear equations given in Example 1. Note that the point (4, 2) is the point of intersection of the two lines.

FIGURE 9.1

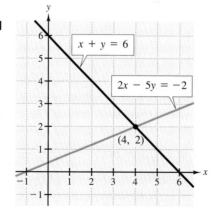

Solving a System of Linear Equations by Graphing

In this chapter we will study three methods for solving a system of two linear equations in two variables. The first method is *solution by graphing*. With this method, we first sketch the lines representing the two equations. Then we try to determine whether the two lines intersect at a point, as illustrated in Example 2.

EXAMPLE 2 ■ Solving a System of Linear Equations by Graphing

Solve the following system by graphing each equation and by locating the point of intersection.

$$x + y = -2 \qquad \text{Equation 1}$$
$$2x - 3y = -9 \qquad \text{Equation 2}$$

Solution

To begin, we write each equation in slope-intercept form. Then, using these forms, we make up two tables of values, as shown in Tables 9.1 and 9.2.

TABLE 9.1   Table of Values for Equation 1

x	-3	-2	-1	0	1	2	3
$y = -x - 2$	1	0	-1	-2	-3	-4	-5

TABLE 9.2   Table of Values for Equation 2

x	-3	-2	-1	0	1	2	3
$y = \frac{2}{3}x + 3$	1	$\frac{5}{3}$	$\frac{7}{3}$	3	$\frac{11}{3}$	$\frac{13}{3}$	5

Using these two tables of values, we sketch the lines shown in Figure 9.2 on the following page. Note that the two lines appear to intersect at the point $(-3, 1)$. To verify this, we substitute the coordinates of the point into each of the two given equations. For instance, by substituting into the first equation, we have

$$-3 + 1 = -2 \qquad \text{Replace } x \text{ by } -3 \text{ and } y \text{ by } 1$$

and by substituting into the second equation, we have

$$2(-3) - 3(1) = -9. \qquad \text{Replace } x \text{ by } -3 \text{ and } y \text{ by } 1$$

Since *both* equations are satisfied, we conclude that the point $(-3, 1)$ is the solution of the system.

FIGURE 9.2

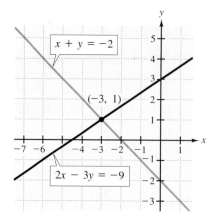

A good question to ask at this point is, Does every system of two linear equations in two variables have a single solution point? This is the same as asking, Do every two lines in a plane *intersect* at a single point? When you think about it, you can see that the answer to the second question is no. In fact, for two lines in a plane, there are three possible relationships.

Number of Points of Intersection of Two Lines

1. The two lines can intersect at a single point. The corresponding system of linear equations has a single solution and is called **consistent**.

2. The two lines can coincide and have infinitely many points of intersection. The corresponding system of linear equations has infinitely many solutions and is called **consistent**.

3. The two lines can be parallel and have no points of intersection. The corresponding system of linear equations has no solution and is called **inconsistent**.

These three possibilities are shown in Figure 9.3. Note that the word *consistent* is used to mean that the system of linear equations has at least one solution, whereas the word *inconsistent* is used to mean that the system of linear equations has no solution.

FIGURE 9.3

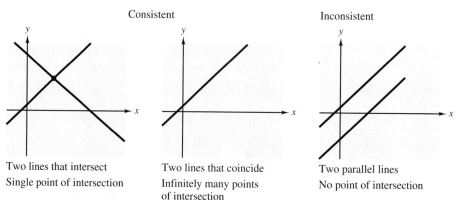

EXAMPLE 3   ■   **A System of Linear Equations with No Solution**

Solve the following system of linear equations.

$$x - y = 2 \qquad \text{Equation 1}$$
$$-3x + 3y = 6 \qquad \text{Equation 2}$$

Solution

We begin by writing each equation in slope-intercept form.

$$y = x - 2 \qquad \text{Slope-intercept form of Equation 1}$$
$$y = x + 2 \qquad \text{Slope-intercept form of Equation 2}$$

From these forms, we can see that the lines representing the two equations are parallel (each has a slope of 1), as shown in Figure 9.4. Therefore, the given system of linear equations has no solution. (The system is inconsistent.)

FIGURE 9.4

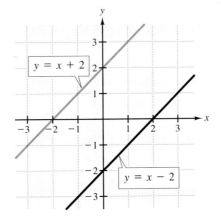

EXAMPLE 4   ■   **A System of Linear Equations with Infinitely Many Solutions**

Solve the following system of linear equations.

$$x - y = 2 \qquad \text{Equation 1}$$
$$-3x + 3y = -6 \qquad \text{Equation 2}$$

Solution

We begin by writing each equation in slope-intercept form.

$$y = x - 2 \qquad \text{Slope-intercept form of Equation 1}$$
$$y = x - 2 \qquad \text{Slope-intercept form of Equation 2}$$

From these forms, we can see that the lines representing the two equations are the same (see Figure 9.5). Therefore, the given system of linear equations has infinitely many solutions. We can describe the solution set by saying that each point on the line $y = x - 2$ is a solution of the system of linear equations.

FIGURE 9.5

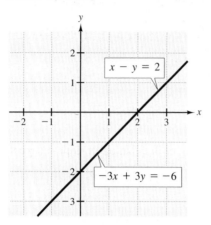

$x - y = 2$

$-3x + 3y = -6$

Note in Examples 3 and 4 that if the two lines representing a system of linear equations have the same slope, then the system must have either no solution or infinitely many solutions. On the other hand, if the two lines have different slopes, then they must intersect at a single point and the corresponding system has a single solution.

EXAMPLE 5   ■   A System of Linear Equations with a Single Solution

Solve the following system of linear equations.

$$2x + y = 4 \qquad \text{Equation 1}$$
$$4x + 3y = 9 \qquad \text{Equation 2}$$

Solution

We begin by writing each equation in slope-intercept form.

$$y = -2x + 4 \qquad \text{Slope-intercept form of Equation 1}$$
$$y = -\frac{4}{3}x + 3 \qquad \text{Slope-intercept form of Equation 2}$$

Since the lines do not have the same slope, we know that they intersect. To find the point of intersection, we sketch both lines in the same rectangular coordinate system, as shown in Figure 9.6. From this sketch, it appears that the solution occurs near the point $\left(\frac{3}{2}, 1\right)$. To check this solution, we substitute the coordinates of the point into each of the two given equations. For instance, by substituting into the first equation, we have

$$2\left(\frac{3}{2}\right) + 1 = 4 \qquad \text{Replace } x \text{ by } \frac{3}{2} \text{ and and } y \text{ by } 1$$

and by substituting into the second equation, we have

$$4\left(\frac{3}{2}\right) + 3(1) = 9. \qquad \text{Replace } x \text{ by } \frac{3}{2} \text{ and } y \text{ by } 1$$

Since *both* equations are satisfied, we conclude that the point $\left(\frac{3}{2}, 1\right)$ is the solution.

FIGURE 9.6

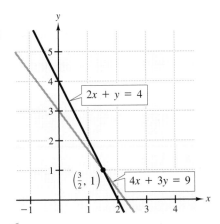

$2x + y = 4$

$\left(\frac{3}{2}, 1\right)$ $4x + 3y = 9$

There are two things you should note in Example 5. First, your success in applying the graphing method of solving a system of linear equations depends on sketching accurate graphs. Second, once you have made a graph and "guessed" at the point of intersection, it is critical that you check to see whether the point you have chosen is actually a solution.

If you have a graphing calculator, the ⎡TRACE⎤ feature on the calculator can help you estimate the point of intersection of two lines.*

EXAMPLE 6 ■ A Graphing Calculator Problem

Use a graphing calculator to approximate the solution of the following system of linear equations.

$$x + 2y = 8 \qquad \text{Equation 1}$$
$$-x + y = 1 \qquad \text{Equation 2}$$

Check your solution in each of the given equations.

Solution

To begin, we solve each equation for y, as follows.

$$y = -\frac{1}{2}x + 4 \qquad \text{Enter as } Y_1 = -(1/2)X + 4$$

$$y = x + 1 \qquad \text{Enter as } Y_2 = X + 1$$

Next, press the ⎡Y=⎤ key, enter the two equations (one as y_1 and the other as y_2), and press ⎡GRAPH⎤. The display screen should look like the one shown in Figure 9.7. From the graph, it appears that the solution is the point (2, 3). To use the calculator to estimate the coordinates of the point, press the ⎡TRACE⎤ key. Then move the cursor until it is as close as possible to the point of intersection. (Because of the limited number of pixels on the display screen, usually it will not be possible to find the exact

*The graphing calculator keystrokes given in this text correspond to the *TI-81* graphing calculator from *Texas Instruments*. For other graphing calculators, the keystrokes may differ.

coordinates of the point. For instance, the calculator may display the solution as $x = 1.9835$ and $y = 3.0248$ when it is actually $x = 2$ and $y = 3$.) After approximating the solution to be $(x, y) = (2, 3)$, we can check the solution as follows.

$$2 + 2(3) = 8 \qquad \text{First equation checks}$$
$$-2 + 3 = 1 \qquad \text{Second equation checks}$$

FIGURE 9.7 $\quad y_1 = \dfrac{-x}{2} + 4, y_2 = x + 1$

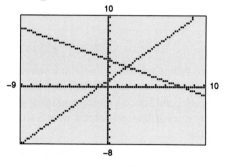

DISCUSSION PROBLEM ■ **You Be the Instructor**

Suppose you are teaching an algebra class and want to create several systems of linear equations for your class to use as practice problems. Write a short paragraph describing how you could create a system of linear equations that has a simple solution such as $(2, -1)$. Use your method to fill in the following blanks so that the given system has $(2, -1)$ as its only solution.

$$3x + 2y = \boxed{} \qquad \text{Equation 1}$$
$$x + \ y = \boxed{} \qquad \text{Equation 2}$$ ■

Warm-Up

The following warm-up exercises involve skills that were covered in earlier sections. You will use these skills in the exercise set for this section.

In Exercises 1–6, sketch the graph of the given equation.

1. $y = -\dfrac{1}{3}x + 6$ **2.** $y = 2(x - 3)$ **3.** $y - 1 = 3(x + 2)$

4. $y + 2 = \dfrac{1}{4}(x - 1)$ **5.** $2x + y = 4$ **6.** $5x - 2y = 3$

In Exercises 7 and 8, find an equation of the line passing through the two points.

7. $(-1, 3)$, $(4, 8)$ **8.** $(2, 6)$, $(5, 1)$

In Exercises 9 and 10, determine the slope of the given line.

9. $3x + 6y = 4$ **10.** $7x - 4y = 10$

9.1   EXERCISES

In Exercises 1–4, determine which ordered pair is a solution of the system of equations.

System	*Points*	*System*	*Points*
1. $x + 3y = 11$	(a) $(2, 3)$	**2.** $3x - y = -2$	(a) $(0, 2)$
$-x + 3y = 7$	(b) $(5, 4)$	$x - 3y = 2$	(b) $(-1, -1)$
3. $2x - 3y = -8$	(a) $(5, -3)$	**4.** $5x - 3y = -12$	(a) $(-3, -1)$
$x + y = 1$	(b) $(-1, 2)$	$x - 4y = 1$	(b) $(3, 1)$

In Exercises 5–10, use the graphs of the linear equations to determine the number of solutions of the given system.

5. $x - y = 1$
    $x + y = 0$

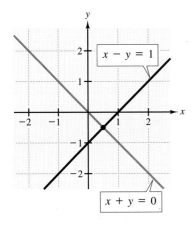

6. $x + 2y = 2$
    $x - y = 2$

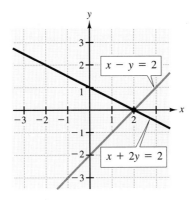

7. $3x - 2y = 0$
    $3x - 2y = -4$

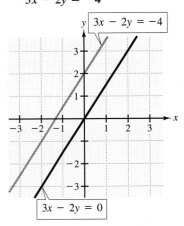
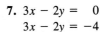

8. $2x + y = 4$
    $-4x - 2y = -8$

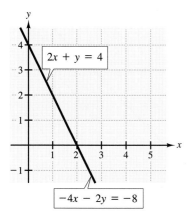

9. $2x - 3y = 6$
    $4x + 3y = 12$

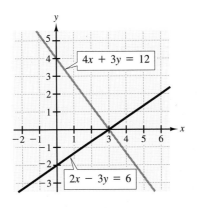

10. $2x - 5y = 10$
     $6x - 15y = 75$

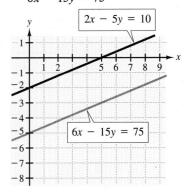

In Exercises 11–30, solve the system of linear equations by graphing.

11. $x - y = 2$
$x + y = 2$

12. $x - y = 0$
$x + y = 4$

13. $3x - 4y = 5$
$x \quad\quad = 3$

14. $5x + 4y = 18$
$y = 2$

15. $4x - 5y = 0$
$6x - 5y = 10$

16. $3x + 2y = -6$
$3x - 2y = 6$

17. $4x - 3y = -6$
$4x - 3y = 0$

18. $3x + 7y = 15$
$x + \frac{7}{3}y = 5$

19. $x + 7y = -5$
$3x - 2y = 8$

20. $-4x + 3y = 10$
$7x + y = 20$

21. $x + 2y = 3$
$x - 3y = 13$

22. $x - 8y = -40$
$-5x + 8y = 8$

23. $4x + 5y = 20$
$\frac{4}{5}x + y = 4$

24. $-x + 3y = 7$
$2x - 6y = 6$

25. $-x + 10y = 30$
$x + 10y = 10$

26. $x = 4$
$y = 6$

27. $2x + 3y = 10$
$y = 3$

28. $-3x + 7y = 25$
$7x + 2y = 70$

29. $-3x + 10y = 15$
$3x - 10y = 15$

30. $9x + 8y = 48$
$-5x + 8y = -8$

In Exercises 31–36, write each equation of the given system in slope-intercept form and use this form to determine the number of solutions of the system. (It is not necessary to sketch the graphs of the equations.)

31. $2x - 3y = -12$
$-8x + 12y = -12$

32. $-5x + 8y = 8$
$7x - 4y = 14$

33. $-2x + 3y = 4$
$2x + 3y = 8$

34. $2x + 5y = 15$
$2x - 5y = 5$

35. $-x + 4y = 7$
$3x - 12y = -21$

36. $3x + 8y = 28$
$-4x + 9y = 1$

In Exercises 37 and 38, the graphs of the two equations appear parallel. Yet, when each equation is written in slope-intercept form, we find that the slopes are not equal and thus their graphs do intersect. Why is the graphical method of solving the system difficult to apply?

37. $x - 200y = -200$
$x - 199y = 198$

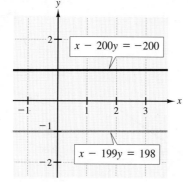

38. $25x - 24y = 0$
$13x - 12y = 24$

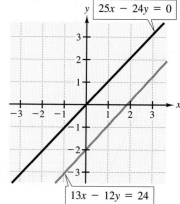

39. *Feed Mixture* A feed mixture consists of two parts of oats for each part of corn (see figure). The total weight of the mixture is 18 tons. If x represents the number of tons of oats and y represents the number of tons of corn, then the mathematical model for the problem is as follows.

$$x \qquad = 2y$$
$$x + y = 18$$

Solve this system graphically.

Oats   Corn

40. *Number Problem* The sum of two numbers is 20 and the difference of the two numbers is 2. Write a system of equations that models this problem and solve the system graphically.

In Exercises 41–44, use a graphing calculator to approximate the solution of the system of equations.

41. $2x + 3y = 8$
     $4x - 2y = 0$

42. $3x - 2y = 2$
     $4x - y = 11$

43. $2x + y = -5$
     $3x + 2y = -4$

44. $5x + 2y = -17$
     $-6x + 3y = 15$

Solving Systems of Linear Equations by Substitution
The Method of Substitution

The Method of Substitution

In this section we look at an algebraic method for solving a system of linear equations—the **method of substitution**. The goal of the method of substitution is to reduce a system of two linear equations in two variables to a single equation in one variable. Example 1 illustrates the basic steps of the method.

EXAMPLE 1 ■ The Method of Substitution

Solve the following system of linear equations.

$$-x + y = 1 \qquad \text{Equation 1}$$
$$2x + y = -2 \qquad \text{Equation 2}$$

Solution

We begin by solving for y in the first equation.

$$y = x + 1 \qquad \text{Solve for } y \text{ in Equation 1}$$

Next, we substitute this expression for y into Equation 2.

$$2x + y = -2 \qquad \text{Equation 2}$$
$$2x + (x + 1) = -2 \qquad \text{Replace } y \text{ by } x + 1$$
$$3x + 1 = -2$$
$$3x = -3$$
$$x = -1 \qquad \text{Solve for } x$$

At this point, we know that the x-coordinate of the solution is -1. To find the y-coordinate, we back-substitute the x-value into the revised Equation 1.

$$y = x + 1 \qquad \text{Revised Equation 1}$$
$$y = -1 + 1 \qquad \text{Replace } x \text{ by } -1$$
$$y = 0 \qquad \text{Solve for } y$$

Thus, the solution is $(-1, 0)$. Check to see that it satisfies both of the original equations. ■

The term **back-substitute** implies that we work backwards. After solving for one of the variables, we substitute that value back into one of the equations in the original (or revised) system to find the value of the other variable.

When using the method of substitution, it does not matter which variable you choose to solve for first. Whether you solve for y first or x first, you will obtain the same solution. You should choose the variable that is easier to work with. For instance, in the system

$$3x - 2y = 1 \qquad \text{Equation 1}$$
$$x + 4y = 3 \qquad \text{Equation 2}$$

it is easier to first solve for x in the second equation. On the other hand, in the system

$$2x + y = 5 \qquad \text{Equation 1}$$
$$3x - 2y = 11 \qquad \text{Equation 2}$$

it is easier to first solve for y in the first equation.

EXAMPLE 2 ■ The Method of Substitution

Solve the following system of linear equations.

$$5x + 7y = 1 \qquad \text{Equation 1}$$
$$x + 4y = -5 \qquad \text{Equation 2}$$

Solution

For this system, it is convenient to begin by solving for x in the second equation.

$$x + 4y = -5 \qquad \text{Given Equation 2}$$
$$x = -4y - 5 \qquad \text{Revised Equation 2}$$

Substituting this expression for x into the first equation produces the following.

$$5x + 7y = 1 \qquad \text{Equation 1}$$
$$5(-4y - 5) + 7y = 1 \qquad \text{Replace } x \text{ by } -4y - 5$$
$$-20y - 25 + 7y = 1$$
$$-13y = 26$$
$$y = -2 \qquad \text{Solve for } y$$

Finally, we back-substitute this y-value into the revised second equation.

$$x = -4y - 5 \qquad \text{Revised Equation 2}$$
$$x = -4(-2) - 5 \qquad \text{Replace } y \text{ by } -2$$
$$x = 3 \qquad \text{Solve for } x$$

Therefore, we conclude that the solution is $(3, -2)$. Check to see that it satisfies both of the original equations. ■

The method of substitution demonstrated in Examples 1 and 2 has the following four steps.

The Method of Substitution	1. Solve one of the equations for one variable in terms of the other.
	2. Substitute the expression obtained in Step 1 into the other equation and solve the resulting equation.
	3. Back-substitute the solution in Step 2 into the expression found in Step 1 to find the other variable.
	4. Check your answer to see that it satisfies both of the original equations.

In Example 3 we use the method of substitution to solve the interest rate problem introduced at the beginning of this chapter.

EXAMPLE 3 ■ **An Application of a System of Linear Equations**

Solve the following system of equations.

$$x + y = 12,000 \qquad \text{Equation 1}$$
$$0.09x + 0.11y = 1,180 \qquad \text{Equation 2}$$

Solution

To begin, it is convenient to multiply both sides of the second equation by 100 to obtain $9x + 11y = 118,000$. This eliminates the need to work with decimals. Then the following steps are performed.

1. Solve for x in Equation 1.

$$x = 12{,}000 - y$$

2. Substitute this expression for x into Equation 2 and solve for y.

$$9(12{,}000 - y) + 11y = 118{,}000$$
$$108{,}000 - 9y + 11y = 118{,}000$$
$$2y = 10{,}000$$
$$y = 5{,}000$$

3. Back-substitute the value $y = 5{,}000$ to solve for x.

$$x = 12{,}000 - 5{,}000 = 7{,}000$$

Therefore, the solution is (7,000, 5,000). From the problem given at the beginning of the chapter, we see that this solution checks because the sum of the two investments is $12,000 and

$$0.09(7{,}000) + 0.11(5{,}000) = 630 + 550 = \$1{,}180. \qquad \blacksquare$$

In the next two examples, we show how the method of substitution identifies systems of equations that have no solution or infinitely many solutions.

EXAMPLE 4 ■ The Method of Substitution: No-Solution Case

Solve the following system of linear equations.

$$x - 3y = 2 \qquad\qquad \text{Equation 1}$$
$$-2x + 6y = 2 \qquad\qquad \text{Equation 2}$$

Solution

1. Solve for x in Equation 1.

$$x - 3y = 2 \qquad\qquad \text{Equation 1}$$
$$x = 3y + 2 \qquad\qquad \text{Solve for } x$$

2. Substitute for x in Equation 2.

$$-2x + 6y = 2 \qquad\qquad \text{Equation 2}$$
$$-2(3y + 2) + 6y = 2 \qquad\qquad \text{Replace } x \text{ by } 3y + 2$$
$$-6y - 4 + 6y = 2$$
$$-4 = 2 \qquad\qquad \text{False statement}$$

Because the substitution resulted in a false statement, we conclude that the given system of linear equations has no solution. (Check the slopes of the graphs of these two equations to see that the lines are parallel.) ■

EXAMPLE 5 ■   The Method of Substitution: Many-Solutions Case

Solve the following system of linear equations.

$$9x + 3y = 15 \qquad \text{Equation 1}$$
$$3x + y = 5 \qquad \text{Equation 2}$$

Solution

1. Solve for y in Equation 2.

$$3x + y = 5 \qquad \text{Equation 2}$$
$$y = -3x + 5 \qquad \text{Solve for } y$$

2. Substitute for y in Equation 1.

$$9x + 3y = 15 \qquad \text{Equation 1}$$
$$9x + 3(-3x + 5) = 15 \qquad \text{Replace } y \text{ by } -3x + 5$$
$$9x - 9x + 15 = 15$$
$$15 = 15 \qquad \text{True statement for any } x$$

This last equation is true for any value of x. This implies that any solution of Equation 2 is also a solution of Equation 1. In other words, the given system of linear equations has infinitely many solutions. The solutions consist of all ordered pairs (x, y) such that $3x + y = 5$. ■

If neither variable has a coefficient of 1 in a system of linear equations, we can still use the method of substitution. It may mean that we have to work with some fractions in the solution steps. This is demonstrated in Examples 6 and 7.

EXAMPLE 6 ■   The Method of Substitution: One-Solution Case

Solve the following system of linear equations.

$$5x + 3y = 18 \qquad \text{Equation 1}$$
$$2x - 7y = -1 \qquad \text{Equation 2}$$

Solution

1. Since neither variable has a coefficient of 1, we can choose to solve for either variable. Let's solve for x in Equation 1.

$$5x + 3y = 18 \qquad \text{Equation 1}$$
$$5x = -3y + 18$$
$$x = -\frac{3}{5}y + \frac{18}{5} \qquad \text{Revised Equation 1}$$

2. Substitute for x in Equation 2 and solve for y.

$$2x - 7y = -1 \qquad \text{Equation 2}$$

$$2\left(-\frac{3}{5}y + \frac{18}{5}\right) - 7y = -1 \qquad \text{Replace } x \text{ by } -\frac{3}{5}y + \frac{18}{5}$$

$$-\frac{6}{5}y + \frac{36}{5} - 7y = -1 \qquad \text{Distributive Property}$$

$$-6y + 36 - 35y = -5 \qquad \text{Multiply both sides by 5}$$

$$-41y = -41$$

$$y = 1 \qquad \text{Solve for } y$$

3. Back-substitute for y into the revised first equation.

$$x = -\frac{3}{5}y + \frac{18}{5} \qquad \text{Revised Equation 1}$$

$$x = -\frac{3}{5}(1) + \frac{18}{5} \qquad \text{Replace } y \text{ by 1}$$

$$x = 3 \qquad \text{Solve for } x$$

4. The solution is $(3, 1)$. Check to see that it satisfies both of the original equations.

■

EXAMPLE 7 ■ **The Method of Substitution: One-Solution Case**

Solve the following system of linear equations.

$$4x + 3y = 1 \qquad \text{Equation 1}$$
$$2x - 3y = 1 \qquad \text{Equation 2}$$

Solution

1. Again, we have a choice of which variable to solve for. Let's solve for y in Equation 1.

$$4x + 3y = 1 \qquad \text{Equation 1}$$

$$3y = -4x + 1$$

$$y = -\frac{4}{3}x + \frac{1}{3} \qquad \text{Revised Equation 1}$$

2. Substitute for y in Equation 2 and solve for x.

$$2x - 3y = 1 \qquad \text{Equation 2}$$

$$2x - 3\left(-\frac{4}{3}x + \frac{1}{3}\right) = 1 \qquad \text{Replace } y \text{ by } -\frac{4}{3}x + \frac{1}{3}$$

$$2x + 4x - 1 = 1$$

$$6x = 2$$

$$x = \frac{1}{3} \qquad \text{Solve for } x$$

3. Back-substitute for x in the revised first equation.

$$y = -\frac{4}{3}x + \frac{1}{3} \qquad\qquad \text{Revised Equation 1}$$

$$y = -\frac{4}{3}\left(\frac{1}{3}\right) + \frac{1}{3} \qquad\qquad \text{Replace } x \text{ by } \tfrac{1}{3}$$

$$y = -\frac{4}{9} + \frac{1}{3}$$

$$y = -\frac{1}{9} \qquad\qquad \text{Solve for } y$$

4. The solution is $\left(\frac{1}{3}, -\frac{1}{9}\right)$. Check to see that it satisfies both of the original equations.

■

DISCUSSION PROBLEM   ■   **Solving a System of Nonlinear Equations**

The method of substitution can be used to solve a nonlinear system of equations. Describe how you would use substitution to solve the following system.

$$x^2 - y = 1 \qquad\qquad \text{Equation 1}$$
$$2x - y = 1 \qquad\qquad \text{Equation 2}$$

How many solutions does this system have? Illustrate your result graphically.   ■

Warm-Up

The following warm-up exercises involve skills that were covered in earlier sections. You will use these skills in the exercise set for this section.

In Exercises 1–6, solve the given equation.

1. $x - (x + 2) = 8$

2. $y + (2y + 3) = 4$

3. $y - 3(4y - 2) = 1$

4. $x + 6(3 - 2x) = 4$

5. $3x + \frac{1}{2}(6x + 5) = \frac{3}{2}$

6. $4y - \frac{4}{5}(3y - 10) = 8$

In Exercises 7–10, solve the system of linear equations by graphing.

7. $-4x + 3y = 8$
$-x + 5y = 2$

8. $2x + y = 3$
$3x - 2y = -6$

9. $x + y = 5$
$2x - 3y = 0$

10. $-2x + 3y = 10$
$5x - 2y = 8$

9.2 EXERCISES

In Exercises 1–10, use the method of substitution to solve the given system of linear equations.

1. $x - y = 0$
$x + y = 2$

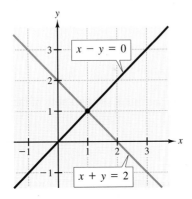

2. $x + y = 1$
$2x - y = 2$

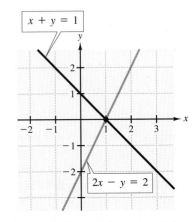

3. $2x + y = 4$
$-x + y = 1$

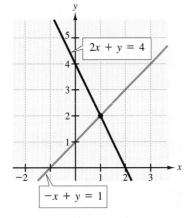

4. $x - y = -5$
$x + 2y = 4$

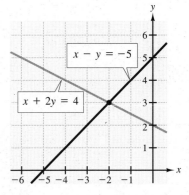

5. $-x + y = 1$
$x - y = 1$

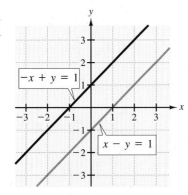

6. $x + 2y = 6$
$x + 2y = 2$

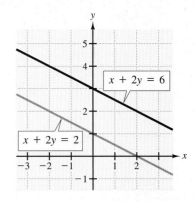

7. $-4x + y = 4$
  $8x - 2y = -8$

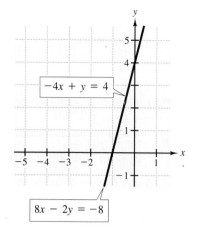

$-4x + y = 4$

$8x - 2y = -8$

8. $4x + 3y = 8$
  $-4x + y = 8$

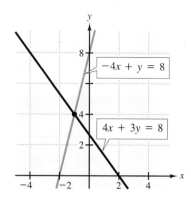

$-4x + y = 8$

$4x + 3y = 8$

9. $2x - y = 2$
  $4x + 3y = 9$

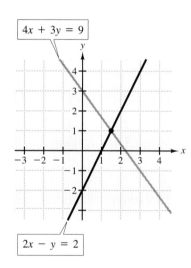

$4x + 3y = 9$

$2x - y = 2$

10. $x - 2y = 0$
  $x + y = 4$

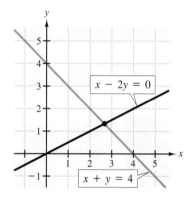

$x - 2y = 0$

$x + y = 4$

In Exercises 11–30, use the method of substitution to solve the given system of linear equations.

11. $x - y = 2$
  $2x + y = 1$

12. $x - 2y = -10$
  $3x - y = 0$

13. $x - y = 0$
  $2x + y = 0$

14. $x - 2y = 0$
  $3x - y = 0$

15. $x - y = 0$
  $5x - 3y = 10$

16. $x + 2y = 1$
  $5x - 4y = -23$

17. $2x - y = -2$
  $4x + y = 5$

18. $-3x + 6y = 4$
  $2x + y = 4$

19. $\dfrac{1}{5}x + \dfrac{1}{2}y = 8$
  $x + y = 20$

20. $\dfrac{1}{2}x + \dfrac{3}{4}y = 10$
  $\dfrac{3}{2}x - y = 4$

21. $-5x + 4y = 14$
  $5x - 4y = 4$

22. $3x - 2y = 3$
  $-6x + 4y = -6$

23. $4x - y = 2$
  $2x - \dfrac{1}{2}y = 1$

24. $x + y = 8$
  $x + y = -1$

25. $5x + 3y = 11$
  $x - 5y = 5$

26. $-3x + y = 4$
  $-9x + 5y = 10$

27. $2x = 5$
$x + y = 1$

28. $3x - y = 0$
$y = 6$

29. $8x + 5y = 100$
$9x - 10y = 50$

30. $x + 4y = 300$
$x - 2y = 0$

In Exercises 31–34, find a system of linear equations with integer coefficients that has the given solution. (*Note:* Each problem has many correct answers.)

31. $(2, 1)$

32. $(4, -3)$

33. $\left(\dfrac{7}{2}, -3\right)$

34. $\left(-\dfrac{1}{2}, 1\right)$

35. When solving a linear system of equations by the method of substitution, how can you recognize that the system has no solution?

36. When solving a linear system of equations by the method of substitution, how can you recognize that the system has infinitely many solutions?

In Exercises 37–40, find the values of a or b such that the given system is inconsistent.

37. $x + by = 1$
$x + 2y = 2$

38. $ax + 3y = 6$
$5x - 5y = 2$

39. $-6x + y = 4$
$2x + by = 4$

40. $6x - 3y = 4$
$ax - y = -2$

41. *Buffet Dinner* A family of six ate a buffet dinner at a restaurant where the price for adults was $11.95 and the price for children was $6.95. The total price for the six meals was $61.70. Find the number of family members charged the adult price.

42. *Number Problem* The sum of two numbers is 40 and the difference of the two numbers is 10. Find the numbers.

SECTION
9.3

Solving Systems of Linear Equations by Elimination
The Method of Elimination • Choosing Methods

The Method of Elimination

In this section we discuss another way to solve a system of linear equations—it is called the **method of elimination**. The key step is to obtain, for one of the variables, coefficients that differ only in sign, so that by *adding* the two equations this variable will be eliminated. For instance, by adding the equations

$$3x + 5y = 7 \qquad \text{Equation 1}$$
$$\underline{-3x - 2y = -1} \qquad \text{Equation 2}$$
$$3y = 6 \qquad \text{Add equations}$$

we eliminate the variable x and obtain a single equation in one variable y.

EXAMPLE 1 ■ The Method of Elimination

Solve the following system of equations.

$$4x + 3y = 1 \qquad \text{Equation 1}$$
$$2x - 3y = 1 \qquad \text{Equation 2}$$

Solution

We begin by noting that the coefficients for y differ only in sign. Therefore, by adding the two equations, we can eliminate y.

$$\begin{array}{ll} 4x + 3y = 1 & \text{Equation 1} \\ \underline{2x - 3y = 1} & \text{Equation 2} \\ 6x \quad\;\; = 2 & \text{Add equations} \end{array}$$

Therefore, $x = \frac{1}{3}$. By back-substituting this value into the first equation, we can solve for y, as follows.

$$\begin{array}{ll} 4x + 3y = 1 & \text{Equation 1} \\ 4\!\left(\dfrac{1}{3}\right) + 3y = 1 & \text{Replace } x \text{ by } \frac{1}{3} \\ 3y = -\dfrac{1}{3} & \\ y = -\dfrac{1}{9} & \text{Solve for } y \end{array}$$

what? ➚

Therefore, the solution is $\left(\frac{1}{3}, -\frac{1}{9}\right)$. Check to see that it satisfies both of the original equations. ∎

Try comparing the solution method used in Example 1 with that used for the same system of equations in Example 7 in the previous section. Which method do you think is easier? Many people find that the method of elimination is more efficient.

To obtain coefficients for one of the variables that differ only in sign, we often need to multiply one or both of the equations by a suitable constant. This is demonstrated in the next two examples.

EXAMPLE 2 ■ **The Method of Elimination**

Solve the following system of linear equations.

$$\begin{array}{ll} 2x - 3y = -7 & \text{Equation 1} \\ 3x + \;\; y = -5 & \text{Equation 2} \end{array}$$

Solution

For this system, we can obtain coefficients of y that differ only in sign by multiplying the second equation by 3.

$$\begin{array}{lll} 2x - 3y = -7 & 2x - 3y = \;\; -7 & \text{Equation 1} \\ \underline{3x + \;\; y = -5} & \underline{9x + 3y = -15} & \text{Multiply Equation 2 by 3} \\ & 11x \quad\;\;\; = -22 & \text{Add equations} \end{array}$$

Thus, we see that $x = -2$. By back-substituting this value of x into the second equation, we can solve for y.

$$3x + y = -5 \qquad \text{Equation 2}$$
$$3(-2) + y = -5 \qquad \text{Replace } x \text{ by } -2$$
$$y = 1 \qquad \text{Solve for } y$$

Therefore, the solution is $(-2, 1)$. Check to see that it satisfies both of the original equations.   ■

We summarize the method of elimination as follows.

The Method of Elimination	1. Obtain coefficients for x (or y) that differ only in sign by multiplying all terms of one or both equations by suitably chosen constants.
	2. Add the equations to eliminate one variable and solve the resulting equation.
	3. Back-substitute the value obtained in Step 2 into either of the original equations and solve for the other variable.
	4. Check your solution in both of the original equations.

EXAMPLE 3   ■   The Method of Elimination

Solve the following system of linear equations.

$$5x + 3y = 6 \qquad \text{Equation 1}$$
$$2x - 4y = 5 \qquad \text{Equation 2}$$

Solution

We can obtain coefficients of y that differ only in sign by multiplying the first equation by 4 and the second equation by 3.

$$5x + 3y = 6 \quad \Longrightarrow \quad 20x + 12y = 24 \qquad \text{Multiply Equation 1 by 4}$$
$$2x - 4y = 5 \quad \Longrightarrow \quad \underline{6x - 12y = 15} \qquad \text{Multiply Equation 2 by 3}$$
$$26x \qquad\quad = 39 \qquad \text{Add equations}$$

From this equation, we can see that $x = \frac{3}{2}$. By back-substituting this value of x into the second equation, we can solve for y, as follows.

$$2x - 4y = 5 \qquad \text{Equation 2}$$
$$2\left(\frac{3}{2}\right) - 4y = 5 \qquad \text{Replace } x \text{ by } \frac{3}{2}$$
$$3 - 4y = 5$$
$$-4y = 2$$
$$y = -\frac{1}{2} \qquad \text{Solve for } y$$

Therefore, the solution is $\left(\frac{3}{2}, -\frac{1}{2}\right)$. Check this solution in both of the original equations.   ■

MATH MATTERS

Magic Squares

A magic square is a square table of consecutive positive integers (from 1 to n) in which each horizontal row, vertical column, and diagonal add up to the same number. Here is an example.

4	3	8
9	5	1
2	7	6

Horizontal: $4 + 3 + 8 = 15$

$9 + 5 + 1 = 15$

$2 + 7 + 6 = 15$

Vertical: $4 + 9 + 2 = 15$

$3 + 5 + 7 = 15$

$8 + 1 + 6 = 15$

Diagonal: $4 + 5 + 6 = 15$

$8 + 5 + 2 = 15$

This engraving by Albrecht Dürer (1471–1528) shows a magic square in the upper right corner.

Can you complete the following magic squares? The magic square on the left has a solution in which each row, column, and diagonal add up to 34, and the magic square on the right has a solution in which each row, column, and diagonal add up to 65. (The answer is given in the back of the text.)

16	3	2	13
5	10	11	8

17	24	1	8	15
23	5	7	14	16

Choosing Methods

To decide which of the three methods (graphing, substitution, or elimination) to use to solve a system of two linear equations, we suggest the following guidelines.

Guidelines for Solving a System of Linear Equations

To decide whether to use the method of graphing, substitution, or elimination, consider the following.

1. The graphing method is useful for approximating the solution and for giving an overall picture of how one variable changes with respect to the other.

2. To find exact solutions, use either substitution or elimination.

3. For systems of equations in which one variable has a coefficient of 1, substitution may be more efficient than elimination.

4. In other cases the elimination method is usually more efficient.

In the next example, note how we can use the method of elimination to determine that a system of linear equations has no solution.

EXAMPLE 4 ■ **The Method of Elimination: No-Solution Case**

Solve the following system of linear equations.

$$2x - 6y = 5 \qquad \text{Equation 1}$$
$$3x - 9y = 2 \qquad \text{Equation 2}$$

Solution

To obtain coefficients of x that differ only in sign, we multiply the first equation by 3 and the second equation by -2.

$$2x - 6y = 5 \implies 6x - 18y = 15 \qquad \text{Multiply Equation 1 by 3}$$
$$3x - 9y = 2 \implies -6x + 18y = -4 \qquad \text{Multiply Equation 2 by } -2$$
$$ 0 = 11 \qquad \text{False statement}$$

Since there are no values of x and y for which $0 = 11$, we conclude that the system is inconsistent and has no solution. ■

Example 5 shows how the method of elimination works with a system that has infinitely many solutions.

EXAMPLE 5 ■ **The Method of Elimination: Many-Solutions Case**

Solve the following system of linear equations.

$$2x - 6y = -5 \qquad \text{Equation 1}$$
$$-4x + 12y = 10 \qquad \text{Equation 2}$$

Solution

To obtain coefficients of x that differ only in sign, multiply the first equation by 2.

$$2x - 6y = -5 \implies 4x - 12y = -10 \qquad \text{Multiply Equation 1 by 2}$$
$$\underline{-4x + 12y = 10} \implies \underline{-4x + 12y = 10} \qquad \text{Equation 2}$$
$$0 = 0 \qquad \text{Add equations}$$

Since the two equations turned out to be equivalent, we conclude that the system has infinitely many solutions. The solution set consists of all ordered pairs (x, y) such that $2x - 6y = -5$. ∎

As our last example in this section, we show how the method of elimination works with a system of linear equations having decimal coefficients.

EXAMPLE 6 ■ **Solving a Linear System Having Decimal Coefficients**

Solve the following system of linear equations.

$$0.02x - 0.05y = -0.38 \qquad \text{Equation 1}$$
$$0.03x + 0.04y = 1.04 \qquad \text{Equation 2}$$

Solution

Because the coefficients in this system have two decimal places, we begin by multiplying each equation by 100. This produces a system in which the coefficients are all integers.

$$2x - 5y = -38 \qquad \text{Revised Equation 1}$$
$$3x + 4y = 104 \qquad \text{Revised Equation 2}$$

Now, to obtain coefficients of x that differ only in sign, we multiply the first equation by 3 and the second equation by -2.

$$2x - 5y = -38 \implies 6x - 15y = -114 \qquad \text{Multiply Equation 1 by 3}$$
$$\underline{3x + 4y = 104} \implies \underline{-6x - 8y = -208} \qquad \text{Multiply Equation 2 by } -2$$
$$-23y = -322 \qquad \text{Add equations}$$

Thus, we find that

$$y = \frac{-322}{-23} = 14.$$

Back-substituting this value into Equation 2 produces the following.

$$3x + 4y = 104 \qquad \text{Equation 2}$$
$$3x + 4(14) = 104 \qquad \text{Replace } y \text{ by } 14$$
$$3x + 56 = 104$$
$$3x = 48$$
$$x = 16 \qquad \text{Solve for } x$$

Therefore, the solution is $(16, 14)$. Check this solution in both of the original equations. ∎

DISCUSSION PROBLEM ■ **A General Solution**

Consider the following *general* system of linear equations.

$$a_1 x + b_1 y = c_1 \qquad \text{Equation 1}$$
$$a_2 x + b_2 y = c_2 \qquad \text{Equation 2}$$

Using the method of elimination, we can determine that the solution is given by

$$x = \frac{c_1 b_2 - c_2 b_1}{a_1 b_2 - a_2 b_1} \qquad \text{and} \qquad y = \frac{a_1 c_2 - a_2 c_1}{a_1 b_2 - a_2 b_1}.$$

What conditions can you place on the coefficients a_1, a_2, b_1, and b_2 so that the system will be guaranteed to have *exactly one* solution? ■

Warm-Up

The following warm-up exercises involve skills that were covered in earlier sections. You will use these skills in the exercise set for this section.

In Exercises 1–4, perform the indicated operations and simplify.

1. $(3x + 2y) - 2(x + y)$ **2.** $(-10u + 3v) + 5(2u - 8v)$

3. $x^2 + (x - 3)^2 + 6x$ **4.** $y^2 - (y + 1)^2 + 2y$

In Exercises 5 and 6, solve the given equation.

5. $3x + (x - 5) = 19$ **6.** $3t - 2(t + 1) = 4$

In Exercises 7–10, determine whether the lines represented by the pair of equations are parallel, perpendicular, or neither.

7. $2x - 3y = -10$
$3x + 2y = 11$

8. $x - 3y = 2$
$6x + 2y = 4$

9. $4x - 12y = 5$
$-2x + 6y = 3$

10. $5x + y = 2$
$3x + 2y = 1$

9.3   EXERCISES

In Exercises 1–10, use the method of elimination to solve the given system of linear equations.

1. $2x + y = 4$
   $x - y = 2$

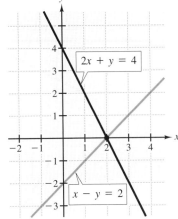

2. $x + 3y = 2$
   $-x + 2y = 3$

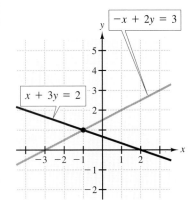

3. $x - y = 0$
   $3x - 2y = -1$

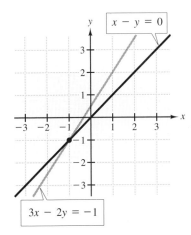

4. $2x - y = 2$
   $4x + 3y = 24$

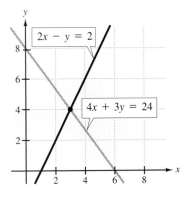

5. $x - y = 1$
   $-2x + 2y = 5$

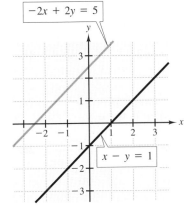

6. $3x + 2y = 2$
   $6x + 4y = 14$

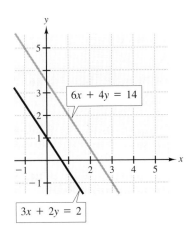

7. $3x - 2y = 6$
$-6x + 4y = -12$

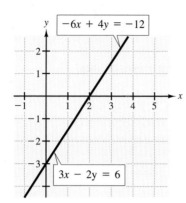

8. $x - 2y = 5$
$6x + 2y = 7$

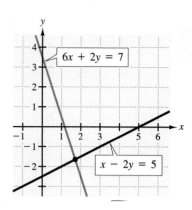

9. $9x - 3y = -1$
$3x + 6y = -5$

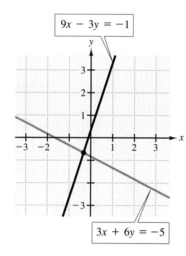

10. $5x + 3y = 18$
$2x - 7y = -1$

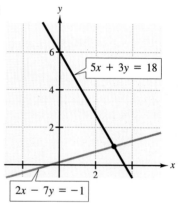

In Exercises 11–30, use the method of elimination to solve the given system of linear equations.

11. $x - y = 4$
$x + y = 12$

12. $-x + 2y = 12$
$x + 6y = 20$

13. $3x - 5y = 1$
$2x + 5y = 9$

14. $x + 2y = 14$
$x - 2y = 10$

15. $x + 7y = 12$
$3x - 5y = 10$

16. $2x + 3y = 18$
$5x - y = 11$

17. $5x + 2y = 7$
$3x - y = 13$

18. $4x + 3y = 8$
$x - 2y = 13$

19. $3x + 2y = 10$
$2x + 5y = 3$

20. $4x + 5y = 7$
$6x - 2y = -18$

21. $6r - 5s = 3$
$-12r + 10s = 5$

22. $\frac{2}{3}x + \frac{1}{6}y = \frac{2}{3}$
$4x + y = 4$

23. $2u + v = 120$
$u + 2v = 120$

24. $5u + 6v = 14$
$3u + 5v = 7$

25. $3a + 3b = 7$
$3a + 5b = 3$

26. $5a + 8b = 11$
$8a + 25b = 31$

27. $0.02x - 0.05y = -0.19$
$0.03x + 0.04y = 0.52$

28. $0.05x - 0.03y = 0.21$
$x + y = 9$

29. $\frac{1}{2}s - t = \frac{3}{2}$
$4s + 2t = 27$

30. $0.2u - 0.1v = 1$
$-0.8u + 0.4v = 3$

In Exercises 31–38, use the more convenient method (substitution or elimination) to solve the given system.

31. $6x + 21y = 132$
$6x - 4y = 32$

32. $-2x + y = 12$
$2x + 3y = 20$

33. $y = 2x - 1$
$y = x + 1$

34. $x + y = 0$
$8x + 3y = 15$

35. $2x - y = 4$
$y = x$

36. $-8x + 11y = 52$
$14x - 9y = 32$

37. $-4x + 3y = 11$
$3x - 10y = 15$

38. $3x - 2y = 0$
$0.2x + 0.8y = 0$

In Exercises 39 and 40, find a system of linear equations with integer coefficients that has the given solution. (*Note:* Each problem has many correct answers.)

39. $\left(6, \frac{4}{3}\right)$

40. $(10, -15)$

41. *Number Problem* The sum of two numbers is 154, and the difference of the numbers is 38. Find the numbers.

42. *Investment* A total of $10,000 is invested in two funds paying 7.5% and 8% simple interest. The combined annual interest for the two funds is $787.50. How much of the $10,000 is invested at each rate?

43. *Ticket Sales* On the first night of a theater production, income from ticket sales was $3,799, and on the second night income was $4,905. The first night 213 student tickets were sold and 632 general-admission tickets were sold. The second night 275 student tickets were sold and 816 general-admission tickets were sold. Find the price of each type of ticket.

44. Answer the given questions for the following system of linear equations.

$x + y = 8$
$2x + 2y = k$

(a) Find the value(s) of k for which the system has an infinite number of solutions.

(b) Find one value of k for which the system has no solutions.

(c) Can the system have a single solution for some value of k? Why or why not?

SECTION 9.4	**Applications of Systems of Linear Equations** *Constructing Systems of Linear Equations* • *Applications of Systems of Linear Equations*

Constructing Systems of Linear Equations

We stated at the beginning of this chapter that systems of linear equations have many applications in science, business, health services, and government. You may be wondering, How can we tell which application can be solved using a system of linear equations? The answer comes from the following considerations.

1. Does the problem involve more than one unknown quantity?

2. Are there two or more equations or conditions to be satisfied?

If one or both of these conditions occur, then the appropriate mathematical model for the problem may be a system of linear equations. Examples 1 and 2 show how to construct such a model.

EXAMPLE 1 ■ Constructing a System of Linear Equations

A total of \$12,000 is invested in two funds paying 9% and 11% simple interest. The combined annual interest for the two funds is \$1,180. How much of the \$12,000 is invested at each rate?

Solution

Notice that this problem has two unknowns: the amount (in dollars) invested at 9% and the amount (in dollars) invested at 11%. As we did in Chapter 4, we begin with a verbal model.

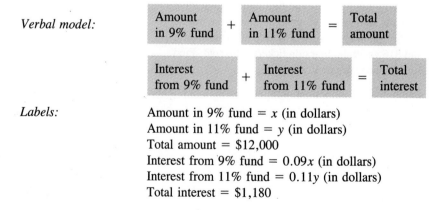

Verbal model:

Amount in 9% fund	+	Amount in 11% fund	=	Total amount

Interest from 9% fund	+	Interest from 11% fund	=	Total interest

Labels:

Amount in 9% fund = x (in dollars)
Amount in 11% fund = y (in dollars)
Total amount = \$12,000
Interest from 9% fund = $0.09x$ (in dollars)
Interest from 11% fund = $0.11y$ (in dollars)
Total interest = \$1,180

System of equations:
$$x + y = 12{,}000 \qquad \text{Equation 1}$$
$$0.09x + 0.11y = 1{,}180 \qquad \text{Equation 2}$$

Recall that we solved this problem in Example 3 in Section 9.2. There we found that the solution is $7,000 in the 9% fund and $5,000 in the 11% fund. ∎

EXAMPLE 2 ∎ **Constructing a System of Linear Equations**

The sum of two numbers is 116. The larger number is one less than twice the smaller number. Find these two numbers.

Solution

In this problem we see that there are two unknowns: the smaller number and the larger number.

Verbal model: Smaller number + Larger number = 116

Larger number = 2 · Smaller number − 1

Labels: Smaller number = x
Larger number = y

System of equations: $x + y = 116$ Equation 1
$y = 2x - 1$ Equation 2

We can solve this system of linear equations (using substitution) as follows.

$$x + y = 116 \qquad \text{Equation 1}$$
$$x + (2x - 1) = 116 \qquad \text{Replace } y \text{ by } 2x - 1$$
$$3x = 117$$
$$x = 39 \qquad \text{Solve for } x$$

Back-substituting this value for x into the second equation produces the following.

$$y = 2x - 1 \qquad \text{Equation 2}$$
$$y = 2(39) - 1 \qquad \text{Replace } x \text{ by } 39$$
$$y = 77 \qquad \text{Solve for } y$$

Therefore, the smaller number is 39 and the larger number is 77. Check this solution in the original statement of the problem. ∎

Applications of Systems of Linear Equations

Many of the problems solved in Chapter 4 using one variable can now be solved using a system of linear equations in two variables. In the next example, we look at a coin problem that was solved using only one variable. (See Example 2 in Section 4.4.)

EXAMPLE 3   ■   A Coin Mixture Problem

A cash register contains $20.90 in dimes and quarters. If there are 119 coins in all, how many of each coin are in the cash register?

Solution

Note that this problem involves two unknowns: the number of dimes and the number of quarters.

Verbal model:
| Number of dimes | + | Number of quarters | = | 119 |

| Value of dimes | + | Value of quarters | = | $20.90 |

Labels: Number of dimes $= x$
Number of quarters $= y$
Value of dimes $= 0.10x$ (in dollars)
Value of quarters $= 0.25y$ (in dollars)

System of equations:
$$x + y = 119 \qquad \text{Equation 1}$$
$$0.10x + 0.25y = 20.90 \qquad \text{Equation 2}$$

To solve this equation, we use the method of elimination. We begin by multiplying the second equation by 100 to eliminate the decimal points. Then, to obtain coefficients that differ only in sign, we multiply the first equation by -10.

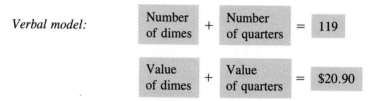

$$x + y = 119 \quad \Longrightarrow \quad -10x - 10y = -1{,}190$$
$$0.10x + 0.25y = 20.90 \quad \Longrightarrow \quad \underline{10x + 25y = 2{,}090}$$
$$15y = 900$$

Therefore, it follows that $y = 60$. By back-substituting this value of y into the first equation, we find that $x = 119 - 60 = 59$. Thus, we conclude that the cash register has 59 dimes and 60 quarters. (Note that this is the same solution we obtained in Example 2 in Section 4.4.)   ■

EXAMPLE 4   ■   A Problem from Geometry

A rectangle is twice as long as it is wide and its perimeter is 132 inches. Find the dimensions of the rectangle using a system of linear equations.

Solution

The two unknowns in this problem are the length and width of the rectangle, as shown in Figure 9.8.

Verbal model:

$$\left(2 \cdot \boxed{\text{Width of rectangle}}\right) + \left(2 \cdot \boxed{\text{Length of rectangle}}\right) = \boxed{\text{Perimeter of rectangle}}$$

$$\boxed{\text{Length of rectangle}} = 2 \cdot \boxed{\text{Width of rectangle}}$$

Labels:

Width of rectangle $= w$ (in inches)
Length of rectangle $= l$ (in inches)
Perimeter of rectangle $= 132$ (inches)

System of equations:

$$2w + 2l = 132 \qquad \text{Equation 1}$$
$$l = 2w \qquad \text{Equation 2}$$

To solve this system of equations, we use the method of substitution, as follows.

$2w + 2l = 132$	Equation 1
$2w + 2(2w) = 132$	Replace l by $2w$
$6w = 132$	
$w = 22$ inches	Solve for w

Back-substituting this value of w into the second equation produces $l = 2w = 2(22) = 44$ inches. Therefore, the rectangle has a width of 22 inches and a length of 44 inches. Check this solution in the original statement of the problem.

FIGURE 9.8

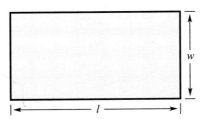

EXAMPLE 5 ■ **Selling Price and Wholesale Cost**

The selling price of a pair of ski boots is $99.20. The markup rate is 55% of the wholesale cost. What is the wholesale cost?

Solution

We could solve this problem using only one unknown (the wholesale cost). However, for the sake of illustration, let's use two unknowns: the wholesale cost and markup (in dollars).

Verbal model:

$$\boxed{\text{Wholesale cost}} + \boxed{\text{Markup (in dollars)}} = \boxed{\text{Selling price}}$$

$$\boxed{\text{Markup (in dollars)}} = 0.55 \cdot \boxed{\text{Wholesale cost}}$$

Labels: Wholesale cost = C (in dollars)
Markup = M (in dollars)
Selling price = $99.20

System of equations: $C + M = 99.20$ Equation 1
$M = 0.55C$ Equation 2

To solve this system of linear equations, we use the method of substitution.

$C + M = 99.20$ Equation 1

$C + 0.55C = 99.20$ Replace M by $0.55C$

$1.55C = 99.20$

$C = \$64$ Solve for C

Therefore, the wholesale cost of the pair of boots is $64.00. (We weren't asked to find the markup, so it is unnecessary to back-substitute to find the value of M.)   ∎

EXAMPLE 6   ■   A Mixture Problem

A company with two stores buys six large delivery vans and five small delivery vans. The first store receives four of the large vans and two of the small vans for a total cost of $160,000. The second store receives two of the large vans and three of the small vans for a total cost of $128,000. What is the cost of each type of van?

Solution

The two unknowns in this problem are the costs of the two types of vans.

Verbal model:

$$\left(4 \cdot \boxed{\text{Cost of large van}}\right) + \left(2 \cdot \boxed{\text{Cost of small van}}\right) = \boxed{\$160,000}$$

$$\left(2 \cdot \boxed{\text{Cost of large van}}\right) + \left(3 \cdot \boxed{\text{Cost of small van}}\right) = \boxed{\$128,000}$$

Labels: Cost of large van = x (in dollars)
Cost of small van = y (in dollars)

System of equations: $4x + 2y = 160,000$ Equation 1
$2x + 3y = 128,000$ Equation 2

To solve this system of linear equations, we use the method of elimination. To obtain coefficients that differ only in sign, we multiply the second equation by -2.

$$4x + 2y = 160,000 \implies 4x + 2y = 160,000$$
$$2x + 3y = 128,000 \implies -4x - 6y = -256,000$$
$$\overline{\; -4y = -96,000}$$

Thus, the cost of each small van is $y = \$24,000$. By back-substituting this value into Equation 1, we can find the cost of each large van.

$$4x + 2y = 160,000 \qquad \text{Equation 1}$$
$$4x + 2(24,000) = 160,000 \qquad \text{Replace } y \text{ by } 24,000$$
$$4x = 112,000$$
$$x = 28,000 \qquad \text{Solve for } x$$

Thus, the cost of each large van is $x = \$28,000$. Check this solution in the original statement of the problem.   ■

EXAMPLE 7   ■   **An Application Involving Two Speeds**

Suppose you take a motorboat trip on a river—18 miles upstream and 18 miles back downstream. You run the motor at the same speed going up and down the river, but because of the current of the river, the trip upstream takes longer than the trip downstream. You don't know the speed of the river's current, but you know that the trip upstream takes one and one-half hours and the trip downstream takes only one hour. From this information, determine the speed of the current.

Solution

One unknown in this problem is the speed of the current. The other unknown is the speed of the boat in still water. To set up a model, we use the fact that the effective speed of the boat (relative to the land) going upstream is $18/\left(1\frac{1}{2}\right) = 12$ miles per hour, and going downstream is $18/1 = 18$ miles per hour. In the following verbal model, note that the current fights against you going upstream but helps you going downstream.

Verbal model:

$$\boxed{\text{Boat speed (still water)}} - \boxed{\text{Speed of current}} = \boxed{\text{Upstream speed}}$$

$$\boxed{\text{Boat speed (still water)}} + \boxed{\text{Speed of current}} = \boxed{\text{Downstream speed}}$$

Labels:

Boat speed in still water $= x$ (in miles per hour)
Current speed $= y$ (in miles per hour)
Upstream speed $= 12$ miles per hour
Downstream speed $= 18$ miles per hour

System of equations:

$$x - y = 12 \qquad \text{Equation 1}$$
$$x + y = 18 \qquad \text{Equation 2}$$

To solve this system of linear equations, we use the method of elimination.

$$
\begin{array}{ll}
x - y = 12 & \text{Equation 1} \\
\underline{x + y = 18} & \text{Equation 2} \\
2x \quad\ = 30 & \text{Add equations}
\end{array}
$$

Thus, the speed of the boat in still water is $x = 15$ miles per hour. To find the speed of the current, we back-substitute this value into Equation 2.

$$
\begin{array}{ll}
x + y = 18 & \text{Equation 2} \\
15 + y = 18 & \text{Replace } x \text{ by 15} \\
y = 3 & \text{Solve for } y
\end{array}
$$

Therefore, the speed of the current is 3 miles per hour. Check this solution in the original statement of the problem. ∎

EXAMPLE 8 ∎ **A Mixture Problem**

A chemist has two different solutions. One is 50% alcohol and 50% water, and the other is 75% alcohol and 25% water. How much of each type of solution should be mixed to obtain 8 liters of solution comprised of 60% alcohol and 40% water?

Solution

The two unknowns in this problem are the amounts of each type of solution.

Verbal model:

$$
\boxed{\text{Liters of 50\% solution}} + \boxed{\text{Liters of 75\% solution}} = \boxed{\text{8 liters of 60\% solution}}
$$

$$
\boxed{\text{Alcohol in 50\% solution}} + \boxed{\text{Alcohol in 75\% solution}} = \boxed{\text{Alcohol in 60\% solution}}
$$

Labels:

Number of liters of 50% solution $= x$
Number of liters of 75% solution $= y$
Number of liters of 60% solution $= 8$
Alcohol in 50% solution $= 0.50x$ (in liters)
Alcohol in 75% solution $= 0.75y$ (in liters)
Alcohol in 60% solution $= 0.60(8) = 4.8$ (in liters)

System of equations:

$$
\begin{array}{ll}
x + \quad y = 8 & \text{Equation 1} \\
0.50x + 0.75y = 4.8 & \text{Equation 2}
\end{array}
$$

To solve this system of linear equations, we use the method of elimination. To obtain coefficients that differ only in sign, we multiply the first equation by -50 and the second equation by 100.

$$\begin{array}{rl} x + y = 8 & \\ 0.50x + 0.75y = 4.8 & \end{array} \quad \Longrightarrow \quad \begin{array}{rl} -50x - 50y = -400 \\ 50x + 75y = 480 \\ \hline 25y = 80 \end{array}$$

Therefore, $y = \frac{80}{25} = 3.2$. Back-substituting this value into the first equation produces $x = 8 - y = 8 - 3.2 = 4.8$. Therefore, the chemist should use 4.8 liters of the 50% solution and 3.2 liters of the 75% solution. Check this answer in the original statement of the problem. ∎

DISCUSSION PROBLEM   ■   A Word Problem with No Solution

Write a short paragraph explaining why the following problem has no solution.

The sum of two numbers is 79, and the difference of the two numbers is 37. One of the numbers is 9 more than the other number. Find the numbers. ■

■

Warm-Up

The following warm-up exercises involve skills that were covered in earlier sections. You will use these skills in the exercise set for this section.

In Exercises 1–6, write an algebraic expression that represents the given statement.

1. The amount of money (in dollars) represented by m nickels and n quarters

2. The amount of income tax due on a taxable income of I dollars (The income tax rate is 13%.)

3. The amount of time required to travel 250 miles at an average speed of r miles per hour

4. The discount on the price of a product with a list price of L dollars and a discount rate of 15%

5. The perimeter of a rectangle of length l and width $l/2$

6. The sum of a number n and two and one-half times that number

In Exercises 7–10, solve the given system of linear equations.

7. $\begin{aligned} x + y &= 25 \\ y &= 10 \end{aligned}$

8. $\begin{aligned} 2x - 3y &= 4 \\ 6x &= -12 \end{aligned}$

9. $\begin{aligned} x + y &= 32 \\ x - y &= 24 \end{aligned}$

10. $\begin{aligned} 2r - s &= 5 \\ r + 2s &= 10 \end{aligned}$

9.4   EXERCISES

Number Problem In Exercises 1–6, find two numbers that satisfy the given requirements.

1. The sum of the numbers is 67, and their difference is 17.

2. The sum of the numbers is 75, and their difference is 15.

3. The sum of the larger number and twice the smaller is 100, and their difference is 10.

4. The sum of the numbers is 46, and the larger number is 2 less than twice the smaller.

5. The sum of the numbers is 132, and the larger number is 6 more than twice the smaller.

6. The sum of three times the smaller number and four times the larger is 225. Nine times the smaller plus two times the larger gives the same sum.

Coin Problems In Exercises 7–12, determine the number of each type of coin.

Number of coins	Types of coins	Value
7. 21	Dimes and quarters	$4.05
8. 21	Dimes and quarters	$2.70
9. 35	Nickels and quarters	$5.75
10. 35	Nickels and quarters	$7.75
11. 44	Nickels and dimes	$3.00
12. 28	Nickels and dimes	$2.40

Dimensions of a Rectangle In Exercises 13–18, find the dimensions of the rectangle that meet the specified conditions.

Perimeter	Relationship Between Length and Width
13. 40 feet	The length is 4 feet greater than the width.
14. 220 inches	The width is 10 inches less than the length.
15. 16 yards	The width is six-tenths of the length.
16. 48 meters	The length is twice the width.
17. 35.2 meters	The length is 120% of the width.
18. 35 feet	The width is 75% of the length.

19. *Wholesale Cost* A watch sells for $108.75. The markup rate is 45% of the wholesale cost. Find the wholesale cost.

20. *Wholesale Cost* The selling price of a cordless phone is $119.91. The markup rate is 40% of the wholesale cost. Find the wholesale cost.

21. *List Price* The sale price of a watch is $35.98. The discount is 30% of the list price. Find the list price.

22. *List Price* The sale price of a microwave oven is $275. The discount is 20% of the list price. Find the list price.

23. *Investment* A combined total of $12,000 is invested in two bonds that pay 10.5% and 12% simple interest. The annual interest is $1,380. How much is invested in each bond?

24. *Investment* A combined total of $8,000 is invested in two bonds that pay 7% and 8.5% simple interest. The annual interest is $635. How much is invested in each bond?

25. *Ticket Sales* Five hundred tickets were sold for a fund-raising dinner. The receipts totaled $3,312.50. Adult tickets were $7.50 each and children's tickets were $4.00 each. How many tickets of each type were sold?

26. *Ticket Sales* A fund-raising dinner was held on two consecutive nights. On the first night 425 adult tickets and 316 children's tickets were sold, for a total of $2,915.00. On the second night 542 adult tickets and 345 children's tickets were sold, for a total of $3,572.50. Find the price of each type of ticket.

27. *Gasoline Mixture* The total cost of 15 gallons of regular unleaded gasoline and 10 gallons of premium unleaded gasoline is $35.50. Premium unleaded gasoline costs $0.20 more per gallon than regular unleaded. Find the price per gallon for each grade of gasoline.

28. *Gasoline Mixture* The total cost of 8 gallons of regular unleaded gasoline and 12 gallons of premium unleaded gasoline is $27.84. Premium unleaded gasoline costs $0.17 more per gallon than regular unleaded. Find the price per gallon for each grade of gasoline.

29. *Nut Mixture* Ten pounds of mixed nuts sell for $5.86 per pound. The mixture is obtained from two kinds of nuts, with one variety priced at $4.25 per pound and the other at $6.55 per pound. How many pounds of each variety of nuts were used in the mixture?

30. *Feed Mixture* How many tons of hay at $110 per ton and $60 per ton must be purchased to have 100 tons of hay with an average value of $75 per ton?

31. *Mixture Problem* How many liters of a 35% alcohol solution must be mixed with a 60% solution to obtain 10 liters of a 50% solution?

32. *Mixture Problem* Ten gallons of 30% acid solution are obtained by mixing a 20% solution with a 50% solution. How many gallons of each solution must be used to obtain the desired mixture?

33. *Average Speed* A van travels for two hours at an average speed of 40 miles per hour. How much longer must the van travel at an average speed of 55 miles per hour so that the average speed for the total trip will be 45 miles per hour?

34. *Average Speed* A van travels for three hours at an average speed of 40 miles per hour. How much longer must the van travel at an average speed of 55 miles per hour so that the average speed for the total trip will be 50 miles per hour?

35. *Airplane Speed* Two planes start from the same airport and fly in opposite directions. The second plane starts one-half hour after the first plane, but its speed is 50 miles per hour faster. Two hours after the first plane starts, the planes are 2,000 miles apart. Find the ground speed of each plane.

36. *Airplane Speed* An airplane flying into a head wind travels 1,800 miles in 3 hours and 36 minutes. On the return flight, the same distance is traveled in 3 hours. Find the speed of the plane in still air and the speed of the wind, assuming that both remain constant through the round-trip.

37. *Number Problem* The sum of the digits of a given two-digit number is 12. If the digits are reversed, the number is increased by 36. Find the number.

38. *Number Problem* The sum of the digits of a given two-digit number is 9. If the digits are reversed, the number is decreased by 45. Find the number.

39. *Best-Fitting Line* The line $y = mx + b$ that best fits the three points $(0, 0)$, $(1, 2)$, and $(2, 2)$ is given by the following system of linear equations.

$$3b + 3m = 4$$
$$3b + 5m = 6$$

(a) Solve the system and find the equation of the best-fitting line.
(b) Plot the three points and sketch the graph of the best-fitting line.

40. *Best-Fitting Line* The line $y = mx + b$ that best fits the three points $(0, 2)$, $(1, 1)$, and $(3, 0)$ is given by the following system of linear equations.

$$3b + 4m = 3$$
$$4b + 10m = 1$$

(a) Solve the system and find the equation of the best-fitting line.
(b) Plot the three points and sketch the graph of the best-fitting line.

CHAPTER 9 SUMMARY

As you review and prepare for a test on this chapter, first try to obtain a global view of what was discussed. Then review the specific skills needed in each category.

Solving Systems of Linear Equations

- *Test* whether a specified point is a solution to a given system of linear equations.

  Section 9.1, Exercises 1–4

- *Solve* a system of linear equations by using graphs to locate the point of intersection.

  Section 9.1, Exercises 11–30

- *Determine* the number of solutions to a system of linear equations from the graphs for the system.

  Section 9.1, Exercises 5–10

- *Solve* a system of linear equations using the method of substitution.

  Section 9.2, Exercises 1–30

- *Solve* a system of linear equations using the method of elimination.

  Section 9.3, Exercises 1–8

Applications of Systems of Linear Equations

- *Create and solve* number problems from verbal statements.

  Section 9.4, Exercises 1–6

- *Create and solve* mixture problems from verbal statements.

  Section 9.4, Exercises 7–12, 27–32

- *Create and solve* consumer problems from verbal statements.

  Section 9.4, Exercises 19–26

- *Create and solve* rate problems from verbal statements.

  Section 9.4, Exercises 33–36

Chapter 9 Review Exercises

In Exercises 1–4, use the graphical method to solve the given system of linear equations.

1. $x + y = 2$
$x - y = 0$

2. $2x = 3(y - 1)$
$y = x$

3. $x - y = 9$
$-x + y = 1$

4. $x + y = -1$
$3x + 2y = 0$

In Exercises 5–8, use the method of substitution to solve the given system of linear equations.

5. $y = 2x$
$y = x + 4$

6. $x = y + 3$
$x = y + 1$

7. $2x - y = 2$
$6x + 8y = 39$

8. $y = -4x + 1$
$y = x - 4$

In Exercises 9–12, use the method of elimination to solve the given system of linear equations.

9. $5x + 4y = 2$
$-x + y = -22$

10. $2x - 5y = 2$
$3x - 7y = 1$

11. $0.2x - 0.1y = 0.07$
$0.4x - 0.5y = -0.01$

12. $2x + y = 0.3$
$3x - y = -1.3$

In Exercises 13–36, use the method of your choice to solve the given system of linear equations.

13. $6x - 5y = 0$
$y = 6$

14. $-x + 2y = 2$
$x = 4$

15. $x - y = 0$
$x - 6y = 5$

16. $-x + y = 4$
$x + y = 4$

17. $6x - 3y = 27$
$-2x + y = -9$

18. $-\dfrac{1}{4}x + \dfrac{2}{3}y = 1$
$3x - 8y = 1$

19. $5x + 8y = 8$
$x - 8y = 16$

20. $-7x + 9y = 9$
$2x + 9y = -18$

21. $2x + 5y = 20$
$4x + 5y = 10$

22. $-3x + 4y = 24$
$-5x + 4y = 8$

23. $-x + 4y = 4$
$x + y = 6$

24. $x + 2y = 2$
$x - 4y = 20$

25. $x + y = 0$
$2x + y = 0$

26. $40x + 30y = 24$
$20x - 50y = -14$

27. $0.2u + 0.3v = 0.14$
$0.4u + 0.5v = 0.20$

28. $12s + 42t = -17$
$30s - 18t = 19$

29. $\dfrac{x}{3} + \dfrac{4y}{7} = 3$
$2x + 3y = 15$

30. $\dfrac{x}{2} - \dfrac{y}{3} = 0$
$3x + 2(y + 5) = 10$

31. $-x + 2y = 1.5$
$2x - 4y = 3$

32. $x - 3y = 5$
$-2x + 6y = 4$

33. $x + 2y = 7$
$2x + y = 8$

34. $2x + 6y = 16$
$2x + 3y = 7$

35. $8x - 4y = 7$
$5x + 2y = 1$

36. $2x - y = -0.1$
$3x + 2y = 1.6$

In Exercises 37 and 38, find a system of linear equations with integer coefficients that has the given solution. (*Note:* Each problem has many correct answers.)

37. $\left(3, \dfrac{8}{3}\right)$

38. $\left(-\dfrac{2}{3}, 5\right)$

39. *Rewriting a Fraction* $(x + 5)/(x^2 + x - 2)$ can be written as the sum of two fractions as follows.

$$\frac{x + 5}{x^2 + x - 2} = \frac{x + 5}{(x - 1)(x + 2)} = \frac{A}{x - 1} + \frac{B}{x + 2}$$

The numbers A and B are the solutions of the system

$$A + B = 1$$
$$2A - B = 5.$$

Solve the system and verify that the sum of the two resulting fractions is the original fraction.

40. *Number of Coins* A group of 15 coins (dimes and quarters) has a value of $2.85. Find the number of each type of coin.

41. *Dimensions of a Rectangle* A rectangular sign has a perimeter of 120 inches. The height of the sign is two-thirds of its width. Find the dimensions of the sign.

42. *Wholesale Cost* The selling price of a VCR is $434. The markup rate is 40% (of the wholesale cost). Find the wholesale cost.

43. *Price per Gallon* Suppose you buy two gallons of gasoline for your lawn mower and five gallons of diesel fuel for your lawn and garden tractor. The total bill is $9.75. Gasoline costs $0.08 more per gallon than diesel fuel. Find the price per gallon of each type of fuel.

44. *Average Speed* A car travels for four hours at an average speed of 50 miles per hour. How much longer must the car travel at an average speed of 65 miles per hour so that the average speed for the total trip will be 55 miles per hour?

45. Consider the following system of linear equations.

$$2x + 3y = 8$$
$$6x + ky = 12$$

(a) Find the value of k for which the system is inconsistent.

(b) Find a value of k for which the system has a unique solution.

(c) Is there a value of k for which the system has an infinite number of solutions? Why or why not?

CHAPTER 9 TEST

Take this test as you would take a test in class. After you are done, check your work with the answers given in the back of the book.

1. Determine which ordered pair is the solution to the system of equations.

$$x - 6y = -19 \qquad \text{(a) } (3, -2)$$
$$4x - 5y = 0 \qquad \text{(b) } (5, 4)$$

2. Determine the number of solutions for the system of equations.

$$3x + 4y = 16$$
$$3x - 4y = 8$$

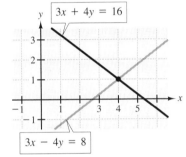

3. Determine the number of solutions for the system of equations.

$$x - 2y = -4$$
$$x - 2y = 2$$

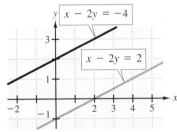

4. Solve the system of equations graphically.

$$x - 2y = -2$$
$$x + y = 4$$

5. Solve the system of equations graphically.

$$2x + y = 4$$
$$x - 2y = -3$$

6. Solve the system of equations by the method of substitution.

$$x + 5y = 10$$
$$4x - 5y = 15$$

7. Solve the system of equations by the method of elimination.

$$x + y = 8$$
$$2x - y = -2$$

8. Solve the system of equations by the method of elimination.

$$7x + 6y = 36$$
$$5x - 4y = 5$$

9. Find the value of a such that the system is inconsistent.

$$ax - 8y = 9$$
$$3x + 4y = 0$$

10. Find a system of linear equations with integer coefficients that has the solution $(-3, 4)$. (The problem has many correct answers.)

11. A rectangle has a perimeter of 40 meters. The length of the rectangle is three times its width. Write a system of linear equations that models the problem. Find the dimensions of the rectangle by solving the system of equations.

12. Twenty liters of 20% acid solution are obtained by mixing a 30% solution and a 5% solution. Write a system of linear equations that models the problem. Solve the system to determine the number of liters of each solution required to obtain the specified mixture.

Radicals

Chapter Overview

We have already studied the rules of algebra as they apply to polynomials and algebraic fractions. In this chapter we extend our study of algebra to include radical expressions such as $\sqrt{x}$ and $\sqrt{x^2 + 1}$. The chapter begins by defining square roots and radicals, then goes on to discuss rules for simplifying radicals and adding, subtracting, multiplying, and dividing radicals. Finally, in Section 10.4 we will see how to solve equations containing radicals.

Square Roots and Radicals

Square Roots • Radicals • Radicals and Calculators

Square Roots

We already know how to find the square of a number (by multiplying the number by itself). For instance, the square of 3 is $3^2 = 9$. We now look at the reverse problem, which is finding a **square root** of a number (by identifying two equal factors of the number). Here are some examples.

Number	Positive Square Root	Equal Factors Check	Negative Square Root	Equal Factors Check
4	2	$(2)(2) = 4$	-2	$(-2)(-2) = 4$
25	5	$(5)(5) = 25$	-2	$(-5)(-5) = 25$
144	12	$(12)(12) = 144$	-12	$(-12)(-12) = 144$

The square of a real number cannot be negative. Thus, as long as we are dealing with real numbers, we do not define the square root of a negative real number. The definition of a square root of a *nonnegative* real number is as follows.

Square Root of a Nonnegative Real Number

If a is a nonnegative real number and b is a real number such that

$$b^2 = a$$

then b is called a **square root** of a.

The real number zero has only one square root, which is also the number zero. All positive real numbers have two square roots: one is called the **positive square root** and the other is called the **negative square root**.

EXAMPLE 1   ■   Finding Square Roots of Numbers

Find all square roots of the following numbers.

(a) 81   (b) 0   (c) -4 (d) $\dfrac{4}{9}$

Solution

(a) There are two different numbers that can be multiplied by themselves to obtain 81. Thus, 81 has two square roots. The positive square root is 9, and the negative square root is -9.

$$\text{Positive square root: } 9 \qquad\qquad \text{Check: } (9)(9) = 81$$
$$\text{Negative square root: } -9 \qquad\qquad \text{Check: } (-9)(-9) = 81$$

(b) There is only one number that can be multiplied by itself to obtain 0. Thus, 0 has only one square root, the number 0.

(c) There is no real number that can be multiplied by itself to obtain −4. Thus, −4 has no square root.

(d) There are two different numbers that can be multiplied by themselves to obtain $\frac{4}{9}$. Thus, $\frac{4}{9}$ has two square roots. The positive square root is $\frac{2}{3}$, and the negative square root is $-\frac{2}{3}$.

$$\text{Positive square root: } \frac{2}{3} \qquad \text{Check: } \left(\tfrac{2}{3}\right)\left(\tfrac{2}{3}\right) = \tfrac{4}{9}$$

$$\text{Negative square root: } -\frac{2}{3} \qquad \text{Check: } \left(-\tfrac{2}{3}\right)\left(-\tfrac{2}{3}\right) = \tfrac{4}{9} \qquad \blacksquare$$

Radicals

In mathematics we use a special symbol, $\sqrt{}$, to denote the nonnegative square root of a number. This symbol is called a **radical sign**. For instance, to denote the positive square root of 4, we can use a radical sign and write

$$\sqrt{4} = 2.$$

Definition of a Radical	If a is a nonnegative real number and b is the *nonnegative* square root of a, then we can write $$\sqrt{a} = b.$$ The number inside the radical sign is called the **radicand**. The entire symbol $\sqrt{a}$ is called a **radical**, and the number b is called the **principal square root** of a.

When used by itself, a radical *always* refers to the principal square root of the radicand. To denote a negative square root, we place a negative sign in front of the radical. For instance, the principal square root of 36 is denoted by $\sqrt{36} = 6$, whereas the negative square root of 36 is denoted by $-\sqrt{36} = -6$.

EXAMPLE 2 ■ Finding the Principal Square Root of a Number

Find the principal square root of each of the following numbers.

(a) 49   (b) 0   (c) −36   (d) $\frac{1}{4}$

no solution   ½

Solution

(a) The number 49 has two square roots, 7 and −7. The principal square root is the positive one.

$$\sqrt{49} = 7 \qquad \text{Principal square root}$$

(b) The number 0 has only one square root, 0. It is the principal square root. That is,

$$\sqrt{0} = 0. \qquad \text{Principal square root}$$

(c) The negative number −36 has no square root, and consequently has no principal square root.

(d) The number $\frac{1}{4}$ has two square roots, $\frac{1}{2}$ and $-\frac{1}{2}$. The principal square root is the positive one.

$$\sqrt{\frac{1}{4}} = \frac{1}{2} \qquad \text{Principal square root} \qquad \blacksquare$$

Integers like 4, 9, 16, and 25, that have integer square roots, are called **perfect squares**. (Rational numbers such as $\frac{1}{4}$, $\frac{1}{9}$, and $\frac{9}{25}$ are also called perfect squares.) The square roots of integers that are not perfect squares are *irrational* numbers. For example, the numbers

$$\sqrt{2}, \qquad \sqrt{3}, \qquad \sqrt{5}, \quad \text{and} \quad \sqrt{6}$$

are all irrational. Remember from Chapter 1 that a rational number is a real number that can be written as the ratio of two integers. (For instance, the numbers 2, $-\frac{3}{2}$, and $\frac{1}{3}$ are rational.) An irrational number is a real number that cannot be written as the ratio of two integers. Thus, when we say that $\sqrt{2}$ is irrational, we are saying that there is no fraction (whose numerator and denominator are integers) that can be multiplied by itself to obtain the number 2.

EXAMPLE 3   ■   Classifying Square Roots as Rational or Irrational

Classify the following numbers as rational or irrational.

(a) $-\sqrt{100}$ (b) $\sqrt{11}$ (c) $\sqrt{\frac{25}{16}}$

Solution

(a) The negative square root of 100 is $-\sqrt{100} = -10$, which is a rational number. (Remember that every integer is a rational number.)

(b) The principal square root of 11, which is $\sqrt{11}$, is irrational because 11 is not a perfect square.

(c) The principal square root of $\frac{25}{16}$ is $\sqrt{\frac{25}{16}} = \frac{5}{4}$, which is a rational number.   ■

Radicals and Calculators

In real-life applications, we often need to use decimal approximations of square roots. Before calculators were available, this was usually done by looking up the decimal approximation in a table of square roots. Today, however, we can use a calculator to easily find decimal approximations for square roots. The key that accomplishes this is labeled $\boxed{\sqrt{}}$ on most calculators. For instance, to find a decimal approximation for $\sqrt{2}$, you could use the following keystrokes.*

Square Root	Keystrokes	Calculator Display	
$\sqrt{2}$	2 $\boxed{\sqrt{}}$	1.4142136	Scientific
$\sqrt{2}$	$\boxed{\sqrt{}}$ 2 $\boxed{\text{ENTER}}$	1.4142136	Graphing

When you are using decimal approximations for irrational numbers, you should remember that the approximation has some round-off error. The more decimal places you list, the more accurate your answer will be. For instance,

$$\sqrt{2} \approx 1.414 \qquad \text{Rounded to three decimal places}$$

is not as accurate as

$$\sqrt{2} \approx 1.4142. \qquad \text{Rounded to four decimal places}$$

EXAMPLE 4 ■ Decimal Approximations of Square Roots

Use a calculator to approximate the following numbers. Round your answers to the indicated number of decimal places.

(a) $\sqrt{3}$ Round to four decimal places.

(b) $-\sqrt{12}$ Round to three decimal places.

(c) $\sqrt{12{,}769}$ Round to two decimal places.

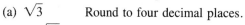

Solution

Square Root	Keystrokes	Calculator Display	Rounded Answer	
(a) $\sqrt{3}$	3 $\boxed{\sqrt{}}$	1.7320508	1.7321	Scientific
$\sqrt{3}$	$\boxed{\sqrt{}}$ 3 $\boxed{\text{ENTER}}$	1.7320508	1.7321	Graphing
(b) $-\sqrt{12}$	12 $\boxed{\sqrt{}}$ $\boxed{+/-}$	-3.4641016	-3.464	Scientific
$-\sqrt{12}$	$\boxed{(-)}$ $\boxed{\sqrt{}}$ 12 $\boxed{\text{ENTER}}$	-3.4641016	-3.464	Graphing
(c) $\sqrt{12{,}769}$	12769 $\boxed{\sqrt{}}$	113	113.00	Scientific
$\sqrt{12{,}769}$	$\boxed{\sqrt{}}$ 12769 $\boxed{\text{ENTER}}$	113	113.00	Graphing

*The graphing calculator keystrokes given in this text correspond to the *TI-81* graphing calculator from *Texas Instruments*. For other graphing calculators, the keystrokes may differ.

When approximating a *negative square root* on a scientific calculator, be sure you press the square root key before pressing the change sign key. For instance, if you used the keystroke sequence 12 $\boxed{+/-}$ $\boxed{\sqrt{}}$, the calculator would think you were asking it to find $\sqrt{-12}$, and it would display some sort of error message. ■

Answers to problems in algebra often involve sums, differences, products, and quotients of integers and square roots. In Example 5 we show how a calculator can be used to approximate this type of expression.

EXAMPLE 5 ■ **Approximating an Expression Involving Square Roots**

Use a calculator to approximate the following expressions.

(a) $1 - \sqrt{2}$ (b) $2 + 3\sqrt{5}$ (c) $\dfrac{1 - 2\sqrt{3}}{4}$

Round your answers to two decimal places.

Solution

(a) The number $1 - \sqrt{2}$ can be approximated using the following keystrokes.

Keystrokes	Display	
1 $\boxed{-}$ 2 $\boxed{\sqrt{}}$ $\boxed{=}$	−0.4142135	Scientific
1 $\boxed{-}$ $\boxed{\sqrt{}}$ 2 $\boxed{\text{ENTER}}$	−0.4142135	Graphing

Therefore, rounded to two decimal places, the answer is

$$1 - \sqrt{2} \approx -0.41.$$

(b) The number $2 + 3\sqrt{5}$ can be approximated using the following keystrokes.

Keystrokes	Display	
2 $\boxed{+}$ 3 $\boxed{\times}$ 5 $\boxed{\sqrt{}}$ $\boxed{=}$	8.7082039	Scientific
2 $\boxed{+}$ 3 $\boxed{\times}$ $\boxed{\sqrt{}}$ 5 $\boxed{\text{ENTER}}$	8.7082039	Graphing

Therefore, rounded to two decimal places, the answer is

$$2 + 3\sqrt{5} \approx 8.71.$$

(c) The number $(1 - 2\sqrt{3})/4$ can be approximated using the following keystrokes.

Keystrokes	Display	
1 $\boxed{-}$ 2 $\boxed{\times}$ 3 $\boxed{\sqrt{}}$ $\boxed{=}$ $\boxed{\div}$ 4 $\boxed{=}$	−0.6160254	Scientific
$\boxed{(}$ 1 $\boxed{-}$ 2 $\boxed{\times}$ $\boxed{\sqrt{}}$ 3 $\boxed{)}$ $\boxed{\div}$ 4 $\boxed{\text{ENTER}}$	−0.6160254	Graphing

Therefore, rounded to two decimal places, the answer is

$$\frac{1 - 2\sqrt{3}}{4} \approx -0.62.$$

■

Even though you can use a calculator to find (or approximate) the square root of any positive real number, we suggest that you memorize the squares of the first fifteen integers $1^2, 2^2, 3^2, 4^2, \ldots, 14^2, 15^2$, as well as the squares of multiples of 10, such as $10^2, 20^2$, and 30^2.

It's easy to make mistakes with a calculator, and knowing the squares of integers will help you estimate square root answers as a check of your calculations. For instance, knowing that $10^2 = 100$ and $20^2 = 400$ can serve as a check when finding $\sqrt{330}$. We know it must lie between

$$10 = \sqrt{100} \quad \text{and} \quad 20 = \sqrt{400}$$

and will be closer to 20 than to 10 because 330 is closer to 400 than to 100. In fact, $\sqrt{330} \approx 18.2$.

EXAMPLE 6   ■   Approximating a Square Root

Give a rough approximation of the following square roots without using a calculator.
(a) $\sqrt{200}$ (b) $\sqrt{110}$

Solution

(a) The number 200 lies between two known perfect squares: $196 = 14^2$ and $225 = 15^2$. Therefore, the square root of 200 must be a number that is between 14 and 15. Since 200 is closer to 196 than to 225, we might roughly approximate $\sqrt{200}$ to be 14.2. (Using a calculator, we find that $\sqrt{200} \approx 14.14$.)

(b) The number 110 lies between two known perfect squares: $100 = 10^2$ and $121 = 11^2$. Therefore, the square root of 110 must be a number that is between 10 and 11. Since 110 is about halfway between 100 and 121, we might roughly approximate $\sqrt{110}$ to be 10.5. (Using a calculator, we find that $\sqrt{110} \approx 10.49$.)   ■

DISCUSSION PROBLEM   ■   Irrationality of $\sqrt{2}$

In this section we mentioned that the square root of 2 is an irrational number. This fact was known to early Greek mathematicians, and a version of their proof that $\sqrt{2}$ is irrational can be found in more advanced textbooks. Because $\sqrt{2}$ is irrational, we know that there is no rational number whose square is 2. However, there are rational numbers whose squares are very close to 2. For instance,

$$\left(\frac{7{,}064}{4{,}995}\right)^2 \approx 2.0000018.$$

Can you find some other rational numbers whose squares are close to 2?   ■

Warm-Up   *The following warm-up exercises involve skills that were covered in earlier sections. You will use these skills in the exercise set for this section.*

In Exercises 1–10, evaluate the given expression.

1. 15^2 **2.** 12^2 **3.** $(-5)^2$ **4.** -5^2

5. $\left(\dfrac{2}{3}\right)^2$ **6.** $\left(\dfrac{4}{5}\right)^2$ **7.** $\dfrac{3}{4}\left(\dfrac{4}{3}\right)^2$ **8.** $\dfrac{5}{6}\left(\dfrac{3}{5}\right)^2$

9. $-\left(\dfrac{1}{3}\right)^2\left(-\dfrac{1}{3}\right)^2$ **10.** $\left(-\dfrac{1}{5}\right)^2\left(\dfrac{1}{5}\right)^2$

10.1 EXERCISES

In Exercises 1–4, fill in the blank with the appropriate real number.

1. $8^2 = 64 \longrightarrow$ Positive square root of 64 is ▢

2. $7^2 = 49 \longrightarrow$ Positive square root of 49 is ▢

3. $(-10)^2 = 100 \longrightarrow$ Negative square root of 100 is ▢

4. $(-13)^2 = 169 \longrightarrow$ Negative square root of 169 is ▢

In Exercises 5–10, find the positive and negative square roots of the real number, if possible. (Do not use a calculator.)

5. 36   **6.** 144   **7.** $\dfrac{9}{16}$ **8.** $\dfrac{4}{25}$ **9.** -16 **10.** -0.09

In Exercises 11–30, evaluate the expression, if possible. (Do not use a calculator.)

11. $\sqrt{100}$ **12.** $\sqrt{64}$ **13.** $-\sqrt{100}$ **14.** $-\sqrt{64}$ **15.** $\sqrt{-100}$ **16.** $\sqrt{-64}$

17. $\sqrt{169}$ **18.** $\sqrt{225}$ **19.** $-\sqrt{\dfrac{1}{9}}$ **20.** $\sqrt{\dfrac{36}{81}}$ **21.** $\sqrt{-\dfrac{1}{25}}$ **22.** $-\sqrt{\dfrac{1}{25}}$

23. $\sqrt{0.04}$ **24.** $\sqrt{0.64}$ **25.** $-\sqrt{0.0009}$ **26.** $\sqrt{0.0144}$ **27.** $\sqrt{32-7}$ **28.** $\sqrt{50+14}$

29. $\sqrt{2\cdot18}$ **30.** $\sqrt{\dfrac{75}{3}}$

In Exercises 31–34, determine whether the square root is rational or irrational.

31. $\sqrt{15}$ **32.** $\sqrt{\dfrac{4}{9}}$ **33.** $\sqrt{400}$ **34.** $\sqrt{50}$

In Exercises 35–50, use a calculator to approximate the expression (if possible). (Round your answer to three decimal places.)

35. $\sqrt{43}$

36. $\sqrt{326}$

37. $\sqrt{632}$

38. $\sqrt{8}$

39. $-\sqrt{517.8}$

40. $-\sqrt{1,250}$

41. $\sqrt{\dfrac{95}{6}}$

42. $-\sqrt{\dfrac{43}{5}}$

43. $-\sqrt{10(324)}$

44. $\sqrt{632(86)}$

45. $16 - \sqrt{92.6}$

46. $27 + \sqrt{32.3}$

47. $\dfrac{9 + \sqrt{45}}{2}$

48. $\dfrac{-3 + \sqrt{1,540}}{2}$

49. $\dfrac{-4 - 3\sqrt{2}}{12}$

50. $\dfrac{-3 + 8\sqrt{-24}}{2}$

51. Complete the following table. (Round your answers to two decimal places.)

x	0	1	2	4	6	8
$\sqrt{x}$			1.414			
x	10	12	14	16	18	20
$\sqrt{x}$						

52. Sketch a graph of the equation $y = \sqrt{x}$ by using the table in Exercise 51. (Remember that negative values of x cannot be used because the square root of a negative number is not a real number.)

In Exercises 53–56, approximate the given square root without using a calculator. Then check your estimate by using a calculator.

53. $\sqrt{55}$

54. $\sqrt{90}$

55. $\sqrt{150}$

56. $\sqrt{125}$

57. Use a calculator to evaluate each of the following.
(a) $\left(\sqrt{8.2}\right)^2$ (b) $\left(\sqrt{142}\right)^2$
(c) $\left(\sqrt{22}\right)^2$ (d) $\left(\sqrt{850}\right)^2$

58. Use the results of Exercise 57 to determine $\left(\sqrt{a}\right)^2$, where a is a nonnegative real number.

59. Find all possible last digits of integers that are perfect squares. (For instance, the last digit of 81 is 1, and the last digit of 64 is 4.)

60. Using the results of Exercise 59, is it possible that 5,788,942,862 is a perfect square?

61. *Floor Space* A square room has 529 square feet of floor space (see figure). What are the dimensions of the room?

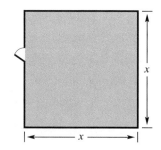

62. *Floor Space* A square room has 729 square feet of floor space. An area carpet covers all of the floor space except for a one-foot border all around the room (see figure). How many square feet are in the carpet? How many square yards are in the carpet?

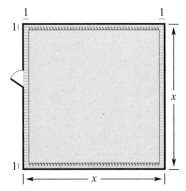

Simplifying Radicals

Simplifying Radicals with Constant Factors • *Simplifying Radicals with Variable Factors* •
Rationalizing Denominators

Simplifying Radicals with Constant Factors

You already know that simplifying algebraic expressions is one of the primary tasks in algebra. In this and the following section, we show how to simplify **radical expressions**.

The first step in rewriting a square root radical in simpler form is to evaluate any perfect square factors. To do this, we need the property of radicals demonstrated by the following two equations.

$$\sqrt{4 \cdot 25} = \sqrt{100} = 10$$
$$\sqrt{4} \cdot \sqrt{25} = 2 \cdot 5 = 10$$

Note that in these two equations we obtain the same result whether we first multiply 4 and 25 and then take the square root, or we first take the square roots of 4 and 25 and then multiply. Thus,

$$\sqrt{4 \cdot 25} = \sqrt{4} \cdot \sqrt{25}.$$

We generalize this property of radicals as follows.

Multiplication Property of Radicals

If a and b are nonnegative real numbers, then
$$\sqrt{a \cdot b} = \sqrt{a} \cdot \sqrt{b}.$$

We can describe this property by saying that the square root of a product is equal to the product of the square roots.

Note how this property is used in Examples 1 and 2. In each of these examples we first factor the radicand so that one of its factors is a perfect square.

EXAMPLE 1 ■ Simplifying Square Roots

Simplify the following radicals.
(a) $\sqrt{18}$
(b) $\sqrt{75}$

Solution

(a) To simplify this radical, we try to factor 18 so that it has a perfect square as one of its factors. Then we use the multiplication property of radicals to write the radical as a product.

$$\sqrt{18} = \sqrt{9 \cdot 2} \qquad \text{9 is perfect square factor}$$
$$= \sqrt{9} \cdot \sqrt{2} \qquad \text{Multiplication property of radicals}$$
$$= 3\sqrt{2} \qquad \text{Simplest form}$$

(b) For this radical, we recognize that 75 can be factored as the product of 3 and the perfect square 25.

$$\sqrt{75} = \sqrt{25 \cdot 3} \qquad \text{25 is perfect square factor}$$
$$= \sqrt{25} \cdot \sqrt{3} \qquad \text{Multiplication property of radicals}$$
$$= 5\sqrt{3} \qquad \text{Simplest form} \qquad \blacksquare$$

EXAMPLE 2   ■   Simplifying Square Roots

Simplify the following radicals.

(a) $\sqrt{63}$ (b) $\sqrt{44}$

Solution

(a) The radicand 63 can be factored as the product of 7 and the perfect square 9. Therefore, we can simplify the square root of 63 as follows.

$$\sqrt{63} = \sqrt{9 \cdot 7} \qquad \text{9 is perfect square factor}$$
$$= \sqrt{9} \cdot \sqrt{7} \qquad \text{Multiplication property of radicals}$$
$$= 3\sqrt{7} \qquad \text{Simplest form}$$

(b) The radicand 44 can be factored as the product of 11 and the perfect square 4. Therefore, we can simplify the square root of 44 as follows.

$$\sqrt{44} = \sqrt{4 \cdot 11} \qquad \text{4 is perfect square factor}$$
$$= \sqrt{4} \cdot \sqrt{11} \qquad \text{Multiplication property of radicals}$$
$$= 2\sqrt{11} \qquad \text{Simplest form} \qquad \blacksquare$$

Some radicands have more than one perfect square factor. In such cases, we try to choose the largest perfect square factor because it minimizes the number of steps needed to write the radical in simplest form. Compare the following two versions of the same problem.

First Solution

$$\sqrt{72} = \sqrt{36 \cdot 2} \qquad \text{36 is largest perfect square factor of 72}$$
$$= \sqrt{36} \cdot \sqrt{2} \qquad \text{Multiplication property of radicals}$$
$$= 6\sqrt{2} \qquad \text{Simplest form}$$

Second Solution

$$\sqrt{72} = \sqrt{9 \cdot 8} \qquad \text{9 is not largest perfect square factor of 72}$$
$$= \sqrt{9} \cdot \sqrt{8} \qquad \text{Multiplication property of radicals}$$
$$= 3\sqrt{8} \qquad \text{Simplify}$$
$$= 3\sqrt{4 \cdot 2} \qquad \text{4 is perfect square factor of 8}$$
$$= 3\sqrt{4} \cdot \sqrt{2}$$
$$= 3 \cdot 2 \cdot \sqrt{2}$$
$$= 6\sqrt{2} \qquad \text{Simplest form}$$

Note that by finding the largest perfect square factor of 72, we saved several steps.

EXAMPLE 3   ■   Radicands with More than One Perfect Square Factor
Simplify the following radicals.
(a) $\sqrt{96}$ (b) $\sqrt{108}$ (c) $\sqrt{288}$ (d) $\sqrt{2,000}$

Solution

(a) The largest perfect square factor of 96 is 16.
$$\sqrt{96} = \sqrt{16 \cdot 6} = \sqrt{16} \cdot \sqrt{6} = 4\sqrt{6}$$

(b) The largest perfect square factor of 108 is 36.
$$\sqrt{108} = \sqrt{36 \cdot 3} = \sqrt{36} \cdot \sqrt{3} = 6\sqrt{3}$$

(c) The largest perfect square factor of 288 is 144.
$$\sqrt{288} = \sqrt{144 \cdot 2} = \sqrt{144} \cdot \sqrt{2} = 12\sqrt{2}$$

(d) The largest perfect square factor of 2,000 is 400.
$$\sqrt{2,000} = \sqrt{400 \cdot 5} = \sqrt{400} \cdot \sqrt{5} = 20\sqrt{5} \qquad ■$$

Simplifying Radicals with Variable Factors

Simplifying radicals that involve *variable* radicands is trickier than simplifying radicals involving only constant radicands. The reason for this can be seen by considering the radical $\sqrt{x^2}$. At first glance, it would appear that this radical simplifies as x. However, in doing so, we have overlooked the possibility that x might be negative. For instance, consider the following.

If $x = 2$, then $\sqrt{x^2} = \sqrt{2^2} = \sqrt{4} = 2 = x$.
If $x = -2$, then $\sqrt{x^2} = \sqrt{(-2)^2} = \sqrt{4} = 2 = |x|$.

In both of these cases, we can conclude that

$$\sqrt{x^2} = |x|.$$

Notice that without knowing whether x is positive, zero, or negative, we *cannot* conclude that $\sqrt{x^2}$ is x.

The Square Root of x^2   If x is a real number, then
$$\sqrt{x^2} = |x|.$$

For the special case in which we know that x is a *nonnegative* real number, we can write $\sqrt{x^2} = x$.

EXAMPLE 4 ■ Simplifying Radicals Involving Variable Factors

Simplify the following radicals.

(a) $\sqrt{25x^2}$

(b) $\sqrt{18a^2}$

(c) $\sqrt{x^4}$

Solution

(a) $\sqrt{25x^2} = \sqrt{25} \cdot \sqrt{x^2} = 5|x|$

(b) $\sqrt{18a^2} = \sqrt{9 \cdot 2 \cdot a^2} = \sqrt{9} \cdot \sqrt{2} \cdot \sqrt{a^2} = 3\sqrt{2}|a|$

(c) This problem is a little different. Note that the absolute value signs do not have to appear in the final simplified version because we know that x^2 cannot be negative.
$$\sqrt{x^4} = \sqrt{(x^2)^2} = |x^2| = x^2 \qquad ■$$

Study the following simplifications of powers of x. Can you justify why absolute values are necessary in some cases and not in others?
$$\sqrt{x^2} = |x|$$
$$\sqrt{x^3} = \sqrt{x^2 \cdot x} = \sqrt{x^2} \cdot \sqrt{x} = x\sqrt{x}$$
$$\sqrt{x^4} = \sqrt{(x^2)^2} = x^2$$
$$\sqrt{x^5} = \sqrt{x^4 \cdot x} = \sqrt{x^4} \cdot \sqrt{x} = x^2\sqrt{x}$$
$$\sqrt{x^6} = \sqrt{(x^3)^2} = |x^3|$$

The reason we don't need to use absolute values in simplifying $\sqrt{x^3} = x\sqrt{x}$ is that the original radical would be undefined if x were negative. Thus, if someone asks us to simplify the radical $\sqrt{x^3}$, we can assume that x must be nonnegative (which means that the absolute value signs are not necessary).

EXAMPLE 5 ■ Simplifying Radicals Involving Variable Factors

Simplify the following radicals.

(a) $\sqrt{16x^3}$

(b) $\sqrt{9a^5}$

Solution

Note that for each of these radicals, we can assume that the variable is nonnegative. (If x were negative, then x^3 would also be negative, and $\sqrt{16x^3}$ would be undefined.)

(a) $\sqrt{16x^3} = \sqrt{16 \cdot x^2 x} = \sqrt{16} \cdot \sqrt{x^2} \cdot \sqrt{x} = 4x\sqrt{x}$

(b) $\sqrt{9a^5} = \sqrt{9 \cdot a^4 a} \cdot = \sqrt{9} \cdot \sqrt{a^4} \cdot \sqrt{a} = 3a^2\sqrt{a}$ ■

Rationalizing Denominators

To simplify a square root having a fractional radicand, we need the property of radicals demonstrated in the following two equations.

$$\sqrt{\frac{64}{16}} = \sqrt{4} = 2$$

$$\frac{\sqrt{64}}{\sqrt{16}} = \frac{8}{4} = 2$$

Since the answers are the same, it follows that

$$\sqrt{\frac{64}{16}} = \frac{\sqrt{64}}{\sqrt{16}}.$$

This illustrates the following general property of radicals.

Division Property of Radicals	If a and b are nonnegative real numbers, and $b \neq 0$, then $$\sqrt{\frac{a}{b}} = \frac{\sqrt{a}}{\sqrt{b}}.$$ We can describe this property by saying that the square root of a quotient is equal to the quotient of the square roots.

Note how this property is used to simplify radicals in Example 6.

EXAMPLE 6 ■ Simplifying Radicals Involving Fractions

Simplify the following radicals.

(a) $\sqrt{\dfrac{21}{4}}$

(b) $\sqrt{\dfrac{45}{25}}$

(c) $\sqrt{\dfrac{48x^4}{3}}$

(d) $\sqrt{\dfrac{3x^2}{12y^4}}$

Solution

(a) $\sqrt{\dfrac{21}{4}} = \dfrac{\sqrt{21}}{\sqrt{4}} = \dfrac{\sqrt{21}}{2}$

(b) $\sqrt{\dfrac{45}{25}} = \dfrac{\sqrt{9 \cdot 5}}{\sqrt{25}} = \dfrac{3\sqrt{5}}{5}$

(c) $\sqrt{\dfrac{48x^4}{3}} = \sqrt{16x^4} = 4x^2$

(d) $\sqrt{\dfrac{3x^2}{12y^4}} = \sqrt{\dfrac{x^2}{4y^4}} = \dfrac{\sqrt{x^2}}{\sqrt{4y^4}} = \dfrac{|x|}{2y^2}$ ∎

In Example 6 note that all of the simplified forms have denominators that are free of radicals. This type of simplification process is called **rationalizing the denominator**. The rationalizing that occurred in Example 6 was easy because (after canceling) the denominators were perfect squares. To rationalize the denominator for a more general fraction, we must do a little more work. The goal is to obtain a perfect square radicand in the denominator by multiplying both the numerator and denominator by the *least* factor that generates such a perfect square. For instance, consider the following fractional expression.

$$\frac{\sqrt{3}}{\sqrt{5}}$$

We can rewrite this expression so that there are no radicals in the denominator by multiplying the numerator and denominator by $\sqrt{5}$, as follows.

$$\frac{\sqrt{3}}{\sqrt{5}} = \frac{\sqrt{3}}{\sqrt{5}} \cdot \frac{\sqrt{5}}{\sqrt{5}} \qquad \text{Multiply numerator and denominator by } \sqrt{5}$$

$$= \frac{\sqrt{15}}{\sqrt{25}} \qquad \text{Denominator has perfect square radicand}$$

$$= \frac{\sqrt{15}}{5} \qquad \text{Simplify}$$

EXAMPLE 7   ∎   Rationalizing Denominators

(a) $\sqrt{\dfrac{13}{3}} = \dfrac{\sqrt{13}}{\sqrt{3}} \cdot \dfrac{\sqrt{3}}{\sqrt{3}} = \dfrac{\sqrt{39}}{3}$

(b) $\sqrt{\dfrac{1}{6}} = \dfrac{\sqrt{1}}{\sqrt{6}} \cdot \dfrac{\sqrt{6}}{\sqrt{6}} = \dfrac{\sqrt{6}}{6}$

(c) $\sqrt{\dfrac{7}{20}} = \dfrac{\sqrt{7}}{\sqrt{20}} \cdot \dfrac{\sqrt{5}}{\sqrt{5}} = \dfrac{\sqrt{35}}{\sqrt{100}} = \dfrac{\sqrt{35}}{10}$ Multiply by $\sqrt{5}/\sqrt{5}$ to obtain perfect square radicand

(d) $\dfrac{12}{\sqrt{18}} = \dfrac{12}{\sqrt{18}} \cdot \dfrac{\sqrt{2}}{\sqrt{2}} = \dfrac{12\sqrt{2}}{\sqrt{36}} = \dfrac{12\sqrt{2}}{6} = 2\sqrt{2}$ Multiply by $\sqrt{2}/\sqrt{2}$ to obtain perfect square radicand   ∎

The three criteria for a radical expression to be in simplest form are summarized as follows.

Simplest Form of a Radical Expression

A (square root) radical expression is in simplest form if the following conditions are met.

1. All possible perfect square factors have been removed from the radicand.

2. All radicands are free of fractions.

3. No denominator contains a radical.

The next two examples show how to rationalize denominators that contain variable factors.

EXAMPLE 8 ■ Rationalizing Denominators Containing Variable Factors

Write the following radicals in simplest form.

(a) $\sqrt{\dfrac{3}{a}}$ (b) $\sqrt{\dfrac{1}{4x^3}}$

Solution

(a) This expression is not in simplest form because the radicand contains a fraction. To simplify the expression, we multiply the numerator and denominator by $\sqrt{a}$, as follows.

$$\sqrt{\frac{3}{a}} = \frac{\sqrt{3}}{\sqrt{a}} \cdot \frac{\sqrt{a}}{\sqrt{a}} = \frac{\sqrt{3a}}{\sqrt{a^2}} = \frac{\sqrt{3a}}{a}$$

Note that absolute value signs are unnecessary here because (from the original radical) we can assume that the variable a cannot be negative.

(b) This expression is not in simplest form because the radicand contains a fraction. To simplify the expression, we could multiply the numerator and denominator by $\sqrt{4x^3}$. But since $\sqrt{4x^3} = 2x\sqrt{x}$, we can accomplish the same thing by simply multiplying the numerator and denominator by $\sqrt{x}$, as follows.

$$\sqrt{\frac{1}{4x^3}} = \frac{\sqrt{1}}{\sqrt{4x^3}} \cdot \frac{\sqrt{x}}{\sqrt{x}} = \frac{\sqrt{x}}{\sqrt{4x^4}} = \frac{\sqrt{x}}{2x^2}$$

■

EXAMPLE 9 ■ Rationalizing Denominators Containing Variable Factors

Write the following radicals in simplest form.

(a) $\sqrt{\dfrac{10x}{8y^5}}$ (b) $\sqrt{\dfrac{12x^2y^2}{5x^5y}}$

Solution

(a) In this case we should reduce the radicand before rationalizing the denominator.

$$\sqrt{\frac{10x}{8y^5}} = \sqrt{\frac{5x}{4y^5}} = \frac{\sqrt{5x}}{\sqrt{4y^5}} \cdot \frac{\sqrt{y}}{\sqrt{y}} = \frac{\sqrt{5xy}}{\sqrt{4y^6}} = \frac{\sqrt{5xy}}{2y^3}$$

(b) Here again, we begin by reducing the radicand.

$$\sqrt{\frac{12x^2y^2}{5x^5y}} = \sqrt{\frac{12y}{5x^3}} = \frac{\sqrt{12y}}{\sqrt{5x^3}} \cdot \frac{\sqrt{5x}}{\sqrt{5x}} = \frac{\sqrt{60xy}}{\sqrt{25x^4}} = \frac{\sqrt{4 \cdot 15xy}}{5x^2} = \frac{2\sqrt{15xy}}{5x^2} \blacksquare$$

DISCUSSION PROBLEM ■ ## Constructing Lengths to Represent Radicals

Consider a right triangle that has two sides of length 1, as shown in Figure 10.1. From the Pythagorean Theorem, we can conclude that the length of the hypotenuse is

$$c = \sqrt{a^2 + b^2} = \sqrt{1^2 + 1^2} = \sqrt{2}.$$

Can you construct a right triangle whose three sides have lengths of 1, 2, and $\sqrt{3}$? (*Note:* The **Pythagorean Theorem** states that the sides of a *right triangle* satisfy the equation $a^2 + b^2 = c^2$, where a and b are the lengths of the legs and c is the length of the hypotenuse.)

FIGURE 10.1

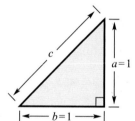

■

Warm-Up

The following warm-up exercises involve skills that were covered in earlier sections. You will use these skills in the exercise set for this section.

In Exercises 1–4, evaluate the square root, if possible. (Do not use a calculator.)

1. $\sqrt{10{,}000}$ **2.** $-\sqrt{196}$ **3.** $-\sqrt{\frac{169}{25}}$ **4.** $\sqrt{-\frac{9}{4}}$

In Exercises 5–10, simplify the given expression.

5. $(2x)^2(2x)^3$ **6.** $(-4x^2y)^3$ **7.** $\frac{32x^3y^2}{2xy^3}$ **8.** $\frac{8y^2z^{-2}}{z^2}$

9. $\frac{9xy^{-2}}{3x^{-2}y}$ **10.** $(u^2v)^3(u^2v)^{-3}$

10.2 EXERCISES

In Exercises 1–4, write the product as a single radical.

1. $\sqrt{2} \cdot \sqrt{7}$ **2.** $\sqrt{5} \cdot \sqrt{19}$ **3.** $\sqrt{11} \cdot \sqrt{10}$ **4.** $\sqrt{35} \cdot \sqrt{3}$

In Exercises 5–8, write the expression as the product of two radicals and simplify.

5. $\sqrt{4 \cdot 15}$ **6.** $\sqrt{16 \cdot 3}$ **7.** $\sqrt{64 \cdot 11}$ **8.** $\sqrt{100 \cdot 3}$

In Exercises 9–20, simplify the radical.

9. $\sqrt{8}$ **10.** $\sqrt{50}$ **11.** $\sqrt{27}$ **12.** $\sqrt{75}$ **13.** $\sqrt{20}$ **14.** $\sqrt{32}$

15. $\sqrt{300}$ **16.** $\sqrt{128}$ **17.** $\sqrt{180}$ **18.** $\sqrt{432}$ **19.** $\sqrt{30{,}000}$ **20.** $\sqrt{800}$

In Exercises 21–30, simplify the radical. (Remember to use absolute value signs if appropriate.)

21. $\sqrt{4x^2}$ **22.** $\sqrt{9x^4}$ **23.** $\sqrt{64x^3}$ **24.** $\sqrt{9z^5}$ **25.** $\sqrt{8a^4}$ **26.** $\sqrt{50b^{11}}$

27. $\sqrt{x^2 y^3}$ **28.** $\sqrt{a^5 b^4}$ **29.** $\sqrt{200x^2 y^4}$ **30.** $\sqrt{128u^4 v^7}$

In Exercises 31–34, write the expression as a single radical.

31. $\dfrac{\sqrt{39}}{\sqrt{15}}$ **32.** $\dfrac{\sqrt{84}}{\sqrt{9}}$ **33.** $\dfrac{\sqrt{152}}{\sqrt{3}}$ **34.** $\dfrac{\sqrt{633}}{\sqrt{5}}$

In Exercises 35–40, simplify the given expression.

35. $\dfrac{\sqrt{54}}{\sqrt{6}}$ **36.** $\dfrac{\sqrt{48}}{\sqrt{8}}$ **37.** $\sqrt{\dfrac{35}{4}}$ **38.** $\sqrt{\dfrac{165}{36}}$ **39.** $\sqrt{\dfrac{100}{11}}$ **40.** $\sqrt{\dfrac{169}{2}}$

In Exercises 41–48, simplify the radical.

41. $\sqrt{\dfrac{3}{25}}$ **42.** $\sqrt{\dfrac{15}{36}}$ **43.** $\sqrt{\dfrac{12x^2}{25}}$ **44.** $\sqrt{\dfrac{8z^3}{y^4}}$ **45.** $\sqrt{\dfrac{18a^3}{2a}}$ **46.** $\sqrt{\dfrac{54b^4}{2b^2}}$

47. $\sqrt{\dfrac{5u^2}{4v^4}}$ **48.** $\sqrt{\dfrac{8y^4}{2z^3}}$

In Exercises 49–62, rationalize the denominator and simplify. (Assume the variables are positive.)

49. $\sqrt{\dfrac{1}{3}}$ **50.** $\sqrt{\dfrac{1}{5}}$ **51.** $\dfrac{1}{\sqrt{3}}$ **52.** $\dfrac{1}{\sqrt{5}}$ **53.** $\dfrac{5}{\sqrt{10}}$ **54.** $\dfrac{7}{\sqrt{14}}$

55. $\dfrac{1}{\sqrt{y}}$ **56.** $\dfrac{1}{\sqrt{z}}$ **57.** $\sqrt{\dfrac{5}{x}}$ **58.** $\sqrt{\dfrac{3}{a}}$ **59.** $\sqrt{\dfrac{2x}{3y}}$ **60.** $\sqrt{\dfrac{20x^2}{5y^2}}$

61. $\dfrac{a^3}{\sqrt{ab}}$ **62.** $\dfrac{3b^2}{\sqrt{6b}}$

In Exercises 63–66, simplify the given radical.

63. $\sqrt{3.2 \times 10^6}$ **64.** $\sqrt{8.5 \times 10^5}$ **65.** $\sqrt{1.8 \times 10^{-7}}$ **66.** $\sqrt{4.4 \times 10^{-4}}$

Period of a Pendulum In Exercises 67 and 68, use the following formula, which gives the period T of a simple pendulum whose length is L. (The period T is measured in seconds and the length is measured in feet.)

$$T = 2\pi \sqrt{\dfrac{L}{32}}$$

67. A trapeze performer is swinging on a trapeze that is 12 feet long. How long will it take the performer to swing from the platform to the "end of the first pendulum," as shown in the figure?

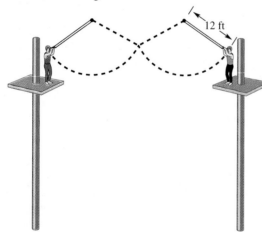

12 ft

68. Find the period of the pendulum shown in the accompanying figure.

Figure for 68

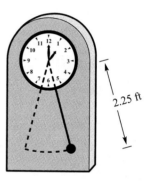

2.25 ft

69. *Area of a Desk Top* Determine the area of the rectangular desk top in the accompanying figure. (Round your answer to two decimal places.)

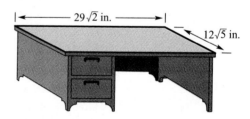

$29\sqrt{2}$ in.

$12\sqrt{5}$ in.

70. *Area of a Triangle* Determine the area of the equilateral triangle in the accompanying figure. (Round your answer to two decimal places.)

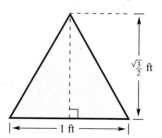

$\dfrac{\sqrt{3}}{2}$ ft

1 ft

Operations with Radical Expressions
Introduction • Combining Radical Expressions • Multiplying Radical Expressions •
Dividing Radical Expressions

Introduction

The rules for operating with radical expressions are similar to those for operating with polynomials. Let's begin with an example that reviews the procedures for combining and multiplying polynomials.

EXAMPLE 1   ■   Combining and Simplifying Polynomial Expressions

(a) To combine like terms in the expression $3x - 7x$, we use the Distributive Property.

$$3x - 7x = (3 - 7)x \qquad \text{Distributive Property}$$
$$= -4x \qquad \text{Simplify}$$

(b) Notice how we combine like terms in the following polynomial expression.

$$2 - 3x + 4x^2 + 2x - 6x^2 + 1$$
$$= 4x^2 - 6x^2 - 3x + 2x + 2 + 1 \qquad \text{Group like terms}$$
$$= (4 - 6)x^2 + (-3 + 2)x + (2 + 1) \qquad \text{Distributive Property}$$
$$= -2x^2 - x + 3 \qquad \text{Simplify}$$

(c) To multiply a monomial by a binomial, we use the Distributive Property, as follows.

$$2x(1 - 3x) = 2x - 6x^2 \qquad \text{Distributive Property}$$

(d) To multiply two binomials, we use the FOIL method, as follows.

$$(x - 2)(2x + 3) = 2x^2 + 3x - 4x - 6 \qquad \text{FOIL}$$
$$= 2x^2 - x - 6 \qquad \text{Combine like terms}$$

(e) To multiply two special binomials that represent the sum and difference of like terms, we use the pattern $(a + b)(a - b) = a^2 - b^2$.

$$(x + 3)(x - 3) = x^2 - 9 \qquad \text{Special product}$$

■

Combining Radical Expressions

Two (square root) radical expressions are **alike** if they have the same radicand. For instance, the expressions $3\sqrt{2}$ and $5\sqrt{2}$ are alike, whereas the expressions

$2\sqrt{3}$ and $2\sqrt{5}$ are not alike. We can combine like radicals by the Distributive Property. For instance, we can simplify the radical expression $2\sqrt{5} + 7\sqrt{5}$ as follows.

$$2\sqrt{5} + 7\sqrt{5} = (2 + 7)\sqrt{5} \qquad \text{Distributive Property}$$
$$= 9\sqrt{5} \qquad \text{Simplify}$$

EXAMPLE 2 ■ Combining Radical Expressions

(a) $2\sqrt{3} - 6\sqrt{3} = (2 - 6)\sqrt{3}$ Distributive Property

       $= -4\sqrt{3}$ Simplify

(b) $\sqrt{7} + 5\sqrt{7} - 2\sqrt{7} = (1 + 5 - 2)\sqrt{7}$ Distributive Property

       $= 4\sqrt{7}$ Simplify

(c) $6\sqrt{6} - \sqrt{3} - 5\sqrt{6} + 2\sqrt{3}$

     $= 6\sqrt{6} - 5\sqrt{6} - \sqrt{3} + 2\sqrt{3}$ Group like terms

     $= (6 - 5)\sqrt{6} + (-1 + 2)\sqrt{3}$ Combine like terms

     $= \sqrt{6} + \sqrt{3}$ Cannot be further simplified

                                                         ■

NOTE: It is important to realize that the expression $\sqrt{a} + \sqrt{b}$ is not equal to $\sqrt{a + b}$. For instance, in Example 2(c), you may have been tempted to add $\sqrt{6} + \sqrt{3}$ and get $\sqrt{9}$ or 3. But remember, we cannot add unlike radicals. Hence, $\sqrt{6} + \sqrt{3}$ cannot be simplified further.

We can simplify expressions involving *variable* radicands in a similar way. Remember that the radicand of a square root cannot be negative. Thus, throughout this section we will assume that all variable expressions that occur as radicands are nonnegative. For instance, in the following expression, we assume that x is nonnegative (even though it is not explicitly stated).

$$3\sqrt{x} - 5\sqrt{x} = (3 - 5)\sqrt{x} \qquad \text{Distributive Property}$$
$$= -2\sqrt{x} \qquad \text{Simplify}$$

EXAMPLE 3 ■ Combining Radical Expressions with Variable Radicands

Simplify the following radical expressions.

(a) $3 + 3\sqrt{x} - \sqrt{x} + \sqrt{4}$ (b) $5\sqrt{y} + y - 2\sqrt{y}$ (c) $3\sqrt{x} + \sqrt{4x}$

Solution

(a) $3 + 3\sqrt{x} - \sqrt{x} + \sqrt{4} = 3 + 3\sqrt{x} - \sqrt{x} + 2$ Simplify: $\sqrt{4} = 2$

                  $= (3 + 2) + \left(3\sqrt{x} - \sqrt{x}\right)$ Group like terms

                  $= (3 + 2) + (3 - 1)\sqrt{x}$ Combine like terms

                  $= 5 + 2\sqrt{x}$ Simplify

(b) $5\sqrt{y} + y - 2\sqrt{y} = (5 - 2)\sqrt{y} + y$ Combine like terms

$= 3\sqrt{y} + y$ Simplify

(c) The radicals in $3\sqrt{x} + \sqrt{4x}$ are not alike as written. However, by writing $\sqrt{4x} = \sqrt{4} \cdot \sqrt{x} = 2\sqrt{x}$, we can simplify the expression as follows.

$$3\sqrt{x} + \sqrt{4x} = 3\sqrt{x} + 2\sqrt{x}$$
$$= (3 + 2)\sqrt{x}$$
$$= 5\sqrt{x}$$ ■

In Example 3(c), note that it is important to simplify individual radicals before combining like radicals. This is demonstrated further in Examples 4 and 5.

EXAMPLE 4 ■ **Simplifying Radical Expressions**

Simplify the following radical expressions.

(a) $2\sqrt{27} + \sqrt{45} - \sqrt{3}$

(b) $2\sqrt{72} - 2\sqrt{32}$

(c) $\sqrt{90} - 2\sqrt{40} + 4\sqrt{10}$

Solution

(a) $2\sqrt{27} + \sqrt{45} - \sqrt{3} = 2\sqrt{9 \cdot 3} + \sqrt{9 \cdot 5} - \sqrt{3}$ Find perfect square factors

$= 6\sqrt{3} + 3\sqrt{5} - \sqrt{3}$ Simplify radicals

$= 5\sqrt{3} + 3\sqrt{5}$ Combine like radicals

(b) $2\sqrt{72} - 2\sqrt{32} = 2\sqrt{36 \cdot 2} - 2\sqrt{16 \cdot 2}$ Find perfect square factors

$= 12\sqrt{2} - 8\sqrt{2}$ Simplify radicals

$= 4\sqrt{2}$ Combine like radicals

(c) $\sqrt{90} - 2\sqrt{40} + 4\sqrt{10} = \sqrt{9 \cdot 10} - 2\sqrt{4 \cdot 10} + 4\sqrt{10}$

$= 3\sqrt{10} - 4\sqrt{10} + 4\sqrt{10}$

$= 3\sqrt{10}$ ■

EXAMPLE 5 ■ **Simplifying Radical Expressions with Variable Radicands**

Simplify the following radical expressions.

(a) $\sqrt{45x} + 2\sqrt{20x}$ (b) $5\sqrt{2x^3} - x\sqrt{8x}$ (c) $\sqrt{50y^5} + \sqrt{32y^5}$

Assume that x and y are nonnegative.

Solution

(a) $\sqrt{45x} + 2\sqrt{20x} = \sqrt{9 \cdot 5x} + 2\sqrt{4 \cdot 5x}$ Factor

$= 3\sqrt{5x} + 4\sqrt{5x}$ Factor out perfect square factors

$= 7\sqrt{5x}$ Combine like terms

(b) $5\sqrt{2x^3} - x\sqrt{8x} = 5\sqrt{2 \cdot x^2 \cdot x} - x\sqrt{4 \cdot 2 \cdot x}$ Factor

$\qquad\qquad\qquad = 5x\sqrt{2x} - 2x\sqrt{2x}$ Factor out perfect square factors

$\qquad\qquad\qquad = (5x - 2x)\sqrt{2x}$ Distributive Property

$\qquad\qquad\qquad = 3x\sqrt{2x}$ Combine like terms

(c) $\sqrt{50y^5} + \sqrt{32y^5} = \sqrt{25y^4 \cdot 2y} + \sqrt{16y^4 \cdot 2y}$ Factor

$\qquad\qquad\qquad = 5y^2\sqrt{2y} + 4y^2\sqrt{2y}$ Factor out perfect square factors

$\qquad\qquad\qquad = 9y^2\sqrt{2y}$ Combine like terms ■

Multiplying Radical Expressions

We can multiply radicals by using the Distributive Property and the FOIL method. In both procedures we also make use of the multiplication property of radicals. Recall from Section 10.2 that the product of two radicals is given by

$$\sqrt{a}\sqrt{b} = \sqrt{ab}$$

where a and b are nonnegative real numbers.

EXAMPLE 6 ■ Multiplying Radicals

(a) $\sqrt{2}\left(1 + \sqrt{3}\right) = \sqrt{2} + \sqrt{2}\sqrt{3}$ Distributive Property

$\qquad\qquad\qquad = \sqrt{2} + \sqrt{6}$ Multiplication property of radicals

(b) $\sqrt{5}\left(\sqrt{15} - \sqrt{5}\right) = \sqrt{5}\sqrt{15} - \sqrt{5}\sqrt{5}$ Distributive Property

$\qquad\qquad\qquad = \sqrt{75} - \sqrt{25}$ Multiplication property of radicals

$\qquad\qquad\qquad = \sqrt{25 \cdot 3} - 5$ Find perfect square factors

$\qquad\qquad\qquad = 5\sqrt{3} - 5$ Simplify ■

EXAMPLE 7 ■ Multiplying Radicals

(a) $\left(\sqrt{7} - 1\right)\left(\sqrt{7} + 3\right) = \overbrace{\sqrt{7 \cdot 7}}^{F} + \overbrace{3\sqrt{7}}^{O} - \overbrace{\sqrt{7}}^{I} - \overbrace{3}^{L}$ FOIL method

$\qquad\qquad\qquad = 7 + (3 - 1)\sqrt{7} - 3$ Combine like radicals

$\qquad\qquad\qquad = 4 + 2\sqrt{7}$ Combine like terms

(b) $\left(2 - \sqrt{6}\right)\left(2 + \sqrt{6}\right) = 2^2 - \left(\sqrt{6}\right)^2$ Special product pattern

$\qquad\qquad\qquad = 4 - 6$

$\qquad\qquad\qquad = -2$ ■

In Example 7(b), the expressions $\left(2 - \sqrt{6}\right)$ and $\left(2 + \sqrt{6}\right)$ are called **conjugates** of each other. The product of two conjugates is the difference of two squares, which is given by the special product pattern: $(a + b)(a - b) = a^2 - b^2$. Here are some other examples.

Expression	*Conjugate*	*Product*
$3 - \sqrt{6}$	$3 + \sqrt{6}$	$(3)^2 - \left(\sqrt{6}\right)^2 = 9 - 6 = 3$
$\sqrt{3} + \sqrt{7}$	$\sqrt{3} - \sqrt{7}$	$\left(\sqrt{3}\right)^2 - \left(\sqrt{7}\right)^2 = 3 - 7 = -4$
$\sqrt{x} - 1$	$\sqrt{x} + 1$	$\left(\sqrt{x}\right)^2 - (1)^2 = x - 1$

MATH MATTERS

Perfect Numbers

A perfect number is a positive integer that is equal to the sum of all of its factors (excluding the number itself). For instance, 6 is a perfect number because the factors of 6 are 1, 2, 3, and 6, and

$$6 = 1 + 2 + 3.$$

The number 28 is also perfect because the factors of 28 are 1, 2, 4, 7, 14, and 28, and

$$28 = 1 + 2 + 4 + 7 + 14.$$

These two perfect numbers were known to Pythagoras in the sixth century B.C. The next two perfect numbers, 496 and 8,128, were discovered 300 years later by Nichomachus of Alexandria, who spent years searching for others. He failed to find any, but this is not surprising because the next one, recorded in a medieval manuscript, is 33,550,336. Until 1952, when a computer was used for the first time, only seven more perfect numbers were found. Today, we know of 30 perfect numbers. Can you show that 496 is a perfect number? (The answer is given in the back of the text.)

Pythagoras

Dividing Radical Expressions

To simplify a *quotient* involving radicals, we rationalize the denominator by multiplying by its conjugate. Note how this is done in the following example.

EXAMPLE 8 ■ Simplifying Quotients Involving Radicals

Simplify the following expression by rationalizing its denominator.

$$\frac{2}{3 - \sqrt{5}}$$

Solution

$$\frac{2}{3 - \sqrt{5}} = \frac{2}{3 - \sqrt{5}} \cdot \frac{3 + \sqrt{5}}{3 + \sqrt{5}} \qquad \text{Multiply by conjugate}$$

$$= \frac{2(3 + \sqrt{5})}{(3)^2 - (\sqrt{5})^2} \qquad \text{Special product}$$

$$= \frac{2(3 + \sqrt{5})}{9 - 5}$$

$$= \frac{2(3 + \sqrt{5})}{4} \qquad \text{Simplify}$$

$$= \frac{3 + \sqrt{5}}{2} \qquad \text{Divide out common factors}$$

EXAMPLE 9 ■ Simplifying Quotients Involving Radicals

Rationalize the denominators and simplify the following.

(a) $\dfrac{5\sqrt{2}}{\sqrt{3} + \sqrt{2}}$ (b) $\dfrac{6}{\sqrt{x} - 2}$

work these out fully

Solution

(a) $\dfrac{5\sqrt{2}}{\sqrt{3} + \sqrt{2}} = \dfrac{5\sqrt{2}}{\sqrt{3} + \sqrt{2}} \cdot \dfrac{\sqrt{3} - \sqrt{2}}{\sqrt{3} - \sqrt{2}}$ Multiply by conjugate

$$= \frac{5\sqrt{6} - 5\sqrt{4}}{(\sqrt{3})^2 - (\sqrt{2})^2}$$

$$= \frac{5\sqrt{6} - 10}{3 - 2}$$

$$= 5\sqrt{6} - 10 \qquad \text{Simplify}$$

(b) $\dfrac{6}{\sqrt{x}-2} = \dfrac{6}{\sqrt{x}-2} \cdot \dfrac{\sqrt{x}+2}{\sqrt{x}+2}$ Multiply by conjugate

$\qquad = \dfrac{6\left(\sqrt{x}+2\right)}{\left(\sqrt{x}\right)^2 - (2)^2}$

$\qquad = \dfrac{6\sqrt{x}+12}{x-4}$ Simplify   ■

EXAMPLE 10 ■ **Dividing Radical Expressions**

Perform the indicated division and simplify your answer.

$$\left(2 - \sqrt{3}\right) \div \left(\sqrt{6} + \sqrt{2}\right)$$

Solution

$\dfrac{2-\sqrt{3}}{\sqrt{6}+\sqrt{2}} = \dfrac{2-\sqrt{3}}{\sqrt{6}+\sqrt{2}} \cdot \dfrac{\sqrt{6}-\sqrt{2}}{\sqrt{6}-\sqrt{2}}$ Multiply by conjugate

$\qquad = \dfrac{2\sqrt{6} - 2\sqrt{2} - \sqrt{18} + \sqrt{6}}{\left(\sqrt{6}\right)^2 - \left(\sqrt{2}\right)^2}$

$\qquad = \dfrac{3\sqrt{6} - 2\sqrt{2} - 3\sqrt{2}}{6-2}$

$\qquad = \dfrac{3\sqrt{6} - 5\sqrt{2}}{4}$ Simplify   ■

DISCUSSION PROBLEM ■ **Closure Property**

Some subsets of real numbers are what we call **closed** with respect to certain operations. For instance, the set of rational numbers is closed with respect to multiplication because the product of any two rational numbers must be another rational number. Show that the set of irrational numbers is *not* closed with respect to multiplication by finding two irrational numbers whose product is rational. Can you find other examples of subsets of real numbers that are closed (or not closed) with respect to multiplication?   ■

> ■ Warm-Up
>
> *The following warm-up exercises involve skills that were covered in earlier sections. You will use these skills in the exercise set for this section.*
>
> In Exercises 1–10, perform the indicated multiplication and simplify the result by combining any like terms.
>
> **1.** $4(2 - x) + 6x$ **2.** $3(2x + 1) + 2(x + 1)$
>
> **3.** $17 - 5(x - 2)$ **4.** $32 - 3(2x + 10)$
>
> **5.** $(x + 6)^2$ **6.** $(2x - 3)^2$
>
> **7.** $(5x - 3)(5x + 3)$ **8.** $(1 - 3x)(1 + 3x)$
>
> **9.** $x(x - 7) - 3x(x + 2)$ **10.** $4x(x + 6) + 3x(10 - 4x)$

10.3 EXERCISES

In Exercises 1–20, simplify the radical expression.

1. $3\sqrt{5} - \sqrt{5}$ **2.** $5\sqrt{6} - 10\sqrt{6}$ **3.** $\frac{2}{5}\sqrt{3} - \frac{6}{5}\sqrt{3}$

4. $\sqrt{15} - \frac{1}{3}\sqrt{15}$ **5.** $\sqrt{3} - 5\sqrt{7} - 12\sqrt{3}$ **6.** $9\sqrt{17} + 7\sqrt{2} - 11\sqrt{17} + \sqrt{2}$

7. $12\sqrt{8} - 3\sqrt{8}$ **8.** $4\sqrt{32} + 2\sqrt{32}$ **9.** $2\sqrt{50} + 12\sqrt{8}$

10. $4\sqrt{27} - \sqrt{75}$ **11.** $5\sqrt{x} - 3\sqrt{x}$ **12.** $3\sqrt{x + 1} + 10\sqrt{x + 1}$

13. $\sqrt{9x} + \sqrt{36x}$ **14.** $\sqrt{64t} - \sqrt{16t}$ **15.** $\sqrt{45z} - \sqrt{125z}$

16. $\sqrt{18u} + \sqrt{8u}$ **17.** $\sqrt{\frac{a}{4}} - \sqrt{\frac{a}{9}}$ **18.** $\sqrt{\frac{v}{36}} - \sqrt{\frac{v}{9}}$

19. $\sqrt{x^3 y} + 4\sqrt{xy}$ **20.** $3t\sqrt{st} - s\sqrt{st}$

In Exercises 21–42, perform the specified multiplication and simplify the product.

21. $\sqrt{3} \cdot \sqrt{27}$ **22.** $\sqrt{5} \cdot \sqrt{15}$ **23.** $\sqrt{7}(1 - \sqrt{2})$

24. $\sqrt{3}(\sqrt{5} - 3)$ **25.** $\sqrt{6}(\sqrt{12} + 8)$ **26.** $\sqrt{2}(\sqrt{14} + 3)$

27. $(\sqrt{2} - 1)(\sqrt{2} + 1)$ **28.** $(\sqrt{7} + 3)(\sqrt{7} - 3)$ **29.** $(\sqrt{10} + \sqrt{5})(\sqrt{10} - \sqrt{5})$

30. $(\sqrt{5} + \sqrt{2})(\sqrt{5} - \sqrt{2})$ **31.** $(\sqrt{13} + 2)^2$ **32.** $(4 - \sqrt{12})^2$

33. $(\sqrt{2} + 1)(\sqrt{3} - 5)$ **34.** $(\sqrt{7} + 6)(\sqrt{2} + 1)$ **35.** $\sqrt{x}(\sqrt{x} + 5)$

36. $\sqrt{x}\left(3 - \sqrt{x}\right)$ 　　　　**37.** $\left(2\sqrt{x} - 3\right)\left(2\sqrt{x} + 3\right)$ 　　　　**38.** $\left(4 - 3\sqrt{t}\right)\left(4 + 3\sqrt{t}\right)$

39. $\left(3 + \sqrt{x}\right)^2$ 　　　　　　　　　　　　**40.** $\left(5 - \sqrt{v}\right)^2$

41. $\left(\sqrt{x} + 1\right)\left(\sqrt{x} - 3\right)$ 　　　　　　　　**42.** $\left(2\sqrt{u} - 3\right)\left(\sqrt{u} - 4\right)$

In Exercises 43–50, determine the conjugate of the expression. Then find the product of the expression and its conjugate.

43. $4 + \sqrt{3}$ 　　　　**44.** $\sqrt{7} - 3$ 　　　　**45.** $\sqrt{15} - \sqrt{7}$ 　　　　**46.** $\sqrt{10} + \sqrt{2}$

47. $\sqrt{x} - 4$ 　　　　**48.** $\sqrt{t} + 5$ 　　　　**49.** $\sqrt{u} - \sqrt{2}$ 　　　　**50.** $\sqrt{a} + \sqrt{3}$

In Exercises 51–60, rationalize the denominator and simplify.

51. $\dfrac{5}{\sqrt{14} - 2}$ 　　　　　　　　　　**52.** $\dfrac{5}{2\sqrt{10} - 5}$

53. $\dfrac{4}{\sqrt{7} - 3}$ 　　　　　　　　　　**54.** $\dfrac{3}{\sqrt{5} + \sqrt{6}}$

55. $\left(\sqrt{5} + 1\right) \div \left(\sqrt{7} + 1\right)$ 　　　　　**56.** $\left(2 - 3\sqrt{7}\right) \div \left(1 + 2\sqrt{7}\right)$

57. $\dfrac{2x}{5 - \sqrt{3}}$ 　　　　　　　　　　**58.** $\dfrac{6}{\sqrt{x} - 1}$

59. $\left(\sqrt{x} - 5\right) \div \left(2\sqrt{x} - 1\right)$ 　　　　**60.** $\left(2\sqrt{t} + 1\right) \div \left(2\sqrt{t} - 1\right)$

In Exercises 61–64, rewrite the expression as a single fraction and simplify.

61. $3 - \dfrac{1}{\sqrt{3}}$ 　　　　　　　　　　**62.** $\dfrac{1}{\sqrt{5}} - 5$

63. $\sqrt{50} - \dfrac{6}{\sqrt{2}}$ 　　　　　　　　**64.** $\dfrac{7}{\sqrt{3}} + \sqrt{12}$

In Exercises 65–68, insert the correct symbol ($<$, $>$, or $=$) between the two real numbers by using a calculator to find the decimal approximation of each expression.

65. $\sqrt{5} + \sqrt{3}$ ░░░ $\sqrt{5 + 3}$ 　　　　　**66.** $\sqrt{5} - \sqrt{3}$ ░░░ $\sqrt{5 - 3}$

67. 5 ░░░ $\sqrt{3^2 + 2^2}$ 　　　　　　　　**68.** 5 ░░░ $\sqrt{3^2 + 4^2}$

69. *A Calculator Experiment* Enter any positive real number in your calculator and repeatedly take the square root. What real number does the display appear to be approaching?

70. Square the real number $3/\sqrt{2}$ and note that the radical is eliminated from the denominator. Is this equivalent to rationalizing the denominator? Why or why not?

71. *The Golden Section* The ratio of the width of the Temple of Hephaestus (see figure) to its height is approximately

$$\frac{w}{h} \approx \frac{2}{\sqrt{5} - 1}.$$

This number is called the **golden section**. Early Greeks believed that the most aesthetically pleasing rectangles were those with sides of this ratio. Rationalize the denominator for this number. Approximate your answer, rounded to two decimal places.

Figure for 71

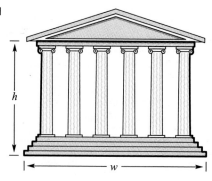

SECTION

10.4

Solving Equations and Applications

Introduction • *Solving Radical Equations* • *Applications of Radicals*

Introduction

We conclude our study of the algebra of radicals by looking at techniques for solving equations involving radicals *and* applications involving radicals. This chapter gives, in miniature, a picture of what algebra is. In other words, the algebra of radicals consists of the following five types of problems.

1. Section 10.1—Definitions of square roots and radicals, and techniques for evaluating radicals

2. Section 10.2—Simplifying radicals and radical expressions

3. Section 10.3—Operations with radicals

4. Section 10.4—Solving equations involving radicals

5. Section 10.4—Applications involving radicals

Think back—didn't the algebra of *linear expressions* have the same five parts? Or how about the algebra of *polynomials*?

Solving Radical Equations

A **radical equation** is an equation that contains one or more radicals with a variable radicand. Here are some examples.

$$\sqrt{x} = 5, \qquad \sqrt{3x - 2} = 7, \quad \text{and} \quad \sqrt{x + 3} = \sqrt{7 - x}$$

Solving radical equations is somewhat like solving equations that contain fractions—we try to get rid of the radicals and obtain a polynomial equation. Then we solve the polynomial equation using the standard procedures. For square root radicals, the following property plays a key role.

Squaring Property of Equality	Let a and b be numbers, variables, or algebraic expressions. If $a = b$, then it follows that $a^2 = b^2$. We call this operation **squaring both sides of an equation**.

You will see in this section that squaring both sides of an equation often introduces extraneous solutions. When you use this procedure, it is critical that you check each solution in the original equation.

To apply the squaring property of equality to solve an equation, we first try to isolate one of the radicals on one side of the equation. Watch how this is done in the examples that follow.

EXAMPLE 1 ■ Solving a Radical Equation Having One Radical

$$\sqrt{x} - 7 = 0 \qquad \text{Given equation}$$
$$\sqrt{x} = 7 \qquad \text{Isolate radical}$$
$$\left(\sqrt{x}\right)^2 = (7)^2 \qquad \text{Square both sides}$$
$$x = 49 \qquad \text{Solution}$$

Check
$$\sqrt{x} - 7 = 0 \qquad \text{Given equation}$$
$$\sqrt{49} - 7 \stackrel{?}{=} 0 \qquad \text{Replace } x \text{ by 49}$$
$$7 - 7 = 0 \qquad \text{Solution checks}$$

Therefore, the equation has one solution: $x = 49$. ■

EXAMPLE 2 ■ Solving a Radical Equation Having One Radical

$$\sqrt{2x + 1} - 2 = 3 \qquad \text{Given equation}$$
$$\sqrt{2x + 1} = 5 \qquad \text{Isolate radical}$$
$$\left(\sqrt{2x + 1}\right)^2 = (5)^2 \qquad \text{Square both sides}$$
$$2x + 1 = 25 \qquad \text{Simplify}$$
$$2x = 24$$
$$x = 12 \qquad \text{Solution}$$

Check

$$\sqrt{2x + 1} - 2 = 3 \qquad \text{Given equation}$$
$$\sqrt{2(12) + 1} - 2 \overset{?}{=} 3 \qquad \text{Replace } x \text{ by } 12$$
$$\sqrt{25} - 2 \overset{?}{=} 3$$
$$5 - 2 = 3 \qquad \text{Solution checks}$$

Therefore, the equation has one solution: $x = 12$. ■

Some radical equations have no solution, as shown in the next example.

EXAMPLE 3 ■ Solving a Radical Equation Having One Radical

$$\sqrt{3x} = -9 \qquad \text{Given equation}$$
$$\left(\sqrt{3x}\right)^2 = (-9)^2 \qquad \text{Square both sides}$$
$$3x = 81$$
$$x = 27 \qquad \text{Trial solution}$$

Check

$$\sqrt{3x} = -9 \qquad \text{Given equation}$$
$$\sqrt{3(27)} \overset{?}{=} -9 \qquad \text{Replace } x \text{ by } 27$$
$$9 \neq -9 \qquad \text{Solution does not check}$$

Therefore, the equation has no solution. ■

NOTE: Another way to see that the equation in Example 3 has no solution is to observe that the right side is negative but the left side *cannot* be negative.

EXAMPLE 4 ■ Solving a Radical Equation Having Two Radicals (on Different Sides)

$$\sqrt{5x + 3} = \sqrt{x + 11} \qquad \text{Given equation}$$
$$\left(\sqrt{5x + 3}\right)^2 = \left(\sqrt{x + 11}\right)^2 \qquad \text{Square both sides}$$
$$5x + 3 = x + 11$$
$$4x = 8$$
$$x = 2 \qquad \text{Solution}$$

Check

$$\sqrt{5x + 3} = \sqrt{x + 11} \qquad \text{Given equation}$$
$$\sqrt{5(2) + 3} \overset{?}{=} \sqrt{2 + 11} \qquad \text{Replace } x \text{ by } 2$$
$$\sqrt{13} = \sqrt{13} \qquad \text{Solution checks}$$

■

EXAMPLE 5   ■   Solving Radical Equations Having Two Radicals (on the Same Side)

(a) $\sqrt{6x-4} - 2\sqrt{4-x} = 0$ .... Given equation

$\sqrt{6x-4} = 2\sqrt{4-x}$ .... Isolate radicals

$6x - 4 = 2^2(4-x)$ .... Square both sides

$6x - 4 = 16 - 4x$

$10x = 20$

$x = 2$ .... Solution

After checking this solution, we can conclude that the solution is $x = 2$. (We leave the checking to you.)

(b) $\sqrt{3x} + \sqrt{2x-5} = 0$ .... Given equation

$\sqrt{3x} = -\sqrt{2x-5}$ .... Isolate radicals

$\left(\sqrt{3x}\right)^2 = \left(-\sqrt{2x-5}\right)^2$ .... Square both sides

$3x = 2x - 5$

$x = -5$ .... Trial solution

A check will show that -5 is not a solution because it yields negative radicands. Therefore, this equation has no solution.   ■

In the next two examples, we show how squaring both sides of an equation can yield a *nonlinear* equation.

EXAMPLE 6   ■   A Radical Equation that Converts to a Nonlinear Equation

$1 - 2x = \sqrt{x}$ .... Given equation

$(1 - 2x)^2 = \left(\sqrt{x}\right)^2$ .... Square both sides

$1 - 4x + 4x^2 = x$ .... Binomial square pattern

$4x^2 - 5x + 1 = 0$ .... Standard form

$(4x - 1)(x - 1) = 0$ .... Factor

$4x - 1 = 0 \implies x = \dfrac{1}{4}$ .... Set first factor equal to 0

$x - 1 = 0 \implies x = 1$ .... Set second factor equal to 0

Try checking each of these solutions. When you do, you will find that $x = \frac{1}{4}$ is a valid solution, but that $x = 1$ is extraneous. Therefore, the equation has only one solution: $x = \frac{1}{4}$.   ■

EXAMPLE 7   ■   A Radical Equation that Converts to a Nonlinear Equation

$$\sqrt{6x + 1} = x - 1 \qquad \text{Given equation}$$
$$\left(\sqrt{6x + 1}\right)^2 = (x - 1)^2 \qquad \text{Square both sides}$$
$$6x + 1 = x^2 - 2x + 1$$
$$0 = x^2 - 8x \qquad \text{Standard form}$$
$$0 = x(x - 8) \qquad \text{Factor}$$
$$x = 0 \qquad \text{Set first factor equal to 0}$$
$$x - 8 = 0 \implies x = 8 \qquad \text{Set second factor equal to 0}$$

A check will show that $x = 0$ is extraneous and $x = 8$ is a valid solution. Therefore, this equation has only one solution: $x = 8$.   ■

Applications of Radicals

A common use of radicals is in applications involving right triangles. Recall that a right triangle is one that contains a right (or 90°) angle, as shown in Figure 10.2. The relationship among the three sides of a right triangle is described by the **Pythagorean Theorem**, which says that if a and b are the lengths of the legs and c is the length of the hypotenuse, then

FIGURE 10.2

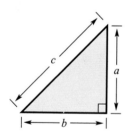

$$c^2 = a^2 + b^2 \qquad \text{Pythagorean Theorem}$$
$$c = \sqrt{a^2 + b^2}.$$

EXAMPLE 8   ■   An Application Involving a Right Triangle

A softball diamond has the shape of a square with 60-foot sides (see Figure 10.3). The catcher is 4 feet behind home plate. How far does the catcher have to throw to second base?

Solution

In Figure 10.3, let x be the hypotenuse of a right triangle with 60-foot sides. Thus, by the Pythagorean Theorem, we have the following equation.

$$x = \sqrt{60^2 + 60^2} \qquad \text{Pythagorean Theorem}$$
$$x = \sqrt{7{,}200}$$
$$x \approx 84.9 \text{ feet}$$

Thus, the distance from home plate to second base is approximately 84.9 feet. Since the catcher is 4 feet behind home plate, the catcher must make a throw of

$$x + 4 \approx 84.9 + 4 = 88.9 \text{ feet.}$$

Check this solution in the original statement of the problem.

FIGURE 10.3

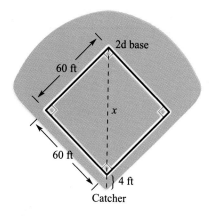

EXAMPLE 9 ■ An Application Involving the Use of Electricity

The amount of power consumed by an electrical appliance is given by the equation

$$I = \sqrt{\frac{P}{R}}$$

where I is the current measured in amps, R is the resistance measured in ohms, and P is the power measured in watts. Find the power used by an electric heater, for which $I = 10$ amps and $R = 15$ ohms.

Solution

$$I = \sqrt{\frac{P}{R}} \qquad \text{Given}$$

$$10 = \sqrt{\frac{P}{15}} \qquad \text{Substitute known values}$$

$$(10)^2 = \left(\sqrt{\frac{P}{15}}\right)^2 \qquad \text{Square both sides}$$

$$100 = \frac{P}{15}$$

$$1{,}500 = P \qquad \text{Solve for } P$$

Therefore, the electric heater uses 1,500 watts of power. Check this solution in the original statement of the problem. ■

EXAMPLE 10   ■   An Application Involving the Velocity of a Falling Object

The velocity of a free-falling object can be determined from the equation

$$v = \sqrt{2gh}$$

where v is the velocity measured in feet per second, $g = 32$ feet per second squared, and h is the distance (in feet) the object has fallen. Find the height from which a rock has been dropped if it strikes the ground with a velocity of 60 feet per second.

Solution

$$v = \sqrt{2gh} \qquad \text{Given equation}$$
$$60 = \sqrt{2(32)h} \qquad \text{Substitute known values}$$
$$(60)^2 = \left(\sqrt{64h}\right)^2 \qquad \text{Square both sides}$$
$$3{,}600 = 64h$$
$$56.25 = h \qquad \text{Solve for } h$$

Thus, the rock has fallen a total of 56.25 feet when it hits the ground, as shown in Figure 10.4. Check this solution in the original statement of the problem.

FIGURE 10.4

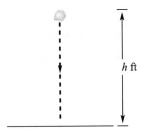

h ft

■

DISCUSSION PROBLEM   ■   Constructing Right Triangles

The ancient Egyptians may have used the following method for constructing a right angle. A rope was marked with 12 equally spaced marks, as shown in Figure 10.5.

FIGURE 10.5

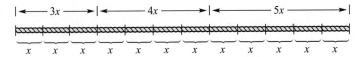

The rope was then stretched tight, as shown in Figure 10.6. Explain why the indicated angle is a right angle. Can you think of any other patterns of marks that could be used to create a right angle in this way?

FIGURE 10.6

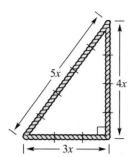

Warm-Up

The following warm-up exercises involve skills that were covered in earlier sections. You will use these skills in the exercise set for this section.

In Exercises 1–6, solve the given equation.

1. $x + 6 = 32$

2. $7x = 18$

3. $6x - 5 = 13$

4. $45 - 8x = 21$

5. $3(x - 5)(x + 11) = 0$

6. $-2(4 - x)(16 + x) = 0$

In Exercises 7–10, simplify the given expression.

7. $\sqrt{500}$

8. $\sqrt{20} - \sqrt{5}$

9. $2\sqrt{18}\sqrt{32}$

10. $\dfrac{2}{\sqrt{2}}$

10.4   EXERCISES

In Exercises 1–30, solve the given equation. (Some of the equations have no solution.)

1. $\sqrt{x} = 10$

2. $\sqrt{x} = 3$

3. $\sqrt{y} - 5 = 0$

4. $\sqrt{t} - 12 = 0$

5. $\sqrt{u} + 3 = 0$

6. $\sqrt{y} + 10 = 0$

7. $\sqrt{a + 3} = 20$

8. $\sqrt{b + 8} = 4$

9. $\sqrt{10x} = 100$

10. $\sqrt{8x} = 4$

11. $\sqrt{3x - 2} = 4$

12. $\sqrt{5x - 1} = 8$

13. $\sqrt{3y + 5} = 2$

14. $\sqrt{5z - 2} = 6$

15. $8 - \sqrt{t} = 3$

16. $12 - \sqrt{s} = 25$

17. $5\sqrt{x + 1} = 6$

18. $2\sqrt{x + 3} = 5$

19. $\sqrt{x^2 + 5} = x + 1$

20. $\sqrt{x^2 - 2} = x - 1$

21. $\sqrt{x + 4} = \sqrt{2x + 1}$

22. $\sqrt{3x + 2} = \sqrt{x + 20}$

23. $\sqrt{5x - 1} = 3\sqrt{x}$

24. $\sqrt{2x + 3} = 2\sqrt{x}$

25. $\sqrt{3t + 11} = 5\sqrt{t}$ **26.** $\sqrt{5 - 4u} = 4\sqrt{u}$ **27.** $\sqrt{x} = \frac{1}{4}x + 1$ **28.** $5\sqrt{x} = x + 4$

29. $\sqrt{6x + 7} = x + 2$ **30.** $\sqrt{4x + 17} = x + 3$

In Exercises 31–36, find the length x of the unknown side of the right triangle. (Round your answer to two decimal places.)

31.

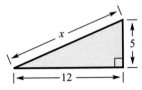

32.

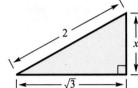

33.

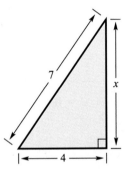

34.

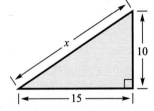

35.

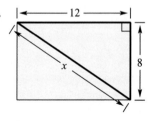

36.

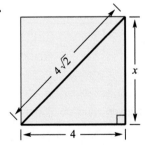

37. *Height of a Ladder* A ladder is 15 feet long, and the bottom of the ladder is 3 feet from the house (see figure). How far does the ladder reach up the side of the house?

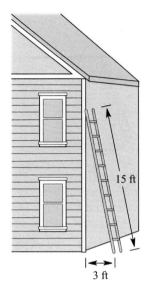

15 ft

3 ft

38. *Length of a Wire* A guy wire on a sailboat mast is attached to the top of the mast and to the deck 15 feet from the base of the mast (see figure). The mast is 40 feet high. How long is the wire?

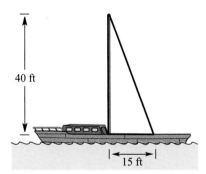

40 ft

15 ft

39. *Distance Between Two Points* Find the length of the line segment connecting the points (1, 2) and (5, 5) in the rectangular coordinate system. (*Hint:* Sketch the line segment as the hypotenuse of a right triangle where the other two sides are parallel to the coordinate axes.)

40. *Distance Between Two Points* Find the length of the line segment connecting the points (2, 5) and (7, 1) in the rectangular coordinate system. (*Hint:* Sketch the line segment as the hypotenuse of a right triangle where the other two sides are parallel to the coordinate axes.)

In Exercises 41–44, use the equation for the velocity of a free-falling object $\left(v = \sqrt{2gh}\,\right)$ as described in Example 10.

41. *Velocity of an Object* An object is dropped from a height of 30 feet. Find the velocity of the object when it strikes the ground.

42. *Velocity of an Object* An object is dropped from a height of 100 feet. Find the velocity of the object when it strikes the ground.

43. *Height of an Object* An object strikes the ground with a velocity of 75 feet per second. Find the height from which the object was dropped.

44. *Height of an Object* An object strikes the ground with a velocity of 100 feet per second. Find the height from which the object was dropped.

Height In Exercises 45 and 46, use the following. The time t in seconds for a free-falling object to fall d feet is given by

$$t = \sqrt{\frac{d}{16}}.$$

45. A construction worker drops a nail from a building and observes it strike a water puddle after approximately three seconds. Estimate the height of the worker.

46. A construction worker drops a nail from a building and observes it hit the ground after approximately five seconds. Estimate the height of the worker.

Pendulum Length In Exercises 47 and 48, use the following. The time *t* in seconds for a pendulum of length *L* feet to move through one complete cycle (its period) is given by

$$t = 2\pi\sqrt{\frac{L}{32}}.$$

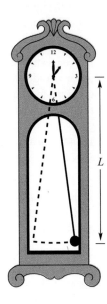

L

47. How long is the pendulum of a grandfather clock that has a period of two seconds? (See figure above.)

48. How long is the pendulum of a mantel clock that has a period of 0.8 seconds?

49. *Demand* The demand equation for a certain product is given by

$$p = 40 - \sqrt{x - 1}$$

where *x* is the number of units demanded per day and *p* is the price per unit. Find the demand when the price is $34.70.

50. *Number of Passengers* An airline offers daily flights between Chicago and Denver. The total monthly cost of these flights is given by

$$C = \sqrt{0.2x + 1}$$

where *C* is measured in millions of dollars and *x* is measured in thousands of passengers. The total cost of the flights for a certain month is 2.5 million dollars. Approximately how many passengers flew that month?

51. *Surface Area* The surface area of a circular cone is given by

$$S = \pi r\sqrt{r^2 + h^2}.$$

Solve this equation for *h*. (See figure.)

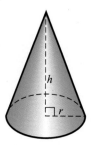

52. *Material in an Ice Cream Cone* Use the formula given in Exercise 51 to approximate the amount of material in an ice cream cone that has a height of 5 inches and a radius of 1.5 inches (see figure). Assume that the material in the cone is $\frac{1}{8}$-inch thick.

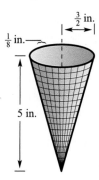

$\frac{3}{2}$ in.

$\frac{1}{8}$ in.

5 in.

53. *Dimensions of a Rectangle* Determine the length and width of a template that has a perimeter of 28 inches and a diagonal length of 10 inches (see figure).

10 in.

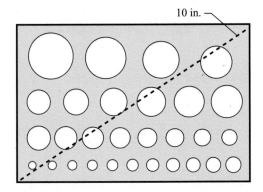

CHAPTER 10   SUMMARY

As you review and prepare for a test on this chapter, first try to obtain a global view of what was discussed. Then review the specific skills needed in each category.

Operations with Radical Expressions

- *Find* the square roots of a number.

  Section 10.1, Exercises 1–10

- *Evaluate* a radical that has a real number radicand.

  Section 10.1, Exercises 11–30

- *Evaluate* a radical expression by using a calculator.

  Section 10.1, Exercises 35–50, 53–57

- *Simplify* a radical with constant factors.

  Section 10.2, Exercises 1–20, 31–48

- *Simplify* a radical with variable factors.

  Section 10.2, Exercises 21–30

- *Rationalize* the denominator of a radical.

  Section 10.2, Exercises 49–62

- *Add and/or subtract* two radicals by combining like radicals.

  Section 10.3, Exercises 1–20

- *Create* the conjugate of a radical expression.

  Section 10.3, Exercises 43–50

- *Multiply* two radicals and simplify the result.

  Section 10.3, Exercises 21–42

- *Divide* two radicals by using the conjugate to rationalize the denominator.

  Section 10.3, Exercises 51–64

Solving Equations that Contain a Radical

- *Solve* a radical equation that contains one radical.

  Section 10.4, Exercises 1–20

- *Solve* a radical equation that contains two radicals.

  Section 10.4, Exercises 21–26

- *Create and solve* a radical equation from a geometric figure.

  Section 10.4, Exercises 31–36

- *Create and solve* a radical equation from verbal statements.

  Section 10.4, Exercises 37–53

Chapter 10 Review Exercises

In Exercises 1–6, evaluate the given square root, if possible. (Do not use a calculator.)

1. $\sqrt{121}$ **2.** $\sqrt{0.09}$ **3.** $\sqrt{-\dfrac{1}{16}}$ **4.** $-\sqrt{\dfrac{64}{9}}$ **5.** $\sqrt{100-36}$ **6.** $\sqrt{16+9}$

In Exercises 7–14, use a calculator to approximate the expression. (Round your answer to two decimal places.)

7. $\sqrt{53}$ **8.** $\sqrt{5{,}335}$ **9.** $\sqrt{\dfrac{7}{8}}$ **10.** $-\sqrt{\dfrac{45}{8}}$

11. $\sqrt{9^2 - 4(2)(7)}$ **12.** $\sqrt{4.5^2 - 4(-3)(8)}$ **13.** $5\sqrt{23.5} - 13.2$ **14.** $\dfrac{-3.4 + \sqrt{12.6}}{2(1.3)}$

In Exercises 15–30, simplify the given radical.

15. $\sqrt{48}$ **16.** $\sqrt{72}$ **17.** $\sqrt{160}$ **18.** $\sqrt{45}$ **19.** $\sqrt{\dfrac{23}{9}}$ **20.** $\sqrt{\dfrac{843}{16}}$

21. $\sqrt{\dfrac{20}{9}}$ **22.** $\sqrt{\dfrac{27}{16}}$ **23.** $\sqrt{36x^4}$ **24.** $\sqrt{81z^2}$ **25.** $\sqrt{4y^3}$ **26.** $\sqrt{100u^5}$

27. $\sqrt{0.04x^2y}$ **28.** $\sqrt{1.44x^2y^3}$ **29.** $\sqrt{32a^3b}$ **30.** $\sqrt{75u^4v^2}$

In Exercises 31–40, rationalize the denominator and simplify.

31. $\sqrt{\dfrac{3}{5}}$ **32.** $\sqrt{\dfrac{7}{10}}$ **33.** $\dfrac{4}{\sqrt{12}}$ **34.** $\dfrac{15}{\sqrt{5}}$ **35.** $\dfrac{3}{\sqrt{x}}$ **36.** $\dfrac{7}{\sqrt{2t}}$

37. $\sqrt{\dfrac{11a}{b}}$ **38.** $\sqrt{\dfrac{4y}{z}}$ **39.** $\dfrac{\sqrt{8x^2}}{\sqrt{2}}$ **40.** $\dfrac{\sqrt{a^2}}{\sqrt{16b^2}}$

In Exercises 41–60, perform the indicated operations and simplify.

41. $7\sqrt{2} + 5\sqrt{2}$ **42.** $15\sqrt{15} - 7\sqrt{15}$ **43.** $3\sqrt{20} - 10\sqrt{20}$

44. $25\sqrt{98} + 2\sqrt{98}$ **45.** $4\sqrt{48} + 2\sqrt{3} - 5\sqrt{12}$ **46.** $12\sqrt{50} - 3\sqrt{8} + \sqrt{32}$

47. $10\sqrt{y+3} - 3\sqrt{y+3}$ **48.** $\sqrt{25x} + \sqrt{49x}$ **49.** $\left(\sqrt{3} + 1\right)^2$

50. $\left(\sqrt{5} - 2\right)^2$ **51.** $\left(\sqrt{7} - 2\right)\left(\sqrt{7} + 2\right)$ **52.** $\left(\sqrt{3} - \sqrt{5}\right)\left(\sqrt{3} + \sqrt{5}\right)$

53. $\left(2\sqrt{3} + 10\right)\left(\sqrt{2} - 3\right)$ **54.** $\left(\sqrt{8} + 2\right)\left(3\sqrt{2} - 1\right)$ **55.** $\sqrt{x}\left(\sqrt{x} + 10\right)$

56. $\sqrt{z}\left(15 - \sqrt{z}\right)$ **57.** $\dfrac{3}{\sqrt{12} - 3}$ **58.** $\dfrac{5}{\sqrt{x} + 2}$

59. $\left(\sqrt{x} - 3\right) \div \left(\sqrt{x} + 3\right)$ **60.** $\left(3\sqrt{s} + 2\right) \div \left(\sqrt{s} + 1\right)$

In Exercises 61–70, solve the given equation. (Some of the equations have no solution.)

61. $\sqrt{y} = 13$

62. $\sqrt{z} = 25$

63. $\sqrt{x} + 2 = 0$

64. $\sqrt{x} - 10 = 0$

65. $\sqrt{t + 3} = 4$

66. $\sqrt{2a - 7} = 15$

67. $\sqrt{2x - 3} = \sqrt{x}$

68. $\sqrt{5x + 3} = \sqrt{x + 15}$

69. $\sqrt{2x} = \frac{1}{2}x + 1$

70. $\sqrt{2x + 7} = x + 4$

71. *Wire Length* A guy wire on a radio tower is attached to the top of the tower and to an anchor 60 feet from the base of the tower (see figure). The tower is 100 feet high. How long is the wire?

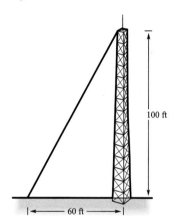

100 ft

60 ft

72. *Height* The time t in seconds for a free-falling object to fall d feet is given by

$$t = \sqrt{\frac{d}{16}}.$$

A child drops a pebble from a bridge and observes it strike the water after approximately 1.5 seconds (see figure). Estimate the height of the bridge.

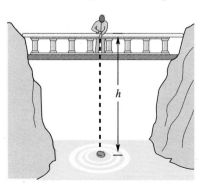

h

73. *Pendulum Length* The time t in seconds for a pendulum of length L feet to move through one complete cycle (its period) is given by

$$t = 2\pi\sqrt{\frac{L}{32}}.$$

How long is the pendulum of a grandfather clock that has a period of 1.75 seconds?

74. *Dimensions of a Rectangle* Determine the length and width of a rectangle that has a perimeter of 70 inches and a diagonal length of 25 inches.

CHAPTER 10 TEST

Take this test as you would take a test in class. After you are done, check your work with the answers given in the back of the book.

1. Evaluate (if possible) without a calculator: $\sqrt{\dfrac{9}{16}}$

2. Evaluate (if possible) without a calculator: $\sqrt{-36}$

3. Use a calculator to approximate $12 - \sqrt{322}$. Round your answer to three decimal places.

4. Simplify: $\sqrt{48}$

5. Simplify: $\sqrt{32x^2y^3}$

6. Simplify: $\dfrac{\sqrt{128}}{\sqrt{2}}$

7. Simplify: $\sqrt{\dfrac{3x^3}{y^4}}$

8. Rationalize the denominator: $\dfrac{5}{\sqrt{15}}$

9. Rationalize the denominator: $\sqrt{\dfrac{8}{t}}$

10. Simplify: $10\sqrt{27} - 7\sqrt{12}$

11. Simplify: $\sqrt{\dfrac{x}{4}} - 3\sqrt{\dfrac{x}{25}}$

12. Multiply and simplify: $\sqrt{2}\left(\sqrt{8} - 5\right)$

13. Multiply and simplify: $\left(\sqrt{6} - 3\right)\left(\sqrt{6} + 5\right)$

14. Multiply and simplify: $\left(\sqrt{5} + 4\right)^2$

15. Determine the conjugate of the real number $\sqrt{3} - 5$ and find the product of the given number and its conjugate.

16. Rationalize the denominator and simplify: $\dfrac{10}{\sqrt{6} + 1}$

17. Solve: $\sqrt{y} = 4$

18. Solve: $2\sqrt{x + 3} = 5$

19. Solve: $2\sqrt{2x} = x + 2$

20. Find the length of the unknown side of the right triangle in the accompanying figure.

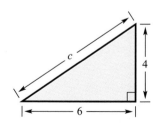

21. The demand equation for a certain product is given by

$$p = 100 - \sqrt{x - 25}$$

where x is the number of units demanded per day and p is the price per unit. Find the demand when the price is $90.

CHAPTER ELEVEN

Quadratic Equations

Chapter Overview

This chapter presents the algebra of quadratic equations. You will learn to solve, graph, and construct quadratic equations. Moreover, this chapter clearly demonstrates that mathematical problem solving is not a cut-and-dried process of matching a single solution method with a given problem type. Rather, mathematical problem solving provides opportunities for you, the problem solver, to be creative and practical in your solution methods.

Solving Quadratic Equations

Introduction • Solving Quadratic Equations by Factoring •
Solving Quadratic Equations by Extracting Square Roots

Introduction

In this chapter we look at second-degree polynomial equations, such as

$$x^2 - 2x - 3 = 0, \qquad 2x^2 + x - 1 = 0, \quad \text{and} \quad x^2 - 5x = 0.$$

We call this type of equation a **quadratic equation** in x.

Definition of a Quadratic Equation in x	A **quadratic equation** in x is an equation that can be written in the standard form $$ax^2 + bx + c = 0$$ where a, b, and c are real numbers with $a \neq 0$.

In Chapter 6 we looked at one method for solving a quadratic equation—factoring. In this chapter we review that method and look at other ways of solving quadratic equations.

Solving Quadratic Equations by Factoring

Remember from Chapter 6 that the first step in solving a quadratic equation by factoring is to write the equation in standard form (with the polynomial on the left side of the equation and zero on the right side). Next, we attempt to factor the left side. Finally, we set each factor equal to zero and solve for x.

EXAMPLE 1 ■ Solving Quadratic Equations by Factoring

Solve the following quadratic equations.

(a) $x^2 + 5x = 14$ (b) $2x^2 = 3 - 5x$

Solution

(a)
$$
\begin{aligned}
x^2 + 5x &= 14 && \text{Given equation} \\
x^2 + 5x - 14 &= 0 && \text{Standard form} \\
(x + 7)(x - 2) &= 0 && \text{Factor} \\
x + 7 &= 0 \implies x = -7 && \text{Set first factor equal to 0} \\
x - 2 &= 0 \implies x = 2 && \text{Set second factor equal to 0}
\end{aligned}
$$

Thus, this quadratic equation has two solutions: -7 and 2. Check these solutions in the original equation.

(b)
$$2x^2 = 3 - 5x \qquad \text{Given equation}$$
$$2x^2 + 5x - 3 = 0 \qquad \text{Standard form}$$
$$(2x - 1)(x + 3) = 0 \qquad \text{Factor}$$
$$2x - 1 = 0 \implies x = \frac{1}{2} \qquad \text{Set first factor equal to 0}$$
$$x + 3 = 0 \implies x = -3 \qquad \text{Set second factor equal to 0}$$

Thus, this quadratic equation has two solutions: $\frac{1}{2}$ and -3. Check these solutions in the original equation. ∎

If the two factors of a quadratic equation in standard form are identical, then the corresponding solution is called a **repeated solution**. This occurs in the next example.

EXAMPLE 2 ■ A Quadratic Equation with a Repeated Solution
Solve the quadratic equation $6x^2 + 4 = 3 + 6x - 3x^2$.

Solution
$$6x^2 + 4 = 3 + 6x - 3x^2 \qquad \text{Given equation}$$
$$9x^2 - 6x + 1 = 0 \qquad \text{Standard form}$$
$$(3x - 1)(3x - 1) = 0 \qquad \text{Factor}$$
$$3x - 1 = 0 \qquad \text{Set factor equal to 0}$$
$$x = \frac{1}{3} \qquad \text{Repeated solution}$$

Thus, this quadratic equation has only one (repeated) solution: $\frac{1}{3}$. Check this solution in the original equation. ∎

EXAMPLE 3 ■ Solving Quadratic Equations by Factoring
Solve the following equations by factoring.
(a) $(x - 5)(x + 2) = 8$ (b) $2x^2 = 32$

Solution
(a) Don't be tricked by this quadratic equation. Even though the left side is factored, the right side is not zero. Thus, we must first rewrite the equation in standard form, and then refactor, as follows.

$$(x - 5)(x + 2) = 8 \qquad \text{Given equation}$$
$$x^2 - 3x - 10 = 8 \qquad \text{Multiply}$$
$$x^2 - 3x - 18 = 0 \qquad \text{Standard form}$$
$$(x - 6)(x + 3) = 0 \qquad \text{Factor}$$

$$x - 6 = 0 \quad \Longrightarrow \quad x = 6 \qquad \text{Set first factor equal to 0}$$
$$x + 3 = 0 \quad \Longrightarrow \quad x = -3 \qquad \text{Set second factor equal to 0}$$

Thus, this equation has two solutions: 6 and -3. Check these solutions in the original equation.

(b)

$$2x^2 = 32 \qquad \text{Given equation}$$
$$2x^2 - 32 = 0 \qquad \text{Standard form}$$
$$2(x^2 - 16) = 0 \qquad \text{Common monomial factor}$$
$$2(x + 4)(x - 4) = 0 \qquad \text{Factor as difference of squares}$$
$$x + 4 = 0 \quad \Longrightarrow \quad x = -4 \qquad \text{Set first factor equal to 0}$$
$$x - 4 = 0 \quad \Longrightarrow \quad x = 4 \qquad \text{Set second factor equal to 0}$$

Thus, this equation has two solutions: -4 and 4. Check these solutions in the original equation.   ■

(Important note)

Can't divide through with a variable because

In Example 3(b) note that it is not necessary to set the *constant* factor 2 equal to zero (because it would produce no solution). Because of this, when an equation in standard form has a constant factor, some people like to divide both sides of the equation by the factor. This procedure is valid because the equations

$$2(x^2 - 16) = 0 \quad \text{and} \quad x^2 - 16 = 0$$

are *equivalent*, which means that both equations have the same solutions. If you use this technique, be sure that you apply it *only* to constant factors. We cannot divide both sides of an equation by a variable factor without risking the loss of one or more solutions. For instance, the equation $x^2 - x = 0$ has *two solutions*: 0 and 1. However, if we divide both sides of the equation by x, we are left with $x - 1 = 0$ which has only one solution. Thus, dividing by x lost the solution $x = 0$.

Solving Quadratic Equations by Extracting Square Roots

There is a nice shortcut for solving quadratic equations of the form

$$u^2 = d$$

where $d > 0$ and u is an algebraic expression. By factoring, we can see that this equation has two solutions.

$$u^2 = d \qquad \text{Given equation}$$
$$u^2 - d = 0 \qquad \text{Standard form}$$
$$\left(u + \sqrt{d}\right)\left(u - \sqrt{d}\right) = 0 \qquad \text{Factor}$$
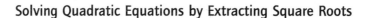
$$u + \sqrt{d} = 0 \quad \Longrightarrow \quad u = -\sqrt{d} \qquad \text{Set first factor equal to 0}$$
$$u - \sqrt{d} = 0 \quad \Longrightarrow \quad u = \sqrt{d} \qquad \text{Set second factor equal to 0}$$

Thus, the equation $u^2 = d$ (where $d > 0$) has two solutions. Since the solutions differ only in sign, we sometimes write both of the solutions together, using a "plus or minus sign."

$$u = \pm\sqrt{d}$$

This form of the solution is read as "u is equal to plus or minus the square root of d." When we solve an equation of the form $u^2 = d$ without going through the steps of factoring, we say that we are **extracting square roots**.

Extracting Square Roots	Let u be a real number, a variable, or an algebraic expression. The equation $u^2 = d$, where $d > 0$ has exactly two solutions:
	$$u = \sqrt{d} \quad \text{and} \quad u = -\sqrt{d}.$$
	These solutions can also be written as $u = \pm\sqrt{d}$.

EXAMPLE 4   ■   **Solving a Quadratic Equation by Extracting Square Roots**

Solve the following quadratic equation by extracting the square roots.

$$4x^2 = 12$$

Solution

$4x^2 = 12$	Given equation
$x^2 = 3$	Divide both sides by 4
$x = \pm\sqrt{3}$	Extract square roots

Thus, the equation has two solutions: $\sqrt{3}$ and $-\sqrt{3}$. Note that $x^2 - 3 = 0$ factors as $(x + \sqrt{3})(x - \sqrt{3}) = 0$, which gives the same two solutions. Check these solutions in the original equation.   ■

EXAMPLE 5   ■   **Solving a Quadratic Equation by Extracting Square Roots**

Solve the following quadratic equation by extracting the square roots.

$$(x - 3)^2 = 7$$

Solution

In this case an extra step is needed after extracting the square roots.

$(x - 3)^2 = 7$	Given equation
$x - 3 = \pm\sqrt{7}$	Extract square roots
$x = 3 \pm\sqrt{7}$	Add 3 to both sides

Thus, the equation has two solutions: $3 + \sqrt{7}$ and $3 - \sqrt{7}$. Check these solutions in the original equation.   ■

EXAMPLE 6 ■   Solving a Quadratic Equation by Extracting Square Roots

Solve the following quadratic equation by extracting the square roots.

$$2(3x + 5)^2 - 16 = 0$$

Solution

In this case some preliminary steps are needed before taking the square roots.

$2(3x + 5)^2 - 16 = 0$	Given equation
$2(3x + 5)^2 = 16$	
$(3x + 5)^2 = 8$	Divide both sides by 2
$3x + 5 = \pm\sqrt{8}$	Extract square roots
$3x = -5 \pm 2\sqrt{2}$	Subtract 5 from both sides
$x = \dfrac{-5 \pm 2\sqrt{2}}{3}$	Divide both sides by 3

Thus, the solutions are

$$\frac{-5 + 2\sqrt{2}}{3} \quad \text{and} \quad \frac{-5 - 2\sqrt{2}}{3}.$$

Check

$2(3x + 5)^2 - 16 = 0$	Given equation
$2\left(3\left[\dfrac{-5 + 2\sqrt{2}}{3}\right] + 5\right)^2 - 16 \overset{?}{=} 0$	Replace x with $(-5 + 2\sqrt{2})/3$
$2([-5 + 2\sqrt{2}] + 5)^2 - 16 \overset{?}{=} 0$	
$2(2\sqrt{2})^2 - 16 \overset{?}{=} 0$	
$2(2^2)(\sqrt{2})^2 - 16 \overset{?}{=} 0$	
$16 - 16 = 0$	Solution checks

We leave the checking of the other solution to you.   ■

Not all quadratic equations have real number solutions. For instance, we know that there is no real number that is the solution of the equation

$$x^2 + 4 = 0$$

because that would imply that there is a real number x such that $x^2 = -4$. (As you know, the square of a real number cannot be negative.) Keep this possibility in mind as you do the exercises for this section.

DISCUSSION PROBLEM ■ **Solving Equations**

A typical third-degree equation can be difficult to solve. However, a third-degree equation that contains no constant term is easy to solve. Why is this? Write a short paragraph explaining how to solve a third-degree equation that contains no constant term. Then apply your technique to the following equations.

(a) $x^3 - x^2 - 2x = 0$ (b) $2x^3 + 4x^2 - 6x = 0$ ■

Warm-Up

The following warm-up exercises involve skills that were covered in earlier sections. You will use these skills in the exercise set for this section.

In Exercises 1–6, factor the given expression.

1. $9x^2 - 25$

2. $4t^2 - 12t + 9$

3. $2x^2 - 8x - 10$

4. $4s^2 - 9$

5. $3x^2 - 11x + 10$

6. $4x^3 - 12x^2 - 16x$

In Exercises 7–10, solve the given equation.

7. $5y + 4 = 0$

8. $2s - 3 = 0$

9. $(x + 7)(2x - 3) = 0$

10. $(5x - 4)(x - 10) = 0$

11.1 EXERCISES

In Exercises 1–30, solve the given equation by factoring.

1. $x^2 - 2x = 0$

2. $t^2 + 5t = 0$

3. $4x^2 + 8x = 0$

4. $25y^2 - 100y = 0$

5. $a^2 - 25 = 0$

6. $v^2 - 100 = 0$

7. $4x^2 - 9 = 0$

8. $16y^2 - 81 = 0$

9. $x^2 - 8x + 16 = 0$

10. $x^2 - 10x + 25 = 0$

11. $x^2 + 4x + 4 = 0$

12. $x^2 + 20x + 100 = 0$

13. $4x^2 + 12x + 9 = 0$

14. $9x^2 - 12x + 4 = 0$

15. $x^2 - 5x + 6 = 0$

16. $x^2 - 7x + 12 = 0$

17. $(x - 3)(x + 1) = 5$

18. $x^2 = 10 - 3x$

19. $(6 + x)(1 - x) = 10$

20. $x^2 + 9x + 14 = 0$

21. $2x^2 = 5x - 2$

22. $3x^2 - 11x + 6 = 0$

23. $4x^2 + 1 = 6x - 4x^2$

24. $12x^2 - 17x + 6 = 0$

25. $5x^2 - 16x - 16 = 0$

26. $3x^2 - 14x - 24 = 0$

27. $u(u - 10) - 6(u - 10) = 0$

28. $x(x + 2) - 3(x + 2) = 0$

29. $3z(z + 20) + 12(z + 20) = 0$

30. $16x(x - 3) - 4(x - 3) = 0$

In Exercises 31–50, solve the quadratic equation by extracting square roots. (Some of the equations have no real number solutions.)

31. $x^2 = 49$

32. $z^2 = 121$

33. $9x^2 = 49$

34. $16z^2 = 121$

35. $u^2 - 100 = 0$

36. $v^2 - 25 = 0$

37. $9u^2 - 100 = 0$

38. $16v^2 - 25 = 0$

39. $9x^2 + 1 = 0$

40. $a^2 + 9 = 0$

41. $(x + 4)^2 = 144$

42. $(y - 7)^2 = 625$

43. $(x - 8)^2 = 0.04$

44. $(x + 2)^2 = 0.09$

45. $4(x + 3)^2 = 25$

46. $9(x - 1)^2 = 16$

47. $(x - 1)^2 = 5$

48. $(x - 10)^2 = 20$

49. $(2x + 5)^2 = 8$

50. $(3x - 7)^2 = 32$

51. *Falling Time* The height h (in feet) of an object dropped from a tower 64 feet high is given by

$$h = 64 - 16t^2$$

where t measures the time in seconds (see figure). How long does it take the object to reach the ground?

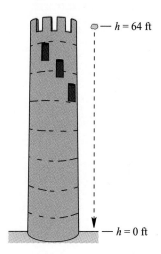

— $h = 64$ ft

— $h = 0$ ft

52. *Falling Time* An object is dropped from a height of 300 feet. Its height (in feet) after t seconds of falling is given by

$$h = 300 - 16t^2.$$

(a) Complete the following table. (Round the times to two decimal places.)

Table for 52

Height, h	Time, t
300	0
250	
200	
150	
100	
50	
0	

(b) Consecutive heights in the table differ by a constant amount of 50 feet. Do the corresponding times differ by a constant amount?

53. *Revenue* The revenue R (in dollars) when x units of a certain product are sold is given by

$$R = x\left(5 - \frac{1}{10}x\right), \quad 0 < x < 25.$$

Determine the number of units that must be sold to produce a revenue of \$60.

54. *Revenue* The revenue R (in dollars) when x units of a certain product are sold is given by

$$R = x\left(100 - \frac{1}{2}x\right), \quad 0 < x < 100.$$

Determine the number of units that must be sold to produce a revenue of \$4,200.

Completing the Square

Introduction   •   Constructing Perfect Square Trinomials   •
Solving Quadratic Equations by Completing the Square

Introduction

Consider the quadratic equation

$$(x - 3)^2 = 7. \qquad\qquad \text{Completed square form}$$

We know from Example 5 of the previous section that this equation has two solutions: $3 + \sqrt{7}$ and $3 - \sqrt{7}$. Now let's suppose that you were given the above equation in its standard form

$$x^2 - 6x + 2 = 0. \qquad\qquad \text{Standard form}$$

How would you solve this equation if you were given only the standard form? You could try factoring, but after attempting to do that you would find that the left side of the equation is not factorable using integer coefficients.

   In this section we look at a technique for rewriting an equation in completed square form. We call this process **completing the square**.

Constructing Perfect Square Trinomials

To complete the square, we must realize that all perfect square trinomials with leading coefficients of one have a similar form. For instance, consider the following perfect square trinomials.

$$x^2 + 2x + 1 = (x + 1)^2$$
$$x^2 - 10x + 25 = (x - 5)^2$$
$$x^2 + 3x + \frac{9}{4} = \left(x + \frac{3}{2}\right)^2$$

In each case note that the constant term of the perfect square trinomial is the square of half of the coefficient of the x-term. That is, a perfect square trinomial with a leading coefficient of one has the following form.

Perfect Square Trinomial = Square of Binomial

$$x^2 + bx + \left(\frac{b}{2}\right)^2 = \left(x + \frac{b}{2}\right)^2$$

$$\underbrace{\qquad\qquad}_{(\text{half})^2}$$

Thus, to complete the square for an expression of the form $x^2 + bx$, we must add $(b/2)^2$ to the expression.

| Completing the Square | To **complete the square** for the expression |

$$x^2 + bx$$

we add $(b/2)^2$, which is the square of half the coefficient of x. Consequently,

$$x^2 + bx + \left(\frac{b}{2}\right)^2 = \left(x + \frac{b}{2}\right)^2.$$

EXAMPLE 1 ■ **Constructing a Perfect Square Trinomial**

What term should be added to the expression

$$x^2 - 4x$$

so that it becomes a perfect square trinomial?

Solution

For this expression, the coefficient of the x-term is -4. By taking half of this and squaring the result, we see that $(-2)^2 = 4$ should be added to the expression to make it a perfect square trinomial.

$$x^2 - 4x + 4 = x^2 - 2(2)x + (-2)^2$$
$$= (x - 2)^2$$ ■

EXAMPLE 2 ■ **Constructing Perfect Square Trinomials**

What terms should be added to the following expressions so that each becomes a perfect square trinomial?
(a) $x^2 + 12x$ (b) $x^2 - 7x$

Solution

(a) For this expression, the coefficient of the x-term is 12. By taking half of this and squaring the result, we see that $6^2 = 36$ should be added to the expression to make it a perfect square trinomial.

$$x^2 + 12x + 36 = x^2 + 2(6)x + 6^2$$
$$= (x + 6)^2$$

(b) For this expression, the coefficient of the x-term is -7. By taking half of this and squaring the result, we see that $\left(-\frac{7}{2}\right)^2 = \frac{49}{4}$ should be added to the expression to make it a perfect square trinomial.

$$x^2 - 7x + \frac{49}{4} = x^2 - 2\left(\frac{7}{2}\right)x + \left(-\frac{7}{2}\right)^2$$
$$= \left(x - \frac{7}{2}\right)^2$$ ■

MATH MATTERS

Calculating with the Number 9

The number 9 has several unusual charac-teristics. For instance, look at the number pattern given by the following multiplica-tions and additions.

$$(1 \times 9) + 2 = 11$$
$$(12 \times 9) + 3 = 111$$
$$(123 \times 9) + 4 = 1{,}111$$
$$(1{,}234 \times 9) + 5 = 11{,}111$$
$$(12{,}345 \times 9) + 6 = 111{,}111$$
$$(123{,}456 \times 9) + 7 = 1{,}111{,}111$$
$$(1{,}234{,}567 \times 9) + 8 = 11{,}111{,}111$$
$$(12{,}345{,}678 \times 9) + 9 = 111{,}111{,}111$$

Try completing the following multiplications and additions. What is the pattern for this set of equations? (The answer is given in the back of the text.)

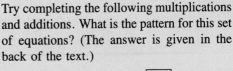

$$(9 \times 9) + 7 = \boxed{}$$
$$(98 \times 9) + 6 = \boxed{}$$
$$(987 \times 9) + 5 = \boxed{}$$
$$(9{,}876 \times 9) + 4 = \boxed{}$$
$$(98{,}765 \times 9) + 3 = \boxed{}$$
$$(987{,}654 \times 9) + 2 = \boxed{}$$
$$(9{,}876{,}543 \times 9) + 1 = \boxed{}$$
$$(98{,}765{,}432 \times 9) + 0 = \boxed{}$$

The first calculating machine (1642) which automatically carried the tens was invented by Blaise Pascal, philosopher and mathematician. It could add figures up to six places.

Solving Quadratic Equations by Completing the Square

We now look at how completing the square can be used to solve a quadratic equation. When using this procedure with an equation, remember that it is essential to preserve the equality. Thus, whatever constant term you add to one side of the equation, you must be sure to add the same constant to the other side of the equation.

EXAMPLE 3 ■ Completing the Square: Leading Coefficient is 1

Solve the equation $x^2 + 10x = 0$ by completing the square.

Solution

$$x^2 + 10x = 0$$ Given equation

$$x^2 + 10x + 5^2 = 25$$ Add $5^2 = 25$ to both sides

$\underbrace{}$
$(\text{half})^2$

$$(x + 5)^2 = 25 \qquad \text{Binomial squared}$$
$$x + 5 = \pm\sqrt{25} \qquad \text{Extract square roots}$$
$$x = -5 \pm 5 \qquad \text{Solve for } x$$
$$x = -5 + 5 \quad \text{or} \quad x = -5 - 5$$
$$x = 0 \quad \text{or} \quad x = -10 \qquad \text{Solutions}$$

Thus, this equation has two solutions: 0 and -10. Check these solutions in the original equation. ■

In Example 3 we used completing the square to solve the quadratic equation simply for the sake of illustration. This particular equation would be easier to solve by factoring. Try reworking the problem by factoring to see that you obtain the same two solutions. In the next example, we look at an equation that cannot be solved by factoring using integer coefficients.

EXAMPLE 4 ■ **Completing the Square: Leading Coefficient is 1**

Solve the following equation by completing the square.

$$x^2 - 4x + 1 = 0$$

Solution

$$x^2 - 4x + 1 = 0 \qquad \text{Given equation}$$
$$x^2 - 4x = -1 \qquad \text{Subtract 1 from both sides}$$
$$x^2 - 4x + (-2)^2 = -1 + 4 \qquad \text{Add } (-2)^2 = 4 \text{ to both sides}$$

$$\underbrace{\qquad}_{(\text{half})^2}$$

$$(x - 2)^2 = 3 \qquad \text{Binomial squared}$$
$$x - 2 = \pm\sqrt{3} \qquad \text{Extract square roots}$$
$$x = 2 \pm \sqrt{3} \qquad \text{Solve for } x$$

Thus, this equation has two solutions: $2 + \sqrt{3}$ and $2 - \sqrt{3}$. Check these solutions in the original equation. ■

If the leading coefficient of a quadratic expression is not 1, we must divide both sides of the equation by this coefficient *before* completing the square, as shown in the following example.

EXAMPLE 5 ■ **Completing the Square: Leading Coefficient is not 1**

Solve the following equation by completing the square.

$$2x^2 + 5x = 3$$

Solution

$$2x^2 + 5x = 3 \qquad \text{Given equation}$$

$$x^2 + \frac{5}{2}x = \frac{3}{2} \qquad \text{Divide both sides by 2}$$

$$x^2 + \frac{5}{2}x + \left(\frac{5}{4}\right)^2 = \frac{3}{2} + \frac{25}{16} \qquad \text{Add } \left(\frac{5}{4}\right)^2 = \frac{25}{16} \text{ to both sides}$$

$$\underbrace{\qquad}_{(\text{half})^2}$$

$$\left(x + \frac{5}{4}\right)^2 = \frac{49}{16} \qquad \text{Binomial squared}$$

$$x + \frac{5}{4} = \pm\frac{7}{4} \qquad \text{Extract square roots}$$

$$x = -\frac{5}{4} \pm \frac{7}{4} \qquad \text{Solve for } x$$

Thus, this equation has two solutions:

$$x = -\frac{5}{4} + \frac{7}{4} = \frac{2}{4} = \frac{1}{2} \quad \text{and} \quad x = -\frac{5}{4} - \frac{7}{4} = -\frac{12}{4} = -3.$$

Check these solutions in the original equation. ∎

NOTE: If you solve a quadratic equation by completing the square and obtain solutions that do not involve radicals, then you could have solved the equation by factoring. For instance, in Example 5 the equation could have been factored as

$$2x^2 + 5x - 3 = 0$$
$$(2x - 1)(x + 3) = 0.$$

In the next example, we use completing the square to solve a quadratic equation that cannot be solved by factoring using integer coefficients.

EXAMPLE 6 ■ Completing the Square: Leading Coefficient is not 1

Solve the following quadratic equation and use a calculator to estimate the solutions to two decimal places.

$$3x^2 - 2x - 4 = 0$$

Solution

$$3x^2 - 2x - 4 = 0 \qquad \text{Given equation}$$
$$3x^2 - 2x = 4 \qquad \text{Add 4 to both sides}$$

$$x^2 - \frac{2}{3}x = \frac{4}{3}$$ Divide both sides by 3

$$x^2 - \frac{2}{3}x + \left(-\frac{1}{3}\right)^2 = \frac{4}{3} + \frac{1}{9}$$ Add $\left(-\frac{1}{3}\right)^2 = \frac{1}{9}$ to both sides

$$\underbrace{\qquad\qquad}_{(\text{half})^2}$$

$$\left(x - \frac{1}{3}\right)^2 = \frac{13}{9}$$ Binomial squared

$$x - \frac{1}{3} = \pm\frac{\sqrt{13}}{3}$$ Extract square roots

$$x = \frac{1}{3} \pm \frac{\sqrt{13}}{3}$$ Solve for x

Thus, this equation has two solutions:

$$x = \frac{1}{3} + \frac{\sqrt{13}}{3} = \frac{1 + \sqrt{13}}{3} \approx 1.54$$

and

$$x = \frac{1}{3} - \frac{\sqrt{13}}{3} = \frac{1 - \sqrt{13}}{3} \approx -0.87.$$

Check these solutions in the original equation.   ∎

The method of completing the square can be used to solve *any* quadratic equation. Moreover, this method will identify those quadratic equations that have no real solutions, as seen in the next example.

EXAMPLE 7   ∎   A Quadratic Equation with No Real Solutions

Use completing the square to show that the following equation has no real solutions.

$$x^2 - 4x + 7 = 0$$

Solution

We begin by writing the equation in completed square form, as follows.

$$x^2 - 4x + 7 = 0$$
$$x^2 - 4x = -7$$
$$x^2 - 4x + 4 = -7 + 4$$
$$(x - 2)^2 = -3$$

Since the square of a real number cannot be negative, we conclude that this equation has no real solution.   ∎

DISCUSSION PROBLEM   ■   Completing the Square

The figure on the left represents the expression $x^2 + 2x$, and the figure on the right represents the expression $(x + 1)^2$. Find the area of the part that was added to the left figure to "complete the square."

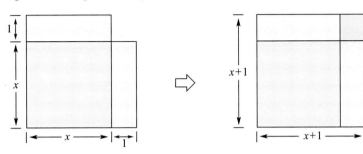

Draw two figures that illustrate completing the square for the expression $x^2 + 6x$.

■

Warm-Up

The following warm-up exercises involve skills that were covered in earlier sections. You will use these skills in the exercise set for this section.

In Exercises 1–4, expand and simplify the given expression.

1. $(x + 2)^2 - 1$ **2.** $(x + 5)^2 + 3$

3. $(u - 8)^2 + 10$ **4.** $(v - 3)^2 - 2$

In Exercises 5–10, solve the given equation.

5. $x^2 = \dfrac{1}{4}$ **6.** $y^2 = \dfrac{9}{16}$

7. $(y - 5)^2 = 36$ **8.** $(z + 2)^2 = 10$

9. $(2x - 1)^2 = 5$ **10.** $(4x + 3)^2 = 7$

11.2   EXERCISES

In Exercises 1–10, determine the constant that must be added to the expression to make it a perfect square trinomial.

1. $x^2 + 10x +$ ▨   **2.** $x^2 + 14x +$ ▨   **3.** $y^2 - 24y +$ ▨   **4.** $y^2 - 8y +$ ▨

5. $t^2 + 3t +$ ▨   **6.** $u^2 + 9u +$ ▨   **7.** $x^2 - \dfrac{4}{5}x +$ ▨   **8.** $y^2 + \dfrac{2}{3}y +$ ▨

9. $y^2 - \dfrac{1}{5}y + $

10. $a^2 - \dfrac{1}{3}a + $

In Exercises 11–20, solve the quadratic equation (a) by completing the square and (b) by factoring.

11. $x^2 - 4x = 0$

12. $x^2 - 2x = 0$

13. $x^2 - 4x + 3 = 0$

14. $x^2 - 8x + 15 = 0$

15. $x^2 + 6x + 8 = 0$

16. $x^2 + 2x - 15 = 0$

17. $x^2 + 2x - 35 = 0$

18. $x^2 + 10x + 16 = 0$

19. $x^2 - 6x - 7 = 0$

20. $x^2 - 6x - 27 = 0$

In Exercises 21–44, solve the quadratic equation by completing the square. (Some of the equations have no real number solution.)

21. $x^2 - 2x - 1 = 0$

22. $x^2 - 6x + 7 = 0$

23. $x^2 + 2x - 1 = 0$

24. $x^2 + 6x + 7 = 0$

25. $u^2 - 4u - 1 = 0$

26. $a^2 - 10a + 15 = 0$

27. $x^2 - 2x + 3 = 0$

28. $x^2 - 6x + 14 = 0$

29. $x^2 - 8x - 2 = 0$

30. $x^2 + 6x - 3 = 0$

31. $y^2 + 14y + 17 = 0$

32. $y^2 + 2y - 26 = 0$

33. $x^2 - x - 3 = 0$

34. $x^2 + 3x + 1 = 0$

35. $t^2 + 5t + 2 = 0$

36. $u^2 - 9u - 5 = 0$

37. $v^2 + 3v - 10 = 0$

38. $z^2 - 7z + 5 = 0$

39. $3x^2 - 6x + 1 = 0$

40. $2x^2 - 4x + 3 = 0$

41. $2x^2 + 6x - 5 = 0$

42. $3x^2 - 12x + 7 = 0$

43. $5y^2 + 5y - 12 = 0$

44. $4z^2 - 3z - 2 = 0$

In Exercises 45–48, solve the equation.

45. $\dfrac{x}{2} + \dfrac{1}{x} = 2$

46. $\dfrac{x}{3} + \dfrac{2}{x} = 4$

47. $\sqrt{2x + 3} = x - 2$

48. $\sqrt{4x + 5} = x - 6$

49. *Number Problem* Find two consecutive positive integers such that the sum of their squares is 85.

50. *Revenue* The revenue R for selling x units of a certain product is given by

$$R = x\left(25 - \dfrac{1}{2}x\right).$$

Find the number of units sold if the revenue is $304.50.

The Quadratic Formula

The Quadratic Formula   •   Solving Quadratic Equations by the Quadratic Formula

The Quadratic Formula

The last technique for solving a quadratic equation involves a formula called the **Quadratic Formula**. To develop this formula, we solve the general quadratic equation $ax^2 + bx + c = 0$ by completing the square.

$$ax^2 + bx + c = 0 \qquad\qquad \text{Standard form, } a \neq 0$$

$$ax^2 + bx = -c \qquad\qquad \text{Subtract } c \text{ from both sides}$$

$$x^2 + \frac{b}{a}x = -\frac{c}{a} \qquad\qquad \text{Divide both sides by } a$$

$$x + \frac{b}{a}x + \underbrace{\left(\frac{b}{2a}\right)^2}_{(\text{half})^2} = -\frac{c}{a} + \left(\frac{b}{2a}\right)^2 \qquad\qquad \text{Complete the square}$$

$$\left(x + \frac{b}{2a}\right)^2 = \frac{b^2 - 4ac}{4a^2} \qquad\qquad \text{Simplify}$$

$$x + \frac{b}{2a} = \pm\sqrt{\frac{b^2 - 4ac}{4a^2}} \qquad\qquad \text{Extract square roots}$$

$$x = -\frac{b}{2a} \pm \frac{\sqrt{b^2 - 4ac}}{2|a|}$$

$$x = \frac{-b \pm \sqrt{b^2 - 4ac}}{2a} \qquad\qquad \text{Solutions}$$

Note that since $\pm 2|a|$ represents the same numbers as $\pm 2a$, we can omit the absolute value sign.

The Quadratic Formula   The solutions of a quadratic equation in the standard form

$$ax^2 + bx + c = 0, \quad a \neq 0$$

are given by the **Quadratic Formula**

$$x = \frac{-b \pm \sqrt{b^2 - 4ac}}{2a}.$$

The expression inside the radical, $b^2 - 4ac$, is called the **discriminant**.

1. If $b^2 - 4ac > 0$, then the equation has two real number solutions.

2. If $b^2 - 4ac = 0$, then the equation has one (repeated) real number solution.

3. If $b^2 - 4ac < 0$, then the equation has no real number solutions.

The Quadratic Formula is one of the most important formulas in algebra, and you should memorize it. We have found that it helps to try to memorize a verbal statement of the rule. For instance, you might try to remember the following verbal statement of the Quadratic Formula: "Minus b, plus or minus the square root of b squared minus $4ac$, all divided by $2a$."

Solving Quadratic Equations by the Quadratic Formula

When using the Quadratic Formula, remember that before the formula can be applied, you must first write the quadratic equation in standard form.

EXAMPLE 1   ■   Using the Quadratic Formula: Two Distinct Solutions

Use the Quadratic Formula to solve the following equation.

$$x^2 + 5x = 14$$

Solution

To begin, we write the equation in standard form, $ax^2 + bx + c = 0$. Then we determine the values of a, b, and c. Finally, we substitute these values into the Quadratic Formula to obtain the solutions.

$$x^2 + 5x = 14 \qquad \text{Given equation}$$

$$x^2 + 5x - 14 = 0 \qquad \text{Standard form with } a = 1, b = 5, c = -14$$

$$x = \frac{-b \pm \sqrt{b^2 - 4ac}}{2a} \qquad \text{Quadratic Formula}$$

$$x = \frac{-5 \pm \sqrt{5^2 - 4(1)(-14)}}{2(1)} \qquad \text{Substitute}$$

$$x = \frac{-5 \pm \sqrt{25 + 56}}{2}$$

$$x = \frac{-5 \pm \sqrt{81}}{2}$$

$$x = \frac{-5 \pm 9}{2} \qquad \text{Solutions}$$

Therefore, the equation has two solutions:

$$x = \frac{-5 + 9}{2} = \frac{4}{2} = 2 \quad \text{and} \quad x = \frac{-5 - 9}{2} = -\frac{14}{2} = -7.$$

Check these solutions in the original equation. (Because the simplified versions of the solutions have no radicals, we know that the equation could have been solved by factoring. To see how this is done, look back at Example 1 in Section 11.1.)   ■

EXAMPLE 2   ■   Using the Quadratic Formula: Two Distinct Solutions

Use the Quadratic Formula to solve the equation $-x^2 - 2x + 4 = 0$.

Solution

This equation is already written in the standard form $ax^2 + bx + c = 0$. However, the leading coefficient is negative. We find that it is cumbersome to have negative leading coefficients, and we suggest multiplying both sides of the equation by -1 (this will produce a positive leading coefficient).

$$-x^2 - 2x + 4 = 0 \qquad\qquad \text{Leading coefficient is negative}$$

$$x^2 + 2x - 4 = 0 \qquad\qquad \text{Standard form with } a = 1, b = 2, c = -4$$

$$x = \frac{-b \pm \sqrt{b^2 - 4ac}}{2a} \qquad\qquad \text{Quadratic Formula}$$

$$x = \frac{-2 \pm \sqrt{2^2 - 4(1)(-4)}}{2(1)} \qquad\qquad \text{Substitute}$$

$$x = \frac{-2 \pm \sqrt{4 + 16}}{2}$$

$$x = \frac{-2 \pm \sqrt{20}}{2}$$

$$x = \frac{-2 \pm 2\sqrt{5}}{2} \qquad\qquad \text{Simplify radical}$$

$$x = \frac{2(-1 \pm \sqrt{5})}{2} \qquad\qquad \text{Divide out common factor}$$

$$x = -1 \pm \sqrt{5} \qquad\qquad \text{Solutions}$$

Therefore, the equation has two solutions:

$$x = -1 + \sqrt{5} \quad \text{and} \quad x = -1 - \sqrt{5}.$$

Check these solutions in the original equation.   ■

EXAMPLE 3   ■   Using the Quadratic Formula: One Repeated Solution

Use the Quadratic Formula to solve the equation $8x^2 - 24x + 18 = 0$.

Solution

We note that this equation has a common factor of 2. To simplify things, we first divide both sides of the equation by 2.

$$8x^2 - 24x + 18 = 0 \qquad\qquad \text{Common factor of 2}$$

$$4x^2 - 12x + 9 = 0 \qquad\qquad \text{Standard form with } a = 4, b = -12, c = 9$$

$$x = \frac{-b \pm \sqrt{b^2 - 4ac}}{2a} \qquad\qquad \text{Quadratic Formula}$$

$$x = \frac{-(-12) \pm \sqrt{(-12)^2 - 4(4)(9)}}{2(4)} \qquad \text{Substitute}$$

$$x = \frac{12 \pm \sqrt{144 - 144}}{8}$$

$$x = \frac{12 \pm \sqrt{0}}{8} = \frac{3}{2} \qquad \text{Repeated solution}$$

Therefore, this quadratic equation has only one solution: $\frac{3}{2}$. Check this solution in the original equation. ■

In the next example, note how the Quadratic Formula can be used to determine that a quadratic equation has no real solution.

EXAMPLE 4 ■ **Using the Quadratic Formula: No Real Solution**

Use the Quadratic Formula to solve the equation $2x^2 - 4x + 5 = 0$.

Solution

$$2x^2 - 4x + 5 = 0 \qquad \text{Standard form with } a = 2, b = -4, c = 5$$

$$x = \frac{-b \pm \sqrt{b^2 - 4ac}}{2a} \qquad \text{Quadratic Formula}$$

$$x = \frac{-(-4) \pm \sqrt{(-4)^2 - 4(2)(5)}}{2(2)} \qquad \text{Substitute}$$

$$x = \frac{4 \pm \sqrt{16 - 40}}{4}$$

$$x = \frac{4 \pm \sqrt{-24}}{4} \qquad \text{Negative discriminant}$$

Since $\sqrt{-24}$ is not a real number, we conclude that the original equation has no real solution. ■

We have now looked at four ways in which to solve quadratic equations.

1. Factoring

2. Extracting square roots

3. Completing the square

4. The Quadratic Formula

Of these four methods, we suggest that you first check to see whether the equation $2x^2 - x - 1 = 0$ is in a form in which you can extract square roots. For instance, it is easy to see that the solutions of $x^2 = 10$ are $x = \pm\sqrt{10}$. Next, we suggest that you try factoring. For instance, by rewriting the equation $2x^2 - x - 1 = 0$ in the factored form $(2x + 1)(x - 1) = 0$, we can see that it has two solutions: $-\frac{1}{2}$ and 1.

If neither of these two methods works, we suggest using the Quadratic Formula (or completing the square).

EXAMPLE 5   ■   Using a Calculator with the Quadratic Formula

Use the Quadratic Formula and a calculator to solve the following equation.

$$1.2x^2 - 17.8x + 8.05 = 0$$

Solution

This quadratic equation is already in standard form. From this form, we see that $a = 1.2$, $b = -17.8$, and $c = 8.05$. Therefore, we can use the Quadratic Formula to solve the equation, as follows.

$$1.2x^2 - 17.8x + 8.05 = 0$$

$$x = \frac{-b \pm \sqrt{b^2 - 4ac}}{2a}$$

$$x = \frac{-(-17.8) \pm \sqrt{(-17.8)^2 - 4(1.2)(8.05)}}{2(1.2)}$$

To evaluate these solutions, we begin by calculating the square root, which we *store* for later use.*

Scientific Calculator Steps	*Display*
17.8 +/− x^2 − 4 × 1.2 × 8.05 = √	16.679329

Graphing Calculator Steps	*Display*
√ (((−) 17.8) x^2 − 4 × 1.2	
× 8.05) ENTER	16.67932852

Storing this result and using the recall key, we find the following two solutions.

$$x \approx \frac{17.8 + 16.67932852}{2.4} \approx 14.366 \qquad \text{Add stored value}$$

$$x \approx \frac{17.8 - 16.67932852}{2.4} \approx 0.467 \qquad \text{Subtract stored value} \qquad ■$$

DISCUSSION PROBLEM   ■   Developing a Formula

At the beginning of this section, we developed the Quadratic Formula by completing the square for the general quadratic equation

$$ax^2 + bx + c = 0.$$

Try reconstructing that development without referring to the text.   ■

*The graphing calculator keystrokes given in this text correspond to the *TI-81* graphing calculator from *Texas Instruments*. For other graphing calculators, the keystrokes may differ.

Warm-Up

The following warm-up exercises involve skills that were covered in earlier sections. You will use these skills in the exercise set for this section.

In Exercises 1–4, solve the quadratic equation by factoring.

1. $x^2 - 7x + 10 = 0$ **2.** $x^2 + 3x - 18 = 0$

3. $2x^2 + 17x - 30 = 0$ **4.** $6x^2 - 17x + 5 = 0$

In Exercises 5 and 6, solve the quadratic equation by completing the square.

5. $x^2 + x - 4 = 0$ **6.** $3x^2 + 6x + 1 = 0$

In Exercises 7–10, simplify the radical.

7. $\sqrt{16 - 4(3)(1)}$ **8.** $\sqrt{9 - 4(-2)(5)}$

9. $\sqrt{36 - 4(2)(-4)}$ **10.** $\sqrt{100 - 4(2)(-6)}$

11.3   EXERCISES

In Exercises 1–4, write the quadratic equation in standard form.

1. $x^2 = 3 - 2x$ **2.** $2x^2 + 3x = 5$ **3.** $x(4 - x) = 10$ **4.** $x(8x + 3) = 2$

In Exercises 5–10, use the discriminant to determine the number of real solutions of the quadratic equation.

5. $2x^2 - 3x - 1 = 0$ **6.** $4x^2 + 4x + 1 = 0$ **7.** $x^2 + 4x + 5 = 0$ **8.** $3x^2 - 2x - 5 = 0$

9. $9x^2 - 12x + 4 = 0$ **10.** $2x^2 + 5x + 6 = 0$

In Exercises 11–20, solve the quadratic equation (a) by the Quadratic Formula, and (b) by factoring.

11. $x^2 - 11x + 30 = 0$ **12.** $x^2 - 7x + 12 = 0$ **13.** $x^2 + 4x + 3 = 0$ **14.** $x^2 + 7x + 10 = 0$

15. $x^2 - 6x + 9 = 0$ **16.** $x^2 + 4x + 4 = 0$ **17.** $4x^2 + 4x + 1 = 0$ **18.** $9x^2 - 12x + 4 = 0$

19. $4x^2 - 19x + 12 = 0$ **20.** $5x^2 - 26x + 5 = 0$

In Exercises 21–40, use the Quadratic Formula to solve the quadratic equation. (Some of the equations have no real number solution.)

21. $8x^2 - 10x + 3 = 0$ **22.** $4x^2 - 13x + 3 = 0$ **23.** $0.5x^2 - 0.8x + 0.3 = 0$

24. $0.04x^2 + 0.04x - 0.03 = 0$ **25.** $2x^2 + 7x + 3 = 0$ **26.** $3x^2 + 11x + 10 = 0$

27. $x^2 - 6x + 7 = 0$

28. $x^2 - 10x + 22 = 0$

29. $t^2 + 5t + 6 = 0$

30. $y^2 + y + 1 = 0$

31. $z^2 + 4z + 4 = 0$

32. $9z^2 + 10z + 4 = 0$

33. $x^2 - 3x - 1 = 0$

34. $u^2 + 5u + 2 = 0$

35. $5x^2 + 2x - 2 = 0$

36. $3z^2 - 4z + 2 = 0$

37. $2y^2 + y + 6 = 0$

38. $0.06t^2 + 0.05t - 0.01 = 0$

39. $0.36s^2 - 0.12s + 0.01 = 0$

40. $0.1x^2 - x - 1.1 = 0$

In Exercises 41–50, solve the quadratic equation by the most convenient method.

41. $x^2 - 625 = 0$

42. $t^2 = 27$

43. $y^2 + 8y = 0$

44. $2y(y - 12) + 3(y - 12) = 0$

45. $(x - 3)^2 - 36 = 0$

46. $4u^2 - 49 = 0$

47. $10x^2 + x - 3 = 0$

48. $x^2 + 14x + 49 = 0$

49. $-2x^2 + 6x + 1 = 0$

50. $6x^2 + 20x + 5 = 0$

In Exercises 51–54, use a calculator to solve the equation. (Round your solution to three decimal places.)

51. $3x^2 - 14x + 4 = 0$

52. $7x^2 + x - 35 = 0$

53. $-0.03x^2 + 2x - 0.5 = 0$

54. $1.7x^2 - 4.2x + 2.1 = 0$

In Exercises 55–58, solve the equation.

55. $\dfrac{x^2}{3} - \dfrac{x}{2} = 1$

56. $\dfrac{x}{3} - \dfrac{1}{x} = 1$

57. $\sqrt{4x + 3} = x - 1$

58. $\sqrt{3x - 2} = x - 2$

59. *Falling Time* A ball is thrown upward with an initial velocity of 20 feet per second from a height that is 100 feet above the level of the water (see figure). The height h (in feet) of the ball t seconds after it is thrown is given by

$$h = -16t^2 + 20t + 100.$$

(a) Find two times when the ball is 100 feet above the water level.

(b) Find the time when the ball strikes the water.

60. *Number Problem* Find two consecutive positive even integers whose product is 224.

Figure for 59

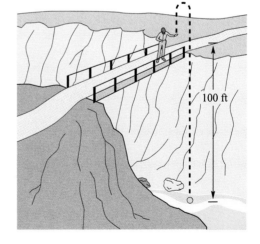

100 ft

SECTION
11.4

Quadratic Functions and Their Graphs
Quadratic Functions • Sketching the Graph of a Quadratic Function

Quadratic Functions

A **quadratic function** of x is a function that can be written in the form

$$y = ax^2 + bx + c, \quad a \neq 0.$$

The graph of a quadratic function is called a **parabola**. Parabolas are cup-shaped. If the coefficient of x^2 is positive, then the parabola opens up, as shown in Figure 11.1. The lowest point on a parabola that opens up is called the **vertex** of the parabola. If the coefficient of x^2 is negative, then the parabola opens down and its vertex is the highest point on the parabola.

FIGURE 11.1   **Parabola Opens Up**   **Parabola Opens Down**

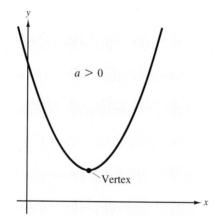

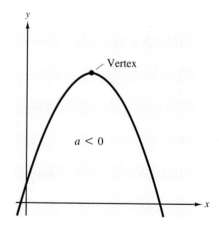

The Leading Coefficient Test for Parabolas

The graph of the quadratic function $y = ax^2 + bx + c$ is called a **parabola**.

1. If $a > 0$, then the parabola opens up.

2. If $a < 0$, then the parabola opens down.

Recall from Section 8.2 that the graph of a function of x must pass the *vertical line test*. (The vertical line test states that no vertical line can intersect the graph of a function of x more than once.) In Figure 11.1 note that both types of parabolas pass the vertical line test.

EXAMPLE 1   ■   Using the Leading Coefficient Test

Determine whether the graphs of the following quadratic functions open up or down.

(a) $y = -x^2 + 2x + 3$ (b) $y = 4x^2 - 1$ (c) $y = 2 - 5x - 3x^2$

Solution

(a) In standard form we see that the leading coefficient of this quadratic function is negative.

$$y = ax^2 + bx + c \qquad \text{Standard form}$$
$$y = -x^2 + 2x + 3 \implies a = -1 \qquad \text{Leading coefficient is negative}$$

Therefore, the graph of the function opens down, as shown in Figure 11.2(a).

(b) In standard form we see that the leading coefficient of this quadratic function is positive.

$$y = ax^2 + bx + c \qquad \text{Standard form}$$
$$y = 4x^2 - 1 \implies a = 4 \qquad \text{Leading coefficient is positive}$$

Therefore, the graph of the function opens up, as shown in Figure 11.2(b).

(c) In standard form we see that the leading coefficient of this quadratic function is negative.

$$y = ax^2 + bx + c \qquad \text{Standard form}$$
$$y = -3x^2 - 5x + 2 \implies a = -3 \qquad \text{Leading coefficient is negative}$$

Therefore, the graph of the function opens down, as shown in Figure 11.2(c).

FIGURE 11.2

(a)
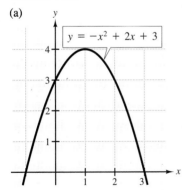

(b)
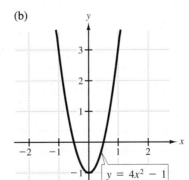

(c)
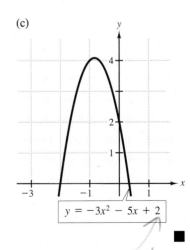

y intercept

Sketching the Graph of a Quadratic Function

In Example 2 we use the standard point-plotting method to sketch the graph of a quadratic function.

EXAMPLE 2 ■ Sketching the Graph of a Quadratic Function by Point Plotting

Find the intercepts of the graph of the function $y = x^2 - 4$. Then sketch the graph of the function and label its intercepts on the graph.

Solution

To find the x-intercepts, we let y be equal to zero and solve the resulting equation for x.

$$x^2 - 4 = 0$$ Let $y = 0$ and solve for x

$$(x + 2)(x - 2) = 0$$ Factor

$$x + 2 = 0 \implies x = -2$$ Set first factor equal to 0

$$x - 2 = 0 \implies x = 2$$ Set second factor equal to 0

From these two solutions, we see that the graph has two x-intercepts. One occurs at the point $(-2, 0)$ and the other occurs at the point $(2, 0)$. To find the y-intercept, we let x equal zero in the original equation, and solve for y. Doing this produces $y = 0^2 - 4 = -4$. Therefore, the y-intercept occurs at the point $(0, -4)$.

To sketch the graph of the function, we make up a table of values, as shown in Table 11.1. (Note that the three intercepts are included in the table.)

TABLE 11.1

x	-3	-2	-1	0	1	2	3
$y = x^2 - 4$	5	0	-3	-4	-3	0	5
Points	$(-3, 5)$	$(-2, 0)$	$(-1, -3)$	$(0, -4)$	$(1, -3)$	$(2, 0)$	$(3, 5)$

After plotting the points, we connect them with a smooth curve, as shown in Figure 11.3. Note that the parabola opens up because the leading coefficient, $a = 1$, is positive.

FIGURE 11.3

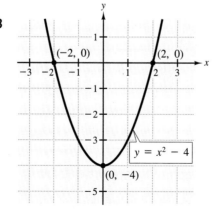

If $a > 0$, then the parabola opens up and the lowest point on the parabola (its vertex) occurs when $x = -b/2a$. If $a < 0$, then the parabola opens down and the highest point on the parabola (its vertex) occurs when $x = -b/2a$. Thus, in either case, the vertex occurs when

$$x = -\frac{b}{2a}. \qquad \text{\textit{x}-value of vertex}$$

We summarize this result as follows.

Vertex of a Parabola

The **vertex** of the parabola given by $y = ax^2 + bx + c$ occurs at the point whose x-coordinate is

$$x = -\frac{b}{2a}.$$

To find the y-coordinate of the vertex, substitute the x-coordinate in the equation $y = ax^2 + bx + c$.

EXAMPLE 3   ■   Finding the Vertex of a Parabola

Find the vertex of each of the following parabolas.

(a) $y = x^2 + 2x - 1$ (b) $y = -x^2 + 3x$

Solution

(a) For this quadratic function, we have $a = 1$ and $b = 2$. Therefore, the x-coordinate of the vertex is

$$x = -\frac{b}{2a} = -\frac{2}{2(1)} = -1.$$

By substituting this value into the original equation, we find the y-coordinate of the vertex to be

$$y = x^2 + 2x - 1 = (-1)^2 + 2(-1) - 1 = 1 - 2 - 1 = -2.$$

Thus, the vertex occurs at the point $(-1, -2)$, as shown in Figure 11.4(a).

(b) For this quadratic function, we have $a = -1$ and $b = 3$. Therefore, the x-coordinate of the vertex is

$$x = -\frac{b}{2a} = -\frac{3}{2(-1)} = \frac{3}{2}.$$

By substituting this value into the original equation, we find the y-coordinate of the vertex to be

$$y = -x^2 + 3x = -\left(\frac{3}{2}\right)^2 + 3\left(\frac{3}{2}\right) = -\frac{9}{4} + \frac{9}{2} = \frac{9}{4}.$$

Thus, the vertex occurs at the point $\left(\frac{3}{2}, \frac{9}{4}\right)$, as shown in Figure 11.4(b).

FIGURE 11.4   (a)

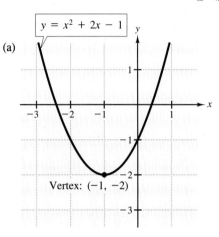

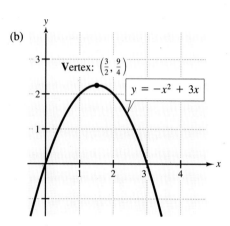

We suggest the following guidelines for sketching the graph of a quadratic function.

Guidelines for Sketching a Parabola	To sketch the parabola given by $y = ax^2 + bx + c$, we suggest the following steps.

1. Use the leading coefficient test to determine whether the parabola opens up or down.

2. Find and plot the y-intercept of the parabola and the vertex.

3. Find and plot the x-intercepts of the parabola (if any).

4. Make up a table of values that includes a few additional points on the parabola.

5. Complete the graph of the parabola with a smooth, cup-shaped curve.

The graph of every quadratic function has exactly one y-intercept. The number of x-intercepts, however, can vary. In the next three examples, we look at parabolas that have two, one, and no intercepts.

EXAMPLE 4   ■   Sketching a Parabola: Two x-intercepts

Sketch a graph of the following quadratic function.

$$y = -2x^2 - x + 6$$

Solution

Leading coefficient test: Since $a = -2$, the parabola opens down.

y-intercept: (0, 6)

Vertex: $x = -\dfrac{b}{2a} = -\dfrac{-1}{2(-2)} = -\dfrac{1}{4}$

$y = -2\left(-\dfrac{1}{4}\right)^2 - \left(-\dfrac{1}{4}\right) + 6 = \dfrac{49}{8}$

$\left(-\dfrac{1}{4}, \dfrac{49}{8}\right)$

x-intercepts: $0 = -2x^2 - x + 6 = -(2x - 3)(x + 2)$

$\left(\dfrac{3}{2}, 0\right)$ and $(-2, 0)$

Table of values:

x	-3	-1	1	2
$y = -2x^2 - x + 6$	-9	5	3	-4
Solution Points	$(-3, -9)$	$(-1, 5)$	$(1, 3)$	$(2, -4)$

We plot the intercepts, the vertex, and the additional points shown in the table of values, and connect the points with a smooth curve, as shown in Figure 11.5.

FIGURE 11.5

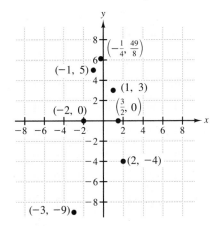
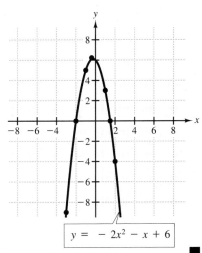

■

EXAMPLE 5   ■   Sketching a Parabola: One *x*-intercept

Sketch a graph of the following quadratic function.

$y = x^2 - 4x + 4$

Solution

Leading coefficient test: Since $a = 1$, the parabola opens up.

y-intercept: (0, 4)

Vertex: $\quad x = -\dfrac{b}{2a} = -\dfrac{-4}{2(1)} = 2$

$\quad\quad\quad\quad y = 2^2 - 4(2) + 4 = 0$

$\quad\quad\quad\quad (2, 0)$

x-intercept: $\quad 0 = x^2 - 4x + 4 = (x - 2)^2$

$\quad\quad\quad\quad (2, 0)$

Table of values:

x	-1	1	3	4	5
$y = x^2 - 4x + 4$	9	1	1	4	9
Solution Points	$(-1, 9)$	$(1, 1)$	$(3, 1)$	$(4, 4)$	$(5, 9)$

We plot the intercepts, the vertex, and the additional points shown in the table of values, and connect the points with a smooth curve, as shown in Figure 11.6.

FIGURE 11.6

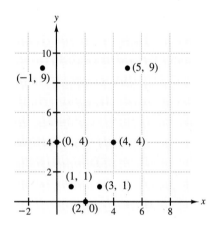 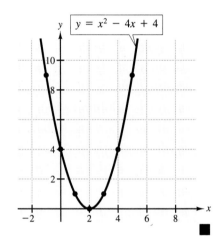

EXAMPLE 6 ■ Sketching a Parabola: No *x*-intercepts

Sketch a graph of the quadratic function $y = x^2 - 6x + 10$.

Solution

When we try to find the *x*-intercepts of this quadratic function, we find that the equation $0 = x^2 - 6x + 10$ has no solution. (Try verifying this by the Quadratic Formula.)

Leading coefficient test: Since $a = 1$, the parabola opens up.

y-intercept: (0, 10)

Vertex: $\quad x = -\dfrac{b}{2a} = -\dfrac{-6}{2(1)} = 3$

$\quad\quad\quad\quad y = 3^2 - 6(3) + 10 = 1$

$\quad\quad\quad\quad (3, 1)$

x-intercept: $\quad 0 = x^2 - 6x + 10 \quad\Longrightarrow\quad$ No real solution

$\quad\quad\quad\quad$ No *x*-intercept

Table of values:

x	1	2	4	5
$y = x^2 - 6x + 10$	5	2	2	5
Solution Points	(1, 5)	(2, 2)	(4, 2)	(5, 5)

We plot the vertex and the additional points shown in the table of values, and connect the points with a smooth curve, as shown in Figure 11.7.

FIGURE 11.7

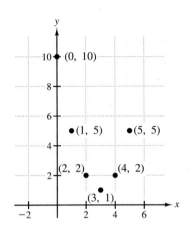

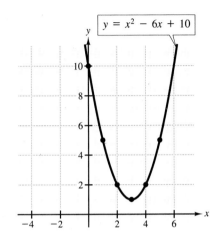

EXAMPLE 7   ■   Graphing Calculator Problem

Use a graphing calculator to sketch the graphs of the following parabolas on the *same* display screen.

(a) $y = x^2$ (b) $y = x^2 + 2$

(c) $y = (x - 3)^2$ (d) $y = (x + 4)^2$

Describe the relationship between the graphs found in parts (b), (c), and (d) and the graph found in part (a).

Solution

To sketch all four graphs on the same screen, press the $\boxed{\text{Y=}}$ key, enter the equations on the lines labeled $y_1 =$, $y_2 =$, $y_3 =$, and $y_4 =$, and press $\boxed{\text{GRAPH}}$. You should obtain the graph shown in Figure 11.8 on the following page.

(a) The graph of $y = x^2$ is a parabola that opens upward and whose vertex is the origin.

(b) The graph of $y = x^2 + 2$ is similar to the graph of $y = x^2$, except that the graph has been shifted two units up.

(c) The graph of $y = (x - 3)^2$ is similar to the graph of $y = x^2$, except that the graph has been shifted three units to the right.

(d) The graph of $y = (x + 4)^2$ is similar to the graph of $y = x^2$, except that the graph has been shifted four units to the left.

FIGURE 11.8

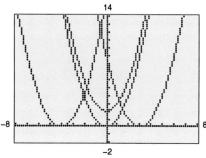

■

DISCUSSION PROBLEM   ■   Finding the Equation of a Parabola

The parabola given by

$$y = (x - 1)(x - 3)$$

has x-intercepts at (1, 0) and (3, 0). The vertex of this parabola is (2, −1). Can you find an equation of a parabola that has the same x-intercepts, but has a vertex at the following points?

(a) (2, 1)   (b) (2, −2)   (c) (2, 2)   ■

Warm-Up

The following warm-up exercises involve skills that were covered in earlier sections. You will use these skills in the exercise set for this section.

In Exercises 1–6, sketch the graph of the equation.

1. $y = 2x$ 　　　　　　　**2.** $y = \frac{1}{3}x$ 　　　　　　　**3.** $y = 2x - 2$

4. $y = \frac{1}{3}x + 2$ 　　　　　**5.** $y = -2x + 5$ 　　　　　**6.** $y = -\frac{1}{3}x + 2$

In Exercises 7–10, solve the equation.

7. $10(x - 3) = 5$ 　　　　　　　　　　**8.** $20 - 3x = 50 - 4x$

9. $x(x + 4) = 0$ 　　　　　　　　　　**10.** $x(x + 4) = 3$

11.4   EXERCISES

In Exercises 1–6, match the equation with the correct graph. [The graphs are labeled (a), (b), (c), (d), (e), and (f).]

1. $y = 5 - x$

2. $y = \frac{1}{2}x - 2$

3. $y = x^2 + 1$

4. $y = -2x^2 + 1$

5. $y = (x - 3)^2$

6. $y = 3 - (x - 3)^2$

(a)

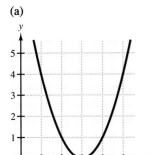

(b)

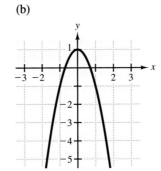

(c)

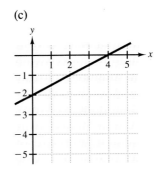

(d)

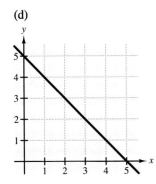

(e)

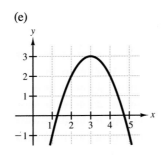

(f)

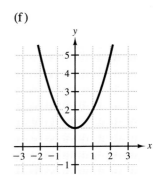

In Exercises 7–12, determine whether the graph of the quadratic function opens up or down.

7. $y = 6x^2 + 2$

8. $y = -3x^2 + x - 3$

9. $y = 4 + 10x - x^2$

10. $y = 1 + 8x(1 - x)$

11. $y = 3 + x(x + 3)$

12. $y = x(x + 1) + 3(x + 1)$

In Exercises 13–20, find the intercepts of the graph of the given function.

13. $y = 16 - x^2$

14. $y = x^2 - 36$

15. $y = x^2 - 2x$

16. $y = x^2 - 4x + 4$

17. $y = x^2 - x - 6$

18. $y = -2x^2 + 5x + 12$

19. $y = x^2 - 2x + 3$

20. $y = x^2 - 2x - 3$

In Exercises 21–26, find the vertex of the parabola.

21. $y = -x^2 + 2$ **22.** $y = 3x^2 - 3$ **23.** $y = x^2 - 4x + 7$ **24.** $y = 1 - 2x - x^2$

25. $y = x(x - 4)$ **26.** $y = 2x(6 - x)$

In Exercises 27–50, sketch the graph of the quadratic function. Identify the vertex of the parabola and the intercepts.

27. $y = x^2 - 1$ **28.** $y = x^2 - 9$ **29.** $y = -x^2 + 1$ **30.** $y = -x^2 + 9$

31. $y = x^2 - 4x$ **32.** $y = x^2 - 6x$ **33.** $y = -x^2 + 4x$ **34.** $y = -x^2 + 6x$

35. $y = (x - 2)^2$ **36.** $y = -(x + 2)^2$ **37.** $y = -x^2 - 2x + 3$ **38.** $y = x^2 - 6x + 8$

39. $y = x^2 + 6x + 8$ **40.** $y = -(x^2 + 4x + 8)$ **41.** $y = x^2 - 4x + 1$ **42.** $y = x^2 + 6x + 11$

43. $y = -(x^2 + 4x + 2)$ **44.** $y = x^2 - 2x - 2$ **45.** $y = \frac{1}{2}x^2 - 4x + 6$ **46.** $y = \frac{1}{3}x^2 - 2x + 4$

47. $y = 2x^2 + 8x + 9$ **48.** $y = 3x^2 - 3$ **49.** $y = -4x^2 + 8x$

50. $y = -(3x^2 + 12x + 10)$

In Exercises 51–54, sketch the graph of the equation and determine the values of x (if any exist) for which the graph is at the specified height y.

51. $y = -x^2 + 3$, $\quad y = 2$ **52.** $y = x^2 - 6x + 6$, $\quad y = 1$

53. $y = \frac{1}{2}x^2 - 4x + 10$, $\quad y = 3$ **54.** $y = -2x^2 + 12x - 14$, $\quad y = 5$

In Exercises 55 and 56, expand the right side of the equation and verify that it has the form $y = ax^2 + bx + c$. Sketch the graph of the equation. Describe how the vertex can be determined from the given completed square form of the equation.

55. $y = (x - 2)^2 - 2$ **56.** $y = -(x - 3)^2 + 1$

In Exercises 57 and 58, complete the square for the right side of the equation. What is the vertex of the parabola?

57. $y = x^2 - 10x + 26$ **58.** $y = x^2 + 8x + 14$

59. *Tossing a Ball* The height y (in feet) of a ball thrown by a child is given by

$$y = -\frac{1}{10}x^2 + 2x + 4$$

where x is the horizontal distance (in feet) from where the ball was thrown (see figure).

Figure for 59

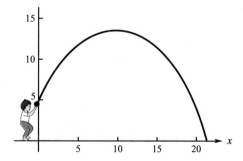

(a) How high was the ball when it left the child's hand? (*Note:* Find y when $x = 0$.)

(b) How high was the ball when it was at its maximum height?

(c) How far from the child was the ball when it struck the ground?

60. *Tossing a Ball* The height y (in feet) of a ball thrown by a child is given by

$$y = -\frac{1}{5}x^2 + 2x + 4$$

where x is the horizontal distance (in feet) from where the ball was thrown (see figure).
  (a) How high was the ball when it left the child's hand? (*Note:* Find y when $x = 0$.)
  (b) How high was the ball when it was at its maximum height?
  (c) How far from the child was the ball when it struck the ground?

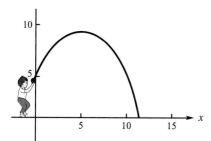

61. Use a graphing calculator to sketch the graphs of the following parabolas on the same screen. Describe the effect on the graph by changing the coefficient of x^2.
  (a) $y = x^2$ (b) $y = \frac{1}{8}x^2$
  (c) $y = -2x^2$ (d) $y = -\frac{1}{6}x^2$

62. Use a graphing calculator to sketch the graphs of the following parabolas on the same screen. Describe the relationship between the graphs.
  (a) $y = x^2$ (b) $y = x^2 + 3$
  (c) $y = x^2 - 3$ (d) $y = x^2 - 8$

63. Use a graphing calculator to sketch the graphs of the following parabolas on the same screen. Describe the relationship between the graphs.
  (a) $y = x^2$ (b) $y = (x - 2)^2$
  (c) $y = (x - 6)^2$ (d) $y = (x + 4)^2$

64. Use the TRACE feature of a graphing calculator to approximate the coordinates of the x- and y-intercepts of the graph of

$$y = 6 - \frac{1}{2}(x - 4)^2.$$

SECTION
11.5

Applications of Quadratic Equations
Review of Problem-Solving Strategy • Applications of Quadratic Equations

Review of Problem-Solving Strategy

We have completed our study of the technical parts of the algebra of quadratic equations. In this section we look at some practical problems involving quadratic equations. To construct equations from verbal problems, we again rely on the problem-solving strategy developed in Chapters 3 and 4. For convenience, we list a summary of the steps involved in this strategy.

Guidelines for Solving Word Problems	1. Search for the *hidden equality*—the two essential expressions said to be equal or known to be equal. (A sketch may be helpful.)
	2. Write a *verbal model* that equates these two essential expressions.
	3. Assign *labels* to the fixed quantities and the variable quantities.
	4. Rewrite the verbal model as an *algebraic equation* using the assigned labels.
	5. *Solve* the algebraic equation.
	6. *Check* to see that your solution satisfies the word problem as stated.

quadratic equation

$$ax^2 + bx + c = 0$$

For word problems that involve quadratic equations, the verbal model may be the *product* of two variable quantities, or the model may be a previously known formula that describes the situation. Watch for these variations in the examples that follow.

Applications of Quadratic Equations

EXAMPLE 1 ■ An Application Involving Area

A picture is 3 inches longer than it is wide and has an area of 108 square inches, as shown in Figure 11.9. What are the outer dimensions of the picture?

Solution

Verbal model: Area of picture = Width · Length

Labels: Picture width = w (in inches)
Picture length = $w + 3$ (in inches)
Area = 108 (in square inches)

Equation:
$$108 = w(w + 3)$$
$$0 = w^2 + 3w - 108$$
$$0 = (w + 12)(w - 9)$$
$$w + 12 = 0 \implies w = -12$$
$$w - 9 = 0 \implies w = 9$$

Of the two possible solutions, we choose the positive value of w and conclude that the width of the picture is

Width of picture = w = 9 inches.

Finally, since the picture is 3 inches longer than it is wide, we see that the length of the picture is

Length of picture = 9 + 3 = 12 inches.

Check these dimensions in the original statement of the problem.

FIGURE 11.9

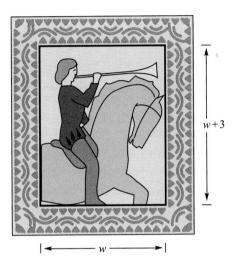

$w + 3$

$\leftarrow w \rightarrow$

■

EXAMPLE 2   ■   An Application Involving the Pythagorean Theorem

An L-shaped sidewalk from Building A to Building B on a college campus is 400 meters long, as shown in Figure 11.10. By cutting diagonally across the grass, students shorten the walking distance to 300 meters. What are the lengths of the two parts of the existing sidewalk?

FIGURE 11.10

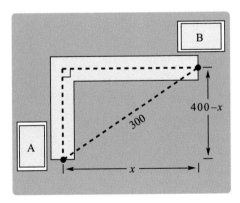

B

$400 - x$

300

A

x

Solution

Common formula: $a^2 + b^2 = c^2$ Pythagorean Theorem

Labels: a = length of one part = x (in meters)
                          b = length of other part = $400 - x$ (in meters)
                          c = length of diagonal = 300 (in meters)

Equation:
$$x^2 + (400 - x^2) = (300)^2$$
$$2x^2 - 800x + 160,000 = 90,000$$
$$2x^2 - 800x + 70,000 = 0$$
$$x^2 - 400x + 35,000 = 0$$

By the Quadratic Formula, we find the solutions of this equation, as follows.

Quadratic Formula

$$x = -b \pm \frac{\sqrt{b^2 - 4ac}}{2a}$$

$$x = \frac{400 \pm \sqrt{(-400)^2 - 4(1)(35,000)}}{2(1)}$$

$$= \frac{400 \pm \sqrt{20,000}}{2}$$

$$= \frac{400 \pm 100\sqrt{2}}{2}$$

$$= 200 \pm 50\sqrt{2}$$

Both solutions are positive, and it does not matter which one we choose. If we let

$$x = 200 + 50\sqrt{2} \approx 270.7 \text{ meters}$$

then the length of the other part is

$$400 - x \approx 400 - 270.7 \approx 129.3 \text{ meters.}$$

Try choosing the other value of x to see that the same two lengths result. ■

EXAMPLE 3 ■ Reduced Rates

A ski club chartered a bus for a ski trip at a cost of $480. In an attempt to lower the bus fare per skier, the club invited nonmembers to go along. After five nonmembers joined the trip, the fare per skier decreased by $4.80. How many club members are going on the bus?

Solution

Verbal model: | Cost per skier | · | Number of skiers | = | 480 |

Labels: Number of ski club members $= x$
Number of skiers $= x + 5$

Original cost per skier $= \dfrac{480}{x}$ (in dollars)

New cost per skier $= \dfrac{480}{x} - 4.80$ (in dollars)

Equation: $\left(\dfrac{480}{x} - 4.80 \right)(x + 5) = 480$

$\left(\dfrac{480 - 4.8x}{x} \right)(x + 5) = 480$

$(480 - 4.8x)(x + 5) = 480x, \quad x \neq 0$

$$480x + 2,400 - 4.8x^2 - 24x = 480x$$
$$-4.8x^2 - 24x + 2,400 = 0$$
$$x^2 + 5x - 500 = 0$$
$$(x + 25)(x - 20) = 0$$
$$x + 25 = 0 \implies x = -25$$
$$x - 20 = 0 \implies x = 20$$

Choosing the positive value of x, we have

$x = 20$ ski club members.

Check this solution in the original statement of the problem. ∎

EXAMPLE 4 ■ **Work Problem**

An office contains two copy machines. Machine B is known to take 12 minutes longer than Machine A to copy the company's monthly report. Working together, it takes 8 minutes to reproduce the report. How long would it take each machine alone to reproduce the report?

Solution

Verbal model: Rate for A + Rate for B = Rate for both

Labels: Both machines: Time = 8 minutes, Rate = $\frac{1}{8}$ job per minute

Machine A: Time = t minutes, Rate = $\frac{1}{t}$ job per minute

Machine B: Time = $t + 12$ minutes, Rate = $\frac{1}{t+12}$ job per minute

Equation:
$$\frac{1}{t} + \frac{1}{t+12} = \frac{1}{8}$$
$$\left(\frac{1}{t} + \frac{1}{t+12}\right)(8t)(t+12) = \frac{1}{8}(8t)(t+12)$$
$$8(t+12) + 8t = t(t+12)$$
$$8t + 96 + 8t = t^2 + 12t$$
$$16t + 96 = t^2 + 12t$$
$$0 = t^2 - 4t - 96$$
$$0 = (t-12)(t+8)$$
$$t - 12 = 0 \implies t = 12$$
$$t + 8 = 0 \implies t = -8$$

We choose the positive value for t and find that

Time for Machine A $= t = 12$ minutes

Time for Machine B $= t + 12 = 24$ minutes.

Check these solutions in the original statement of the problem. ■

DISCUSSION PROBLEM ■ **What Is Algebra?**

Now that you are completing this course, suppose someone asks you what algebra is. Write a short paper describing how you would answer this question. ■

Warm-Up

The following warm-up exercises involve skills that were covered in earlier sections. You will use these skills in the exercise set for this section.

In Exercises 1–8, solve the given equation.

1. $3(x - 2) = 0$ **2.** $3x(x - 2) = 0$

3. $2n + (n + 2) = 30$ **4.** $2n(n + 2) = 30$

5. $2(x + 8)^2 = 200$ **6.** $t^2 + 3t - 1 = 0$

7. $t + \dfrac{2}{t} = 3$ **8.** $\sqrt{3s + 4} = s$

In Exercises 9 and 10, sketch the graph of the quadratic function.

9. $y = 2 - 2x - x^2$ **10.** $y = (x - 4)^2$

[handwritten: quadratic equation $= ax^2 + by + c = 0$]
[handwritten: Quadratic formula $= x = \dfrac{-b \pm \sqrt{b^2 - 4ac}}{2 \cdot 2a}$]

11.5 EXERCISES

In Exercises 1–6, find two positive integers satisfying the given requirement. (*Note:* Recall that even and odd integers can be represented by $2n$ and $2n + 1$, respectively.)

1. The product of two consecutive integers is 132.   **2.** The product of two consecutive integers is 420.

3. The product of two consecutive even integers is 168.   **4.** The product of two consecutive odd integers is 143.

5. The sum of the squares of two consecutive integers is 113.   **6.** The sum of the squares of two consecutive integers is 481.

In Exercises 7–10, find the time necessary for an object to fall to ground level from an initial height of h_0 feet if its height h at any time t in seconds is given by $h = h_0 - 16t^2$.

7. $h_0 = 1,600$

8. $h_0 = 400$

9. $h_0 = 550$ (approximate height of the Washington Monument)

10. $h_0 = 1,350$ (approximate height of the World Trade Center)

In Exercises 11–20, find the dimensions of the rectangle and complete the table where l and w represent the length and width of a rectangle, respectively.

Width	Length	Perimeter	Area
11. $0.6l$	l	64 in.	
13. w	$2w$		50 ft^2
15. $\frac{1}{4}l$	l		100 in.^{2}
17. w	$w + 4$	56 km	
19. $l - 10$	l		75 m^2

Width	Length	Perimeter	Area
12. w	$1.5w$	75 m	
14. w	$1.2w$		1,440 cm^2
16. $\frac{2}{3}l$	l		24 in.^{2}
18. $l - 5$	l	18 ft	
20. w	$w + 6$		160 ft^2

21. *Storage Area* A rectangular region in a retail lumberyard is to be fenced for the storage of treated lumber. The region will be fenced on three sides, and the fourth side will be bounded by the existing building of the business (see figure). The area of the region to be fenced is 3,750 square feet, and 175 feet of fencing is available. Find the dimensions of the region.

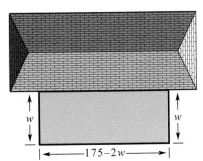

22. *Dimensions of a Conduit* An open-topped rectangular conduit for carrying water in a manufacturing process is made by folding up the edges of a sheet of aluminum 36 inches wide (see figure). The area of a cross section of the conduit must be 162 square inches. Find the dimensions of a cross section of the conduit.

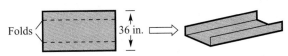

23. *Dimensions of a Triangle* The height of a triangle is one-third its base and the area of the triangle is 24 square inches. Find the dimensions of the triangle.

24. *Dimensions of a Triangle* The height of a triangle is three times its base and the area of the triangle is 864 square inches. Find the dimensions of the triangle.

25. *Delivery Distance* You are asked to deliver pizza to Offices B and C (see figure), and you are required to keep a log of all the mileages between stops. You forget to look at the odometer at stop B, but after getting to stop C you record the total distance traveled from the pizza shop as 14 miles. Also, the return distance from C to A is 10 miles. The route forms a right triangle, and the distance from A to B is greater than the distance from B to C. Find the distance from A to B.

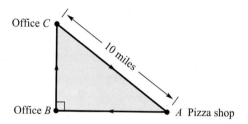

26. *Dimensions of a Rectangle* The perimeter of a rectangle is 68 inches and the length of the diagonal is 26 inches. Find the dimensions of the rectangle.

27. *Ticket Prices* A service organization paid $100 for a block of tickets to a ball game. The block contained five more tickets than the organization needed for its members. By inviting five more people to attend (and share in the cost), the organization lowered the price per ticket by $1. How many people are going to the game?

28. *Ticket Prices* A science club charters a bus to attend a science fair at a cost of $360. In an attempt to lower the bus fare per member, the club invites nonmembers to go along. After 10 nonmembers join the trip, the fare per person is decreased by $3. How many people are going on the excursion?

29. *Air Speed* An airline runs a commuter flight between two cities that are 480 miles apart. If the average speed of the planes could be increased by 20 miles per hour, the travel time would be decreased by 20 minutes. What airspeed is required to obtain this decrease in travel time?

30. *Average Speeds* A truck traveled the first 200 miles of a trip at one speed and the last 225 miles at an average speed of five miles per hour less. The entire trip took 10 hours. What were the two average speeds?

31. *Work Rate* Working together, two people can complete a task in four hours. One person takes six hours longer than the other. Working alone, how long would it take each to do the task?

32. *Work Rate* An office contains two printers. Printer B is known to take five minutes longer than Printer A to produce the company's monthly financial report. Working together, it takes three minutes to produce the report. How long would it take each printer to produce the report?

CHAPTER 11   SUMMARY

As you review and prepare for a test on this chapter, first try to obtain a global view of what was discussed. Then review the specific skills needed in each category.

Solving Quadratic Equations

■ *Solve* a quadratic equation by factoring.   Section 11.1, Exercises 1–30

■ *Solve* a quadratic equation by extracting square roots.   Section 11.1, Exercises 31–50

■ *Create* a perfect square trinomial.   Section 11.2, Exercises 1–10

■ *Solve* a quadratic equation by completing the square.   Section 11.2, Exercises 11–44

■ *Determine* the number of solutions to quadratic equations using the   Section 11.3, Exercises 5–10
discriminant of the Quadratic Formula.

■ *Solve* a quadratic equation using the Quadratic Formula.   Section 11.3, Exercises 11–40

■ *Use* a calculator to evaluate the Quadratic Formula.   Section 11.3, Exercises 51–54

■ *Create and solve* a quadratic equation from verbal statements.   Section 11.5, Exercises 11–24

Sketching Graphs of Quadratic Functions

■ *Identify* the upward or downward orientation of a parabola by using   Section 11.4, Exercises 1–12
the leading coefficient test.

■ *Find* the intercepts as an aid to sketching a parabola.   Section 11.4, Exercises 13–20

■ *Find* the vertex of a parabola.   Section 11.4, Exercises 21–26

■ *Sketch* the graph of a quadratic function using the vertex and intercepts   Section 11.4, Exercises 27–50
as aids.

Chapter 11 Review Exercises

In Exercises 1–10, solve the quadratic equation by factoring, if possible.

1. $x^2 + 10x = 0$

2. $u^2 - 12u = 0$

3. $5z(z + 1) - 8(z + 1) = 0$

4. $x(2x - 15) + 4(2x - 15) = 0$

5. $4y^2 - 25 = 0$

6. $8z^2 - 32 = 0$

7. $9y^2 - 30y + 25 = 0$

8. $x^2 + x + \dfrac{1}{4} = 0$

9. $x^2 - x - 56 = 0$

10. $5x^2 - 10x - 15 = 0$

In Exercises 11–16, solve the quadratic equation by extracting square roots, if possible.

11. $x^2 = 625$

12. $a^2 = 98$

13. $y^2 + 8 = 0$

14. $y^2 - 8 = 0$

15. $(x - 15)^2 = 400$

16. $(x + 2)^2 = 0.01$

In Exercises 17–22, solve the quadratic equation by the method of completing the square, if possible.

17. $x^2 - 6x - 1 = 0$

18. $x^2 + 10x + 12 = 0$

19. $x^2 - x - 1 = 0$

20. $t^2 + 3t + 1 = 0$

21. $2y^2 + 10y + 5 = 0$

22. $3x^2 - 2x + 1 = 0$

In Exercises 23–28, solve the quadratic equation by using the Quadratic Formula, if possible.

23. $y^2 + y - 42 = 0$

24. $x^2 - x - 20 = 0$

25. $2y^2 + y - 42 = 0$

26. $2x^2 - x - 20 = 0$

27. $0.3t^2 - 2t + 1 = 0$

28. $-u^2 + 3.1u + 5 = 0$

In Exercises 29–40, solve the equation by the most convenient method.

29. $v^2 = 250$

30. $x^2 - 45x = 0$

31. $-x^2 + 3x + 70 = 0$

32. $4x^2 + 4x + 1 = 0$

33. $(x - 12)^2 - 169 = 0$

34. $50 - (x - 6)^2 = 0$

35. $c^2 - 6c + 6 = 0$

36. $c^2 - 6c + 5 = 0$

37. $y^2 + y + 1 = 0$

38. $0.5y^2 + 0.75y - 2 = 0$

39. $\dfrac{1}{x} + \dfrac{1}{x + 1} = \dfrac{1}{2}$

40. $x = \sqrt{4x + 5}$

In Exercises 41–50, sketch the graph of the quadratic function. Identify the vertex of the parabola and any intercepts.

41. $y = (x - 4)^2$

42. $y = -(x - 4)^2$

43. $y = 3 - (x - 4)^2$

44. $y = (x - 4)^2 - 2$

45. $y = -x^2 + 3x$

46. $y = x^2 - 10x$

47. $y = \frac{1}{5}x^2 - 2x + 4$

48. $y = \frac{1}{3}(x^2 - 4x + 6)$

49. $y = \frac{1}{4}(4x^2 - 4x + 3)$

50. $y = 2x^2 + 4x + 5$

51. *Number Problem* Find two consecutive positive integers whose product is 240.

52. *Number Problem* Find two consecutive positive integers such that the sum of their squares is 365.

53. *Falling Time* The height h in feet of an object above the ground is given by $h = 48 - 16t^2$ where t is the time in seconds. Find the time when the object strikes the ground.

54. *Falling Time* The height h in feet of an object above the ground is given by $h = -16t^2 + 48t + 160$ where t is the time in seconds. Find the time when the object strikes the ground.

55. *Maximum Area* The perimeter of a rectangle of length l and width w is 40 feet (see figure).

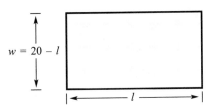

$w = 20 - l$

l

(a) Show that $w = 20 - l$.

(b) Show that the area A is given by $A = lw = l(20 - l)$.

(c) Complete the following table by using the equation from part (b).

l	2	4	6	8	10	12	14	16	18
A									

(d) Sketch the graph of the function $A = l(20 - l)$.

(e) What are the dimensions of the rectangle when its area is maximum?

56. *Dimensions of a Triangle* The height of a triangle is one and one-half times its base and the area of the triangle is 60 square inches. Find the dimensions of the triangle.

57. *Ticket Sales* A Little League baseball team paid $72 for a block of tickets to a ball game. The block contained three more tickets than the team needed for its members. By inviting three more people to attend (and share in the cost), the team lowered the price per ticket by $1.20. How many people are going to the game?

58. *Average Speeds* A train traveled the first 165 miles of a trip at one speed, and the last 300 miles at an average speed that was five miles per hour greater. The entire trip took eight hours. What were the two average speeds?

59. *Work Rate* Working together, two people can complete a task in 10 hours. Working alone, one person takes 4 hours longer than the other to complete the task. Working alone, how long would it take each to complete the task?

60. *Work Rate* Working together, two people can complete a task in 15 hours. Working alone, one person takes 2 hours longer than the other to complete the task. Working alone, how long would it take each to complete the task?

CHAPTER 11 TEST

Take this test as you would take a test in class. After you are done, check your work with the answers given in the back of the book.

1. Solve by extracting square roots: $x^2 - 144 = 0$

2. Solve by factoring: $x(x + 3) - 10(x + 3) = 0$

3. Solve by factoring: $6x^2 + x - 15 = 0$

4. Solve by completing the square: $t^2 - 6t + 7 = 0$

5. Solve by completing the square: $3z^2 + 9z + 5 = 0$

6. Solve by using the Quadratic Formula: $x^2 - x - 3 = 0$

7. Solve by using the Quadratic Formula: $2u^2 + 4u + 1 = 0$

8. Solve: $\dfrac{1}{x + 1} - \dfrac{1}{x - 2} = 1$

9. Solve: $\sqrt{2x} = x - 1$

10. Find the x-intercepts of the graph of $y = x^2 - 8x + 12$.

11. Sketch the graph and give the coordinates of the vertex of the parabola $y = x^2 - 4x$.

12. The product of two consecutive positive integers is 420. Find the integers.

13. The height of a triangle is three times the length of its base. The area of the triangle is 54 square inches. Find the dimensions of the triangle.

14. The length of a rectangle is 10 inches greater than the width of the rectangle. The area of the rectangle is 96 square inches. Find its dimensions.

15. Together, two people can complete mowing the school lawn in six hours. When working alone, it takes one person five hours longer than the other. Find the time required for each to mow the lawn alone.

16. Amtrak runs a train between two cities that are 360 miles apart. If the average speed could be increased by $7\frac{1}{2}$ miles per hour, the travel time would be decreased by 40 minutes. What average speed is required to obtain this decrease in travel time?

CUMULATIVE TEST: CHAPTERS 8–11

Take this test as you would take a test in class. After you are done, check your work with the answers given in the back of the book.

1. Determine whether the ordered pairs are solution points of the equation $9x - 4y + 36 = 0$.
   (a) $(-1, -1)$ (b) $(8, 27)$ (c) $(-4, 0)$ (d) $(3, -2)$

2. Find the x- and y-intercepts (if any) of the graph of the equation $y = x^2 - 4x + 3$.

3. Plot the points $(-7, 0)$ and $(2, 6)$ and find the slope of the line through the points.

4. The slope of a line is $-\frac{1}{4}$ and a point on the line is $(2, 1)$. Find the coordinates of a second point on the line. (There are many correct answers to this problem.)

5. Find an equation of the line through the point $\left(0, -\frac{3}{2}\right)$ with slope $m = \frac{5}{6}$.

6. Sketch the graph of the equation $3x - 2y + 8 = 0$.

7. Sketch the half-plane determined by the inequality $(y - 1) \geq 2(x + 3)$.

8. Solve the following system of equations graphically.
$$x + 5y = 0$$
$$7x + 5y = 30$$

9. Solve the following system of equations by the method of substitution.
$$x - 2y = -3$$
$$x + 5y = 4$$

10. Solve the following system of equations by the method of elimination.
$$2x + y = 4$$
$$4x - 3y = 3$$

11. Simplify: $\sqrt{300x^3}$

12. Rationalize the denominator: $\sqrt{\dfrac{5}{2x}}$

13. Rationalize the denominator: $\dfrac{5}{\sqrt{8} - \sqrt{3}}$

14. Expand and simplify: $\left(\sqrt{6} + 1\right)^2$

15. Solve: $3\sqrt{2x - 3} = 5$ **16.** Solve: $4x^2 - 5x - 6 = 0$ **17.** Solve: $2x^2 + 2x - 5 = 0$

18. Sketch the graph of the quadratic function $y = -x^2 + 8x$. Label the intercepts and the vertex of the graph.

19. A sales representative is reimbursed $125 per day for lodging and meals, plus $0.35 per mile driven. Write a linear equation giving the daily cost C to the company in terms of x, the number of miles driven. Find the cost for a day when the representative drives 70 miles.

20. The total cost of 10 gallons of regular unleaded gasoline and 12 gallons of premium unleaded gasoline is $31.88. Premium unleaded gasoline costs $0.20 more per gallon than regular unleaded gasoline. Find the price per gallon of each grade of gasoline.

Chapter 1

SECTION 1.1   *(page 8)*

1. $2 < 5$ **3.** $-4 < -1$ **5.** $-2 < 1.5$
7. $-4.5 < -3$
9. $\frac{1}{3} < 4$ **11.** $4 > -\frac{7}{2}$

13. $-8 > -10$ **15.** $-4.6 < 1.5$

17. $\frac{7}{16} < \frac{5}{8}$ **19.** $0 > -\frac{7}{16}$

21. $|2| = 2$ **23.** $|-10| = 10$ **25.** -5 **27.** 3
29. 7 **31.** 3.4 **33.** $\frac{7}{2}$ **35.** -4.09 **37.** -23.6
39. $|-15| = |15|$ **41.** $|-4| > |3|$
43. $|32| < |-50|$ **45.** $|\frac{3}{16}| < |\frac{3}{2}|$
47. $-|-48.5| < |-48.5|$ **49.** $|-3.1| > -|-3.1|$
51.   **53.**

55. False, $|0| = 0$ **57.** True   **59.** True

SECTION 1.2   *(page 17)*

Warm-Up   *(page 17)*

1. $-2.5 > -4$ **2.** $\frac{3}{16} < \frac{3}{8}$

3. $-3.1 < 2.7$ **4.** $4.3 > -1$

5. -0.75 **6.** 25.2 **7.** $|\frac{7}{2}| = |-3.5|$
8. $|\frac{3}{4}| > -|0.75|$ **9.** $|-25| > -|20|$
10. $0 < |-3|$

1. 9 **3.** -2 **5.** -1 **7.** 0 **9.** -1 **11.** 30
13. 6 **15.** 16 **17.** 32 **19.** $1,997$
21. $12 + (-9) = 3$ **23.** $-10 + (13) = 3$
25. $4 + (4) = 8$ **27.** 35 **29.** -13 **31.** $1,500$
33. -610 **35.** 8 **37.** -12 **39.** 5
41. $2 + 2 + 2 = 6$
43. $(-3) + (-3) + (-3) + (-3) = -12$ **45.** 210
47. -32 **49.** 54 **51.** $-9,920$ **53.** -90
55. $30,814$ **57.** 3 **59.** -6 **61.** 27
63. Division by zero is undefined.   **65.** -29 **67.** 32
69. 110 **71.** 47 **73.** 86 **75.** $1,045$ **77.** $12°$ F
79. $7,000$ feet   **81.** $\$6,000$
83.

Day	Daily Gain or Loss
Tues	$+ 21$
Wed	$+16$
Thur	-17
Fri	-28

85. Negative

SECTION 1.3   *(page 32)*

Warm-Up   *(page 32)*

1. 0 **2.** 3 **3.** 8 **4.** 6 **5.** -12
6. 17 **7.** 2 **8.** 2 **9.** 6 **10.** 1

1. $\frac{3}{5}$ **3.** $\frac{3}{5}$ **5.** $\frac{1}{4}$ **7.** $\frac{5}{16}$ **9.** $\frac{3}{2}$ **11.** $\frac{3}{5}$
13. $\frac{3}{8}$ **15.** $\frac{14}{11}$ **17.** $-\frac{1}{2}$ **19.** $\frac{5}{6}$ **21.** $\frac{9}{16}$
23. $-\frac{1}{24}$ **25.** $-\frac{17}{16}$ **27.** $-\frac{121}{12}$ **29.** $\frac{4}{3}$ **31.** $\frac{5}{6}$
33. $\frac{3}{8}$ **35.** $-\frac{3}{8}$ **37.** $\frac{21}{20}$ **39.** $\frac{27}{40}$ **41.** $\frac{1}{2}$
43. $-\frac{8}{27}$ **45.** 5 **47.** $-\frac{16}{3}$ **49.** 0.75
51. 0.5625 **53.** $0.\overline{6}$ **55.** $0.58\overline{3}$ **57.** $0.\overline{45}$
59. 2.27 **61.** -1.90 **63.** -57.02 **65.** 39.08
67. 1.57 **69.** $\frac{3}{10}$ **71.** 36% **73.** $\$30,600$
75. $\$1.202$ **77.** $\$623.68$ **79.** False

A1

SECTION 1.4 *(page 40)*

Warm-Up *(page 40)*

1. -64 **2.** 75   **3.** -7 **4.** 0   **5.** $\frac{1}{8}$
6. $-\frac{1}{2}$ **7.** $\frac{3}{4}$ **8.** $\frac{14}{3}$ **9.** $-\frac{6}{5}$ **10.** $\frac{14}{5}$

1. 2^5 **3.** $\left(-\frac{1}{4}\right)^3$ **5.** $(-3)(-3)(-3)(-3)(-3)(-3)$
7. $(9.8)(9.8)(9.8)$ **9.** 9   **11.** -125 **13.** -125
15. $\frac{1}{64}$ **17.** $\frac{8}{27}$ **19.** -1.728 **21.** 4   **23.** 13
25. 210   **27.** 0.0084
29. Division by zero is undefined.   **31.** True
33. False   **35.** 28   **37.** 8   **39.** $-\frac{11}{2}$ **41.** 17
43. -52 **45.** -32 **47.** $\frac{7}{3}$ **49.** 12   **51.** 18
53. 9   **55.** -1 **57.** $\frac{5}{6}$ **59.** $\frac{5}{4}$ **61.** $\frac{1}{20}$
63. 1.19   **65.** 836.94
67. $4 \cdot 10^2 = 4 \cdot 10 \cdot 10 = 400$
      $4 \cdot 10 \cdot 4 \cdot 10 = 4 \cdot 4 \cdot 10 \cdot 10 = 1,600$
69. $-3^2 = -(3 \cdot 3) = -9$
      $(-3)(-3) = 9$
71. $4 - (6 - 2) = 4 - 4 = 0$
      $4 - 6 - 2 = -2 - 2 = -4$
73. $2^2 < 2^4$ **75.** $\frac{3}{4} > \left(\frac{3}{4}\right)^2$ **77.** 36
79. $V = 7^3$ cubic inches   **81.** 0.07, \$31,500

SECTION 1.5 *(page 52)*

Warm-Up *(page 52)*

1. 13   **2.** 28   **3.** 9,300,000   **4.** 0.0066
5. 2   **6.** -4 **7.** 35   **8.** -30 **9.** 10
10. 8

1. Commutative Property of Multiplication
3. Commutative Property of Addition
5. Additive Identity Property
7. Associative Property of Addition
9. Additive Inverse Property
11. Associative Property of Multiplication
13. Multiplicative Inverse Property
15. Distributive Property
17. Distributive Property
19. Associative Property of Addition
21. $6x + 6 \cdot 2$ **23.** $5 + y$ **25.** $6(xy)$

27. $(3x + 2y) + 5$ **29.** $5(v + u)$
31. (a) -50 (b) $\frac{1}{50}$ **33.** (a) $-2x$ (b) $\frac{1}{2x}$
35. (a) $-(x + y)$ (b) $\frac{1}{x + y}$ **37.** (a) $-ab$
(b) $\frac{1}{ab}$ **39.** $(10 + 8) + 2$ **41.** $x + (3 + 2)$
43. $2(3 \cdot 4)$ **45.** $(2 \cdot 3)y$ **47.** $3 \cdot 6 + 3 \cdot 10 = 48$
49. $3 \cdot 2x + 3 \cdot 4 = 6x + 12$
51. $\frac{2}{3} \cdot 9z + \frac{2}{3} \cdot 24 = 6z + 16$
53. $x \cdot 3 + x \cdot x = 3x + x^2$

55. $3 + 10(x + 1)$
       $= 3 + 10x + 10$ Distributive Property
       $= 3 + 10 + 10x$ Commutative Property of Addition
       $= (3 + 10) + 10x$ Associative Property of Addition
       $= 13 + 10x$ Addition of real numbers
57. $5(x + 3) = 5x + 5 \cdot 3$ **59.** $\frac{8}{0}$ is undefined.
61. No

CHAPTER 1 REVIEW EXERCISES *(page 55)*

1. $-\frac{1}{10} < 4$ **3.** $-3 > -7$

5. $10.6 > -3.5$ **7.** 8.5   **9.** -8.5

11. $|-84| = |84|$ **13.** $\left|\frac{3}{10}\right| > -\left|\frac{4}{5}\right|$ **15.** 100
17. 11   **19.** 240   **21.** -268 **23.** -38 **25.** 45
27. 1,500   **29.** $-2,358$ **31.** -18 **33.** Undefined
35. 0   **37.** 7   **39.** $\frac{2}{5}$ **41.** $\frac{3}{4}$ **43.** $\frac{1}{9}$ **45.** $\frac{103}{96}$
47. $\frac{17}{8}$ **49.** $-\frac{1}{12}$ **51.** 1   **53.** $\frac{2}{3}$ **55.** $\frac{6}{7}$
57. $\frac{10}{3}$ **59.** 13.1   **61.** 21   **63.** 343   **65.** -343
67. $\frac{81}{625}$ **69.** 160   **71.** 54   **73.** $\frac{37}{8}$
75. Commutative Property of Addition
77. Multiplicative Inverse Property
79. Distributive Property
81. Associative Property of Addition
83. Multiplicative Identity Property
85. \$3.52   **87.** \$0.065   **89.** 0

CHAPTER 1 TEST *(page 57)*

1. $-|-2| < -\frac{3}{5}$ **2.** -4 **3.** 10   **4.** -160
5. -8 **6.** $\frac{17}{24}$ **7.** $\frac{3}{10}$ **8.** $\frac{7}{12}$ **9.** -27
10. -64 **11.** $-\frac{4}{9}$ **12.** 33   **13.** $\frac{1}{8}$

14. Distributive Property
15. Multiplicative Inverse Property
16. Associative Property of Addition
17. 15 feet   **18.** 128 cubic feet
19.

Chapter 2

SECTION 2.1   *(page 65)*

Warm-Up   *(page 65)*

1. -10 **2.** $-\frac{5}{3}$ **3.** 20   **4.** 42   **5.** 5
6. 12   **7.** Distributive Property
8. Commutative Property of Multiplication
9. Associative Property of Addition
10. Multiplicative Inverse Property

1. $3x^2$, 5   **3.** $\frac{5}{3}$, $-3y^3$ **5.** $2x$, $-3y$, 1
7. $3(x + 5)$, 10   **9.** $4x$, $\dfrac{x + 2}{3}$ **11.** -6 **13.** $\frac{1}{2}$
15. $\frac{3}{4}$ **17.** $-\frac{3}{2}$ **19.** $-\frac{1}{3}$ **21.** $2u^4$ **23.** $(2u)^4$
25. a^3b^2 **27.** $4x^3y^2$ **29.** $3^3(x - y)^2$
31. $2 \cdot 2 \cdot x \cdot x \cdot x \cdot x$ **33.** $4 \cdot y \cdot y \cdot z \cdot z \cdot z$
35. $a^2 \cdot a^2 \cdot a^2 = a \cdot a \cdot a \cdot a \cdot a \cdot a$
37. $5 \cdot x \cdot x \cdot x \cdot x \cdot x \cdot x \cdot x$ **39.** $(ab)(ab)(ab)$
41. $(-2y)(-2y)(-2y)$ **43.** $(x + y)(x + y)$ **45.** u^6
47. $5x^7$ **49.** $-8x^3$ **51.** $-6ab^7$ **53.** $(x - 2y)^4$
55. $-72x^7 + 5x^2$ **57.** $-54x^3 + 5x^2$ **59.** $a^{10}b^{11}$

61. Not equal   **63.** Not equal   **65.** $P\left(1 + \dfrac{r}{4}\right)^4$

67. 10^4 **69.** 8,390
71.

t	2^t
1	\$ 0.02
5	0.32
10	10.24
20	10,485.76
25	335,544.32

73. 1,728 cubic inches = 1 cubic foot

SECTION 2.2   *(page 73)*

Warm-Up   *(page 72)*

1. $(3z)^4$ **2.** $8^3x^3y^2$
3. $3 \cdot 3 \cdot x \cdot x \cdot x \cdot x \cdot y \cdot y \cdot y$
4. $5 \cdot uv \cdot uv \cdot uv \cdot uv$ **5.** v^5 **6.** u^6
7. $4x^6$ **8.** $8y^5$ **9.** $5z^7$ **10.** $(a + 3)^7$

1. Commutative Property of Addition
3. Multiplicative Identity Property
5. Associative Property of Addition
7. Multiplicative Inverse Property
9. Distributive Property
11. Additive Inverse and Additive Identity Properties
13. $(x + 1) - (x + 1) = 0$ **15.** $v(2) = (2)v$
17. $(t + 5)(t - 2) = t(t - 2) + 5(t - 2)$

19. $5x\left(\dfrac{1}{5x}\right) = 1$, $x \neq 0$ **21.** $-10x + 5y$

23. $3x + 6$ **25.** $x^2 + x^2y + xy^2$ **27.** $3x^2y + 3x$
29. $-u + v$ **31.** $6x^2$, 6; $-3xy$, -3; y^2, 1
33. $4x^3$, 4; $-3x^2$, -3; $2x$, 2
35. $0.12x$, 0.12; $0.36x^2y$, 0.36; $-1.40y$, -1.40
37. $16t^3$, $3t^3$; 4, -5 **39.** $6x^2y$, $-4x^2y$ **41.** $-2y$

43. $-2x + 4y$ **45.** $x^2 - xy + 4$ **47.** $2\left(\dfrac{1}{x}\right) + 8$

49. $3x^2 + 2x^2y + 3xy^2 + y^2$
51. Variable factors are not alike. $x^2y \neq xy^2$
53. Unequal. $3(x - 4) = 3x - 12$ **55.** Equal
57. 24   **59.** 416   **61.** 39.9
63. $a(b + c) = ab + ac$

SECTION 2.3   *(page 82)*

Warm-Up   *(page 81)*

1. $-8x + 20$ **2.** $30t - 40$ **3.** $-2x^2y + xy^3$
4. $-xz^2 + 2y^2z$ **5.** $-9 + 6x$ **6.** $-10x - 8z$
7. $11s - 5t$ **8.** $-x^2 + 1$ **9.** $x - 4$
10. $3x^2y - 5xy - xy^2$

1. $-12x$ **3.** $6x^2$ **5.** $2x$ **7.** $-6x^5$
9. $-24x^4y^4$ **11.** $2x + 3$ **13.** $-2m + 40$
15. $2x + 44$ **17.** $8x + 26$ **19.** $2x - 17$
21. $3x^2 + 5x$ **23.** $4t^2 - 11t$ **25.** $\dfrac{x}{3}$ **27.** $\dfrac{7z}{5}$

29. $-\dfrac{11x}{12}$ **31.** x

33. (a) 0   (b) 7   **35.** (a) -5 (b) 25
37. (a) 6   (b) 0   **39.** (a) 3   (b) -20
41. (a) 17   (b) 4
43. (a) 0   (b) Division by zero is undefined.
45. (a) $\frac{15}{2}$ (b) 10   **47.** (a) 175   (b) 140
49. (a)

x	-1	0	1	2	3	4
$3x - 2$	-5	-2	1	4	7	10

(b) 3   (c) $\frac{2}{3}$ **51.** (a) $5x^2$ (b) 1,125 square feet
53. 9,375 square feet

SECTION 2.4 *(page 93)*

Warm-Up *(page 92)*

1. $-9x + 11y$ **2.** $8v - 4$ **3.** $-y^4 + 2y^2$
4. $10t - 4t^2$ **5.** 7   **6.** 0   **7.** 0
8. Division by zero is undefined.   **9.** 4   **10.** 9

---

1. $x + 5$ **3.** $x - 6$ **5.** $2x$ **7.** $\dfrac{x}{4}$ **9.** $\dfrac{x}{3}$
11. $8 + 5x$ **13.** $3x + 5$ **15.** $10(x + 4)$
17. $|16 - x|$ **19.** $x^2 + 1$
21. A number decreased by 10
23. A number is tripled, and the product is increased by 2.
25. The product of 3 and the sum of a number and 2
27. The sum of a number and 1 is divided by 2.
29. The square of a number increased by 5
31. $(x + 3)x = x^2 + 3x$ **33.** $(x - 9)3 = 3x - 27$
35. $x^2 + (x + 1) = x^2 + x + 1$ **37.** $0.10n$
39. $0.06L$ **41.** $\dfrac{100}{r}$ **43.** $15 + 2n$
45. s^2 square inches   **47.** $5w$
49. $n + (n + 1) + (n + 2) = 3n + 3$
51. $(2n + 1) + (2n + 3) = 4n + 4$
53. (a)

n	0	1	2	3	4	5
$3n + 2$	2	5	8	11	14	17
Differences		3	3	3	3	3

(b) The coefficient of n determines the difference.

CHAPTER 2 REVIEW EXERCISES *(page 96)*

1. $4x$, 4; $-\frac{1}{2}x^3$, $-\frac{1}{2}$ **3.** y^2, 1; $-10yz$, -10; $\frac{2}{3}z^2$, $\frac{2}{3}$
5. $(5z)^3$ **7.** $a^2(b - c)^2$ **9.** x^6 **11.** $-2t^6$
13. $-5x^3y^4$ **15.** $-64y^7$
17. Commutative Property of Multiplication
19. Associative Property of Addition
21. Multiplicative Inverse Property   **23.** $4x + 12y$
25. $-10u + 15v$ **27.** $8x^2 + 5xy$ **29.** $a - 3b$
31. $-2a$ **33.** $11p - 3q$ **35.** $x^2 + 2xy + 4$
37. $3(x - y) + 3xy$ **39.** $5u - 10$ **41.** $5s - r$
43. $10z - 1$ **45.** $4y - 2x$ **47.** (a) 5   (b) 5
49. (a) 4   (b) -2 **51.** $\frac{2}{3}x + 5$ **53.** $2x - 10$
55. $50 + 7x$ **57.** $\dfrac{x + 10}{8}$ **59.** $x^2 + 64$

61. A number plus 3
63. A number decreased by 2 is divided by 3.
65. $0.28I$ **67.** $w^2 + 3w$ square feet
69. $(2n - 1) + (2n + 1) + (2n + 3) = 6n + 3$
71. $9x^2$
73. (a)

n	0	1	2	3	4	5
$3n + 2$	2	6	12	20	30	42
Differences		4	6	8	10	12
Differences			2	2	2	2

(b) Third row: Entries increase by 2
    Fourth row: Constant 2

CHAPTER 2 TEST *(page 99)*

1. $2x^2$, 2; $-7xy$, -7; $3y^3$, 3   **2.** $x^3(x + y)^2$ **3.** c^8
4. t^8 **5.** $-10u^4v$
6. Associative Property of Multiplication
7. Commutative Property of Addition
8. Additive Inverse Property   **9.** $3x + 24$
10. $-3y + 2y^2$ **11.** $-a - 7b$ **12.** $8u - 8v$
13. $4z - 4$ **14.** $18 - 2t$ **15.** 25   **16.** -15
17. Not possible; division by zero   **18.** $\dfrac{1}{5}n + 2$
19. $2w - 4$ **20.** $3n + 2m$

Chapter 3

SECTION 3.1 *(page 107)*

Warm-Up *(page 107)*

1. $-8y^6$ **2.** $12a^3b$ **3.** $x + 10$ **4.** $-9x$
5 (a) 0   **(b)** 4   **6. (a)** 1   **(b)** $\frac{1}{5}$
7. $2x + 4$ **8.** $2(x - 10)$ **9.** $(x - 5)^2$
10. $\dfrac{x + 25}{2}$

1. (a) Solution   **(b)** Not a solution
3. (a) Not a solution   **(b)** Solution
5. (a) Not a solution   **(b)** Solution   **7. (a)** Solution
(b) Not a solution   **9. (a)** Solution
(b) Not a solution   **11. (a)** Solution   **(b)** Solution
13. (a) Solution   **(b)** Not a solution

15.

$5x + 12 = 22$	Given equation
$5x + 12 - 12 = 22 - 12$	Subtract 12 from both sides
$5x = 10$	Combine like terms
$\dfrac{5x}{5} = \dfrac{10}{5}$	Divide both sides by 5
$x = 2$	Solution

17.

$\frac{2}{3}x = 12$	Given equation
$\frac{3}{2}\left(\frac{2}{3}x\right) = \frac{3}{2}(12)$	Multiply both sides by 3/2
$x = 18$	Solution

19.

$2(x - 1) = x + 3$	Given equation
$2x - 2 = x + 3$	Distributive Property
$-x + 2x - 2 = -x + x + 3$	Subtract x from both sides
$x - 2 = 3$	Combine like terms
$x - 2 + 2 = 3 + 2$	Add 2 to both sides
$x = 5$	Solution

21. $x + 6 = 94$ **23.** $3{,}650 + x = 4{,}532$
25. $x + 12 = 45$ **27.** $4(x + 6) = 100$
29. $2x - 15 = \dfrac{x}{3}$ **31.** $2n + (2n + 2) + (2n + 4) = 18$
33. $2l + 2\left(\frac{1}{3}l\right) = 96$ **35.** $3r + 25 = 160$
37. $P + 0.25P = 375$ **39.** $750{,}000 - 3D = 75{,}000$
41. $4r + 24 = 200$

SECTION 3.2 *(page 118)*

Warm-Up *(page 117)*

1. 3   **2.** -40 **3.** $\frac{3}{2}$ **4.** $\frac{7}{6}$ **5.** $-\frac{1}{16}$
6. $\frac{8}{5}$ **7.** Solution   **8.** Solution
9. Solution   **10.** Not a solution

1.

$$5(-3) + 15 = -15 + 15 = 0$$
$$5(-3) + 15 - 15 = -15 + 15 - 15 = 0 - 15$$
$$5(-3) = -15$$
$$\frac{5(-3)}{5} = \frac{-15}{5}$$
$$-3 = -3$$

3.

$$-2(-4) + 5 = 8 + 5 = 13$$
$$-2(-4) + 5 - 5 = 8 + 5 - 5 = 13 - 5$$
$$-2(-4) = 8$$
$$\frac{-2(-4)}{-2} = \frac{8}{-2}$$
$$-4 = -4$$

5.

$8x - 2 = 20$	Given equation
$8x - 2 + 2 = 20 + 2$	Add 2 to both sides
$8x = 22$	Combine like terms
$\dfrac{8x}{8} = \dfrac{22}{8}$	Divide both sides by 8
$x = \dfrac{11}{4}$	Isolate x

7.

$10 - 4x = -6$	Given equation
$10 - 4x - 10 = -6 - 10$	Subtract 10 from both sides
$-4x = -16$	Combine like terms
$\dfrac{-4x}{-4} = \dfrac{-16}{-4}$	Divide both sides by -4
$x = 4$	Isolate x

9. 3   **11.** -6 **13.** $\frac{3}{2}$ **15.** $\frac{2}{3}$ **17.** 2   **19.** 2
21. $\frac{1}{3}$ **23.** -2 **25.** 1   **27.** 0   **29.** No solution
31. $\frac{2}{5}$ **33.** 0   **35.** $\frac{2}{3}$ **37.** 30   **39.** $\frac{5}{3}$ **41.** $\frac{1}{3}$
43. $\frac{5}{6}$ **45.** No solution   **47.** 50   **49.** 5
51. 7.71   **53.** 8.99   **55.** 30   **57.** 35, 37
59. 36 ft   **61.** $9,750   **63.** $312.50   **65.** $12\frac{1}{4}$ ft

67.

t	Width	Length	Area
1	300	300	90,000
1.5	240	360	86,400
2	200	400	80,000
3	150	450	67,500
4	120	480	57,600
5	100	500	50,000

SECTION 3.3   *(page 127)*

Warm-Up   *(page 127)*

1. -6 **2.** 5   **3.** 35   **4.** -8 **5.** 14
6. 0   **7.** 0   **8.** $-\frac{1}{3}$ **9.** 2   **10.** $\frac{5}{2}$

1. 5   **3.** -10 **5.** 2   **7.** -5 **9.** -7
11. -4 **13.** $\frac{8}{5}$ **15.** 1   **17.** 9   **19.** 3
21. $-\frac{3}{2}$ **23.** $\frac{5}{2}$ **25.** 0   **27.** $-\frac{2}{5}$ **29.** $-\frac{10}{3}$
31. $\frac{32}{5}$ **33.** 10   **35.** 4   **37.** $\frac{1}{7}$ **39.** $\frac{5}{6}$
41. 4.8 hrs   **43.** 97   **45.** 25 quarts 10%
47. $1\frac{1}{3}$ quarts   **49.** $x = 4$ ft

SECTION 3.4   *(page 137)*

Warm-Up   *(page 137)*

1. $-\frac{1}{2} > -7$ **2.** $-\frac{1}{3} < -\frac{1}{6}$ **3.** $-2 > -3$
4. $-6 > -\frac{13}{2}$ **5.** $-\frac{5}{2}$ **6.** 24   **7.** 36
8. $\frac{1}{2}$ **9.** 2   **10.** 6

1. (a) Yes   **(b)** No   **(c)** Yes   **(d)** No
3. (a) Yes   **(b)** No   **(c)** No   **(d)** Yes
5. (c)   **6.** (b)   **7.** (f)   **8.** (e)   **9.** (d)
10. (a)
11. $t \geq 5$ **13.** $x \leq 2$

15. $x < 3$ **17.** $x > -4$

19. $n < 3$ **21.** $x \leq 18$

23. $x < 4$ **25.** $x > 6$

27. $x \geq 4$ **29.** $x > \frac{1}{2}$

31. $x > \frac{11}{3}$ **33.** $y > 1$

35. $z \leq -1$ **37.** $y > \frac{7}{8}$

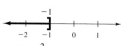   

39. $x > -\frac{2}{5}$ **41.** $x \geq -\frac{10}{3}$

43. $-1 < x < 3$ **45.** $-\frac{9}{2} < x < \frac{15}{2}$

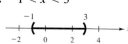   

47. $-16 < x < 8$ **49.** $-\frac{3}{4} < x < -\frac{1}{4}$

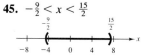

51. $x \geq 0$ **53.** $y > -6$ **55.** $z \geq 3$
57. x is not more than 10.
59. y is greater than $-\frac{3}{2}$ and no more than 5.
61. Mars is farther from the Sun than Mercury.
63. $m < 24{,}062.50$
65. 0 min $\leq t \leq 12.4$ min (or 12 min if portion of minute is charged as a whole minute)   **67.** 1 mi $\leq d \leq$ 5 mi

CHAPTER 3 REVIEW EXERCISES   *(page 141)*

1. (a) Not a solution   **(b)** Solution
3. (a) Not a solution   **(b)** Solution   **5. (a)** Solution
(b) Not a solution   **7. (a)** Not a solution
(b) Solution   **9. (a)** Solution   **(b)** Solution   **11.** 5

13. 4   **15.** 3   **17.** $\frac{4}{3}$ **19.** 20   **21.** $\frac{5}{3}$ **23.** 20
25. 7   **27.** $\frac{1}{3}$ **29.** -1 **31.** 7.99   **33.** 224.31
35. $x \geq 2$ **37.** $x < 3$

39. $x \geq 2$ **41.** $x > 10$

43. $t > 4$ **45.** $y \leq -1$

47. $-6 < x \leq 6$ **49.** $-7 < x < -1$

51. $x + \dfrac{1}{x} = \dfrac{37}{6}$ **53.** $6x - \frac{1}{2}(6x) = \frac{1}{2}(6x) = 24$
55. $z \geq 10$ **57.** $8 < y < 12$ **59.** $V < 12$

CHAPTER 3 TEST   *(page 143)*

1. (a) Not a solution   **(b)** Solution   **2.** $\frac{21}{4}$ **3.** 7
4. 10   **5.** 1   **6.** 11.03
7. $x \leq 4$

8. $x < -6$

9. $-1 < x \leq 2$

10. $-1 \leq x < 5$

11. $y \leq 10$ **12.** $t \geq 4$ **13.** 50   **14.** $200
15. 2 hours

Chapter 4

SECTION 4.1   *(page 154)*

Warm-Up   *(page 153)*
1. $\frac{3}{5}$ **2.** $\frac{6}{7}$ **3.** $\frac{11}{41}$ **4.** $\frac{2}{3}$ **5.** $\frac{4}{3}$ **6.** 4
7. $\frac{8}{5}$ **8.** $\frac{15}{2}$ **9.** 10   **10.** $\frac{32}{5}$

1. $\frac{3}{2}$ **3.** $\frac{1}{4}$ **5.** $\frac{3}{8}$ **7.** $\frac{3}{4}$ **9.** $\frac{2}{1}$ **11.** $\frac{12}{1}$
13. $\frac{3}{2}$ **15.** $0.0445   **17.** $0.0645
19. 32-ounce jar   **21.** 12   **23.** 50   **25.** $\frac{10}{3}$
27. 16   **29.** $\frac{1}{2}$ **31.** 27   **33.** $\frac{14}{5}$ **35.** $\frac{3}{16} = 0.1875$
37. $22\frac{2}{9}$ gal   **39.** 250 blocks   **41.** 15 pints
43. $1,142.31   **45.** 22,691 votes   **47.** 245 miles
49. 46 min, 40 sec   **51.** $\frac{5}{2}$ **53.** $\frac{15}{2}$ **55.** 6 ft 8 in.
57. $5,268.87   **59.** $0.51   **61.** $\frac{100}{49}$

SECTION 4.2   *(page 164)*

Warm-Up   *(page 163)*
1. -100 **2.** 4   **3.** $-10,000$ **4.** 900
5. $-\frac{1}{100}$ **6.** $\frac{1}{4}$ **7.** 200   **8.** 40   **9.** $\frac{1}{5}$
10. $\frac{2}{5}$

1.

Percent	40%
Parts out of 100	40
Decimal	0.40
Fraction	$\frac{2}{5}$

3.

Percent	7.5%
Parts out of 100	7.5
Decimal	0.075
Fraction	$\frac{3}{40}$

5.

Percent	63%
Parts out of 100	63
Decimal	0.63
Fraction	$\frac{63}{100}$

7.

Percent	15.5%
Parts out of 100	15.5
Decimal	0.155
Fraction	$\frac{31}{200}$

9.

Percent	60%
Parts out of 100	60
Decimal	0.60
Fraction	$\frac{3}{5}$

11. 0.125   **13.** 2.50   **15.** 7.5%   **17.** 62%
19. 80%   **21.** 35%   **23.** 37.5%   **25.** 41.67%
27. 45   **29.** 77.52   **31.** 0.42   **33.** 176
35. 2,100   **37.** 2,200   **39.** 132   **41.** 360
43. 72%   **45.** 12.5%   **47.** $\frac{2}{3}$%   **49.** 500%
51. $425   **53.** $71\frac{2}{3}$%   **55.** 10,210 eligible voters
57. $20,772.73   **59.** 500 points
61. Attendance: 42.22%
Advertising: 12.06%
Concessions: 11.69%
TV and radio: 34.03%
63. Bored: 108
Moving: 57
Redecorating: 48
New furniture: 45
Other: 42
65. 25% reduction

SECTION 4.3   *(page 175)*

Warm-Up   *(page 174)*

1. 14   **2.** 4   **3.** −3   **4.** 4   **5.** 15.5
6. 30   **7.** 200%   **8.** $66\frac{2}{3}$%   **9.** 455
10. 250

Cost, Selling Price, Markup, Markup Rate

1. $26.97, $49.95, $22.98, 85.2%
3. $69.29, $125.98, $56.69, 81.8%
5. $13,250.00, $15,900.00, $2,650.00, 20%
7. $40.98, $74.38, $33.40, 81.5%
9. $107.97, $199.96, $91.99, 85.2%

List Price, Sale Price, Discount, Discount Rate

11. $39.95, $29.95, $10.00, 25%
13. $189.99, $159.99, $30.00, 15.8%
15. $119.96, $59.98, $59.98, 50%
17. $23.69, $18.95, $4.74, 20%
19. $995.00, $695.00, $300.00, 30.2%

21. 2 hrs   **23.** 18%   **25.** 7 min, $1.18
27. $960.70   **29.** $87, $1,537, $1,037   **31.** $575
33. $35,714.29   **35.** 12.5%   **37.** 10.75 hrs
39. Local store at $47.99   **41.** $60

SECTION 4.4   *(page 188)*

Warm-Up   *(page 188)*

1. 4   **2.** −1   **3.** −2   **4.** 1   **5.** 9
6. 10   **7.** $\frac{2}{5}$ **8.** $\frac{10}{3}$ **9.** −9   **10.** $-\frac{11}{5}$

1. 80 15¢-stamps; 20 30¢-stamps
3. 8 nickels; 12 dimes
5. Candidates A and B: 250 each; Candidate C: 500
7. 100 children's tickets; 300 adult tickets
9.

Oats x	Corn $100 - x$	Price per Ton of the Mixture
0	100	$450
20	80	420
40	60	390
60	40	360
80	20	330
100	0	300

(a) Decreases   **(b)** Decreases
(c) Average of the two prices
11. 4 roses   **13.** Solution 1: 25 gal, Solution 2: 75 gal
15. Solution 1: 5 qts, Solution 2: 5 qts   **17.** 1.14 gallons

Distance, d,   Rate, r,   Time, t

19. 165 mi, 55 mi/hr, 3 hrs
21. 500 km, 90 km/hr, $\frac{50}{9}$ hr
23. 5,280 ft, 2,112 ft/sec, $\frac{5}{2}$ sec

25. $\frac{3}{17}$ hr ≈ 10.6 min   **27.** 28 mi   **29.** 1,153.85 mi/hr
31. 20 min   **33.** 45.2 mi/hr   **35.** 1 hr, 5 mi
37. $\frac{6}{5}$ hr

39. (a) $\dfrac{1}{h}t + \dfrac{1}{3h/2}t = 1$ **(b)**

$$\frac{1}{h}t + \frac{2}{3h}t = 1$$
$$3t + 2t = 3h$$
$$5t = 3h$$
$$t = \frac{3h}{5}$$

h	1	2	3	4	5
t	$\frac{3}{5}$	$\frac{6}{5}$	$\frac{9}{5}$	$\frac{12}{5}$	3

41. 13   **43.** 15   **45.** 262, 263   **47.** 54, 56
49. 8, 9, 10   **51.** 15, 75

SECTION 4.5 *(page 199)*

Warm-Up *(page 199)*

1. $15x$ **2.** $-8x^2$ **3.** $\dfrac{x^2}{2}$ **4.** $2x$

5. $30x - 5y$ **6.** $15 - 7s$ **7.** $9x + 4$
8. $z^2 + 9z$ **9.** 5 **10.** -2

1. 40 mi/hr **3.** $12\pi \approx 37.70$ m^3 **5.** \$1,450
7. $\dfrac{2A}{b}$ **9.** $\dfrac{E}{I}$ **11.** $\dfrac{V}{wh}$ **13.** $\dfrac{S}{1+R}$ **15.** $\dfrac{A-P}{Pt}$
17. $\dfrac{2A - ah}{h}$ **19.** $\dfrac{\pi h^3 + 3V}{3\pi h^2}$ **21.** $\dfrac{2(h - tv_0)}{t^2}$
23. $\dfrac{L + d - a}{d}$ **25.** $\dfrac{a - S}{L - S}$ **27.** 24 in.^{2}
29. 96 in.^{3} **31.** 784 sq ft **33.** \$540 **35.** 11%
37. \$230 **39.** \$15,975
41. \$10,000 at 8.5%; \$15,000 at 10%

CHAPTER 4 REVIEW EXERCISES *(page 203)*

1. $\frac{1}{8}$ **3.** $\frac{4}{3}$ **5.** $\frac{7}{2}$ **7.** 4 **9.** 10
11. \$1,686.67 **13.** 214 mi **15.** $\frac{4}{3}$
17.

Percent	Parts out of 100	Decimal	Fraction
35%	35	0.35	$\frac{7}{20}$

19. 20 **21.** 400 **23.** 60% **25.** 3.5%
27. \$114.75 **29.** \$181.35 **31.** 16.8%
33. 13 dimes; 17 quarters **35.** 9 hrs, 22 min
37. 3,500 mi **39.** $\frac{30}{11}$ hr **41.** 48, 51, 54
43. 30 ft by 26 ft **45.** \$475 **47.** \$285,714.29
49. \$15,000 at 8%; \$5,000 at 10.5% **51.** $t = \dfrac{A - P}{Pr}$

CHAPTER 4 TEST *(page 206)*

1. $\frac{5}{9}$ **2.** $\frac{15}{2}$ **3.** 6 **4.** 150 miles
5. (a) 37.5% (b) 0.375 **6.** 42 **7.** 1,200
8. 36% **9.** 40% **10.** 8 dimes, 12 quarters
11. 4 roses **12.** 48 mi/hr **13.** $\frac{36}{7}$ hr
14. 30, 31, 32 **15.** $R = \dfrac{S - C}{C}$ **16.** \$480
17. \$6,250

CUMULATIVE TEST: CHAPTERS 1–4 *(page 207)*

1. $-\frac{3}{4} < \left|-\frac{7}{8}\right|$ **2.** 1,200 **3.** $-\frac{11}{24}$ **4.** $-\frac{25}{12}$
5. 8 **6.** $3^3(x + y)^2$ **7.** $15x^7$ **8.** a^8b^7
9. $-2x^2 + 6x$ **10.** $7x^2 - 6x - 2$
11. Associative Property of Addition **12.** 6 **13.** $\frac{52}{3}$
14. 5
15. $-5 \le x < 1$

16. $\frac{3}{4}$ **17.** \$2,968.42 **18.** \$920 **19.** 246, 248
20. \$8,000 at 7.5%; \$4,000 at 9%

Chapter 5

SECTION 5.1 *(page 214)*

Warm-Up *(page 214)*

1. $10x - 10$ **2.** $12 - 8z$ **3.** $-2 + 3x$
4. $-50x + 75$ **5.** $10x, -3y, 4$
6. $-2r, 8s, -6$ **7.** $\frac{3}{4}$ **8.** -8 **9.** $5x - 2y$
10. $\frac{1}{6}x + 8$

Polynomial, Standard Form, Degree, Leading Coefficient

1. $2x - 3$, $2x - 3$, 1, 2
3. $9 - 2y^4$, $-2y^4 + 9$, 4, -2
5. $8x + 2x^5 - x^2 - 1$, $2x^5 - x^2 + 8x - 1$, 5, 2
7. 10, 10, 0, 10
9. $v_0t - 16t^2$ (v_0 is a constant), $-16t^2 + v_0t$, 2, -16
11. Trinomial **13.** Binomial **15.** Monomial
17. Not a polynomial **19.** Polynomial **21.** $5x^3 - 10$
23. $6x^2$ **25.** $-x^6 + 3x^4 + x^2$ **27.** $3x^2 + 2$
29. $8 + 2x^3$ **31.** $y^4 + 4$ **33.** $4x^2 + 2x + 2$
35. $4x^2 + 8$ **37.** $4z^2 - z - 2$ **39.** 1
41. $-x^2 - 2x + 2$ **43.** $-2x^3$ **45.** $4t^3 - 3t^2 + 15$
47. $5x^3 - 6x^2 + x - 3$ **49.** $3x^3 + 4x + 10$
51. $x^2 - 2x + 2$ **53.** $-3x^3 + 1$ **55.** $-u^2 + 5$
57. $-2x - 20$ **59.** $3x^3 - 2x + 2$
61. $2x^4 + 9x + 2$ **63.** $8x^3 + 29x^2 + 11$
65. $12z + 8$ **67.** $4t^2 + 20$ **69.** $6v^2 + 90v + 30$
71. $2x^2 - 2x$

SECTION 5.2 *(page 225)*

Warm-Up   *(page 224)*

1. $(-3)^4$ **2.** x^3 **3.** $\frac{4}{5} \cdot \frac{4}{5} \cdot \frac{4}{5} \cdot \frac{4}{5}$

4. $(4.5)(4.5)(4.5)(4.5)(4.5)$ **5.** $x^2 \cdot x^2 \cdot x^2$

6. $y^4 \cdot y^4$ **7.** $7x - 8$ **8.** $-2y + 14$

9. $-4z + 12$ **10.** $-5u - 5$

---

1. $-2x^2$ **3.** $\frac{5}{2}x^2$ **5.** $4t^3$ **7.** $6b^3$ **9.** x^8

11. $-27t^3$ **13.** $18x^3$ **15.** $12x^5 - 6x^4$

17. $3y - y^2$ **19.** $-x^3 + 4x$ **21.** $6t^2 - 15t$

23. $3x^3 - 6x^2 + 3x$ **25.** $-12x - 12x^3 + 24x^4$

27. $2x^3 - 4x^2 + 16x$ **29.** $30x^3 + 12x^2$

31. $x^2 + 7x + 12$ **33.** $6x^2 - 7x - 5$

35. $x^2 + 3xy + 2y^2$ **37.** $-x^2 + 17x$ **39.** $2st - 4t^2$

41. $x^3 - 8$ **43.** $x^4 - 5x^3 - 2x^2 + 11x - 5$

45. $x^3 - 3x^2 - 16x + 6$ **47.** $x^3 + 27$

49. $x^4 - x^2 + 4x - 4$

51. $x^5 + 5x^4 - 3x^3 + 8x^2 + 11x - 12$

53. $x^3 + 3x^2 + x - 1$ **55.** $4u^3 + 4u^2 - 5u - 3$

57. $x^2 - 4$ **59.** $4u^2 - 9$ **61.** $4x^2 - 9y^2$

63. $x^2 + 12x + 36$ **65.** $a^2 - 4a + 4$

67. $9x^2 - 12x + 4$ **69.** $4x^2 - 20xy + 25y^2$

71. $64 - 48z + 9z^2$

73. $x^2 + y^2 + 2xy + 2x + 2y + 1$

75. $u^2 + v^2 - 2uv + 6u - 6v + 9$ **77.** $8x$

79. $(a + b)(a + b) = a^2 + 2ab + b^2$

$$
\begin{array}{r}
a^2 + 2ab + b^2 \\
\times \qquad\qquad a + b \\
\hline
a^2b + 2ab^2 + b^3 \\
a^3 + 2a^2b + ab^2 \qquad\quad \\
\hline
a^3 + 3a^2b + 3ab^2 + b^3
\end{array}
$$

81. $x^3 + 6x^2 + 12x + 8$ **83.** (a) $6w$ (b) $2w^2$

85. $x(x + 10) = x^2 + 10x$ **87.** $x^2 + ax + bx + ab$

89. (a) $x^2 - 1$ (b) $x^3 - 1$ (c) $x^4 - 1$

(d) $x^5 - 1$

SECTION 5.3 *(page 233)*

Warm-Up   *(page 233)*

1. $\frac{2}{3}$ **2.** $\frac{1}{8}$ **3.** $\frac{2}{5}$ **4.** $\frac{25}{6}$ **5.** $-10x^5$

6. $12y^3 - 4y$ **7.** $4z^2 - 1$ **8.** $2x^2 + 3x - 20$

9. $x^2 + 14x + 49$ **10.** $x^3 + 1$

---

1. x^3 **3.** $2y^2$ **5.** $\frac{1}{z^3}$ **7.** 1 **9.** $\frac{4^4}{x^2}$ **11.** $-\frac{x}{2}$

13. $4z^2$ **15.** $\frac{8b}{3}$ **17.** $-\frac{11y}{2}$ **19.** $\frac{3s^3}{2r^2}$ **21.** $\frac{1}{2z}$

23. $z + 1$ **25.** $z - 3$ **27.** $5x - 2$

29. $-5z^2 - 2z$ **31.** $8z^2 + 3z - 2$ **33.** $-4x + 3$

35. $m^2 + 3 - \frac{4}{m}$ **37.** $5x + 3$ **39.** $x - 2$

41. $x + 5$ **43.** $2z - 1$ **45.** $y + 2$ **47.** $6t + 1$

49. $3x - 1$ **51.** $x^2 + 2x + 4$ **53.** $x^2 + 2x - 3$

55. $7 - \frac{11}{x + 2}$ **57.** $x - 3 + \frac{18}{x + 3}$

59. $4x - 1 - \frac{2}{x + 1}$ **61.** $x + 2 + \frac{2}{2x + 3}$

63. $2z + 4 + \frac{4}{2z - 1}$ **65.** $x^2 - 2x + 5 + \frac{3}{x - 2}$

67. $3t^2 + t + 1 - \frac{4}{t + 2}$ **69.** $x^3 + x^2 + x + 1$

71. $2x$ **73.** $5uv$ **75.** Not valid   **77.** Valid

SECTION 5.4 *(page 241)*

Warm-Up   *(page 241)*

1. 64 **2.** -32 **3.** $\frac{4}{9}$ **4.** -64 **5.** 7

6. 144 **7.** x^5 **8.** ab **9.** u^3v **10.** y^2z^5

---

1. $\frac{1}{9}$ **3.** $-\frac{1}{64}$ **5.** 16 **7.** $\frac{9}{16}$ **9.** $\frac{9}{4}$ **11.** x^5

13. $\frac{1}{y^6}$ **15.** $\frac{1}{x^2}$ **17.** $\frac{1}{y^6}$ **19.** $\frac{1}{s^2}$ **21.** $\frac{1}{4x^4}$

23. $\frac{1}{b^5}$ **25.** $\frac{1}{9x^4y^2}$ **27.** $\frac{a^6}{64b^9}$ **29.** $-2x^3$ **31.** 1

33. $\frac{10}{x}$ **35.** $\frac{x^2}{9z^4}$ **37.** 1 **39.** $-\frac{81}{16}$

41. $1{,}090{,}000$ **43.** 6.21 **45.** 0.00852

47. 0.0867 **49.** 8003.05 **51.** 9.3×10^7

53. 1.637×10^9 **55.** 4.35×10^{-4}

57. 4.392×10^{-3} **59.** 1.6×10^7 **61.** 4.984×10^{12}

63. 3.0981×10^6 **65.** 3.35544×10^{32}

67. 1.15743×10^{-22} **69.** 523.456

71. 4.4688×10^{14} mi   **73.** 0.006 gal   **75.** 1

CHAPTER 5 REVIEW EXERCISES *(page 245)*

1. $\frac{1}{2}$ **3.** $-3x + 6$ **5.** $16x^4$ **7.** $x^2 + 4x + 4$

9. $1 - \frac{x}{7}$ **11.** $3x - 1$ **13.** $-x^2 + 2x$

15. $-t^2 + 2t$ **17.** $-5x^3 - 5x - 2$

19. $7y^2 - y + 6$ **21.** $2x^2 + 8x$ **23.** $2x^2 + 2x - 12$

25. $2x^3 + 13x^2 + 19x + 6$ **27.** $u^2 - 6u + 5$

29. $x + 3$ **31.** $\frac{7}{3}x$ **33.** $x + 2$

35. $8x + 5 + \dfrac{2}{3x - 2}$ **37.** $2x^2 + 4x + 3 + \dfrac{5}{x - 1}$

39. $x^2 - 2$ **41.** $x^2 + 6x + 9$ **43.** $16x^2 - 56x + 49$

45. $u^2 - 36$ **47.** $6t^2 - t - 1$

49. $a^2 - b^2 - 2a + 1$ **51.** $\frac{1}{16}$ **53.** $\frac{1}{36}$ **55.** $\frac{125}{27}$

57. 9×10^6 **59.** 3.7×10^4 **61.** $\dfrac{x^4}{y^6}$ **63.** $\dfrac{1}{t^3}$

65. $\dfrac{25}{y^2}$ **67. (a)** $4x - 6$ **(b)** $x^2 - 3x$ **69.** 0.15 ft

CHAPTER 5 TEST *(page 247)*

1. Degree: 4; Leading coefficient: -3

2. $2z^2 - 3z + 15$ **3.** $7u^3 - 1$ **4.** $-6x^2 + 12x$

5. $-y^2 + 8y + 3$ **6.** $9x^3$ **7.** $6a^3 - 15a$

8. $10b^2 + b - 3$ **9.** $2z^3 + z^2 - z + 10$

10. $x^2 - 10x + 25$ **11.** $4x^2 - 9$

12. $6x^2 + 9x$ square units   **13.** $\dfrac{2a}{3}$ **14.** $3x + 5$

15. $x^2 + 2x + 3$ **16.** $2x^2 + 4x - 3 - \dfrac{2}{2x + 1}$

17. $\frac{3}{8}$ **18.** $\dfrac{x^4}{9y^6}$ **19.** $384{,}000{,}000$

20. 1.013×10^5 **21.** 2.25×10^{10}

Chapter 6

SECTION 6.1 *(page 255)*

Warm-Up *(page 255)*

1. -20 **2.** -42 **3.** $24x - 36$

4. $28 - 21x$ **5.** $-60 + 42x$ **6.** $-2y^2 - 2y$

7. $-3t^2 - 6t$ **8.** $-8x^2y^2 - 24xy$ **9.** $4 - x^2$

10. $x^2 + 8x + 16$

1. 6   **3.** 2   **5.** z^2 **7.** $2x$ **9.** u^2v **11.** $3yz^2$

13. $14a^2b^2$ **15.** 1   **17.** $3(x + 1)$ **19.** $6(z - 1)$

21. $8(t - 2)$ **23.** $-5(5x + 2)$ **25.** $6(4y^2 - 3)$

27. $x(x + 1)$ **29.** $u(25u - 14)$ **31.** $2x^3(x + 3)$

33. No common factor   **35.** $xy(3xy - 1)$

37. $-rs^2(10r^2 + 7)$ **39.** $8a^3b^3(2 + 3a)$

41. $10abc(1 + a)$ **43.** $4(3x^2 + 4x - 2)$

45. $25(4 + 3z - 2z^2)$ **47.** $3x^2(3x^2 + 2x + 6)$

49. $(x - 3)(x + 5)$ **51.** $(s + 10)(t - 8)$

53. $a(b + 2)(a - 1)$ **55.** $z^2(z + 5)(z^2 + 5z + 1)$

57. $2y^2(y - 8)^2(y - 4)$ **59.** $5uv^2(uv^2 + u + 1)$

61. $(x + 10)(x + 1)$ **63.** $(y - 4)(y + 2)$

65. $(x + 2)(x^2 + 1)$ **67.** $(t - 3)(t^2 + 2)$

69. $(z + 3)(z^2 - 2)$ **71.** $-5(2x - 1)$

73. $-(x^2 - 2x - 4)$ **75.** $\frac{1}{4}(2x + 3)$

77. $\frac{1}{16}(14x + 5)$ **79.** $\frac{1}{5}(10y - 1)$ **81.** $44 - h$

83. $P(1 + rt)$

SECTION 6.2 *(page 263)*

Warm-Up *(page 263)*

1. $-4x + 24$ **2.** $18x + 48$ **3.** $y^2 + 2y$

4. $-a^3 + a^2$ **5.** $x^2 - 7x + 10$

6. $t^2 + 9t + 18$ **7.** $u^2 - 5u - 24$

8. $v^2 - 7v + 6$ **9.** $z^2 - 2z - 3$

10. $x^2 + 3x - 28$

1. $x + 1$

3. $a - 2$

5. $y - 5$

7. $z - 2$

9. $(x + 12)(x + 1)$ **11.** $(x + 4)(x + 2)$
$(x - 12)(x - 1)$
$(x + 6)(x + 2)$
$(x - 6)(x - 2)$
$(x + 4)(x + 3)$
$(x - 4)(x - 3)$

13. $(x - 8)(x - 5)$ **15.** $(x - 4)(x - 3)$

17. $(x - 3)(x + 2)$ **19.** $(x + 5)(x - 3)$

21. $(x + 10)(x - 7)$ **23.** $(x - 9)(x - 8)$

25. $(x + 15)(x + 4)$ **27.** $(x + 2y)(x - y)$

29. $(x + 5y)(x + 3y)$ **31.** $(a + 5b)(a - 3b)$

33. $(x - 9z)(x + 2z)$ **35.** $3(x + 5)(x + 2)$

37. $x(x - 10)(x - 3)$ **39.** $8(x - 3)(x + 1)$

41. $x(x + 3y)(x + 2y)$ **43.** $2xy(x + 3y)(x - y)$

45. ±16, ±8   **47.** ±11, ±4, ±1
49. ±37, ±20, ±15, ±13, ±12   **51.** 2, −4
53. 5, −7   **55.** 8, −10
57. $(x + 3)(x + 1)$

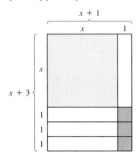

59. $(x + 3)(x + 2)$

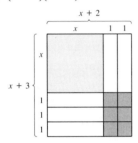

MATH MATTERS: PRIME NUMBERS   *(page 272)*

The first three prime numbers that are greater than or equal to 1,000 are 1,009, 1,013, and 1,019.

---

5. $2y − 9$
7. $4z − 1$
9. $(5x + 12)(x + 1)$ $(5x − 12)(x − 1)$
$(5x + 1)(x + 12)$ $(5x − 1)(x − 12)$
$(5x + 6)(x + 2)$ $(5x − 6)(x − 2)$
$(5x + 2)(x + 6)$ $(5x − 2)(x − 6)$
$(5x + 4)(x + 3)$ $(5x − 4)(x − 3)$
$(5x + 3)(x + 4)$ $(5x − 3)(x − 4)$
11. $(2x + 3)(x + 1)$ **13.** $(2x − 3)(x + 1)$
15. $(2y − 1)(y − 1)$ **17.** $(5a − 2)(3a + 4)$
19. $(3u − 2)(6u + 1)$ **21.** $(5t + 6)(2t − 3)$
23. Irreducible   **25.** $(5m − 3)(3m + 5)$
27. $(8z − 5)(2z − 3)$ **29.** Irreducible
31. $(3x + 1)(x + 2)$ **33.** $(2x + 3)(x − 1)$
35. $(3x + 4)(2x − 1)$ **37.** $(5x − 2)(3x − 1)$
39. $(8x − 3)(2x + 1)$ **41.** $−(2x − 3)(x + 1)$
43. $(1 + 6x)(1 − 10x)$ **45.** $−(6x + 5)(x − 2)$
47. $x(x − 3)$ **49.** $(u − 3)(u + 9)$
51. $(v + 7)(v − 6)$ **53.** $2(3x − 2)(x + 2)$
55. Irreducible   **57.** $y^2(3 − 2y)(5 + y)$
59. $v^2(u + 3)(u − 1)$ **61.** ±11, ±13, ±17, ±31
63. ±1, ±4, ±11   **65.** −1, −7, −10, −22
67. −13, 7, 8, −8, 3
69. $(2x + 1)(x + 2)$

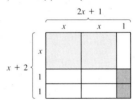

71. $l = 2x + 3$

SECTION 6.3   *(page 273)*

Warm-Up   *(page 273)*

1. $\frac{1}{3}x + \frac{5}{9} = \frac{1}{9}(3x + 5)$
2. $\frac{5}{8}x − \frac{3}{2} = \frac{1}{8}(5x − 12)$ **3.** $6(x + 2)$
4. $4(x − 3)$ **5.** $xy(x − y)$
6. $3x(2x^2 − x + 3)$ **7.** $(x + 7)(x − 6)$
8. $(x + 7)(x + 6)$ **9.** $x(x − 7)(x + 6)$
10. $2y(x + 8)(x − 4)$

---

1. $5x + 3$
3. $5a − 3$

SECTION 6.4   *(page 283)*

Warm-Up   *(page 283)*

1. $−12x^2 + 36x$ **2.** $−6z^2 + 15z$
3. $x^2 − 100$ **4.** $y^2 − a^2$ **5.** $x^2 + 10x + 25$
6. $9x^2 − 12x + 4$ **7.** $10(x^2 + 7)$
8. $4x(4 − x)$ **9.** $(3x − 2)(x − 1)$
10. $(2x + 1)(x − 1)$

---

1. $(x + 6)(x − 6)$ **3.** $\left(x + \frac{1}{2}\right)\left(x − \frac{1}{2}\right)$
5. $(4y + 3)(4y − 3)$ **7.** $(7 + 3y)(7 − 3y)$

9. $-z(10 + z)$　**11.** $2(x + 6)(x - 6)$
13. $2(2 + 5x)(2 - 5x)$　**15.** $(x^2 + 1)(x + 1)(x - 1)$
17. $(y^2 + 9)(y + 3)(y - 3)$　**19.** $(x - 2)^2$
21. $(z + 3)^2$　**23.** $(2t + 1)^2$　**25.** $(5y - 1)^2$
27. $\left(b + \frac{1}{2}\right)^2$　**29.** $(x - 3y)^2$　**31.** $(2y + 5z)^2$
33. ± 20　**35.** ± 2　**37.** ± 4　**39.** 9　**41.** 4
43. $(x - 2)(x^2 + 2x + 4)$　**45.** $(y + 4)(y^2 - 4y + 16)$
47. $(1 - 2t)(1 + 2t + 4t^2)$
49. $(2t - 1)(4t^2 + 2t + 1)$　**51.** $x^2(x - 4)$
53. $(x - 1)^2$　**55.** $(1 - 2x)^2$　**57.** $2x(2 - x)(1 + x)$
59. $(9x + 1)(x + 1)$　**61.** $(2x - 1)(4x + 1)$
63. $(1 + x^2)(5 - x)$　**65.** $x(x^2 + 1)(x - 4)$
67. $(7 + z)(3 - z)$　**69.** $2(t - 2)(t^2 + 2t + 4)$
71. $u(u^2 + 2u + 3)$　**73.** $(x^2 + 9)(x + 3)(x - 3)$
75. $(1 + x^2)(1 + x)(1 - x)$　**77.** 441　**79.** 3,599
81. $(x + 3)^2 + 1^2$　**83.** $(x + 10)$ in. $\times$ $(x + 10)$ in.

SECTION 6.5　*(page 291)*

Warm-Up　*(page 291)*

1. 12　**2.** -3　**3.** $\frac{4}{7}$　**4.** 6　**5.** $x(3x + 7)$
6. $(2x + 5)(2x - 5)$　**7.** $-(x - 15)(x - 7)$
8. $(x + 9)(x - 2)$　**9.** $(2x + 3)(5x - 1)$
10. $(6x - 1)(x - 12)$

1. 0, 5　**3.** 2, 3　**5.** $-1, 2$　**7.** $\frac{5}{2}, -\frac{1}{3}$
9. 0, 3, -25　**11.** 4, -4　**13.** 3, -3　**15.** 1, -1
17. $-2, 8$　**19.** 0, $-\frac{1}{2}$　**21.** 8, -2　**23.** 4, -2
25. 1　**27.** -7　**29.** $\frac{3}{2}$　**31.** 3, $-\frac{1}{2}$　**33.** $-\frac{5}{3}, 1$
35. $-2, 7$　**37.** $-\frac{3}{2}, 1$　**39.** 0, -5　**41.** 8, 2
43. 0, $-2, -3$　**45.** 0, $\frac{3}{2}, -4$　**47.** 3, $-3, 2$
49. 2, -2　**51.** 8, 9　**53.** $w = 9$ in.; $l = 12$ in.
55. 10 sec

57. (a) $V = lwh$　**(b)**
　　　$V = (x)(x)(2)$
　　　$V = 2x^2$

x	2	4	6	8
V	8	32	72	128

(c) 14 in. $\times$ 14 in.

59. $x(ax + b) = 0$
　　$x = 0, -\dfrac{b}{a}$

CHAPTER 6 REVIEW EXERCISES　*(page 295)*

1. $5x^2(1 + 2x)$　**3.** $4a(2 - 3a^2)$
5. $-6(x + 1)(3x - 1)$　**7.** $(a + 10)(a - 10)$
9. $(u + v + 2)(u + v - 2)$　**11.** $(x - 4)^2$
13. $(3s + 2)^2$　**15.** $2x + 5$　**17.** $3x + 2$
19. $x - 1$　**21.** $(x - 7)(x + 4)$
23. $(3x + 2)(2x + 1)$　**25.** $3u(2u + 5)(u - 2)$
27. $-4a(2a + 1)^2$　**29.** $(5x + 2y)(2x + y)$
31. $st(s + t)(s - t)$　**33.** $(2x - 1)(x - 1)$
35. $(3 - 2t)(9 + 6t + 4t^2)$　**37.** $(x^2 + 1)(x + 2)$
39. $(x + 2)(x - 2)(x - 4)$　**41.** $\pm 6, \pm 10$
43. $\pm 2, \pm 5, \pm 10, \pm 23$
45. $\pm 4, \pm 7, \pm 11, \pm 17, \pm 28, \pm 59$　**47.** 5, 8, 9
49. 2, $-6, -16, -30$　**51.** 0, $\frac{3}{2}$　**53.** 9, -9
55. 6　**57.** $-1, \frac{3}{4}$　**59.** 3, 4　**61.** 3　**63.** 12, 14

CHAPTER 6 TEST　*(page 297)*

1. $7x^2(1 - 2x)$　**2.** $(z + 7)(z - 3)$
3. $(t - 5)(t + 1)$　**4.** $(2x - 1)(3x - 4)$
5. $3y(y + 5)(2y + 5)$　**6.** $(2 + 5v)(2 - 5v)$
7. $(2x - 5)^2$　**8.** $-(z + 5)(z + 13)$
9. $(x + 2)(x + 3)(x - 3)$　**10.** $8x - 9$　**11.** $-4, \frac{3}{2}$
12. 0, 2　**13.** $\frac{2}{3}, -3$　**14.** 6, -6　**15.** 36
16. 7 in. by 12 in.　**17.** $\frac{5}{4}$ sec

Chapter 7

SECTION 7.1　*(page 305)*

Warm-Up　*(page 304)*

1. -9　**2.** 48　**3.** -21　**4.** -18　**5.** $\frac{2}{3}$
6. $\frac{3}{7}$　**7.** $\frac{6}{7}$　**8.** $\frac{7}{11}$　**9.** $x(x^2 - 3x + 4)$
10. $x^2(5x + 4)(2x - 1)$

1. $x \neq 4$　**3.** $x \neq -2$　**5.** All real x　**7.** $t \neq \pm 5$
9. $x \neq 2, -1$　**11.** $3x$　**13.** $(x + 1)$　**15.** $(x + 2)$
17. $(x - 2)$　**19.** $(x + 2)$　**21.** $\dfrac{x}{3}$　**23.** $2y, \ y \neq 0$
25. $\dfrac{3x}{2}, \ x \neq 0$　**27.** $x, \ x \neq 0, x \neq -1$
29. $\frac{1}{2}, \ x \neq 5$　**31.** $-\frac{1}{2}, \ x \neq 5$　**33.** $\dfrac{3y}{y + 1}, \ x \neq 0$

35. $\dfrac{y-4}{3}$, $y \neq -4$ **37.** $\dfrac{1}{a+2}$ **39.** $\dfrac{x}{x-5}$

41. $\dfrac{y+2}{y+5}$, $y \neq 2$ **43.** $-\dfrac{1}{x-2}$, $x \neq 3$

45. $\dfrac{x-2}{x+1}$, $x \neq -10$ **47.** $\dfrac{x(x+3)}{x-2}$, $x \neq -2$

49. $x^2 + 1$, $x \neq 2$

51. (a) The average cost per unit equals $\dfrac{3{,}000 + 7.50x}{x}$.

(b) Domain $= (1, 2, 3, \ldots)$ **(c)** \$37.50
53. $0 \leq p < 1$

SECTION 7.2   *(page 312)*

Warm-Up   *(page 311)*

1. $\frac{1}{4}$ **2.** $\frac{3}{8}$ **3.** $\frac{10}{63}$ **4.** $-\frac{15}{112}$ **5.** $\frac{15}{2}$
6. $-\frac{7}{30}$ **7.** $x(2x+1)(2x-1)$
8. $(3x+2)(3x-2)$ **9.** $(5x-7)(3x+2)$
10. $(2x-7)^2$

---

1. x^2 **3.** ab **5.** $(2-x)$ **7.** $\dfrac{3x}{2}$, $x \neq 0$

9. 3, $x \neq 0, y \neq 0$ **11.** 32, $x \neq -\frac{2}{3}$
13. 1, $a \neq -2$ **15.** -1, $r \neq 1$

17. $\dfrac{1}{5(x-2)}$, $x \neq 1$ **19.** $-\dfrac{x(x+7)}{(x+1)}$, $x \neq 9$

21. $\dfrac{r+1}{r}$, $r \neq 1$ **23.** $\dfrac{(t-3)}{(t+3)(t-2)}$, $t \neq -2$

25. $\dfrac{xy(x+2y)}{(x-2y)}$ **27.** $\dfrac{(x-y)}{x(x+y)^2}$, $x \neq -2y$

29. $3(a+1)^2(a-1)$, $a \neq 0, a \neq 1$ **31.** $\dfrac{3}{4x}$

33. $\frac{3}{2}$, $x \neq -4$
35. $xy(x+y)$, $x \neq 0, y \neq 0, x \neq -y$
37. $2x^2$, $x \neq 0$ **39.** $\frac{1}{6}$, $a \neq -5$

41. $\dfrac{2x+3}{2x-3}$, $x \neq \pm 2$

43. $xy(x+y)$, $x \neq 0, y \neq 0, x \neq -y$

45. $3x(x+y)$, $x \neq 0$ **47.** $\dfrac{8}{45x}$

49. $(x+2)(x+1)$, $x \neq -1, x \neq -2$ **51. (a)** 0
(b) Division by zero   **(c)** $\frac{2}{3}$ **(d)** $\frac{1}{18}$

53. (a) $\dfrac{1}{12}$ min   **(b)** $\dfrac{x}{12}$ min   **(c)** $\dfrac{8}{3}$ min

55. $\dfrac{3xy}{8}$

SECTION 7.3   *(page 323)*

Warm-Up   *(page 322)*

1. 1 **2.** $\frac{1}{4}$ **3.** $\frac{3}{5}$ **4.** $-\frac{1}{2}$ **5.** $\frac{9}{14}$ **6.** $\frac{19}{48}$
7. $-\frac{4}{75}$ **8.** $\frac{146}{35}$ **9.** $\frac{11}{24}$ **10.** $-\frac{7}{18}$

---

1. $\dfrac{5}{x}$ **3.** y **5.** $\dfrac{14}{3a}$ **7.** $\frac{1}{3}$ **9.** 1 **11.** $\dfrac{x+5}{x-1}$

13. $\dfrac{y+1}{y-1}$ **15.** $2x^3$ **17.** $3x(x+5)$

19. $x(x^2-4)$ **21.** $\dfrac{x+5}{3x-6}, \dfrac{30}{3x-6}$

23. $\dfrac{2(x + 3)}{x^2(x + 3)}, \dfrac{5x}{x^2(x + 3)}$

25. $\dfrac{(x - 8)(x - 4)}{(x - 4)^2(x - 4)}, \dfrac{9x(x + 4)}{(x - 4)^2(x + 4)}$ **27.** $\dfrac{1 - 3x}{5x}$

29. $\dfrac{5z + 6}{z^2}$ **31.** $0, \; x \ne 3$ **33.** $\dfrac{2x + 5}{x - 5}$

35. $\dfrac{6x + 13}{x + 3}$ **37.** $\dfrac{3}{(x + 2)(x - 1)}$ **39.** $\dfrac{x + 2}{x(x - 4)}$

41. $\dfrac{4x - 9}{(x + 3)(x - 3)}$ **43.** $\dfrac{7v + 8}{v(v + 4)}$

45. $-\dfrac{x^2 - 2x + 1}{x(x^2 + 1)}$ **47.** $\dfrac{x - 4}{(x + 2)(x - 1)(x - 2)}$

49. $\dfrac{3x}{(x - 3)^2}$ **51.** $\dfrac{x^2 - 1}{x(x^2 + 1)}$ **53.** $\dfrac{4x^2 + 2x - 1}{x^2(x + 1)}$

55. $\dfrac{x^3 + 3x + 6}{(x + 1)(x - 1)}$ **57.** $\dfrac{y - x}{xy}$ **59.** $\dfrac{5u - 2v}{(u - v)^2}$

61. $\dfrac{x}{2}, \; x \ne 0$ **63.** $\dfrac{y + 3}{y^2}$ **65.** $\dfrac{1}{y}, \; x \ne y$

67. $\dfrac{z^2 - 4}{1 - 4z}, \; z \ne 0$ **69.** $\dfrac{1}{x}, \; x \ne -1$ **71.** $\dfrac{15t}{56}$ hr

73. $\dfrac{11x}{60}$

75. $\overbrace{\dfrac{8x}{72}}^{x_1} \; \overbrace{\dfrac{9x}{72}}^{x_2} \; \overbrace{\dfrac{10x}{72}}^{x_3} \; \dfrac{11x}{72} \; \dfrac{12x}{72}$

77. $\dfrac{R_1 R_2}{R_1 + R_2}$

SECTION 7.4 *(page 333)*

Warm-Up *(page 333)*

1. $\frac{8}{5}$ **2.** $-\frac{15}{2}$ **3.** 2 **4.** $\frac{1}{8}$ **5.** $-10, 4$
6. $-\frac{3}{2}, 9$ **7.** 0, 8 **8.** 4 **9.** $-8, 7$
10. 0, 5

1. (a) Not a solution **(b)** Solution
(c) Not a solution **(d)** Solution **3. (a)** Solution
(b) Not a solution **(c)** Not a solution
(d) Not a solution **5.** 2 **7.** $\frac{2}{5}$ **9.** $\frac{32}{5}$ **11.** 50
13. 5 **15.** $\frac{1}{4}$ **17.** 13 **19.** 12 **21.** $\frac{3}{4}$ **23.** 4
25. 7 **27.** 2 **29.** 3 **31.** No solution
33. No solution **35.** 5 **37.** ± 3 **39.** ± 5
41. -2 **43.** $-2, \frac{3}{2}$ **45.** $-5, 4$ **47.** $\frac{1}{2}, 2$
49. $-1, 3$ **51.** 3 **53.** $66\frac{2}{3}$ mi/hr **55.** 12 partners

57. 75%
59.

Person #1	Person #2	Together
4 days	4 days	2 days
4 hours	6 hours	$\frac{12}{5}$ hours
4 hours	$2\frac{1}{2}$ hours	$\frac{20}{13}$ hours

61. 50 mi/hr; 60 mi/hr **63.** 10 consecutive hits
65. Truck: 50 mi/hr, Car: 60 mi/hr

CHAPTER 7 REVIEW EXERCISES *(page 337)*

1. All real numbers except $x = 5$
3. All real numbers except $t = 1$ and $t = 2$

5. $\dfrac{x}{3y}, \; x \ne 0$ **7.** $\frac{3}{4}, \; b \ne 2$ **9.** $-4, \; x \ne y$

11. $\dfrac{2y}{y + 3}$ **13.** $\dfrac{3x^3}{4y^2}$ **15.** $\dfrac{z(z - 1)}{5}, \; z \ne -1$

17. $-\frac{1}{3}, \; v \ne 0, v \ne 1$ **19.** $4x, \; x \ne 0$ **21.** $\dfrac{3x^2 y}{4}$

23. $50y, \; y \ne 0$ **25.** $\dfrac{u}{u - 3}, \; u \ne 0, u \ne -3$

27. $\dfrac{x(x + 1)}{x - 8}, \; x \ne 1, x \ne -1$

29. $\dfrac{(x + 1)(y - 1)}{y + 1}, \; x \ne 0, y \ne 0, y \ne 1$ **31.** $\frac{1}{4}$

33. $\dfrac{x + 4}{x + 2}$ **35.** $\frac{5}{48}$ **37.** $-\dfrac{1}{(x + 1)(x + 2)}$

39. $\dfrac{x^3 - x + 3}{(x + 2)(x - 1)}$ **41.** $\dfrac{x + 1}{x(x^2 + 1)}$

43. $\dfrac{2x^2 - 3x + 2}{(x + 2)(x - 2)^2}$

45. $\dfrac{x(x + 1)}{(x + 2)}, \; x \ne 0, x \ne 1, x \ne -1$

47. $\dfrac{x^2}{x - 1}, \; x \ne 0$ **49.** $-\dfrac{1}{xy(x + y)}, \; x \ne y$

51. -8 **53.** 5 **55.** $\frac{1}{2}$ **57.** $3, -4$ **59.** $2, -6$
61. 54 mi/hr **63.** 4 farmers **65.** $\frac{40}{9}$ min

CHAPTER 7 TEST *(page 339)*

1. All real numbers except $x = 10$ **2.** $x(x + 1)$

3. $\dfrac{8x}{x + 1}, \; x \ne 0$ **4.** $\dfrac{x + 8}{x + 5}, \; x \ne 8$ **5.** $\dfrac{18}{x^2}$

6. $\dfrac{(x + 2)(x - 2)}{x^2}, \; x \ne -2$ **7.** $\dfrac{5}{6x}$

8. $-\dfrac{1}{t}$, $t \neq 5$ **9.** $\dfrac{x^3}{(x-3)^8}$, $x \neq 0$

10. $36x^3(x+3)^2$ **11.** $\dfrac{9u+8}{3u^2}$ **12.** $\dfrac{-3(2x+3)}{x+2}$

13. $\dfrac{2}{(x+1)^2}$ **14.** $\dfrac{2x}{4x+1}$, $x \neq 0$

15. (a) Not a solution   **(b)** Solution
(c) Not a solution   **(d)** Solution   **16.** $\dfrac{9}{2}$ **17.** 4

18. $\dfrac{1}{2}$, 2   **19. (a)** $\dfrac{1}{80}$ min   **(b)** $\dfrac{x}{80}$ min   **(c)** $\dfrac{1}{5}$ min

20. 6 partners

CUMULATIVE TEST: CHAPTERS 5–7 *(page 340)*

1. $-5x^2 + 5$ **2.** $-42z^4$ **3.** $3x^2 - 7x - 20$
4. $25x^2 - 9$ **5.** $25x^2 + 60x + 36$ **6.** $x + 12$

7. $x + 1 + \dfrac{2}{x-4}$ **8.** $\dfrac{81}{16}$ **9.** $\left(\dfrac{2}{x}\right)^2$

10. $2u(u-3)$ **11.** $(x+2)(x-6)$ **12.** $x(x+4)^2$

13. $(x-2)(x+2)^2$ **14.** $\dfrac{5}{x+5}$, $x \neq 5$

15. $\dfrac{c+10}{c^2}$, $c \neq 1$ **16.** $\dfrac{3c^2}{4(c-1)}$, $c \neq 0$

17. $\dfrac{2(x+3)}{x^2-4}$ **18.** $\dfrac{2(a^2-1)}{a+2}$, $a \neq 0$ **19.** $-\dfrac{3}{5}$, 3

20. $\dfrac{2}{5}$ **21.** 50 mi/hr
22. New employee: 9 hrs, Experienced employee: 4.5 hrs

Chapter 8

SECTION 8.1 *(page 350)*

Warm-Up *(page 349)*

1. $-4 < 3$ **2.** $\dfrac{8}{3} > 2$

3. $-2 > -6$ **4.** $-8 < 0$

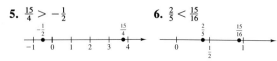

5. $\dfrac{15}{4} > -\dfrac{1}{2}$ **6.** $\dfrac{2}{5} < \dfrac{15}{16}$

7. 1   **8.** 4   **9.** $\dfrac{9}{2}$ **10.** 5

1.   **3.**

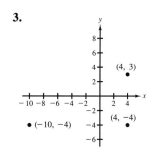

5.

7. (a) Quadrant II   **(b)** Quadrant IV
9. A: (5, 2), B: (−3, 4), C: (2, −5), D: (−2, −2)

11.   **13.**

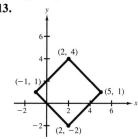

15.

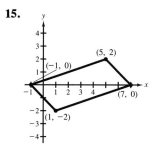

17. Quadrants II or III   **19.** Quadrants II or IV
21.

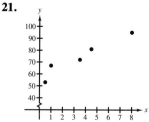

23.

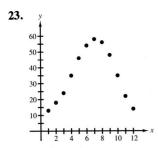

25. **(a)** Solution point   **(b)** Not a solution point
(c) Not a solution point   **(d)** Solution point
27. **(a)** Solution point   **(b)** Solution point
(c) Not a solution point   **(d)** Solution point
29. **(a)** Not a solution point   **(b)** Solution point
(c) Solution point   **(d)** Not a solution point
31.

x	-2	0	2	4	6
$y = 3x - 4$	-10	-4	2	8	14

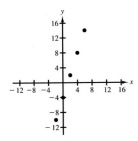

33.

x	-5	$\frac{3}{2}$	5	10	20
$y = -\frac{3}{2}x + 5$	$\frac{25}{2}$	$\frac{11}{4}$	$-\frac{5}{2}$	-10	-25

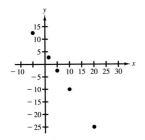

35.

x	$y = 35x + 5,000$
100	8,500
150	10,250
200	12,000
250	13,750
300	15,500

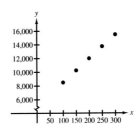

37. The point reflects about the y-axis.

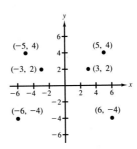

39. 1,100   **41.** 450   **43.** $2,700   **45.** $850
47. 35%

SECTION 8.2   *(page 365)*

Warm-Up   *(page 364)*

1.   **2.**

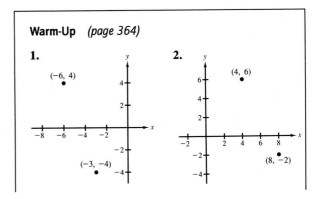

3.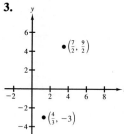

4.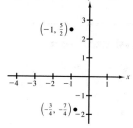

5. $\frac{42}{5}$ **6.** 24   **7.** $y = 14$ **8.** $y = -7$
9. $x = 12$ **10.** $x = 4$

1.

x	y	(x, y)
-2	11	$(-2, 11)$
-1	10	$(-1, 10)$
0	9	$(0, 9)$
1	8	$(1, 8)$
2	7	$(2, 7)$

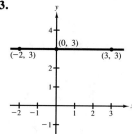

3.

x	y	(x, y)
-2	3	$(-2, 3)$
0	2	$(0, 2)$
2	1	$(2, 1)$
4	0	$(4, 0)$
6	-1	$(6, -1)$

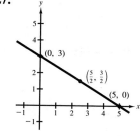

5. $(1, 0), (0, -1)$ **7.** $(2, 0), (0, 4)$
9. $(-40, 0), (0, 15)$ **11.** $(4, 0), (-4, 0), (0, -16)$
13. $(0, -4)$ **15.** $(0, 3)$ **17.** $\left(-\frac{1}{2}, 0\right), (1, 0), (0, -1)$

19.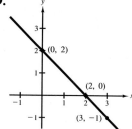

21.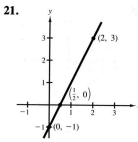

23.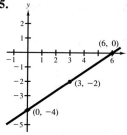

25.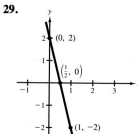

27.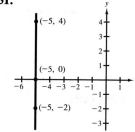

29.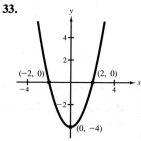

31.

33.

35.

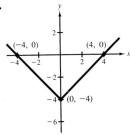

37. y is a function of x.   **39.** y is a function of x.
41. y is *not* a function of x.

43. y is a function of x.   **45.** y is a function of x.

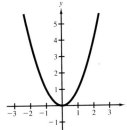

47. y is *not* a function of x.   **49.** $y = 35t$

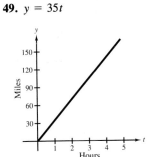

MATH MATTERS: THE SPEEDS OF ANIMALS
(page 372)

A speed of one mile per four minutes is equal to

$$\frac{1 \text{ mile}}{4 \text{ minutes}} = \frac{1 \text{ mile}}{4 \text{ minutes}} \cdot \frac{60 \text{ minutes}}{1 \text{ hour}} = \frac{15 \text{ miles}}{1 \text{ hour}}.$$

SECTION 8.3   *(page 375)*

Warm-Up   *(page 375)*

1. $y = 4 - 3x$ **2.** $y = x$ **3.** $y = x + 5$
4. $y = x + 4$ **5.** $y = 5 - x$ **6.** $y = x - 5$
7. $y = \frac{2}{3} - \frac{2}{3}x$ **8.** $y = \frac{4}{5}x + \frac{2}{5}$
9. $y = \frac{5}{4} - \frac{3}{4}x$ **10.** $y = 2 - \frac{2}{3}x$

1.

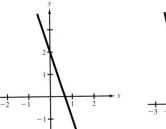

3.

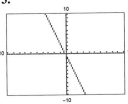

5.

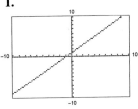

7.

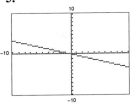

9.

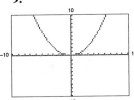

11.

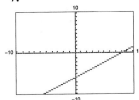

13.

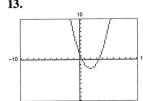

15.

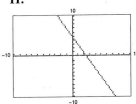

17.

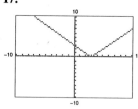

19.

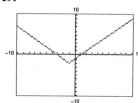

21. $y = \frac{3}{2}x - 2$

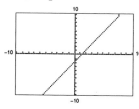

23. $y = -\frac{1}{2}x^2 + 2$

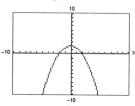

25.

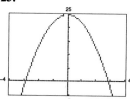

27.

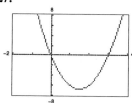

29.

```
RANGE
Xmin = -15
Xmax = 15
Xscl = 1
Ymin = -10
Ymax = 10
Yscl = 1
Xres = 1
```

31.

```
RANGE
Xmin = -5
Xmax = 20
Xscl = 5
Ymin = -5
Ymax = 20
Yscl = 5
Xres = 1
```

33. 2 x-intercepts **35.** (b) **36.** (c) **37.** (d)
38. (a)

39. Triangle

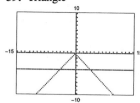

41. Square

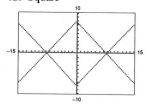

43. $(-0.3, 0.0)$
$(3.3, -0.1)$
(Rounded to one
decimal place)

45.

SECTION 8.4 *(page 389)*

Warm-Up *(page 388)*

1. $\frac{1}{2}$ **2.** $\frac{1}{3}$ **3.** $-\frac{9}{2}$ **4.** $-\frac{13}{3}$ **5.** $\frac{4}{5}$
6. $-\frac{8}{7}$ **7.** $y = \frac{1}{3}(2x - 5)$ **8.** $y = -2x$
9. $y = 3x - 1$ **10.** $y = \frac{2}{3}x + 5$

1. 1 **3.** 0 **5.** $-\frac{1}{3}$ **7.** (a) L_2 (b) L_3
(c) L_1

9. $m = \frac{5}{4}$; rises

11. $m = 2$; rises

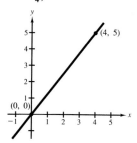
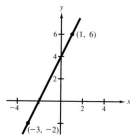

13. $m = -\frac{3}{4}$; falls

15. Undefined slope; vertical

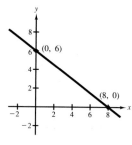
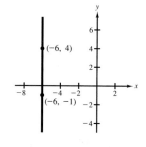

17. $m = 0$; horizontal

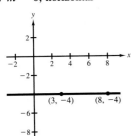

19. $m = -\frac{18}{17}$; falls

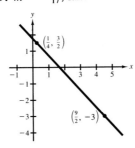

45.

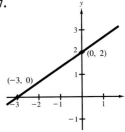

47.

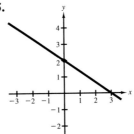

21. $m = -\frac{25}{32}$; falls

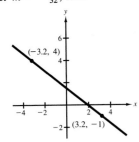

23. $m = \frac{7}{3}$; rises

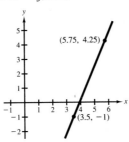

49.

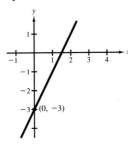

51. $y = 2x - 3$

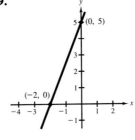

25. $m = \dfrac{0}{a - 4} = 0$

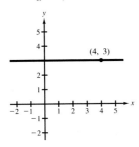

53. $y = -x$

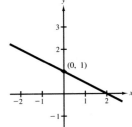

55. $y = -\frac{1}{2}x + 1$

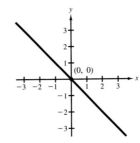

27. $x = -2$ **29.** $y = -\frac{25}{4}$ **31.** $(0, 1), (1, 1)$
33. $(2, -4), (3, -2)$ **35.** $(1, -1), (2, -3)$
37. $(-1, 2), (2, 4)$ **39.** $(-8, 0), (-8, -1)$

57. $y = \frac{3}{4}x + \frac{1}{2}$

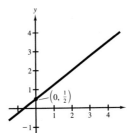

59. $y = 3$

41.

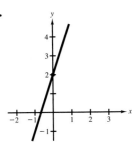

43.

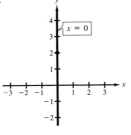

61. Perpendicular   **63.** Parallel

65. Parallel

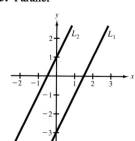

67. Perpendicular

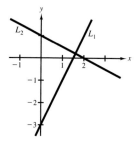

69.

Year	83	84	85	86	87	88	89
Slope	0.05	0.07	0.18	0.18	0.26	0.35	

(a) 83–84   **(b)** 88–89
71. The slope

SECTION 8.5   *(page 401)*

Warm-Up   *(page 400)*

1. $-\frac{1}{4}$ **2.** $\frac{15}{14}$ **3.** Undefined   **4.** $\frac{3}{2}$
5.   **6.**

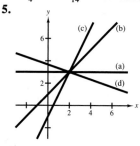   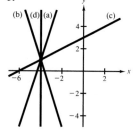

7. $y = -\frac{2}{5}x + \frac{48}{5}$ **8.** $y = \frac{4}{3}x - \frac{29}{3}$
9. $y = -2x + 7$ **10.** $y = x + 3$

---

1. 5   **3.** $\frac{2}{3}$ **5.** $\frac{3}{8}$ **7.** 2   **9.** $\frac{3}{2}$
11. $y = -2x$ **13.** $y = 3x - 2$

15. $y = -6x + \frac{3}{2}$

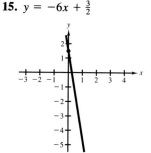

17. $x - y = 0$

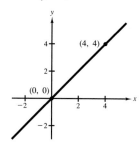

19. $3x + 4y - 18 = 0$

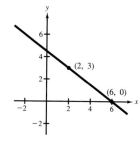

21. $x - 3y + 12 = 0$

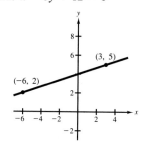

23. $y - 3 = 0$

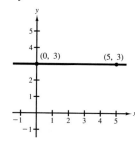

25. $4x - y - 11 = 0$

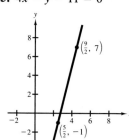

27. $x - 2y - 8 = 0$ **29.** $2x + y = 0$
31. $x + 3y - 4 = 0$ **33.** $x - 6 = 0$
35. $8x - 6y - 17 = 0$ **37.** $3x + 5y - 10 = 0$
39. $x + y - 5 = 0$ **41.** $8x - 7y - 12 = 0$
43. $x + 2y - 3 = 0$ **45.** $x + 8 = 0$
47. $6x + 5y - 9 = 0$
49. (a) $2x - y - 3 = 0$ **(b)** $x + 2y - 4 = 0$
51. (a) $3x + 4y + 2 = 0$ **(b)** $4x - 3y + 36 = 0$
53. (a) $y = 0$ **(b)** $x + 1 = 0$
55. (a) ii; $m = -10$; $\frac{-\$10.00}{1 \text{ week}}$ **(b)** iii; $m = 1.5$; $\frac{\$1.50}{1 \text{ unit}}$
(c) i; $m = 0.25$; $\frac{\$0.25}{1 \text{ mile}}$ **(d)** iv; $m = -100$; $\frac{-\$100}{1 \text{ year}}$
57. $W = 2,000 + 0.02S$ **59.** $S = L - 0.2L = 0.8L$
61. $y = 70 + 4t$ **63. (a)** $x = 82 - \frac{1}{15}p$
(b) 45 units   **(c)** 49 units

65. Perpendicular

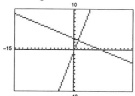

67.

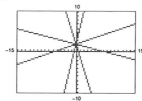

SECTION 8.6 *(page 410)*

Warm-Up *(page 409)*

1.

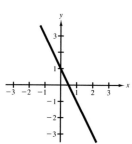

2.

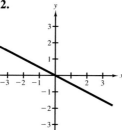

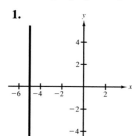

4.

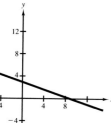

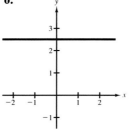

6.

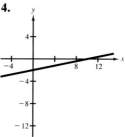

7. (a) Not a solution point **(b)** Solution point
9. (a) Not a solution point **(b)** Solution point

8. (a) Solution point **(b)** Not a solution point
10. (a) Solution point **(b)** Not a solution point

---

1. (a) Not a solution point **(b)** Solution point
(c) Solution point **(d)** Not a solution point
3. (a) Solution point **(b)** Solution point
(c) Solution point **(d)** Not a solution point
5. Dashed **7.** Solid **9.** (d) **10.** (e) **11.** (a)
12. (f) **13.** (c) **14.** (b)
15. $y \geq 3$ **17.** $x > \frac{3}{2}$

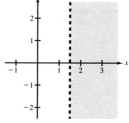 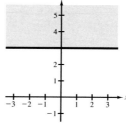

19. $y > x$ **21.** $y \leq x - 2$

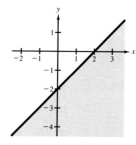 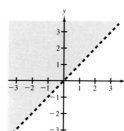

23. $y > x - 2$ **25.** $y \geq \frac{2}{3}x + \frac{1}{3}$

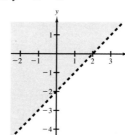 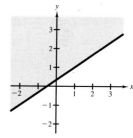

27. $y > -2x + 10$ **29.** $y < \frac{3}{2}x + 3$

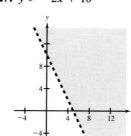 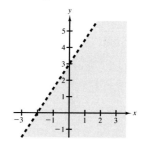

31. $y \geq -2x + 6$

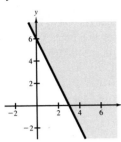

33. $y < -\frac{5}{2}x + \frac{5}{2}$

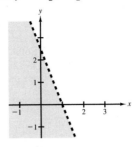

35. $y \leq \frac{1}{3}x + \frac{5}{3}$

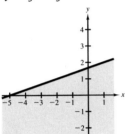

37. $y < \frac{1}{2}x + 1$

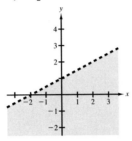

39. $y < -\frac{4}{3}x + 4$

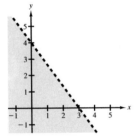

41. $y \geq -\frac{2}{3}x + 2$ **43.** $y > \frac{2}{3}x + 2$ **45.** $y \leq 2x$

47. $6x + 4y \geq 120$
   x: Grocery store hours
   y: Hours spent mowing lawns
   (5, 30), (20, 10), (10, 15)

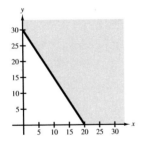

49. $T + \frac{3}{2}C \leq 12$
   T: tables
   C: chairs
   (*T, C*): (5, 4), (2, 6), (0, 8)

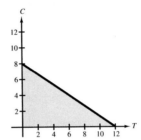

51.

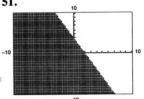

53.

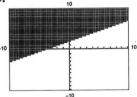

CHAPTER 8 REVIEW EXERCISES   *(page 414)*

1.

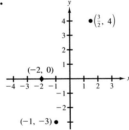

3. *y* is *not* a function of *x*. (−4, 0), (0, 2), (0, −2)
5. *y* is a function of *x*. (0, 0), (4, 0)
7. (0, 4), (8, 0)   **9.** (0, 3), $\left(-\frac{3}{2}, 0\right)$

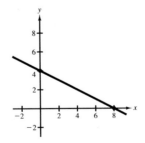

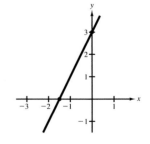

11. $(3, 0)$, $(-3, 0)$, $(0, 3)$ **13.** $(0, 1)$

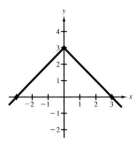 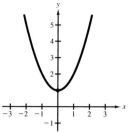

15. $(4, 0)$, $(0, 16)$

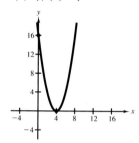

17. $\frac{5}{12}$ **19.** $\frac{2}{7}$ **21.** Undefined   **23.** $-\frac{4}{3}$ **25.** $\frac{7}{2}$
27. Two sides are perpendicular

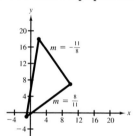

29. Opposite sides are parallel

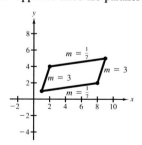

31. $(1, 3)$, $(2, 1)$ **33.** $(6, 6)$, $(10, 9)$
35. $2x - y - 9 = 0$ **37.** $4x + y - 6 = 0$
39. $4x - 5y - 6 = 0$ **41.** $8x + 3y - 37 = 0$
43. $x - 3 = 0$ **45.** $x + 2y + 4 = 0$ **47.** $y - 8 = 0$

49. $12x - 42y - 1 = 0$
51. (a) $2x + 3y - 7 = 0$ **(b)** $3x - 2y + 9 = 0$
53. (a) $8x + 6y - 29 = 0$ **(b)** $24x - 32y + 113 = 0$
55. $x \geq 2$ **57.** $y < -2x + 1$

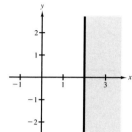 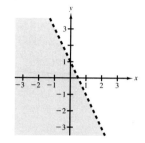

59. $y \geq \frac{1}{4}x + \frac{1}{2}$

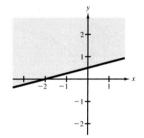

61. (a) 16 ft/sec   **(b)** 1.5 sec   **(c)** -16 ft/sec
63. $A = x(12 - x)$

CHAPTER 8 TEST *(page 417)*
1.

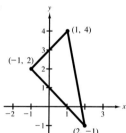

2. Quadrants I or III

3.

x	$y = 0.75x + 4$
2	$ 5.50
4	7.00
6	8.50
8	10.00
10	11.50
12	13.00

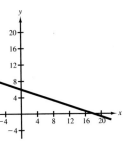

4. $(-4, 0), (0, 3)$ **5.** $(-1, 0), (2, 0), (0, -2)$
6.   **7.**

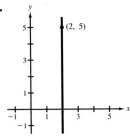

8. 4   **9.** $(-2, 2), (-1, 0)$
10.

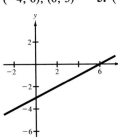

11. $\frac{3}{14}$ **12.** $3x + 8y - 48 = 0$
13. $5x - 4y - 6 = 0$ **14.** $-\frac{5}{3}$ **15. (a)** Solution
(b) Solution   **(c)** Solution   **(d)** Solution
16.   **17.**

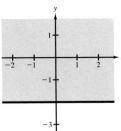

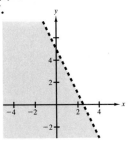

18. $V = -800t + 4{,}200 = \$1{,}800$

Chapter 9
SECTION 9.1 *(page 427)*

Warm-Up *(page 426)*

1.

2.

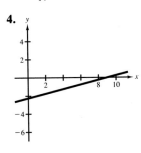

3.

4.

5.

6.

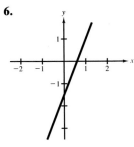

7. $x - y + 4 = 0$ **8.** $5x + 3y - 28 = 0$
9. $-\frac{1}{2}$ **10.** $\frac{7}{4}$

1. $(2, 3)$ **3.** $(-1, 2)$ **5.** Single solution
7. No solutions   **9.** Single solution   **11.** $(2, 0)$
13. $(3, 1)$ **15.** $(5, 4)$ **17.** No solutions
19. $(2, -1)$ **21.** $(7, -2)$

23. Infinitely many solutions   **25.** $(-10, 2)$
27. $\left(\frac{1}{2}, 3\right)$ **29.** No solutions
31. $y = \frac{2}{3}x + 4$ **33.** $y = \frac{2}{3}x + \frac{4}{3}$
$y = \frac{2}{3}x - 1$ $y = -\frac{2}{3}x + \frac{8}{3}$
No solutions   Single solution
35. $y = \frac{1}{4}x + \frac{7}{4}$, Infinitely many solutions
37. Slopes of the two lines are nearly the same.
39. $x = 12, y = 6$ **41.** $(1, 2)$ **43.** $(-6, 7)$

SECTION 9.2   *(page 436)*

Warm-Up   *(page 435)*

1. No solution   **2.** $\frac{1}{3}$ **3.** $\frac{5}{11}$ **4.** $\frac{14}{11}$
5. $-\frac{1}{6}$ **6.** 0   **7.** $(-2, 0)$ **8.** $(0, 3)$
9. $(3, 2)$ **10.** $(4, 6)$

1. $(1, 1)$ **3.** $(1, 2)$ **5.** No solutions
7. Infinitely many solutions   **9.** $\left(\frac{3}{2}, 1\right)$ **11.** $(1, -1)$
13. $(0, 0)$ **15.** $(5, 5)$ **17.** $\left(\frac{1}{2}, 3\right)$ **19.** $\left(\frac{20}{3}, \frac{40}{3}\right)$
21. No solutions   **23.** Infinitely many solutions
25. $\left(\frac{5}{2}, -\frac{1}{2}\right)$ **27.** $\left(\frac{5}{2}, -\frac{3}{2}\right)$ **29.** $(10, 4)$
31. $x - 2y = 0, x + y = 3$
33. $2x - y = 10, 4x + 3y = 5$
35. You end with a contradictory equation that shows the equality of two unequal real numbers.
37. $b = 2$ **39.** $b = -\frac{1}{3}$ **41.** 4 adults

MATH MATTERS: MAGIC SQUARES   *(page 441)*

Two possible completed magic squares are as follows.

16	3	2	13
5	10	11	8
9	6	7	12
4	15	14	1

17	24	1	8	15
23	5	7	14	16
10	12	19	2	22
4	6	25	21	9
11	18	13	20	3

SECTION 9.3   *(page 445)*

Warm-Up   *(page 444)*

1. x **2.** $-37v$ **3.** $2x^2 + 9$ **4.** -1
5. 6   **6.** 6   **7.** Perpendicular
8. Perpendicular   **9.** Parallel   **10.** Neither

1. $(2, 0)$ **3.** $(-1, -1)$ **5.** No solutions
7. Infinitely many solutions   **9.** $\left(-\frac{1}{3}, -\frac{2}{3}\right)$
11. $(8, 4)$ **13.** $(2, 1)$ **15.** $(5, 1)$ **17.** $(3, -4)$
19. $(4, -1)$ **21.** No solutions   **23.** $(40, 40)$
25. $\left(\frac{13}{3}, -2\right)$ **27.** $(8, 7)$ **29.** $\left(6, \frac{3}{2}\right)$ **31.** $(8, 4)$
33. $(2, 3)$ **35.** $(4, 4)$ **37.** $(-5, -3)$
39. $x + 3y = 10, x - 3y = 2$ **41.** 96, 58
43. Student tickets: \$3.00, General admission: \$5.00

SECTION 9.4   *(page 456)*

Warm-Up   *(page 455)*

1. $0.05m + 0.10n$ **2.** $0.13I$ **3.** $\dfrac{250}{r}$
4. $0.15L$ **5.** $3l$ **6.** $3.5n$ **7.** $(15, 10)$
8. $\left(-2, -\frac{8}{3}\right)$ **9.** $(28, 4)$ **10.** $(4, 3)$

1. 42, 25   **3.** 40, 30   **5.** 90, 42
7. 8 dimes, 13 quarters   **9.** 15 nickels, 20 quarters
11. 28 nickels, 16 dimes   **13.** 8 ft by 12 ft
15. 3 yd by 5 yd   **17.** 8 m by 9.6 m   **19.** \$75
21. \$51.40   **23.** \$4,000 at 10.5%, \$8,000 at 12%
25. 375 adult, 125 children
27. Unleaded: \$1.34/gal, Premium: \$1.54/gal
29. 3 lb of \$4.25 nuts, 7 lb of \$6.55 nuts
31. 4L of 35% solution, 6L of 60% solution   **33.** 1 hour
35. 550 mi/hr, 600 mi/hr   **37.** 48
39. (a) $y = x + \frac{1}{3}$ **(b)**

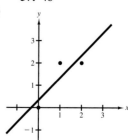

CHAPTER 9 REVIEW EXERCISES *(page 459)*

1. $(1, 1)$ **3.** No solutions **5.** $(4, 8)$ **7.** $\left(\frac{5}{2}, 3\right)$

9. $(10, -12)$ **11.** $\left(\frac{3}{5}, \frac{1}{2}\right)$ **13.** $(5, 6)$

15. $(-1, -1)$ **17.** Infinitely many solutions

19. $\left(4, -\frac{3}{2}\right)$ **21.** $(-5, 6)$ **23.** $(4, 2)$ **25.** $(0, 0)$

27. $\left(-\frac{1}{2}, \frac{4}{5}\right)$ **29.** $(-3, 7)$ **31.** No solutions

33. $(3, 2)$ **35.** $\left(\frac{1}{2}, -\frac{3}{4}\right)$

37. $x + 3y = 11,\ 3x - 3y = 1$

39. $\dfrac{2}{x - 1} - \dfrac{1}{x + 2} = \dfrac{2x + 4 - x + 1}{(x - 1)(x + 2)} = \dfrac{x + 5}{x^2 + x - 2}$

41. 24 inches by 36 inches

43. \$1.45/gal: gasoline, \$1.37/gal: diesel fuel

45. (a) 9 **(b)** 1

(c) No. By solving the equations for y, it can be seen that the slopes are equal but the y-intercepts are not equal. Therefore, the two equations are not equivalent.

29. 6 **31.** Irrational **33.** Rational **35.** 6.557

37. 25.140 **39.** -22.755 **41.** 3.979

43. -56.921 **45.** 6.377 **47.** 7.854 **49.** -0.687

51.

x	$\sqrt{x}$	x	$\sqrt{x}$
0	0	10	3.16
1	1	12	3.46
2	1.41	14	3.74
4	2	16	4
6	2.45	18	4.24
8	2.83	20	4.47

53. 7.416 **55.** 12.247 **57. (a)** 8.2 **(b)** 142

(c) 22 **(d)** 850 **59.** 0, 1, 4, 5, 6, 9

61. 23 ft by 23 ft

CHAPTER 9 TEST *(page 461)*

1. (a) Not a solution **(b)** Solution **2.** One

3. Zero **4.** $(2, 2)$ **5.** $(1, 2)$ **6.** $(5, 1)$

7. $(2, 6)$ **8.** $\left(3, \frac{5}{2}\right)$ **9.** -6

10. $x + 2y = 5$
$\quad\ \ x - \ \ y = -7$

11. $2l + 2w = 40$
$\quad\ \ l - 3w = 0$
$\quad\quad$ 15m by 5m

12. $\quad\quad\quad x + y = 20$
$\quad\ 0.30x + 0.05y = 0.20(20)$
$\quad$ 12L 30% solution
$\quad$ 8L 5% solution

Chapter 10

SECTION 10.1 *(page 469)*

Warm-Up *(page 469)*

1. 225 **2.** 144 **3.** 25 **4.** -25 **5.** $\frac{4}{9}$

6. $\frac{16}{25}$ **7.** $\frac{4}{3}$ **8.** $\frac{3}{10}$ **9.** $-\frac{1}{81}$ **10.** $\frac{1}{625}$

1. 8 **3.** -10 **5.** $6, -6$ **7.** $\frac{3}{4}, -\frac{3}{4}$

9. Not possible **11.** 10 **13.** -10

15. Not possible **17.** 13 **19.** $-\frac{1}{3}$

21. Not possible **23.** 0.2 **25.** -0.03 **27.** 5

SECTION 10.2 *(page 479)*

Warm-Up *(page 478)*

1. 100 **2.** -14 **3.** $-\frac{13}{5}$ **4.** Not possible

5. $32x^5$ **6.** $-64x^6y^3$ **7.** $\dfrac{16x^2}{y}$ **8.** $\dfrac{8y^2}{z^4}$

9. $\dfrac{3x^3}{y^3}$ **10.** 1

1. $\sqrt{14}$ **3.** $\sqrt{110}$ **5.** $\sqrt{4} \cdot \sqrt{15} = 2\sqrt{15}$

7. $\sqrt{64} \cdot \sqrt{11} = 8\sqrt{11}$ **9.** $2\sqrt{2}$ **11.** $3\sqrt{3}$

13. $2\sqrt{5}$ **15.** $10\sqrt{3}$ **17.** $6\sqrt{5}$ **19.** $100\sqrt{3}$

21. $2|x|$ **23.** $8x\sqrt{x}$ **25.** $2a^2\sqrt{2}$ **27.** $|x|y\sqrt{y}$

29. $10|x|y^2\sqrt{2}$ **31.** $\sqrt{\frac{13}{5}}$ **33.** $\sqrt{\frac{152}{3}}$ **35.** 3

37. $\frac{1}{2}\sqrt{35}$ **39.** $\dfrac{10}{\sqrt{11}} = \dfrac{10\sqrt{11}}{11}$ **41.** $\dfrac{\sqrt{3}}{5}$

43. $\dfrac{2|x|\sqrt{3}}{5}$ **45.** $3|a|$ **47.** $\dfrac{|u|\sqrt{5}}{2v^2}$ **49.** $\dfrac{\sqrt{3}}{3}$

51. $\dfrac{\sqrt{3}}{3}$ **53.** $\dfrac{\sqrt{10}}{2}$ **55.** $\dfrac{\sqrt{y}}{y}$ **57.** $\dfrac{\sqrt{5x}}{x}$

59. $\dfrac{\sqrt{6xy}}{3|y|}$ **61.** $\dfrac{a^2\sqrt{ab}}{b}$ **63.** $\sqrt{3.2} \times 10^3$

65. $3\sqrt{2} \times 10^{-4}$ **67.** $\dfrac{\pi\sqrt{6}}{2}$ sec **69.** 1,100.47 sq in.

MATH MATTERS: PERFECT NUMBERS   *(page 485)*

To show that 496 is a perfect number, we first find all factors of 496 (excluding 496 itself). These are 1, 2, 4, 8, 16, 31, 62, 124, and 248. The sum of these factors is

$$1 + 2 + 4 + 8 + 16 + 31 + 62 + 124 + 248 = 496.$$

Therefore, 496 is a perfect number.

SECTION 10.3   *(page 488)*

Warm-Up   *(page 488)*

1. $8 + 2x$ **2.** $8x + 5$ **3.** $27 - 5x$
4. $2 - 6x$ **5.** $x^2 + 12x + 36$
6. $4x^2 - 12x + 9$ **7.** $25x^2 - 9$ **8.** $1 - 9x^2$
9. $-2x^2 - 13x$ **10.** $54x - 8x^2$

1. $2\sqrt{5}$ **3.** $-\frac{4}{5}\sqrt{3}$ **5.** $-11\sqrt{3} - 5\sqrt{7}$
7. $18\sqrt{2}$ **9.** $34\sqrt{2}$ **11.** $2\sqrt{x}$ **13.** $9\sqrt{x}$
15. $-2\sqrt{5z}$ **17.** $\frac{1}{6}\sqrt{a}$ **19.** $(|x| + 4)\sqrt{xy}$ **21.** 9
23. $\sqrt{7} - \sqrt{14}$ **25.** $6\sqrt{2} + 8\sqrt{6}$ **27.** 1   **29.** 5
31. $4\sqrt{13} + 17$ **33.** $\sqrt{3} + \sqrt{6} - 5\sqrt{2} - 5$
35. $x + 5\sqrt{x}$ **37.** $4x - 9$ **39.** $x + 6\sqrt{x} + 9$
41. $x - 2\sqrt{x} - 3$
43. For the conjugate $4 - \sqrt{3}$, the product is 13.
45. For the conjugate $\sqrt{15} - \sqrt{7}$, the product is 8.
47. For the conjugate $\sqrt{x} + 4$, the product is $x - 16$.
49. For the conjugate $\sqrt{u} + \sqrt{2}$, the product is $u - 2$.

51. $\dfrac{\sqrt{14} + 2}{2}$ **53.** $-2(\sqrt{7} + 3)$

55. $\dfrac{\sqrt{35} - \sqrt{5} + \sqrt{7} - 1}{6}$

57. $\dfrac{(\sqrt{3} + 5)x}{11}$

59. $\dfrac{2x - 9\sqrt{x} - 5}{4x - 1}$

61. $\dfrac{9 - \sqrt{3}}{3}$ **63.** $2\sqrt{2}$ **65.** $\sqrt{5} + \sqrt{3} > \sqrt{5 + 3}$

67. $5 > \sqrt{3^2 + 2^2}$ **69.** 1   **71.** $\dfrac{\sqrt{5} + 1}{2} \approx 1.62$

SECTION 10.4   *(page 497)*

Warm-Up   *(page 497)*

1. 26   **2.** $\frac{18}{7}$ **3.** 3   **4.** 3   **5.** 5, -11
6. 4, -16 **7.** $10\sqrt{5}$ **8.** $\sqrt{5}$ **9.** 48
10. $\sqrt{2}$

1. 100   **3.** 25   **5.** No solution   **7.** 397
9. 1,000   **11.** 6   **13.** $-\frac{1}{3}$ **15.** 25   **17.** $\frac{11}{25}$
19. 2   **21.** 3   **23.** No solution   **25.** $\frac{1}{2}$ **27.** 4
29. $-1, 3$ **31.** 13   **33.** 5.74   **35.** 14.42
37. $\sqrt{216} \approx 14.7$ ft   **39.** 5   **41.** 43.82 ft/sec
43. 87.89 ft   **45.** 144 ft   **47.** 3.24 ft   **49.** 29 units
51. $\dfrac{\sqrt{S^2 - \pi^2 r^4}}{\pi r}$ **53.** 8 in. by 6 in.

CHAPTER 10 REVIEW EXERCISES   *(page 502)*

1. 11   **3.** Not possible   **5.** 8   **7.** 7.28
9. 0.94   **11.** 5   **13.** 11.04   **15.** $4\sqrt{3}$
17. $4\sqrt{10}$ **19.** $\frac{1}{3}\sqrt{23}$ **21.** $\frac{2}{3}\sqrt{5}$ **23.** $6x^2$
25. $2y\sqrt{y}$ **27.** $0.2|x|\sqrt{y}$ **29.** $4|a|\sqrt{2ab}$
31. $\dfrac{\sqrt{15}}{5}$ **33.** $\dfrac{2\sqrt{3}}{3}$ **35.** $\dfrac{3\sqrt{x}}{x}$ **37.** $\dfrac{\sqrt{11ab}}{|b|}$
39. $2|x|$ **41.** $12\sqrt{2}$ **43.** $-14\sqrt{5}$ **45.** $8\sqrt{3}$
47. $7\sqrt{y} + 3$ **49.** $2\sqrt{3} + 4$ **51.** 3
53. $2\sqrt{6} - 6\sqrt{3} + 10\sqrt{2} - 30$ **55.** $x + 10\sqrt{x}$
57. $2\sqrt{3} + 3$ **59.** $\dfrac{(\sqrt{x} - 3)^2}{x - 9}$ **61.** 169
63. No solution   **65.** 13   **67.** 3   **69.** 2
71. $20\sqrt{34} \approx 116.6$ ft   **73.** 2.48 ft

CHAPTER 10 TEST *(page 504)*

1. $\frac{3}{4}$ **2.** Not possible **3.** -5.944 **4.** $4\sqrt{3}$

5. $4|x|y\sqrt{2y}$ **6.** 8 **7.** $\frac{x}{y^2}\sqrt{3x}$ **8.** $\frac{\sqrt{15}}{3}$

9. $\frac{2\sqrt{2t}}{t}$ **10.** $16\sqrt{3}$ **11.** $-\frac{\sqrt{x}}{10}$ **12.** $4 - 5\sqrt{2}$

13. $2\sqrt{6} - 9$ **14.** $21 + 8\sqrt{5}$

15. For the conjugate $\sqrt{3} + 5$, the product is -22.

16. $2(\sqrt{6} - 1)$ **17.** 16 **18.** $\frac{13}{4}$

19. 2 **20.** $2\sqrt{13} \approx 7.21$ **21.** 125 units

Chapter 11

SECTION 11.1 *(page 511)*

Warm-Up *(page 511)*

1. $(3x + 5)(3x - 5)$ **2.** $(2t - 3)^2$

3. $2(x - 5)(x + 1)$ **4.** $(2s + 3)(2s - 3)$

5. $(3x - 5)(x - 2)$ **6.** $4x(x - 4)(x + 1)$

7. $-\frac{4}{5}$ **8.** $\frac{3}{2}$ **9.** $-7, \frac{3}{2}$ **10.** $\frac{4}{5}, 10$

---

1. $0, 2$ **3.** $0, -2$ **5.** $5, -5$ **7.** $\frac{3}{2}, -\frac{3}{2}$ **9.** 4

11. -2 **13.** $-\frac{3}{2}$ **15.** $2, 3$ **17.** $-2, 4$

19. $-1, -4$ **21.** $2, \frac{1}{2}$ **23.** $\frac{1}{2}, \frac{1}{4}$ **25.** $4, -\frac{4}{5}$

27. $6, 10$ **29.** $-4, -20$ **31.** $7, -7$ **33.** $\frac{7}{3}, -\frac{7}{3}$

35. $10, -10$ **37.** $\frac{10}{3}, -\frac{10}{3}$ **39.** No real solution

41. $8, -16$ **43.** $8.2, 7.8$ **45.** $-\frac{11}{2}, -\frac{1}{2}$

47. $1 + \sqrt{5}, 1 - \sqrt{5}$ **49.** $\frac{-5 \pm 2\sqrt{2}}{2}$ **51.** 2 sec

53. 20 units

MATH MATTERS: CALCULATING WITH THE NUMBER 9 *(page 515)*

The pattern for the given multiplication table is as follows.

$$
\begin{aligned}
(9 \times 9) + 7 &= 88 \\
(98 \times 9) + 6 &= 888 \\
(987 \times 9) + 5 &= 8{,}888 \\
(9{,}876 \times 9) + 4 &= 88{,}888 \\
(98{,}765 \times 9) + 3 &= 888{,}888 \\
(987{,}654 \times 9) + 2 &= 8{,}888{,}888 \\
(9{,}876{,}543 \times 9) + 1 &= 88{,}888{,}888 \\
(98{,}765{,}432 \times 9) + 0 &= 888{,}888{,}888
\end{aligned}
$$

SECTION 11.2 *(page 519)*

Warm-Up *(page 519)*

1. $x^2 + 4x + 3$ **2.** $x^2 + 10x + 28$

3. $u^2 - 16u + 74$ **4.** $v^2 - 6v + 7$ **5.** $\frac{1}{2}, -\frac{1}{2}$

6. $\frac{3}{4}, -\frac{3}{4}$ **7.** $11, -1$

8. $-2 + \sqrt{10}, -2 - \sqrt{10}$

9. $\frac{1}{2} + \frac{\sqrt{5}}{2}, \frac{1}{2} - \frac{\sqrt{5}}{2}$

10. $-\frac{3}{4} + \frac{\sqrt{7}}{4}, -\frac{3}{4} - \frac{\sqrt{7}}{4}$

---

1. 25 **3.** 144 **5.** $\frac{9}{4}$ **7.** $\frac{4}{25}$ **9.** $\frac{1}{100}$

11. $0, 4$ **13.** $1, 3$ **15.** $-2, -4$ **17.** $5, -7$

19. $7, -1$ **21.** $1 + \sqrt{2}, 1 - \sqrt{2}$

23. $-1 + \sqrt{2}, -1 - \sqrt{2}$ **25.** $2 + \sqrt{5}, 2 - \sqrt{5}$

27. No real solution **29.** $4 + 3\sqrt{2}, 4 - 3\sqrt{2}$

31. $-7 + 4\sqrt{2}, -7 - 4\sqrt{2}$ **33.** $\frac{1 \pm \sqrt{13}}{2}$

35. $\frac{-5 \pm \sqrt{17}}{2}$ **37.** $2, -5$ **39.** $\frac{3 \pm \sqrt{6}}{3}$

41. $\frac{-3 \pm \sqrt{19}}{2}$ **43.** $\frac{-5 \pm \sqrt{265}}{10}$ **45.** $2 \pm \sqrt{2}$

47. $3 + 2\sqrt{2}$ **49.** $6, 7$

SECTION 11.3 *(page 526)*

Warm-Up *(page 526)*

1. $2, 5$ **2.** $3, -6$ **3.** $\frac{3}{2}, -10$ **4.** $\frac{1}{3}, \frac{5}{2}$

5. $\frac{-1 \pm \sqrt{17}}{2}$ **6.** $\frac{-3 \pm \sqrt{6}}{3}$ **7.** 2 **8.** 7

9. $2\sqrt{17}$ **10.** $2\sqrt{37}$

---

1. $x^2 + 2x - 3 = 0$ **3.** $-x^2 + 4x - 10 = 0$ **5.** 2

7. 0 **9.** 1 **11.** $5, 6$ **13.** $-1, -3$ **15.** 3

17. $-\frac{1}{2}$ **19.** $4, \frac{3}{4}$ **21.** $\frac{1}{2}, \frac{3}{4}$ **23.** $1, \frac{3}{5}$

25. $-3, -\frac{1}{2}$ **27.** $3 \pm \sqrt{2}$ **29.** $-2, -3$ **31.** -2

33. $\frac{3 \pm \sqrt{13}}{2}$ **35.** $\frac{-1 \pm \sqrt{11}}{5}$

37. No real solutions **39.** $\frac{1}{6}$ **41.** ± 25 **43.** $0, -8$

45. $-3, 9$ **47.** $\frac{1}{2}, -\frac{3}{5}$ **49.** $\frac{3 \pm \sqrt{11}}{2}$

51. 4.361, 0.306 **53.** 0.251, 66.416 **55.** $\dfrac{3 \pm \sqrt{57}}{4}$

57. $3 + \sqrt{11}$ **59. (a)** 0 sec, $\frac{5}{4}$ sec **(b)** 3.20 sec

SECTION 11.4 *(page 537)*

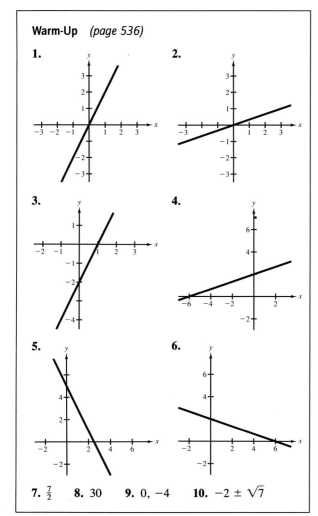

Warm-Up *(page 536)*

1.

2.

3.

4.

5.

6.

7. $\frac{7}{2}$ **8.** 30 **9.** 0, −4 **10.** $-2 \pm \sqrt{7}$

1. (d) **2.** (c) **3.** (f) **4.** (b) **5.** (a) **6.** (e)
7. Up **9.** Down **11.** Up
13. (4, 0), (−4, 0), (0, 16) **15.** (0, 0), (2, 0)
17. (3, 0), (−2, 0), (0, −6) **19.** (0, 3) **21.** (0, 2)
23. (2, 3) **25.** (2, −4)

27.

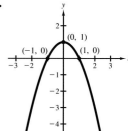

29.

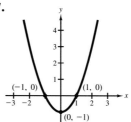

31.

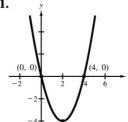

33.

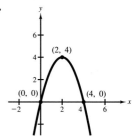

35.

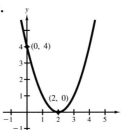

37.

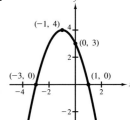

39.

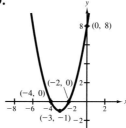

41.

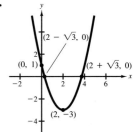

43.

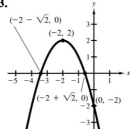

45.

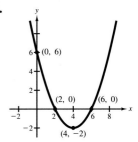

47.

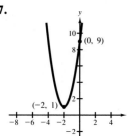

49.

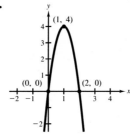

51. 1, −1

53. $4 \pm \sqrt{2}$

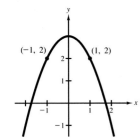

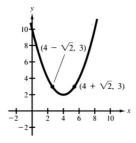

55. $y = x^2 - 4x + 2$. If the equation is in the form $y = (x - b)^2 + c$, then the vertex is (b, c).

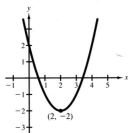

57. $y = (x - 5)^2 + 1$, (5, 1)   **59. (a)** 4 ft   **(b)** 14 ft
(c) $10 + 2\sqrt{35} \approx 21.8$ ft

61. (b) Parabola is wider. For the same x, y is $\frac{1}{8}$ what it was in the original equation.
(c) Parabola opens downward. Graph is more narrow; y is −2 times what it was in the original equation.
(d) Parabola opens downward. Graph is wider.

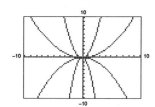

63. (b) Graph is shifted 2 units to the right.
(c) Graph is shifted 6 units to the right.
(d) Graph is shifted 4 units to the left.

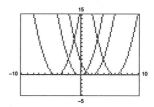

SECTION 11.5 *(page 544)*

Warm-Up *(page 544)*

1. 2   **2.** 0, 2   **3.** $\frac{28}{3}$ **4.** 3, −5
5. 2, −18   **6.** $\dfrac{-3 \pm \sqrt{13}}{2}$ **7.** 1, 2   **8.** 4
9.   **10.**

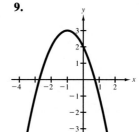

1. 11, 12   **3.** 12, 14   **5.** 7, 8   **7.** 10 sec
9. 5.86 sec

Width, Length, Perimeter, Area

11. 12 in., 20 in., 64 in., 240 in.²
13. 5 ft, 10 ft, 30 ft, 50 ft²
15. 5 in., 20 in., 50 in., 100 in.²
17. 12 km, 16 km, 56 km, 192 km²
19. 5 m, 15 m, 40 m, 75 m²

21. 50 ft by 75 ft   or   37.5 ft by 100 ft
23. Base: 12 in. Height: 4 in.   **25.** 8 mi
27. 25 people   **29.** 180 mi/hr   **31.** 6 hr and 12 hr

CHAPTER 11 REVIEW EXERCISES *(page 548)*

1. 0, −10   **3.** −1, $\frac{8}{5}$ **5.** $\pm\frac{5}{2}$ **7.** $\frac{5}{3}$ **9.** −7, 8
11. ±25   **13.** No real solution

15. $-5, 35$ **17.** $3 \pm \sqrt{10}$ **19.** $\dfrac{1 \pm \sqrt{5}}{2}$

21. $\dfrac{-5 \pm \sqrt{15}}{2}$ **23.** $6, -7$ **25.** $\dfrac{-1 \pm \sqrt{337}}{4}$

27. $\dfrac{10 \pm \sqrt{70}}{3}$ **29.** $\pm 5\sqrt{10}$ **31.** $-7, 10$

33. $-1, 25$ **35.** $3 \pm \sqrt{3}$ **37.** No real solution

39. $\dfrac{3 \pm \sqrt{17}}{2}$

(c)

l	A
2	36
4	64
6	84
8	96
10	100
12	96
14	84
16	64
18	36

41.

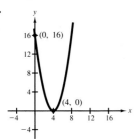

43.

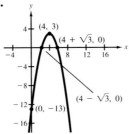

45.

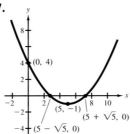

47.

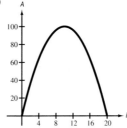

(d)

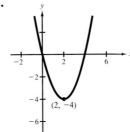

(e) 10 ft by 10 ft

57. 15 people   **59.** 18.20 hrs, 22.20 hrs

49.

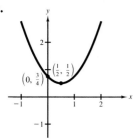

CHAPTER 11 TEST   *(page 550)*

1. $12, -12$ **2.** $10, -3$ **3.** $\dfrac{3}{2}, -\dfrac{5}{3}$ **4.** $3 \pm \sqrt{2}$

5. $\dfrac{-9 \pm \sqrt{21}}{6}$ **6.** $\dfrac{1 \pm \sqrt{13}}{2}$ **7.** $\dfrac{-2 \pm \sqrt{2}}{2}$

8. No real solution   **9.** $2 + \sqrt{3}$ **10.** $(6, 0), (2, 0)$

11.

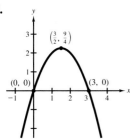

51. $15, 16$ **53.** $\sqrt{3}$ sec

55. (a) $2l + 2w = 40$
$l + w = 20$
$w = 20 - l$

(b) $A = lw$
$w = 20 - l$
$A = l(20 - l)$

12. 20, 21   **13.** Base: 6 in. Height: 18 in.
14. 16 in. by 6 in.   **15.** 10 hrs, 15 hrs
16. 67.5 mi/hr

CUMULATIVE TEST: CHAPTERS 8–11 *(page 551)*

1. (a) Not a solution   **(b)** Solution   **(c)** Solution
(d) Not a solution   **2.** $(1, 0), (3, 0), (0, 3)$
3.

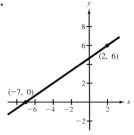

$m = \frac{2}{3}$

4. $(6, 0)$ **5.** $5x - 6y - 9 = 0$
6.

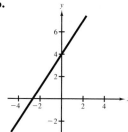

7.

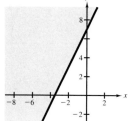

8. $(5, -1)$ **9.** $(-1, 1)$ **10.** $\left(\frac{3}{2}, 1\right)$ **11.** $10x\sqrt{3x}$
12. $\dfrac{\sqrt{10x}}{2x}$ **13.** $\sqrt{8} + \sqrt{3}$ **14.** $7 + 2\sqrt{6}$
15. $\dfrac{26}{9}$ **16.** $2, -\frac{3}{4}$ **17.** $\dfrac{-1 \pm \sqrt{11}}{2}$
18.

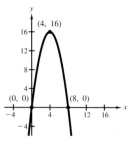

19. $C = 0.35x + 125$, $149.50
20. Unleaded: $1.34, Premium: $1.54

Graphing Linear Equations

A **linear equation in two variables** is an equation of first degree in both variables.

The **graph of an equation** is the set of all points in the rectangular coordinate system whose coordinates are solutions of the equation.

The **graph of a linear equation in two variables** is a straight line.

To find the **x-intercepts** (where the graph intersects the x-axis), let y be zero and solve the equation for x.

To find the **y-intercepts** (where the graph intersects the y-axis), let x be zero and solve the equation for y.

Slope, $m = \dfrac{y_2 - y_1}{x_2 - x_1} = \dfrac{\text{Change in } y}{\text{Change in } x}$.

Slope-intercept form of the equation of a line:
$y = mx + b$, m is the slope, and $(0, b)$ is the y-intercept.

Point-slope form of the equation of a line:
$y - y_1 = m(x - x_1)$, m is the slope, and (x_1, y_1) is a point on the line.

Graphs of Systems of Linear Equations

The **solution** of a system of linear equations is an ordered pair (a, b) that satisfies each of the equations. A system of linear equations may have no solution, exactly one solution, or infinitely many solutions.

Graphs of Quadratic Equations

Standard form of a quadratic equation:
$ax^2 + bx + c = 0$, a, b, and c are real numbers with $a \neq 0$. A quadratic equation can be solved by factoring, extracting square roots, completing the square, or using the Quadratic Formula.

Extracting square roots: If $u^2 = d$, where $d > 0$, then $u = \pm\sqrt{d}$.

Quadratic Formula: $x = \dfrac{-b \pm \sqrt{b^2 - 4ac}}{2a}$

Discriminant: $b^2 - 4ac$
If $b^2 - 4ac > 0$, then the equation has two real number solutions.
If $b^2 - 4ac = 0$, then the equation has one (repeated) real number solution.
If $b^2 - 4ac < 0$, then the equation has no real number solutions.

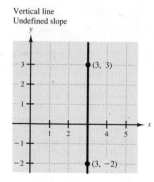

Vertical line
Undefined slope

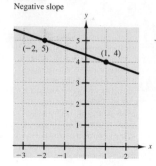

Line falls
Negative slope

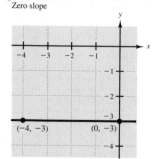

Horizontal line
Zero slope

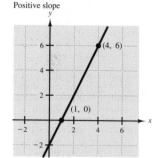

Line rises
Positive slope

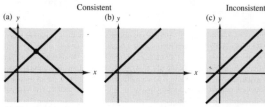

(a) Two lines that intersect at a single point
(b) Two lines that coincide with infinitely many points of intersection
(c) Two parallel lines with no point of intersection

The graph of $y = ax^2 + bx + c$, $a \neq 0$, is called a **parabola** which opens up if $a > 0$ and opens down if $a < 0$.

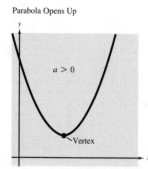

Parabola Opens Up

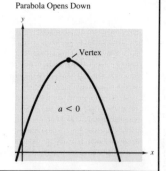

Parabola Opens Down